D1603947

GASEOUS ELECTRONICS

VOLUME I

Electrical Discharges

Contributors

P. BOGEN
SANBORN C. BROWN
L. M. CHANIN
ALAN GARSCADDEN
A. GOLDMAN
M. GOLDMAN
E. HINTZ
JOHN H. INGOLD
A. D. MACDONALD
E. PFENDER
S. J. TETENBAUM
ALAN J. TOEPFER
GEROLD YONAS

GASEOUS ELECTRONICS

Edited by MERLE N. HIRSH

Science and Mathematics Division
University of Minnesota
Morris, Minnesota

H. J. OSKAM

Department of Electrical Engineering
University of Minnesota
Minneapolis, Minnesota

VOLUME I
Electrical Discharges

ACADEMIC PRESS New York San Francisco London 1978

A Subsidiary of Harcourt Brace Jovanovich, Publishers

ACADEMIC PRESS, INC.
111 Fifth Avenue, New York, New York 10003

United Kingdom Edition published by
ACADEMIC PRESS, INC. (LONDON) LTD.
24/28 Oval Road, London NW1 7DX

Library of Congress Cataloging in Publication Data

Main entry under title:

Gaseous electronics.

Includes bibliographies.
1. Electric discharges through gases. I. Hirsh, Merle N. II. Oskam, Hendrik Jan.
QC711.G25 537.5 77-6598
ISBN 0-12-349701-9 (v. 1)

PRINTED IN THE UNITED STATES OF AMERICA

Contents

List of Contributors

Numbers in parentheses indicate the pages on which the authors' contributions begin.

P. BOGEN (453), Institut für Plasmaphysik der Kernforschungsanlage Jülich GmbH, Association Euratom-KFA, Jülich, Federal Republic of Germany

SANBORN C. BROWN (1), Research Laboratory of Electronics, Massachusetts Institute of Technology, Cambridge, Massachusetts 02139

L. M. CHANIN (109, 133), Department of Electrical Engineering, University of Minnesota, Minneapolis, Minnesota 55455

ALAN GARSCADDEN (65), High Power Branch, Air Force Aero Propulsion Laboratory, Wright-Patterson AFB, Ohio 45433

A. GOLDMAN (219), Laboratoire de Physique des Décharges du Centre National de la Recherche Scientifique, École Supérieure d'Electricité, Gif-sur-Yvette, France

M. GOLDMAN (219), Laboratoire de Physique des Décharges du Centre National de la Recherche Scientifique, École Supérieure d'Electricité, Gif-sur-Yvette, France

E. HINTZ (453), Institut für Plasmaphysik der Kernforschungsanlage Jülich GmbH, Association Euratom-KFA, Jülich, Federal Republic of Germany

JOHN H. INGOLD (19), Lighting Business Group, General Electric Company, Cleveland, Ohio 44112

A. D. MACDONALD (173), Lockheed Palo Alto Research Laboratory, Palo Alto, California 94304

E. PFENDER (291), Heat Transfer Division, Department of Mechanical Engineering, University of Minnesota, Minneapolis, Minnesota 55455

S. J. TETENBAUM (173), Lockheed Palo Alto Research Laboratory, Palo Alto, California 94304

ALAN J. TOEPFER (399), Fusion Research Department, Sandia Laboratories, Albuquerque, New Mexico 87115

GEROLD YONAS (399), Fusion Research Department, Sandia Laboratories, Albuquerque, New Mexico 87115

Preface

Throughout the long history of gaseous electronics, we have witnessed the parallel growth of basic understanding of the physical processes occurring in ionized gases, and of the application of that understanding to a variety of practical problems. A number of technological developments since the end of World War II, most notably ultrahigh vacuum techniques and microwave diagnostics, engendered a remarkable increase in the rates of growth of both aspects of the field. In view of the increasing importance of many of these applications to critical problems facing our society, we have decided to develop a multivolume treatise directed toward the intelligent application of gaseous electronics principles and devices to those problems.

The treatise will consist ultimately of about a dozen volumes. The first three volumes will concentrate on general principles, at a level more appropriate to the applications engineer than is now found in the basic physics oriented texts. Thus, Volume I describes a number of representative types of electrical discharges; Volume II will consider the microprocesses resulting in the growth and decay of charged particles and of excited species in the discharge; Volume III will be directed toward the interface between the ionized gas "plasma" and surfaces. Then will follow a number of books devoted to the practical applications of electrical discharges in various fields, such as gaseous electronics applications of corona discharges, plasma chemistry, the physics of the ionosphere, and thermonuclear fusion.

This first volume describes our best current understanding of various kinds of discharges. Following a historical introduction to the field in general, we consider the glow discharge (Chapter 2), high frequency and microwave discharges (Chapter 3), corona discharges (Chapter 4), arcs and torches (Chapter 5), electron beam produced plasmas (Chapter 6), and shock wave induced plasmas (Chapter 7). These treatments of the various kinds of discharge include macroscopic manifestations, such as I–V characteristics and qualitative phenomena, as well as an attempt to describe the underlying phenomena in terms of microscopic processes. The level of the treatment has been directed to the applications-oriented reader as much as possible.

This volume would not have been possible were it not for the diligent efforts and wholehearted cooperation of our contributing authors, to whom we express our appreciation. We also wish to thank the members of the staff of Academic Press for their constant support and encouragement throughout this venture.

Chapter 1

A Short History of Gaseous Electronics

SANBORN C. BROWN
RESEARCH LABORATORY OF ELECTRONICS
MASSACHUSETTS INSTITUTE OF TECHNOLOGY
CAMBRIDGE, MASSACHUSETTS

Physics has few prophets, but Sir William Crookes was one who merited that distinction. After a lifetime of working with his "vacuum" tubes, focusing his mysterious rays on glowing ores, driving mica paddle wheels with them, trying to determine their characteristics long before electrons or ions in the modern sense were discovered, long before the concepts of mobility, diffusion, attachment, or recombination had been thought of as even applicable to his studies, he told an audience at a meeting of the British Association for the Advancement of Science in 1879:

> So distinct are these phenomena from anything which occurs in air or gas at the ordinary tension, that we are led to assume that we are here brought face to face with Matter in a Fourth state or condition, a condition so far removed from the State of gas as a gas is from a liquid.
>
> In studying this Fourth state of Matter we seem at length to have within our grasp and obedient to our control the little indivisible particles which with good warrant are suppose to constitute the physical basis of the universe . . . We have actually touched the border land where Matter and Force seem to merge into one another, the shadowy realm between Known and Unknown, which for me has always had peculiar temptations. I venture to think that the greatest scientific problems of the future will find their solution in this Border Land, and even beyond; here it seems to me, lie Ultimate Realities, subtle, far-reaching, wonderful (Crookes, 1879a).

Crookes was by no means speaking from ignorance when he suggested a fourth state of matter although he was at the time just emerging from a scientific shadow brought on by his enthusiastic and wholehearted espousal

ISBN 0-12-349701-9

of spiritualism and communication with the spirit world. He had made his reputation mostly as a chemist, the discoverer of the element thallium, and as editor of the *Chemical News*. As a member of the Royal Society of London he had vitriolic encounters with Fellows of that society over their refusal to accept his papers on "mysterious forces and apparitions" during the years of 1870–1874. An aggressive entrepreneur, he was making his living as a director of the Native Guano Company converting London sewage into saleable manure when he invented and named his famous radiometer which caused an immense sensation throughout both the scientific and popular worlds.

While experimenting with his "electric radiometer" in 1878, Crookes noticed that if the vanes were negatively charged, a dark space surrounded each vane which rotated with the vane, and he started out (with the aid of a most skillful glassblower, Charles Gimingham) on his long series of glow discharge experiments and the design of his many "Crookes tubes."

An important technical breakthrough that allowed progress to be made in the middle of the nineteenth century was the discovery of the induction coil by Heinrich Daniel Ruhmkorff in 1851. Up until that time the necessary voltage was supplied by charging up Leyden jars or from current supplied by batteries. Incidentally, the Leyden jar was not discovered in Leyden but in Pomerania by Ewald Georg von Kleist in 1745. It was called the Leyden jar because Professor Pieter van Musschenbroek of Leyden almost killed himself experimenting with it and ended his published account with the statement that he would not receive such a shock again even if offered the whole kingdom of France. For such "unworthy sentiments" he was severely rebuked by Joseph Priestley who in his "History of Electricity" compared the "cowardly professor" with the "truly philosophic heroism" of one Georg Wilhelm Richman, a Swedish physicist who was one of the first unfortunate victims of an attempt to reproduce Franklin's kite experiment.

On the subject of victims, the Ruhmkorff coil was also considered to be a very dangerous piece of laboratory equipment, and warnings in laboratory textbooks such as the following were not at all uncommon:

> The physiological effects obtained by means of a Ruhmkorff coil are very surprising. The discharge from a fair-sized coil is sufficient to kill a small animal. Great caution in handling the terminal wires is therefore needful when the coil is in action, especial care being necessary that both wires are not touched simultaneously (Dunman, 1891, p. 158).

Up until the time of Ruhmkorff, the principal electrical discharge studied had been the low voltage arc since this could be maintained by battery power. Although it had been discovered as early as 1808 by Humphry Davy (using a voltaic pile that Davy and Count Rumford had designed at the Royal Institution in London), little knowledge of a fundamental nature came out of these studies, and the arc was looked on more as a novelty than as a gateway to knowledge. This type of discharge received its name "arc" from Davy (1821). Atmospheric pressure sparks and breakdown experiments at lower pressures could be initiated by electrostatic machines, but the deliverable current was so small that steady-state experiments were most difficult and measuring the parameters of transient phenomena was almost beyond the technical instrumentation of the day. The study of the tremendous electrical discharges of lightning (after identification by Benjamin Franklin in 1752), most unfortunately, proved fatal to more physicists than it enlightened.

Another important technique that did much to provide experimental advancement was developed by an itinerant glass blower by the name of Heinrich Geissler, who founded a shop in Bonn for making scientific apparatus. In 1858 Geissler developed a technique for fusing platinum electrodes directly into glass, and "Geissler" tubes became the show pieces of late nineteenth century demonstration physics lectures. Subsequently, Geissler became a professor of physics at the University of Bonn and his most famous student Johann Wilhelm Hittorf carried on for a generation beyond him.

Cathode rays were discovered by Julius Plücker and named by him "Kathodenstrahlen." He showed that these rays made glass, mica, and other substances fluoresce, and he investigated the effects of magnetic fields on cathode rays. Plücker (1859) postulated that the fluorescence came from currents of electricity which flowed from the cathode to the tube walls and then retraced their paths. Hittorf (1869), discovered that the rays cast shadows and further work on these shadows was carried on by Eugene Goldstein (1876), who showed that a small body near a large cathode cast a shadow, an important observation since it meant that the rays came in a definite direction from the cathode, and he advanced the theory that they were waves in the ether. William Crookes (1879b), on the other hand, championed the theory that these rays were charged particles shot out at right angles from the cathode at high velocity. He showed that when these rays hit vanes mounted as in a radiometer, the vanes were set in motion. Crookes and Hittorf entered into a spirited controversy as to whether this effect was a purely mechanical momentum transfer as maintained by Crookes, or was the effect due to secondary thermal effects as believed by Hittorf. These two, Crookes and Hittorf, laid the foundations for some of the most spectacular discoveries in physics, yet two more different men can hardly be imagined.

Crookes, the son of a well-to-do English businessman, had little formal training, never graduated from a university, and drifted from job to job as an industrial chemist. Early in his life he became an expert in spectroscopy, which led him to the discovery of thallium. Crookes was much interested in photography and for a while was editor of the *Journal of the Photographic Society of London*. He started, and for many years was the editor of, a chemical weekly called *Chemical News*. He carried this on in parallel with a great deal of chemical consulting at which he was highly successful both technically and financially. His early experiments on the passage of electricity in gases was stimulated by his interest in trying to invent a practical electric light. He experimented with carbon and mercury arcs and also the "electric egg" (see Fig. 1), but in the absence of a cheap source of power, batteries being what were available, he turned to the chemical "lime light" as being much more practical. He had an aggressive, flamboyant, and quarrelsome disposition and could shift in a single week from experiments on psychic forces, to the atomic weight of thallium, and then to the production of fertilizer from rotten fish. As he became wealthy in his own right, he could

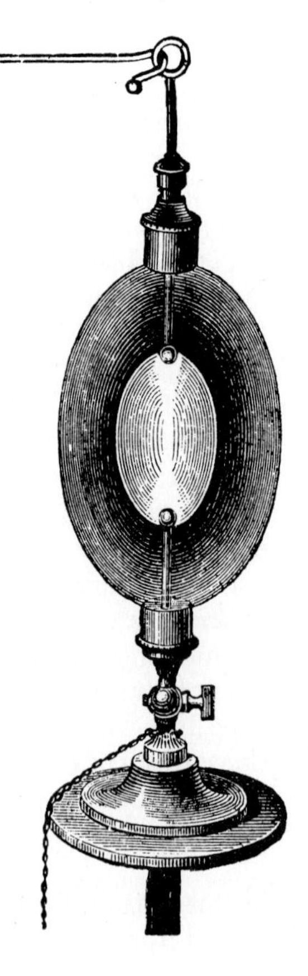

Fig. 1. The electric egg. The influence of air pressure on the electrical discharge was studied by attaching the electrodes to a source of electric power and observing the changes in size, shape, and color of the discharge.

afford to hire an excellent glassblower for his private laboratory. That, plus his own technical skill at producing a good vacuum was largely responsible for his success with all types of "Crookes" tubes.

Hittorf, on the other hand, was almost a caricature of a pedantic German physics professor. He went through the full range of German university education, then taught in a small-town university with a heavy teaching load, stealing time to do research; he strained and skimped to carry on his studies. Always half-starved himself, Hittorf laboriously built his own equipment, with which he was always having trouble. There is a story, perhaps apocryphal, which somehow typifies his bad luck. To understand its full import it should be remarked that he evacuated his tubes with a Toepler pump, a long and slow process. In 1874 McLeod had introduced his famous gauge and shortly thereafter, Toepler devised a pump which worked like a McLeod gauge except the top measuring bore was attached to the tube to be evacuated, so that by patient and repeated raisings and lowerings of the mercury column the pressure in the tube could be reduced.

Hittorf was trying to determine if there was an ultimate length to the positive column. Week after week his discharge tube grew as he added meter after meter, sealing it off, pumping it down, and testing it as he pushed his Ruhmkorff induction coil to higher and higher voltage. His tube went all the way across the room, turned and curved back, turned again until his whole laboratory seemed full of thin glass tubing. It was summer, and as he sweated away with his Toepler pump, he opened the window to make it bearable. Suddenly, from outside came the howl of a pack of dogs in full pursuit, and flying through the open window came a terrified cat, to land, feet spread for the impact, in the middle of the weeks and weeks of labor. "Until an unfortunate accident terminated my experiment," Hittorf wrote, "the positive column appeared to extend without limit."

Careful studies by Jean Baptiste Perrin (1895), which showed that the cathode rays carried a negative charge and that this negative electrification followed the same course as the rays producing the fluorescence on the glass, and measurements of the heat developed by cathode rays made by Gustav Heinrich Wiedemann (1891), were aimed very specifically at discovering the fundamental nature of cathode rays, but they never really did. As late as 1893, J. J. Thomson explained the phenomenon as chemical in nature, analogous to electrolysis. He wrote:

> The passage of electricity through a gas, as well as through an electrolyte . . . is accompanied and affected by chemical changes; also that chemical decomposition is not to be considered merely as an accidental attendant on the electrical discharges, but as an essential feature of the discharge without which it could not occur (Thomson, 1893, p. 189).

The equivalence of cathode rays and the phenomena of electrolytis was so close in most physicists' minds that the same nomenclature was used for both. In 1834 Michael Faraday had named "that surface at which the current leaves the decomposing body" the "cathode" from the Greek *κάθοδος* meaning the "road down," and the other electrode "anode" from *ἄνοδος*, the "road up." These designations were universally adopted for the gas discharge tubes as well.

Faraday went further in his paper on electrolysis to turn to the Greek word "to go" (*ἰόν*) and wrote:

> I propose to distinguish such bodies by calling those *anions* which go to the anode . . . and those passing to the cathode, *cations;* and when I have occasion to speak of these together, I shall call them *ions* (Faraday, 1839, p. 663).

Although 50 years later, the equivalence of electrolytes and gas discharges was generally accepted, Faraday did not think of conductors of electricity in gases as particles. He considered electrical phenomena in terms of stresses and strains in the medium. Electrical force was transmitted in an analogous way to the elastic deformation of a rod transmitting action down its length. In "vacuum" tubes, electrical force was transmitted by stresses and strains through the "ether." Faraday's concepts were brilliantly reinforced by the theoretical work of Maxwell and the experiments of Hertz, and few were the dissenters.

In terms of our modern nomenclature, the word "electron" was introduced years before the particle was discovered. Professor G. Johnstone Stoney (1891) was studying the atomic theory of electricity in electrolytes and measured the charge in 1874. In 1891 he suggested that these charges be called "electrons." But he very clearly wrote that these were not particles—just charges; they had no mass, no inertia. Sir Oliver Lodge (1906), called them "electric ghosts."

While Crookes and Hittorf were concentrating on trying to understand the electric "rays," many other researchers were making studies on optical and continuum effects. "Geissler" tubes were extensively used for spectroscopic light sources and one of the main thrusts of physics research concentrated on trying to understand the origin of spectral lines. It was during this kind of study that Lord Rayleigh (1906) introduced what we now call the "plasma frequency." J. J. Thomson had proposed the watermelon model of the atom, where electrons were imbedded in a positive jelly (like seeds in a watermelon) and Lord Rayleigh calculated the frequencies at which such electrons would oscillate when displaced from their equilibrium position. Rayleigh was searching for an explanation of spectral series and he was able to explain the Balmer series for hydrogen using this model. It did not work for any other atom, and Rutherford's hard-core nucleus was ultimately the successful model. Nevertheless, Rayleigh's theory still lives on when we calculate plasma frequencies and Langmuir's plasma oscillations.

Geissler and Crookes tubes were called "vacuum" tubes, although up to 1880 they were evacuated to a "torricelli" vacuum or by a piston pump whose ultimate limit was about 0.1 Torr. With the introduction of the Toepler pump the pressure in tubes could be reduced to around 10^{-2} Torr. Gaede invented the mechanical rotary oil pump in 1907, and in 1915 he invented the mercury diffusion pump, so that vacua gradually got better as time went on.

A few illustrations of the kind of studies that were carried out are instructive. A Crookes tube to demonstrate the effect of pressure on the discharge is shown in Fig. 1. It was called the "electric egg." In 1880 de la Rue and Müller worked out the details of what we now call the Paschen curves, and in 1884 Hittorf publicized a tube to demonstrate the effect, a sketch of which is shown in Fig. 2. The generation of heat produced by cathode rays was shown in a tube sketched in Fig. 3, the effect of magnetic fields in Fig. 4, shadows on the glass in Fig. 5, and the mechanical effect on a paddle wheel in Fig. 6.

For years W. de la Rue and H. W. Müller worked to assemble the world's largest battery (14,400 cells of zinc and silver electrodes in an NH_4Cl electrolyte) and their studies of the optical behavior of discharges were widely acclaimed. A much reproduced illustration in a paper they published in 1878 is shown in Fig. 7. Striations had been discovered by Abria in 1848 and for 50 years thereafter, physicists studied these and other complicated visual changes in hopes of learning something of the nature of the "electrical

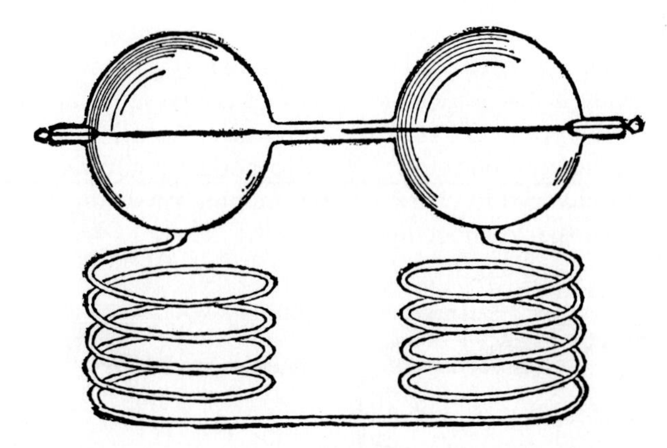

Fig. 2. Demonstration of the Paschen effect. When the pressure of the gas in the bulbs was reduced, the discharge took place through the long tube and not across the small gap between the metal electrodes.

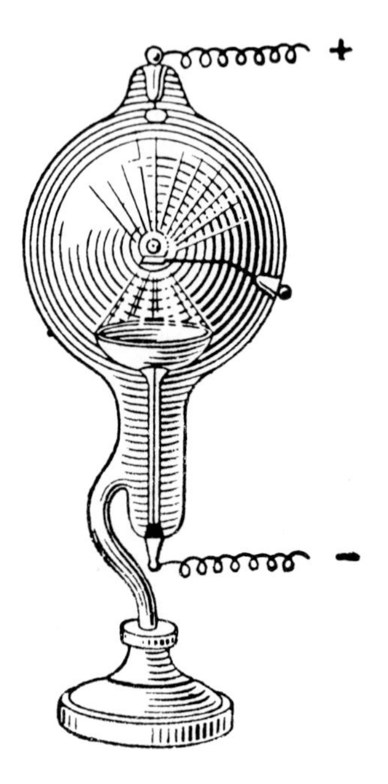

Fig. 3. The heating effects of cathode rays. The negative electrode was concave and a piece of platinum foil was placed at the center of the sphere of which the cathode was a portion. Where the cathode rays came to focus, the foil was heated to white heat.

fluid." The behavior of the glowing discharges as a function of pressure, electrode geometry, and type of gas, filled the literature for many years and a nomenclature emerged, some of which is still used today. This is indicated in Fig. 8.

Starting at the negative electrode (the cathode), a very thin "Aston dark space" is usually visible although sometimes obscured by the cathode glow which took on different names depending on the pressure in the tube. At low pressure this was often called the "cathode layer" and sometimes two or three of these layers were visible. At higher pressures and voltages the cathode surface appeared covered with a velvety coating which was universally called the "cathode glow." If the cathode was covered with an oxide layer, the glow changed its characteristic (we now know, due to sputtering), and it was called the "cathode light."

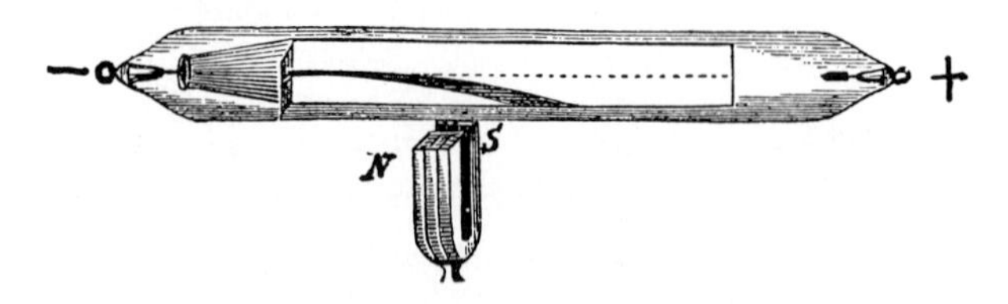

Fig. 4. The deflection of cathode rays by a magnetic field.

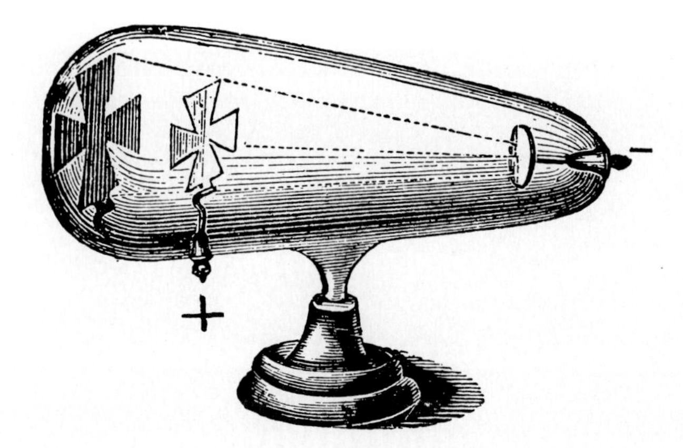

Fig. 5. Cathode rays travel in straight lines. The positive electrode was a cross cut from an aluminum sheet. A dark shadow was cast against a bright green background of the glass excited by the cathode ray bombardment. The cross was on a horizontal hinge so that it could be tipped out of the way, and the shadow disappeared.

The next dark space was often called the "cathode dark space" although in England it was more often known as the "Crookes dark space" and in Germany the "Hittorf dark space." The next glowing region had two names, either the "negative glow" or the "negative column." The next comparatively nonluminous space was called the "second negative dark space" or more commonly the "Faraday dark space." This was of very variable length and sometimes was not observed at all. The name "positive column" seemed to agree with everyone, although the way it broke up into "striations" or "striae" was subject to a great deal of investigation. The "anode glow" and the "anode dark space" complete the nomenclatural description.

It is an interesting historical circumstance that very little fundamental knowledge emerged from the intense study given to this discharge "vacuum"

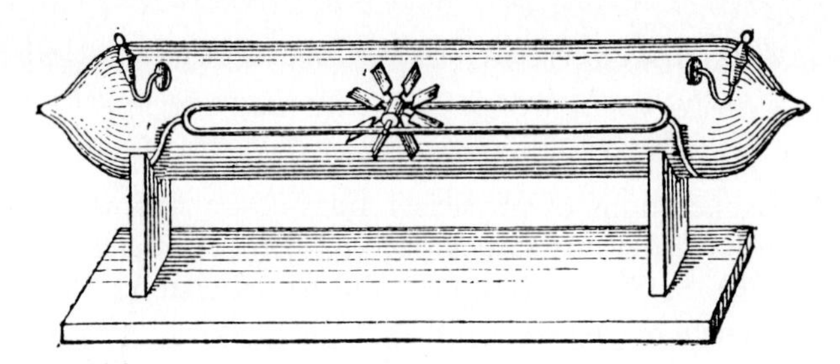

Fig. 6. The railway tube. Cathode rays were observed to produce a mechanical force on bodies which they hit. In the tube, a small wheel with mica paddles could be made to roll from cathode to anode along a glass track.

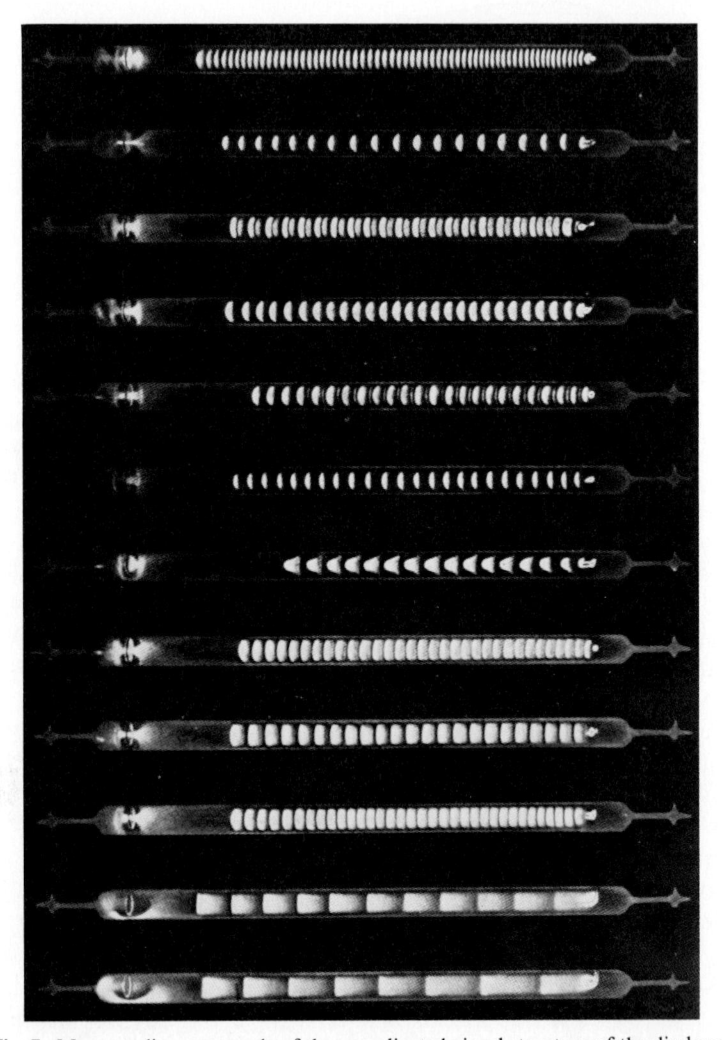

Fig. 7. Many studies were made of the complicated visual structure of the discharge.

tube. Several generations of physicists amassed an impressive amount of data on the behavior of the discharges and their characteristics, but until the nature of the "rays" involved began to be understood, the fourth state of matter remained an enigma.

The search for the nature of electricity has taken a long and circuitous route. The ancient Greeks had discovered that amber rubbed with various things had the property of attracting light objects, but it was not until the

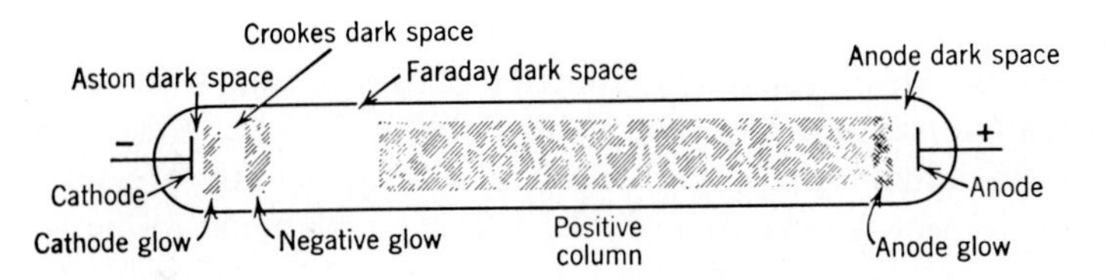

Fig. 8. Nomenclature for various parts of an electrical discharge.

early 17th century when William Gilbert, the English physician and physicist, published his treatise "De Magnete" (Gilbert, 1600) that any consistent study of electrical phenomena was reported. It was Gilbert who introduced the term "electricity" by calling a glass rod rubbed with silk "electrified," that is "amberized." In 1733 the French chemist Charles François de Cisternay Du Fay found that sealing wax rubbed with cat's fur took on an opposite charge from glass rubbed with silk and is therefore credited with discovering two kinds of electricity which Benjamin Franklin defined as "positive" and "negative." There then followed about 100 years of speculation as to the nature of the "electrical fire." There were one-fluid theories and two-fluid theories, but no experiments to decide between them. Michael Faraday made his remarkable discovery of the laws of electrolysis in 1833. These laws were immediately applied to electricity passing through gases, but so strongly held at the time was the concept that electricity was not a substance of any kind, that these experiments had no effect on establishing the atomic nature of electricity. Faraday wrote convincingly of electrical force being transmitted in the same way as the elastic deformation of a solid rod. In a vacuum it was the "ether" that transmitted the electrical stresses and strains. James Clerk Maxwell put these ideas into mathematical form and when in 1887, Heinrich Hertz showed experimentally that electrical waves traveled through space as predicted by the Faraday–Maxwell theory, proof of the "ether-stress' theory was complete.

For two generations before 1900, most physicists differentiated two different kinds of electricity: electrolytic conduction by some kind of atom of electricity, and metallic conduction as an electrical stress and strain. During the eighteenth century there were many reported studies of the leakage of charge from electrically insulated bodies, but the most careful one was carried out by Coulomb in 1785. Coulomb took into account leakage over the strings that insulated his charged body and concluded that there was an actual passage of electrical charge through the air which he attributed to molecules of the air becoming charged by contact with the body. Since they would be charged with the same sign as the body, they were then repelled out into space.

Seventy-five years after Coulomb's careful experiment, Matteucci (1850) measured the rate of leakage as a function of pressure and showed that it was less when the pressure was lower; and in 1900, C. T. R. Wilson concluded that the maximum leakage was roughly proportional to the pressure. Wilson also showed that the maximum leakage was proportional to the volume of the enclosing vessel which was inconsistent with Coulomb's theory of charged molecules and led the way to the acceptance that there was some kind of "atom of electricity" involved in the process. It was not until Wilhelm Conrad Röntgen, using a common type of Crookes tube, made the remarkable discovery of x rays in 1895, and Antoine Henri Becquerel's discovery of radioactive radiation from uranium a year later, that the particular nature of the electrical "fluid" began to be accepted. The new fact that was common to both of these discoveries was that a gas could be made a conductor by these mysterious rays.

Röntgen's discovery was completely fortuitous. At his laboratory at Würzburg he was studying the nature of the fluorescent glow on materials struck by cathode rays, a phenomenon often demonstrated with a Crookes tube of the form illustrated in Fig. 9. One of the materials which responded

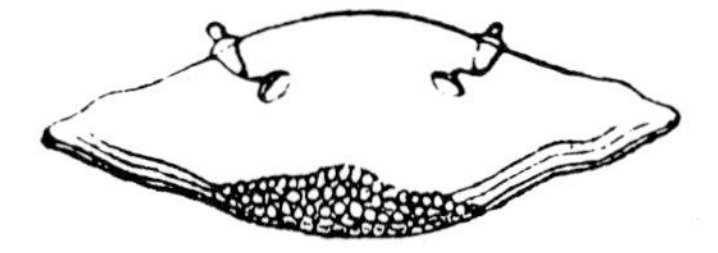

Fig. 9. The fluorescence of a large number of chemicals, ores, and rocks bombarded by cathode rays were demonstrated in this type of Crookes tube.

particularly well to this bombardment was barium platinocyanide. Röntgen was trying to determine whether the commonly observed fluoresce of the glass walls of the Crookes tube would have any effect in barium platinocyanide outside the tube. He chose a form of Crookes tube which was called a "focus tube," shown in Fig. 10, to maximize bouncing of the cathode rays off the anode onto the glass walls, and hence, the wall fluorescence. Having made screens coated with this material, he was encouraged to discover that in fact it did, and his screens glowed brightly when the Crookes tube was on. His Nobel prize-winning move was to cover up his Crookes tube with black paper to make sure that it was the glass fluorescence that was making his

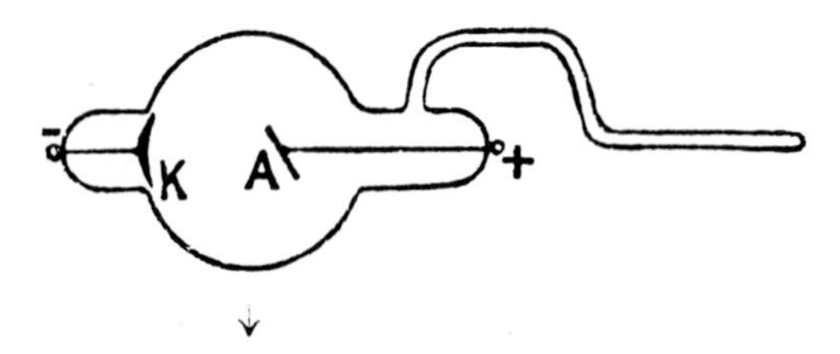

Fig. 10. A Crooke; "focus tube" of the type Röntgen used to study fluorescence and x-rays.

barium platinocyanide glow, and to his astonishment the black paper made no difference whatever. It took him almost no time at all to discover that his x rays could "discharge electrified bodies" and also could expose a photographic film, the later effect leading to a curious kind of priority claim from a physicist working with a Crookes tube in Oxford. Frederick Smith had already found that photographic plates stored in a box near the tube kept getting fogged. Rather than investigating, Smith told his assistant to keep the plates somewhere else!

Immediately on hearing of Röntgen's discovery, J. J. Thomson at the Cavendish Laboratory found that he too had a Crookes tube of the right geometry to give out x rays and he set to work with the help of his student, Ernest Rutherford, to study the nature of gas made a conductor by these new rays. (Thomson and Rutherford, 1896). They arranged their experiment so that the x rays would pass through a metal box attached to a long metal tube. At the far end of the box was an insulated wire attached to a quadrant electrometer. It was a simple apparatus, but a tricky experiment to carry out. One can get a feeling about this from their acknowledgment at the end of their paper:

> In conclusion, we desire to thank Mr. E. Everett for the assistance he has given us in these experiments. The period during which a bulb gives out Röntgen rays at a uniform rate is not a long one, and as most of our experiments required the rate of emission to be constant, they have entailed the use of a very large number of bulbs, all of which have been made by Mr. Everett (Thomson and Rutherford, 1896, p. 392).

Also suggestive of their experimental difficulties was the comment of another Cavendish physicist of this era, the younger Lord Rayleigh, who is quoted as saying of the "old quadrant electrometer:" "I do not know who designed it, but . . . I suspect that it was primarily the Devil."

In spite of obvious problems with technique, Thomson and Rutherford carried out one of the most important experiments of the age. With a pair of common household fireplace bellows they blew the air exposed to x rays along the tube and measured its conductivity with the electrometer. They showed that the electrified gas could be separated from the natural gas since it lost its charge if blown through glass wool, or a strong electric field. The gas did not obey Ohm's law, but as the potential difference across it increased, the current reached a saturation value and they concluded that the Röntgen rays produced positive and negative charged particles which tended to recombine. They wrote the equation for this loss in the form that has been used ever since:

$$\frac{dn}{dt} = -\alpha n^2 \qquad \text{or} \qquad \frac{1}{n} - \frac{1}{N} = \alpha t.$$

Their basic conclusion was "that the analogy between a dilute solution of an electrolyte and gas exposed to the Röntgen rays holds through a wide range of phenomena."

Thomson and Rutherford measured the current–voltage characteristics of the conducting gas and related that to the charged particle velocity. They measured for the first time the mobility of both positive and negative particles in air, hydrogen chloride, "coal gas," and mercury vapor. Another one of J. J. Thomson's students, John S. Townsend, measured the loss of charged particles as they were blown through a small-diameter metal tube, and developed a theory for ions based on Maxwell's (1867) "Dynamical Theory of Gases." Townsend (1900), calculated the charged particle diffusion coefficients for the same gases studied by Rutherford. Faraday had shown that the quantity of electricity transferred in electrolysis could be measured in terms of ne where n represented the number of molecules per cubic centimeter and e was the charge. With μ measured by Rutherford and D by Townsend the Cavendish group proved that

$$ne = (\mu/D)p$$

gave almost the same value for ionized gases as for liquid electrolytes.

At the same time that his students were showing that ne in the ionized gas was like that found in an electrolyte, J. J. Thomson was measuring the ratio e/m for cathode rays by deflecting a narrow beam of these rays in electric and magnetic fields. Not only did he find that his measured values were independent of the gas used and the metal of the electrodes in his tube, but most remarkable was the conclusion that his e/m was almost a thousand time greater than for a charged hydrogen ion in the electrolysis of a liquid (Thomson, 1897). He had discovered a subatomic particle, a thousand times lighter than hydrogen, and he called it a "corpuscle." Lord Kelvin wanted it to be called an "electrion" and it was the Dutch physicist H. A. Lorentz who suggested expanding Stoney's "electron" to include the particle itself, and this was the name that stuck.

The electron was so esoteric a concept, of so little importance except to those who loved it, that for years after its discovery, the toast at the annual Cavendish Laboratory dinners was: "The electron: may it never be of any use to anybody."

At this point in history, Thomson's two brilliant students left the Cavendish Laboratory and went their separate ways. Rutherford turned his attention to radioactivity and Townsend moved to Oxford to set up what became to be the most productive research laboratory in the world in the field of "electricity in gases."

With the clear recognition of the nature of the elementary charged particles: the electron, the positive ion, and the negative ion, J. S. Townsend became the acknowledged leader in gas discharge research. He and his students dominated the field for a generation, amassing an impressive volume of data, coefficients, and parameters to characterize the "passage of electricity through gases." Townsend based his explanations on the behavior of the "average electron" and the "average ion," and a very mechanistic picture of the whole field evolved. Great emphasis was placed on laws of similitude, dimensionless and proper variables, and coefficients defined in terms of the motion of individual particles.

In the late 1920s and the early 1930s, progress toward a fundamental understanding of the physics of the process was slowed, not only by the limits of the "average electron" concept to lead to new insights, but also by an experimental limitation to the purity of both gas and electrode surfaces, the seriousness of which was completely unanticipated. The best available vacua were obtained by mercury vapor pumps and pressure was read from mercury manometers or McLeod gauges. The resulting mercury contamination dominated the discharge characteristics to a point where correlation with such fundamental constants as ionization potentials, spectroscopically determined energy levels, and surface work functions appeared to have little promise.

More phenomenological studies, however, were pursued with vigor, and Loeb (1955) and his students, among others, described the details of an encyclopedic array of sparks, arcs, glows, coronas, pulses, and discharges which greatly increased the store of knowledge of the total field.

In 1935 the now classic paper on "Velocity Distributions for Elastically Colliding Electrons" was published by Morse, Allis, and Lamar (1935), which clearly showed the superiority of the electron energy distribution function approach over the average electron method for dealing with the particle motions in a discharge. Since that time, W. P. Allis, his students, and many other theoretical physicists have capitalized on the energy distribution concept, and particularly with the advent of electronic computers, the whole orientation of the field is polarized in this direction.

In the matter of the mercury contamination, which for so many years confused the results of experiments designed to relate directly atomic parameters with gas discharge behavior, the initial high-purity work was done in the Philips Laboratories in Eindhoven (Druyvesteyn and Penning, 1940). Very rapidly thereafter, the techniques for ultrahigh vacuum and better than spectroscopically pure gases were developed, and a real match between theory and experiments became possible.

The name of the field of gaseous electronics has changed drastically several times in the last one hundred years. In writing a sequel to James Clerk Maxwell's "Treatise on Electricity and Magnetism," J. J. Thomson introduced a chapter on "The Passage of Electricity Through Gases" with the sentence:

> The importance which Maxwell attached to the study of the phenomena attending the passage of electricity through gases, as well as the fact that there is no summary in English text books of the very extensive literature on this subject, leads me to think that a short account of recent researches on this kind of electric discharge may not be out of place in this volume (Thomson, 1893, p. 189).

For many years the field was designated as the "passage of electricity through gases" or "electrical discharges in gases." Gradually this was shortened to "gas discharge" physics. With the rise in popularity of the field of electronics during and after the second world war, the field changed its name again to "gaseous electronics." At the present time, when a great deal of attention is focused on the use of ionized media in controlled thermonuclear reactors, the specific type of electrical discharge, which is defined as a plasma, has tended to be corrupted to include the whole field; and by some the term "plasma physics" is used synonymously with "gaseous electronics."

Lewi Tonks published an account of Irving Langmuir's introduction of the word "plasma" in 1928:

> Langmuir came into my room in the General Electric Research Laboratory one day and said "Say, Tonks, I'm looking for a word. In these gas discharges we call the region in the immediate neighborhood of the wall or an electrode a sheath, and that seems to be quite appropriate; but what should we call the main part of the discharge? . . . there is complete space-charge neutralization. I don't want to invent a word, but it must be descriptive of this kind of region as distinct from a sheath. What do you suggest?"
>
> My reply was classic: "I'll think about it, Dr. Langmuir."
>
> The next day Langmuir breezed in and announced, "I know what we'll call it! We'll call it the plasma." The image of blood plasma immediately came to mind: I think Langmuir even mentioned blood (Tonks, 1967, p. 857).

But Langmuir was not one to be deterred by other people's concepts even in the definition of words, and as a good student of Greek, he had studied his etymology. He was impressed by the obvious characteristic that the glowing discharge molded itself to any shape into which the tube was formed, so he chose the Greek word πλάσμα which means "to mold;" the word came into general use when research problems of controlled thermonuclear fusion involved large numbers of physicists who began to refer to their working "fluid" as a Langmuir plasma.

Even a short history of gaseous electronics is never finished since progress is being made all of the time, new concepts and new terms are constantly being added, and the main focus of attention of workers in the field is always changing and expanding. It is, however, useful to realize that we are building on a long tradition and much of our language in the field predated our present understanding of the physics of the terms we use. Some of our concepts grew so naturally out of the state of the art of the times that it is difficult to pinpoint their origin. Some, on the other hand, can be given specific dates and originators. To summarize this brief history, Table I highlights some of these landmarks of progress toward our present understanding of the field of gaseous electronics.

TABLE I

DISCOVERIES IN GASEOUS ELECTRONICS

Date	Concept	Originator
1600	Electricity	Gilbert
1742	Sparks	Desaguliers
1808	Diffusion	Dalton
1808	Arc (discharge)	Davy
1817	Mobility	Faraday
1821	Arc (name)	Davy
1834	Cathode and anode	Faraday
1834	Ions	Faraday
1848	Striations	Abria
1860	Mean free path	Maxwell
1876	Cathode rays	Goldstein
1879	Fourth state of matter	Crookes
1880	Paschen curve	la Rue and Müller
1889	Maxwell–Boltzmann distribution	Nernst
1891	Electron (charge)	Stoney
1895	X rays	Röntgen
1897	[Cyclotron] frequency	Lodge
1898	Ionization	Crookes
1899	Transport equations	Townsend
1899	Energy gain equations	Lorentz
1901	Townsend coefficients	Townsend
1905	Diffusion of charged particles	Einstein
1906	Electron (particle)	Lorentz
1906	[Plasma] frequency	Rayleigh
1914	Ambipolar diffusion	Seeliger
1921	Ramsauer effect	Ramsauer
1925	Debye length	Debye and Hückel
1928	Plasma	Langmuir
1935	Velocity distribution functions	Allis

REFERENCES

Abria, A. (1848). *Ann. Chim.* **7,** 477.
Coulomb, C. A. (1785). *Mém. Acad. Sci.,* 612
Crookes, Sir William (1879a). Lecture on "Radiant Matter" delivered to the Brit. Assoc. Adv. Sci., Sheffield, August 22:
Crookes, Sir William (1879b). *Phil. Trans.* Pt. I, p. 152.
Davy, H. (1821). *Phil. Trans.* **111,** 427.
de la Rue, W., and Müller, H. W. (1878). *Phil. Trans.* **169,** 155.
Druyvesteyn, M. J., and Penning, F. M. (1940). *Rev. Mod. Phys.* **12,** 87.
Dunman, T. (1891). "A Short Text Book of Electricity and Magnetism," p. 158. Ward, Lock, Bowden & Co., London.
Faraday, M. (1839). *Res. Elec.* 663.
Gilbert, W. (1600). "De Magnete, Magneticisque Corporibus." Petrus Short, London.
Goldstein, E. (1876). *Berl. Monat.* 283.
Hittorf, J. W. (1869). *Pogg. Ann.* **136,** 8.
Hittorf, J. W. (1884). *Wied. Ann.* **21,** 96.
Lodge, O. (1906). "Electrons," p. 36. George Bell and Sons, London.
Loeb, L. B. (1955). "Basic Processes of Gaseous Electronics." Univ. of California Press, Berkeley.
Matteucci, C. (1850). *Ann. Chim. Phys.* **28,** 390.
Maxwell, J. C. (1867). *Phil. Trans.* **A157,** 49.
Morse, P. M., Allis, W. P., and Lamar, E. S. (1935). *Phys. Rev.* **48,** 412.
Perrin, J. B. (1895). *Co. Re. Acad. Sci. Paris.* **121,** 1130.
Plücker, J. (1859). *Pogg. Ann.* **107,** 77.
Rayleigh, Lord (1906). *Phil. Mag.* **11,** 117.
Röentgen, W. C. (1895). *Sitzungsber. Würzburger Phys. Med. Gesellschaftswiss. Jahrb.*
Stoney, G. J. (1891). *Sci. Trans. Roy. Dublin Soc.* **4,** 563.
Thomson, J. J. (1893). "Recent Researches in Electricity and Magnetism," p. 189. Oxford Univ. Press, London and New York.
Thomson, J. J. (1897). *Phil. Mag.* **44,** 298.
Thomson, J. J., and Rutherford, E. (1896). *Phil. Mag.* **42,** 392.
Tonks, L. (1967). *Am. J. Phys.* **35,** 857.
Townsend, J. S. (1900). *Phil. Trans.* **A193,** 129.
Wiedemann, G. H. (1891). *Sitzungsber. Phys. Med. Soz. Erlangen.*
Wilson, C. T. R. (1900). *Proc. Cambridge Phil. Soc.* **11,** 32.
Wilson, C. T. R. (1901). *Proc. Roy. Soc. London* **68,** 151.

Chapter 2

Glow Discharges at DC and Low Frequencies

Part 2.1 Anatomy of the Discharge

JOHN H. INGOLD
LIGHTING BUSINESS GROUP
GENERAL ELECTRIC COMPANY
CLEVELAND, OHIO

I. GENERAL REMARKS

The aim of this chapter is to describe the behavior of glow discharges when the applied voltage is dc or low frequency ac. One of the first problems encountered in pursuing this aim is defining "glow discharge." For example, there are arc discharges, Townsend discharges, and self-sustained and non-self-sustained discharges. According to Webster (Gove, 1961), a glow discharge is "a silent luminous electrical discharge without sparks through a gas." Now the word "discharge" means "the equalization of a difference of electric potential between two points" in the electrical sense (Gove, 1961). This meaning implies that a glow discharge is a transient phenomenon. However, this chapter is devoted mainly to steady-state dc glow discharges, with only a brief excursion into the realm of steady-state ac glow discharges maintained by a periodically varying applied voltage.

ISBN 0-12-349701-9

How is a glow discharge distinguished from an arc discharge? There is one school of thought that says that it is a glow discharge if it has a large cathode fall of potential, and an arc discharge if it has a small cathode fall of potential. Another says that to be an arc, a discharge must have nearly equal electron and heavy particle temperatures. The former definition distinguishes between a glow and an arc on the basis of electrode-related phenomena. When it comes to afterglows, however, no distinction can be made between arcs and glows on the basis of electrode-related phenomena. Consequently, the latter definition seems more appropriate to this author and will be used in this work.

In its simplest form, a glow discharge can be established by passing electric current through a gas between two electrodes, all of which may or may not be contained inside an insulating envelope. A simple circuit that can be used to establish a glow discharge is shown diagrammatically in Fig. 1. The time evolution of current I and discharge voltage V_d after the switch S is closed is depicted schematically in Fig. 2, provided that the supply voltage V_0 is greater than the breakdown voltage of the discharge tube. For a length of time t_c after the switch is closed, I is negligibly small and V_d is equal to V_0. After breakdown, which occurs for practical purposes at $t = t_c$, the current and voltage settle down to their steady-state values, which are determined by the equation

$$V_d = V_0 - IR.$$

The discharge voltage V_d depends on I and certain properties of the discharge

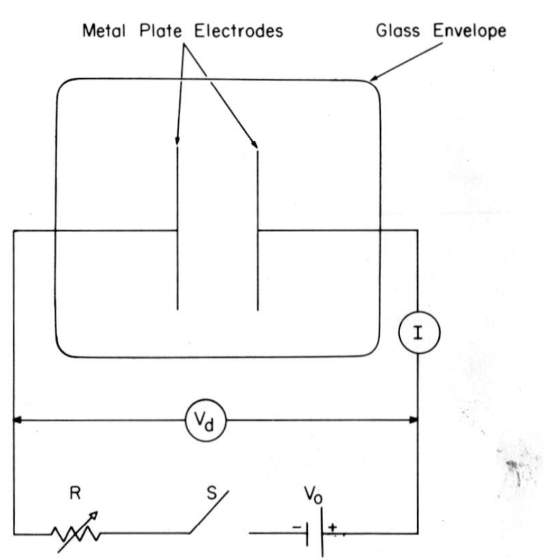

Fig. 1. Circuit used to establish glow discharge.

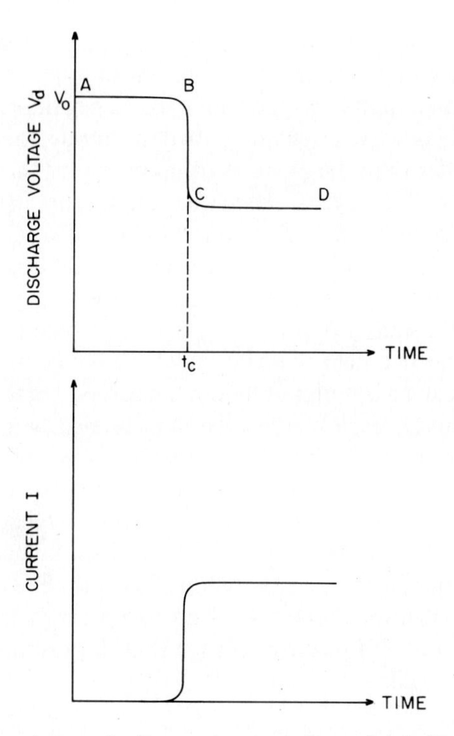

Fig. 2. Time evolution of voltage and current after switch in Fig. 1 is closed.

tube, such as gas type, gas pressure, electrode material, and electrode temperature. It is one of the main purposes of this chapter to quantify this voltage–current relationship. Before doing so, however, it is appropriate to describe qualitatively the visual and electrical phenomena that are observed. This description serves the dual purpose of orienting the reader and introducing some of the terminology that will be used here.

Part *AB* of the curve V_d versus t in Fig. 2 is called the dark discharge region. The reason is obvious—there is little or no light detected by the human eye during this part of the discharge. The light intensity grows very rapidly during the *BC* part of the curve, and a steady glow from the discharge which may occupy most of the discharge tube, is observed during the *CD* part of the curve. Visual examination of the discharge reveals the following characteristics. When the length of the discharge tube is many times the lateral dimensions of the tube, then most of the tube is filled with what appears to be a glow of uniform light intensity. This part of the discharge is called the positive column. However, there are regions near each electrode where the light intensity is not uniform. At the negative electrode, called the cathode, there may be several alternately light and dark layers, depending on the gas pressure and current. It is sufficient to note here that the cathode region has three main parts: cathode dark space, negative glow, and Faraday dark space. At the positive electrode, called the anode, there may be a region of intensity brighter than the positive column—called the anode glow—and there is usually a dark layer—called the anode dark space—between the end of the positive column and the anode glow. The relative intensities of these regions are shown in Fig. 3a. According to von Engel (1965), the glow discharge derives its name from the negative glow.

The potential distribution in a long discharge tube is roughly as shown in Fig. 3b. The electric field is much higher in the cathode region because electron multiplication is required to join the cathode to the positive column. Right at the cathode, most of the current is carried by ions flowing to the cathode; in the rest of the discharge, including the negative glow and the Faraday dark space, nearly all of the current is carried by electrons. The Faraday dark space is dark because there is little or no accelerating electric field in this region. No field is required in this region because of the "overproduction" of ions in the negative glow. The Faraday dark space can be shortened and sometimes removed altogether by providing a sink where ions can recombine with electrons faster than by ordinary ambipolar diffusion to faraway walls.

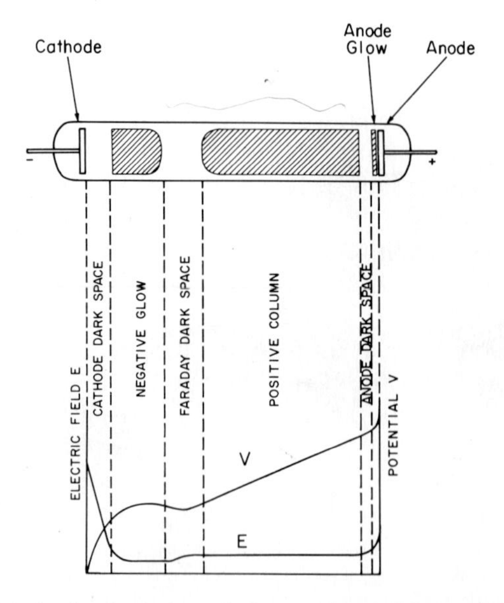

Fig. 3. Potential distribution in long discharge tube, showing variation of emitted light intensity.

The electrical characteristics of a glow discharge in a short tube without a positive column can be described conveniently by referring to the voltage–current curve shown in Fig. 4. The voltage–current curve can be divided into three main regions, as shown in the figure. At extremely low currents, on the order of microamperes and less, the discharge is a Townsend discharge, with little or no visible light emanating from the discharge tube. The electric field distribution between the electrodes is constant at very low values of current. When the current reaches the neighborhood of 0.1–1 μA, however, the field configuration begins to change due to the onset of space-charge distortion. The main difference between regions I and II is that the Townsend discharge in region II is now confined to a short distance from the cathode, whereas in region I it occupied the entire tube. Region II can be subdivided into the normal glow region (low current) and the abnormal glow region (high current); this is a cathode-related distinction. In the normal glow region, the cathode fall is nearly independent of current. The reason that the voltage begins to rise in the abnormal glow region as the current is increased is that the cathode fall must increase to supply the additional current. When the current is allowed to increase even further, a phenomenon customarily known as the glow-to-arc transition (GAT) takes place. Once again, this is a cathode-related distinction, for the behavior of the negative glow, and of the positive column, if any, is the same in region III as in region II. What happens during GAT is that the cathode fall undergoes a transformation from a cold-cathode discharge depending primarily

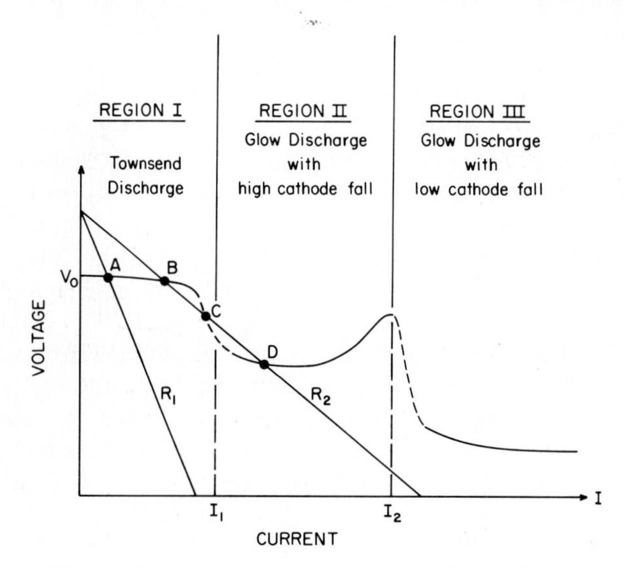

Fig. 4. Schematic voltage–current curve of glow discharge.

on Townsend's first and second ionization processes, to a hot cathode discharge in which thermionic emission plays an important role. The discharge in region III is not an arc discharge according to the definition given at the beginning of this chapter; nevertheless, workers in this field refer to the discharge in region III as an arc discharge.

The voltage–current curve in Fig. 4 shows schematically what happens when the resistance R in the circuit shown in Fig. 1 is varied. The load lines in Fig. 4 correspond to two different values of R: $R_1 > R_2$. The points where the load lines cross the voltage–current curve of the discharge represent possible operating points. For example, when $R = R_1$, then the discharge will operate with voltage and current indicated by point A. When the resistance R is decreased to R_2, then the discharge operates with voltage and current indicated by point B. However, if the resistance R_2 were to be approached from below, then the discharge would operate at the voltage and current given by point D. Point C is an unstable operating point, and cannot be reached in the steady state with this simple circuit.

II. CATHODE REGION

Of the three main parts of the glow discharge, the cathode fall region is probably the least understood from a theoretical point of view. Unlike the positive column, which is uniform in the direction of current flow, the cathode fall is highly nonuniform in the direction of current flow, due primarily to electron multiplication, which is required for joining the cathode to the positive column. Because the cathode fall, negative flow, and Faraday dark space are related to cathode properties, whereas the positive column is independent of the cathode, it is helpful in the interests of simplicity to discuss the cathode fall, negative glow, and Faraday dark space together by assuming that they form a diode, as shown in Fig. 1. The generic voltage–current curve for a self-sustaining discharge in this diode is shown in Fig. 5. This curve can be traced out by varying the load resistance R shown in Fig. 1. When V_d in Fig. 1 is less than V_b in Fig. 5, there is no discharge. When V_d reaches V_b, however, a small current on the order of 10^{-9} A will flow, if a temporary initial source of electrons has been provided. This low-current region is the dark discharge referred to in Section I. As the current is increased further either by increasing V_0 or by decreasing R, V_d begins to decrease as shown in Fig. 5. This decrease is due to the onset of space-charge distortion of the electric field in the gas diode. For a further increase in current, the voltage levels out at a value commonly called the normal glow voltage V_n. The normal glow region usually extends from about 10^{-6} to 10^{-4} or 10^{-3} A/cm^2. It is a region characterized by little change in voltage and a

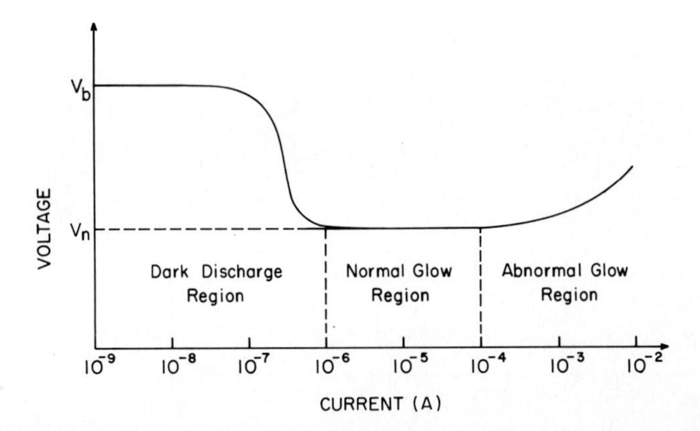

Fig. 5. Voltage–current curve of self-sustaining glow discharge (no positive column).

large change in current. Finally, however, as the current is increased even further, the voltage across the discharge begins to increase once more. By custom, this region of the glow discharge is called the abnormal glow. To help clarify the differences between these three main regions of the glow discharge, schematic potential diagrams thought to be typical of each are shown in Fig. 6. In summary, the dark discharge is characterized by a uniform electric field, the normal glow by high field at the cathode and low field at the anode, and the abnormal glow by a higher field at the cathode, which increases with current. The physical processes thought to account for this behavior will now be discussed.

The discharge tube shown schematically in Fig. 1 has electrodes at each end to permit electric current to flow to and from the actual discharge. The electrode at the negative end of the tube is called the cathode; its purpose is to provide electrons to maintain the discharge. There are several physical processes by which a cathode can emit electrons. Direct emission processes include thermionic emission and field emission, while indirect processes, usually called secondary emission, consist of ejection of electrons from the cathode by energetic particles, such as ions, electrons, and photons. For a cold discharge tube with applied voltages up to a few thousand volts, the direct processes are negligible. Furthermore, at the cathode where there is a fairly large positive potential gradient, there are no electrons impinging on the cathode. Therefore, the physical processes most likely to cause emission of electrons from the cathode are ion and photon bombardment. Only secondary emission by ion bombardment will be discussed in this chapter.

The large positive potential gradient in front of the cathode is necessary to accelerate the few electrons ejected from the cathode by impinging ions

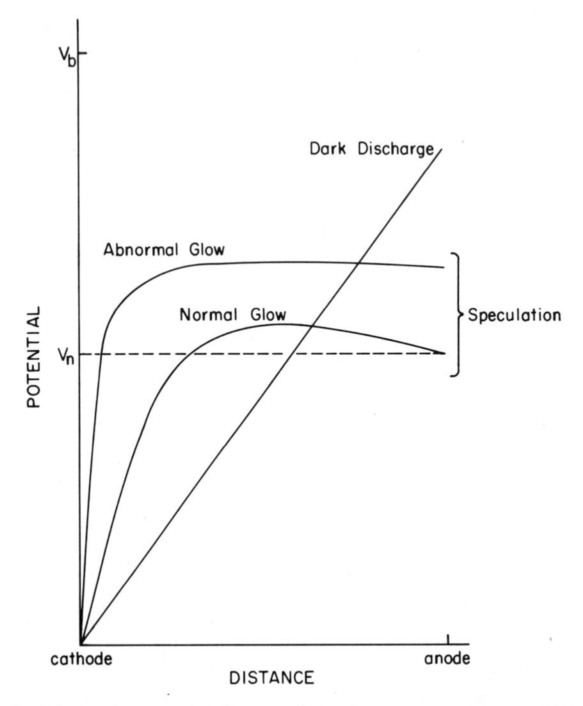

Fig. 6. Schematic potential diagram for voltage–current curve of Fig. 5.

to energies high enough to create additional electrons by ionizing the gas in front of the cathode. The nature of the potential gradient in front of the cathode is such that the electrons are accelerated away from the cathode, whereas ions are accelerated toward the cathode, where they can eject more electrons.

One of the earliest theoretical treatments of the self-sustaining glow discharge cathode fall current–voltage relationship was given by von Engel and Steenbeck (1934). They assumed that the electric field was linear in the cathode dark space, and that the ion flow to the cathode was mobility limited. Their analysis is an extension of Townsend's (1915) dark discharge theory, which will be reviewed here. Townsend reasoned that the growth of electron current could be described by a simple exponential increase with distance away from the cathode. In other words, the increase in electron current should be proportional to the current. This assumption leads to the relation

$$d\Gamma_e/dx = \alpha\Gamma_e, \tag{1}$$

where Γ_e is electron current density expressed in electrons per unit area per unit time, x is distance, and α is known as Townsend's first ionization coefficient. For constant α, this equation integrates to give

$$\Gamma_e(x) = \Gamma_e(0)e^{\alpha x}, \tag{2}$$

where $\Gamma_e(0)$ is the value of Γ_e at the cathode located at $x = 0$. Townsend reasoned further that electrons were produced at the cathode by ions that had been accelerated through the cathode fall. He related the outgoing electrons current density $\Gamma_e(0)$ to the incoming current density $\Gamma_p(0)$ by the relation

$$\Gamma_e(0) = -\gamma\Gamma_p(0) + \Gamma_0, \tag{3}$$

where γ is known as Townsend's second ionization coefficient and Γ_0 is electron emission from the cathode caused by an agent external to the discharge. By Eq. (2), the electron current density at $x = d$, where d denotes the distance separating the discharge cathode and anode, is given by the relation

$$\Gamma_e(d) = \Gamma_e(0)e^{\alpha d}. \tag{4}$$

Because a new ion is created each time an electron is created, the ion current density at the cathode is related to the ion current density at $x = d$ by the relation

$$\Gamma_p(0) = \Gamma_p(d) - \Gamma_e(0)(e^{\alpha d} - 1). \tag{5}$$

When $\Gamma_p(0)$ is eliminated from Eq. (5) by means of Eq. (3), the result can be expressed by

$$\Gamma_p(d) = \frac{\Gamma_0}{\gamma} - \frac{\Gamma_e(0)}{\gamma}[1 - \gamma(e^{\alpha d} - 1)] \tag{6}$$

Because the anode does not emit ions, $\Gamma_p(d) = 0$. In this case, Eq. (6) can be combined with Eq. (4) to give

$$\Gamma_e(d) = \frac{\Gamma_0 e^{\alpha d}}{1 - \gamma(e^{\alpha d} - 1)}, \tag{7}$$

which is the usual expression for the current growth in a Townsend discharge.

In a self-sustained glow discharge, however, there is no external agent to produce electron emission from the cathode; consequently, $\Gamma_0 = 0$. But if $\Gamma_0 = 0$, there is no current to the anode by Eq. (7), unless the denominator of Eq. (7) is also zero. This is the case of a self-sustained discharge, with the maintenance condition

$$\alpha d = \ln(1 + 1/\gamma). \tag{8}$$

Because α depends on the electric field, Eq. (8) defines the electric field required to maintain the discharge in the self-sustaining condition for the dark discharge. In the normal and abnormal glow conditions, however, the electric field is variable. Therefore, it is customary to replace α by $\bar{\alpha}$, where $\bar{\alpha}$ is defined by the relation

$$\bar{\alpha} = \frac{1}{d_c} \int_0^{d_c} \alpha \, dx \tag{9}$$

so that the maintenance condition becomes

$$\bar{\alpha} d_c = \ln(1 + 1/\gamma) \tag{10}$$

in which d_c is the thickness of the cathode fall region. Definition of the thickness of the cathode fall region is rather tenuous, but most workers in the field tend to define the cathode fall thickness as the distance from the cathode point in the discharge where the electric field is zero. Equation (10) forms the basis of most of the presently accepted theories of the cathode fall, even though its derivation depends on the assumption that no ions enter the cathode fall region from the negative glow.

In addition to its dependence on electric field strength, the first ionization coefficient α is a property of the gas. The second ionization coefficient γ is also a property of the gas (through its dependence on the kind of ion), as well as the electrode material. Therefore, it is apparent that the cathode fall current–voltage relationship is a property of both gas and electrode. On the other hand, the cathode fall does not depend on the properties of the positive column or of the anode. In fact, if the cathode is moved while the anode position is held constant, the negative glow moves with the cathode as though attached to it.

In addition to the maintenance condition expressed by Eq. (10), an equation that relates current to the cathode fall voltage and thickness is needed. In principle, this equation should take into account the possibility that the ion flow to the cathode can be either mobility limited or inertia limited, depending on whether the ion mean free path is small or large compared with the cathode fall thickness. It is generally assumed that the ion current to the cathode is mobility limited, in which case the high pressure space-charge-limited current–voltage relationship should be used. When the ion mobility is independent of the electric field, this equation is

$$\Gamma_p(0) = (9\varepsilon\mu_p/8e)V_c^2/d_c^3, \tag{11}$$

where ε is permittivity of free space, μ_p is ion mobility, e is electron charge, and V_c is cathode fall voltage (Cobine, 1958).

Equations (10) and (11), along with an empirically (usually) determined relation for α in terms the electric field E* and pressure p, uniquely determine the cathode fall voltage and current. The first ionization coefficient is usually expressed in terms of E and p by the relation

$$\alpha/p = A\exp(-Bp/E), \tag{12}$$

where A and B are empirically determined constants characteristic of the gas. It is customary to express cathode fall thickness in terms of the parameter pd_c, and the current density in terms of the parameter J/p^2. Following this custom, it can be shown that the simple picture outlined above gives the following cathode fall relations:

$$(\bar{\alpha}/p)pd_c = \ln(1 + 1/\gamma), \tag{13}$$

$$V_c = \int_0^{pd_c} (E/p)\, d(px), \tag{14}$$

$$J/p^2 = (1 + \gamma)(9\varepsilon\mu_p p/8)V_c{}^2/(pd_c)^3, \tag{15}$$

where α/p is given by Eq. (12), and $\bar{\alpha}$ is defined above.

Equations (13)–(15) are similar, if not identical, to the early theory proposed by Von Engel and Steenbeck. They assumed a linear electric field in the cathode fall region, i.e.,

$$E = E_0(1 - x/d_c),$$

where E_0 is the value of E at the cathode. When this relation is substituted into Eq. (12) and the result integrated, the expression for $\bar{\alpha}$ is found to be

$$\bar{\alpha}d_c = \alpha_0 d_c \exp(Bp/E_0)E_2(Bp/E_0),$$

where α_0 is the value of α at $x = 0$ and $E_2(u)$ is the exponential integral defined by the equation

$$E_n(u) = \int_1^\infty e^{-tu}t^{-n}\, dt.$$

To illustrate the simplified picture outlined above, the cathode fall current–voltage curve for an argon discharge will be calculated by means of the equations given above, except that a slightly different equation for α, due to Ward (1962) will be used. This relation is

$$\alpha/p = A\exp[-B(p/E)^{1/2}],$$

* In this section, the electric field is defined by the relation $E = +dV/dx$.

where $A = 29.22\ \mathrm{cm^{-1}\ Torr^{-1}}$ and $B = 26.64\ \mathrm{V^{1/2}\ cm^{-1/2}\ Torr^{-1/2}}$. For this kind of field-dependence for α, the form of $\bar{\alpha}$ is expressed by the relation

$$\bar{\alpha}d_c = 2\alpha_0 d_c \exp[B(p/E_0)^{1/2}]E_3[B(p/E_0)^{1/2}].$$

The calculated results for the variation of J/p^2 with V_c for $\mu_p p = 10^3\ \mathrm{cm^2\ Torr\ V^{-1}\ sec^{-1}}$ are shown as solid curves in Fig. 7. The parameter in Fig. 7 is γ, and the dashed curve represents the experimental results of Guntershulze (1930; averaged by Ward). These curves show qualitative agreement between theory and experiment for values of J/p^2 greater than about $10^{-5}\ \mathrm{A\ cm^{-2}\ Torr^{-2}}$. However, for values of J/p^2 less than this amount, the theoretical curves show that V_c begins to increase, which is not observed experimentally. More will be said about this later.

After the pioneering work by Von Engel and Steenbeck, Druyvesteyn (1938) made extensive comparisons between their model and his own meaurements. His findings are summarized in Fig. 8, which shows the universal theoretical curve derived by Von Engel and Steenbeck and the corresponding experimental results of Druyvesteyn. Note that neon is the only gas that follows the theoretical curve, while the other rare gases deviate significantly

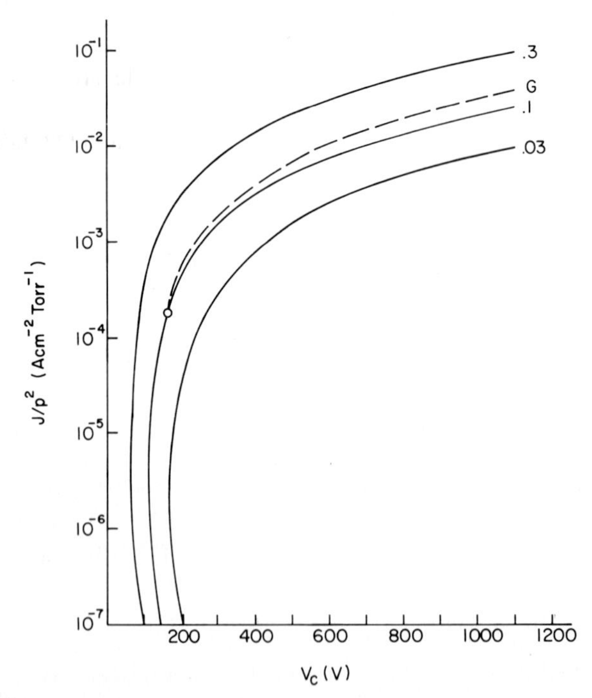

Fig. 7. Current density J/p^2 vs. cathode fall voltage V_c for simple model.

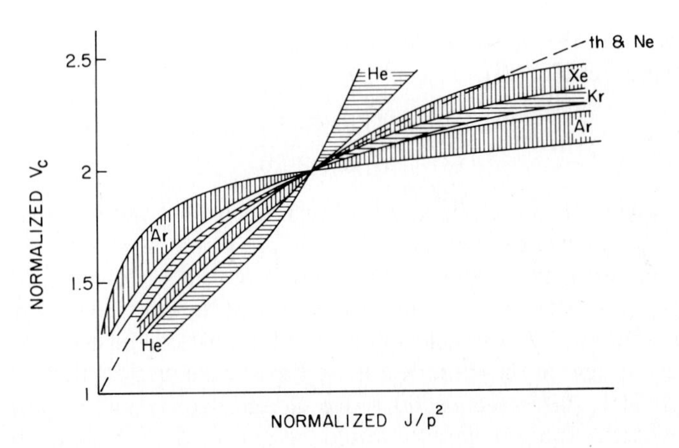

Fig. 8. Druyvesteyn's comparison of simple model with measurement.

from the universal curve. This deviation is probably due to the simplified model used by Von Engel and Steenbeck.

This original theory and subsequent refinements and improvements were summarized in a long review article by Druyvesteyn and Penning (1940), in another by Francis (1956), and in the book by Weston (1968). Weston concludes that a better understanding of the cathode fall region is now available, but he doubts that present theoretical current voltage characteristics are more accurate than the early ones. The better understanding stems from steadily improving experimental procedures used to measure cathode fall properties. It is only by having well outgassed discharge tubes, very pure gases, and very clean electrodes that reproducible results can be obtained for measurements of α and γ. For pure inert gases, γ generally ranges from 0.01 to 0.3, and the normal glow value of V_c from 100 to 300 V. Weston's book presents a fairly complete survey of the experimentally determined values of these quantities, and the reader who is interested in the experimental measurements should consult it.

One controversy that has not been satisfactorily resolved at the present time has to do with the role of the negative glow in the establishment of a self-sustaining discharge. The specific question has to do with the origin of the ions that strike the cathode and liberate electrons to maintain the discharge. The maintenance condition embodied in Eq. (10) is based on the assumption that all of the ions that reach the cathode are produced in the cathode fall, i.e., there are no ions entering the cathode fall region from the negative glow. This is a natural, but questionable, extension of Townsend's breakdown theory to a steady-state discharge. Druyvesteyn and Penning (1940) have pointed out that if a significant number of ions enter the cathode fall region from the negative glow, the maintenance condition is altered to the following [cf. Eq. (8)]:

$$\bar{\alpha} d_c = \ln \frac{1 + 1/\gamma}{1 + \delta}, \tag{16}$$

where δ is defined as the ratio $J_p(d_c)/J_e(d_c)$. According to this equation, the value of $\bar{\alpha} d_c$ required to maintain the discharge would be significantly lower if $J_p(d_c)$ amounted to an appreciable fraction of $J_e(d_c)$. However, Little and Von Engel argue that there are no ions entering the cathode fall region from the negative glow. Their conclusion is based in part on the assumption of a linear electric field in the cathode fall. By Poisson's equation, they show that a linear field requires a constant space charge density in the cathode fall region, whereas the ion density would be expected to decrease from the negative glow to cathode if most of the ions originated in the negative glow. This argument does not appear to be conclusive, because almost any slowly varying field would appear approximately linear to the experimentalist. For example, the Child–Langmuir space charge equation gives the space variation of the electric potential as $x^{4/3}$, and the high-pressure space charge equation as $x^{3/2}$. All of these functional forms appear quite similar, as shown in Fig. 9. Weston does not mention ion current from the negative glow in his book; therefore, he apparently agrees with the position taken by Little and von Engel on this matter. In addition, Weston cites numerical calculations by Ward (1962) that show an approximately linear field at the cathode. What should be examined, however, is the spatial variation of charge density in the cathode fall to establish the validity of the assumption of a linear field.

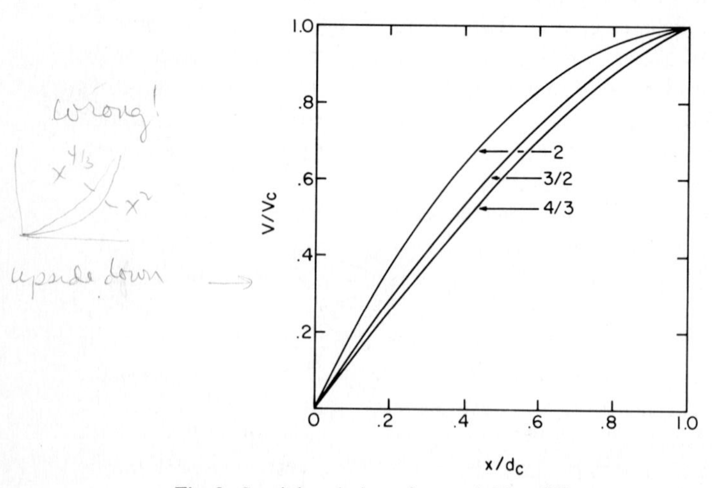

Fig. 9. Spatial variation of potential for different space charge models.

Another weakness of the presently accepted models of the cathode fall region has to do with the concept of α. In Equation (1), α denotes the number of ionizing collisions per centimeter of path length traversed by an electron. This Townsend coefficient is usually measured in swarm-type experiments with uniform electric fields at low electron density to avoid space charge distortion. Great care is taken in these measurements to ensure that the electrons "are in equilibrium with the field," which means that the electron energy distribution is not changing with position. These conditions are probably not met in the cathode fall region of a cold-cathode glow discharge, especially in the abnormal glow regime where the value of E/p can reach 10^4 V cm^{-1} Torr^{-1} over a distance of 10^{-2} cm. In such a situation, the electron energy is probably changing throughout the cathode fall, and the energy distribution is probably highly peaked in the forward direction in velocity space, so that the ionization probability depends not only on the value of E/p, but also on the distance through which the ionizing electron has been accelerated. This means that α is not a unique function of E/p, but depends on the thickness of the cathode fall as well. Consequently, theoretical models based on the concept of a constant, slightly anisotropic velocity distribution for the electrons in the cathode fall may not be valid.

The effects of (a) ion current to the cathode from the negative glow and (b) variable α in the cathode fall can be investigated qualitatively with the help of a simplified model. The main feature of the model is the assumption of a monoenergetic beam for the electron energy distribution. For a monoenergetic beam of electrons, Townsend's first ionization coefficient is given by the relation

$$\alpha = n_a Q_i(\xi), \tag{17}$$

where n_a is gas density and $Q_i(\xi)$ the cross section for ionization of the gas atoms by electrons of energy ξ. The energy balance equation for the electrons in the cathode fall is

$$\frac{d}{dx}(\Gamma_e \xi) = \Gamma_e E - \alpha \Gamma_e V_i,$$

where $\Gamma_e E$ is the energy gained from the electric field, and $\alpha \Gamma_e V_i$ the energy lost due to ionization of gas atoms with ionization potential V_i. By Eq. (1), this equation can be put in the form

$$(d\xi/dx) + \alpha\xi = E - V_i, \tag{18}$$

which can be solved numerically when the energy dependence of the cross section and spatial dependence of the electric field are specified. For example, above threshold, the ionization cross section for argon follows approximately the functional form

$$Q_i(\xi) = C(\xi - V_i)/(1 + D\xi^{3/2}), \tag{19}$$

where C and D are constants that have the values 2.63×10^{-17} cm^2 V^{-1} and 7×10^{-3} V$^{-3/2}$, respectively, up to several hundred volts of electron energy (Kieffer, 1969). Numerical solutions to Eqs. (18) and (19) for a constant electric field are shown in Fig. 10 for the condition $V(0) = V_i$. [The solution for $V(x)$ for energies smaller than V_i is trivial.] The results are shown as electron energy measured in volts plotted against distance-measured units of $\lambda_i = 1/n_a C V_i = 0.067$ cm for argon at a gas density of 3.53×10^{16} cm^{-3} (1 Torr pressure). The parameter is E/p, and the curves have been terminated at the distance at which the maintenance condition embodied in Eq. (16) is satisfied, i.e., at the end of the cathode fall, where $x = d_c$. To evaluate this maintenance condition, it was assumed that every ion produced in the negative glow by primary electrons from the cathode fall region was returned to the cathode, and that the number of such ions returning per unit time was given by the relation

$$\Gamma_p(d_c) = -\Gamma_e(0)\exp(\bar{\alpha} d_c)(\xi_c/V_i - 1).$$

Note that the quantity $\xi_c/V_i - 1$ is the maximum number of ion pairs that one electron of energy ξ_c can produce in the negative glow. The average

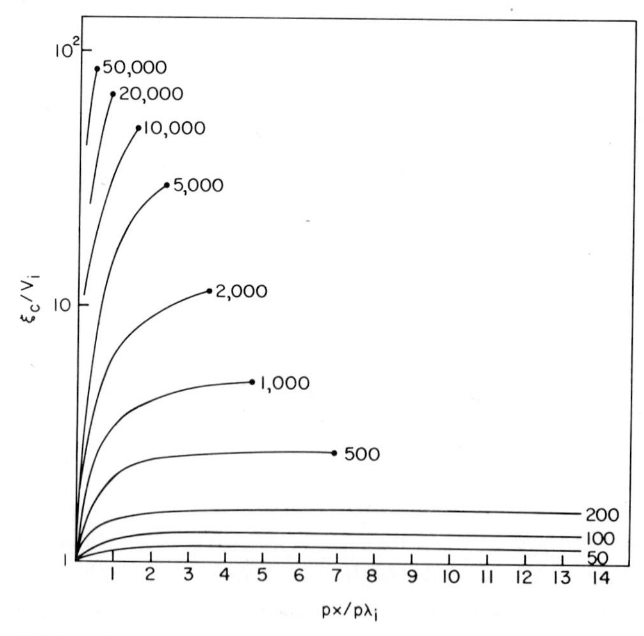

Fig. 10. Spatial dependence of electron energy for electron beam in cathode fall (theoretical).

energy ξ_c of the electrons as they enter the negative glow is denoted by the circled points in Fig. 10. According to these curves, the average electron energy is not constant anywhere in the cathode fall for $E/p \geq 1000$ V cm^{-1} Torr^{-1}. Consequently, the constant electron energy assumption may not be valid in the cathode fall region in the abnormal glow.

The corresponding current–voltage curves are shown in Fig. 11 for several values of Townsend's second ionization coefficient γ. These curves were generated by getting the cathode fall thickness measured in units of λ_i from Fig. 10, and then generating pairs of values of V_c and J/p^2 from Eq. (14) and either Eq. (15) or the inertia-limited space charge equation

$$J/p^2 = (1 + \gamma)(4\varepsilon/9)(2e/m_i)^{1/2} V_c^{3/2}/(pd_c)^2, \tag{20}$$

according to which gave the *smaller* current for the same values of V_c and pd_c. This is necessary because at very high values of E/p the high-pressure space charge equation [Eq. (15)] can give higher values of current than does the vacuum space charge equation, whereas the latter equation should always be viewed as giving the upper limit to the current density for a given pair of values of V_c and pd_c. This approximate procedure is not as accurate as the

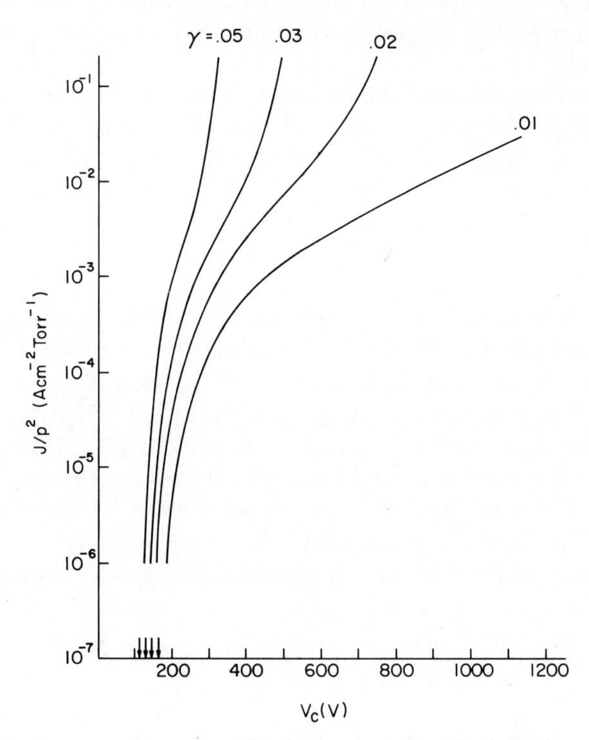

Fig. 11. Current–voltage curves for electron beam in cathode fall (theoretical).

treatment given by Ingold (1969) for the transition from inertia-limited flow to mobility-limited flow, but it is adequate for the treatment presented here. Because current–voltage curves measured in discharges with planar electrodes seldom show steeply rising portions like those in Fig. 11 (cf. Fig. 7), it can be concluded that ion current given to the cathode from the negative glow is small. Perhaps the curves in Fig. 11 are more indicative of hollow-cathode glow discharges, where the ions must return to the cathode rather than diffuse to the anode or to the confining walls.

One interesting feature of the curves shown in Fig. 11 is their behavior at low current density. Instead of V_c increasing at low current as given by earlier work (cf. Fig. 7) V_c approaches asymptotically to the limit

$$V_c - V_i = 2V_i \ln(1 + 1/\gamma).$$

Therefore, the present model predicts a normal glow voltage given by the relation

$$V_n = V_i[2 \ln(1 + 1/\gamma) + 1].$$

On the basis of the present model, the so-called normal glow voltage must be interpreted as the minimum voltage required to ensure that the electron energy ξ_c is greater than the ionization potential V_i, for otherwise the glow would extinguish.

In theory, the mathematical formulation allows J/p^2 to decrease indefinitely, while V_c remains equal to V_n and pd_c increases indefinitely. In practice, however, what happens in a planar discharge is that J/p^2, V_c, and pd_c remain equal to J_n/p^2, V_n, and pd_n, respectively, while the discharge contracts radially to allow the total current to decrease. Consequently, this model does not explain all of the observed characteristics of the normal glow discharge.

A serious shortcoming of this model has to do with the assumption of a monoenergetic beam for the distribution of electron energy in the cathode fall. Unless the thickness of the cathode fall is less than the mean free path for elastic collisions between electrons and atoms, the electron energy distribution cannot be beamlike. Instead, it is more likely to be a spherical shell with its center displaced in the direction of current flow. This is the approximate shape of the distribution to be expected when electron momentum but not electron energy, is randomized.

In mathematical form, the one-dimensional distribution function $f(x,u,v)$ is

$$f(x,u,v) = \frac{n(x)}{2\pi w(x)} \delta\{[u - u_0(x)]^2 + v^2 - w^2(x)\},$$

where the electron density $n(x)$ is given by the relation

$$n(x) = 2\pi \int_{-\infty}^{\infty} \int_{0}^{\infty} f(x,u,v)v \, dv \, du.$$

The random speed $w(x)$ represents the radius of the spherical shell, and the drift speed $u_0(x)$ represents the displacement of the center of the spherical shell from the origin in velocity space. This form for the distribution function is probably a very good approximation in the near-cathode region of the CDS, and perhaps not so good near the NG. Nevertheless, it is instructive to investigate the consequences of such a model.

Plane geometry is assumed for the CDS, so that pertinent velocity moments are the following:

Average x component of velocity:

$$\bar{u} = u_0.$$

Average x component of momentum flow in x direction:

$$\overline{nmu^2} = nm(u_0{}^2 + w^2/3).$$

Average energy:

$$\overline{\tfrac{1}{2}nm(u^2 + v^2)} = \tfrac{1}{2}nm(w^2 + u_0{}^2).$$

Average x component of energy flow in x direction:

$$\overline{\tfrac{1}{2}nm(u^2 + v^2)u} = nu_0(\tfrac{1}{2}mu_0{}^2 + \tfrac{5}{6}mw^2),$$

where m is electron mass. The three equations that determine $n(x)$, $u_0(x)$, and $w(x)$, are the following:

Particle balance:

$$\frac{d}{dx}(nu_0) = \nu_i n.$$

Momentum balance:

$$\frac{d}{dx}[nm(u_0{}^2 + w^2/3)] = enE - m\nu_m nu_0.$$

Energy balance:

$$\frac{d}{dx}[nu_0(\tfrac{1}{2}mu_0{}^2 + \tfrac{5}{6}mw^2)] = enu_0E - 2\frac{m}{M}\nu_m e\xi n - eV_i\nu_i n,$$

where V_i is ionization potential, e electron charge, ν_i the frequency of electron impact ionization, ν_m the frequency of elastic momentum transfer collisions between electrons and atoms, and $e\xi = \frac{1}{2}m(w^2 + u_0{}^2)$ the total average energy per electron.

Two features of the balance equations should be pointed out:

(a) When the electron temperature is defined by the relation

$$\tfrac{3}{2}kT = \tfrac{1}{2}mw^2,$$

then the left-hand sides of the momentum and energy balance equations become

$$\frac{d}{dx}(nmu_0{}^2 + nmkT), \qquad \frac{d}{dx}[nu_0(\tfrac{1}{2}mu_0{}^2 + \tfrac{5}{2}kT)],$$

which are identical to the results found for a shifted Maxwellian distribution (Chapman and Cowling, 1960).

(b) These results differ from those published previously for the same delta function distribution (Kozlov and Khvesyuk, 1971).

When n is eliminated from the momentum and energy balance equations by means of the particle balance equation, two equations remain to be solved for w and u_0. In the near-cathode region where the effects of ionization and elastic energy loss can be neglected, these equations can be solved analytically in the case of a constant momentum transfer collision frequency ν_m. This solution shows that the drift velocity u_0 saturates at the value $u_0 = 0.6\mu E$ at a short distance pd_1 from the cathode. The value of pd_1 is given by the relation

$$pd_1 = \frac{0.72(\mu p)(E/p)}{\nu_m/p},$$

where $\mu = e/m\nu_m$ is the electron mobility. On the other hand, the energy $\frac{1}{2}mu_0{}^2 + \frac{5}{6}mw^2$ increases linearly with distance away from the cathode. Consequently, the distance pd_2 required for the electron energy to reach the ionization potential of the gas is

$$pd_2 = \left[\frac{5}{3}\frac{eV_i/m}{(\mu p)^2(E/p)^2} - \frac{3}{25}\right]\frac{(\mu p)(E/p)}{\nu_m/p},$$

and the voltage V_2 at this point is simply $(E/p)(pd_2)$, which is about $\frac{5}{3}V_i$.

Numerical solutions of the complete equations show that the distance pd_3 from the cathode at which the total energy saturates is given approximately by the relation

$$pd_3 = \left[\frac{5}{3}\frac{e\xi/m}{(\mu p)^2(E/p)^2} - \frac{3}{25}\right]\frac{(\mu p)(E/p)}{\nu_m/p}$$

provided that departure from normal glow conditions is not severe. (The saturated values of u_0 and w are obtained by solving the three balance equations with $du_0/dx = dw/dx = 0$.)

To satisfy the maintenance condition given by Eq. (16), it is necessary that

$$\ln[(1 + 1/\gamma)/(1 + \delta)] = \bar{\alpha} d_c \approx \alpha(d_c - d_3),$$

where α is evaluated at the saturated value of electron energy. Typical results calculated for argon with $\nu_m = 6.62 \times 10^9\ \text{sec}^{-1}$ are given in Table I for $\gamma = 0.1$ and $\delta = 0$.* Note that V_c appears to approach a minimum value of about 127 V, which is interpreted here to be the normal value V_n. The values of V_c vs. pd_c listed in Table I compare favorably with measured values

TABLE I

CALCULATED RESULTS FOR ARGON

Saturated energy (eV)	pd_2 (Torr-cm)	pd_3 (Torr-cm)	pd_c (Torr-cm)	α/p ($\text{cm}^{-1}\text{Torr}^{-1}$)	E/p ($\text{V cm}^{-1}\text{Torr}^{-1}$)	V_c (V)
15.8	0.316	0.320	1.543	1.96	82	127
16	0.222	0.228	1.1	2.77	117	128
20	0.06	0.077	0.349	8.82	423	147
30	0.027	0.052	0.219	14.35	892	195

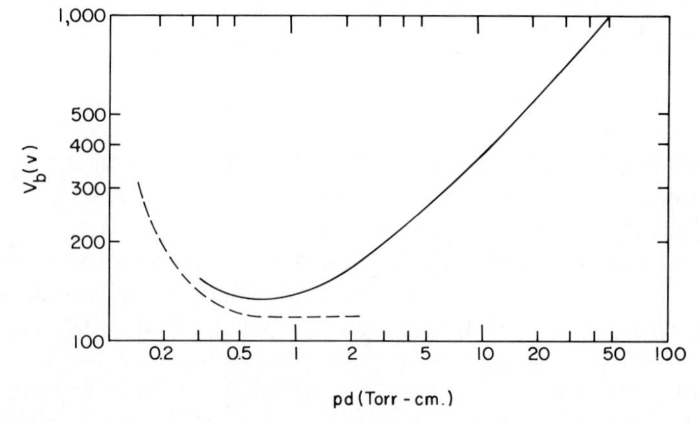

Fig. 12. Paschen curves for argon. —, measured by Frouws (CIPIG, 3, Venice, 1957); ---, calculated for spherical shell electron energy distribution.

* This value of ν_m for electrons in argon is approximately valid for electron energies greater than about 15 eV.

of breakdown voltage vs. *pd* to the left of the Paschen minimum, as shown in Fig. 12 (Weston, 1968). The reason for disagreement to the right of the Paschen minimum is thought to be as follows: The present model is based on the assumption that "tail" electrons are not important in determining the ionization rate, i.e., the average energy of the "bulk" electrons is greater than the ionization potential of the gas. This assumption is no longer valid on the right side of the Paschen minimum because E/p is getting smaller and smaller as *pd* increases. Consequently, tail electrons are important in determining the ionization rate of the right side of the Paschen minimum, and in the present model, there are no tail electrons.

III. POSITIVE COLUMN

The positive column is probably the best understood of the three main parts of the glow discharge. This understanding is undoubtedly due to the axial and azimuthal symmetry of the positive column, which allows a one-dimensional theoretical analysis to be made. In other words, all of the measurable quantities—particle densities, current densities, particle temperatures—vary with radial position only. The analysis is simplified even further at low current densities where negligible power is dissipated in the positive column by the applied electric field: the particle drift velocities and temperatures can be assumed constant independent of radial position. Then only the particle densities remain to be determined. Furthermore, when the fractional ionization is small and there are no chemical reactions taking place, then the density of neutral particles in the ground state can be assumed constant independent of radial position. Finally, when there is only one ionic species, then there are only two particle densities—electrons n_e and ions n_p—for which the radial dependence must be determined. In principle, this determination can be made by solving the Boltzmann equations for the electron energy distribution functions (EEDF) and the electrons and ions. In practice, the solution of the Boltzmann equations is so complicated that various approximations are usually made. One of the simplest approximations is called the method of moments, which consists of taking various velocity moments of the Boltzmann equation and retaining only those moments that have simple physical interpretations. This method has its limitations, but it is a useful one for illustrating the main features of the positive column. The first, second, and third velocity moments of the Boltzmann equation give rise to the equations of conservation of mass, momentum, and energy for each species. To first order in the gradients of the macroscopic variables, which are density, mean velocity, and temperature, these equations in the case of cylindrical symmetry are of the following form:

$$\frac{1}{r}\frac{d}{dr}(r\Gamma_{er}) = S_i, \qquad \Gamma_{er} = n_e u_e, \qquad \Gamma_{ez} = n_e v_e, \tag{21}$$

$$\frac{d}{dr}(n_e e\theta_e + n_e m_e u_e^2) - en_e E_r = -m_e \nu_{em}\Gamma_{er}, \qquad en_e E_z = m_e \nu_{em}\Gamma_{ez}, \tag{22}$$

$$n_e(e/m_e\nu_{em})E_z^2 = 3\frac{m_e}{m_a}\nu_{em}e(\theta_e - \theta_g)n_e - eV_i S - \sum_j eV_j S_j, \tag{23}$$

where n is particle density, u the radial component of mean velocity; v the axial component of mean velocity; S_i the rate of production of charged particles expressed in units of particles per unit volume per unit time; θ temperature expressed in units of volts; ν_m the frequency of elastic momentum transfer collisions between charged particles and neutrals; the subscript a refers to neutral particles and e to electrons; and V_j is the energy level in volts of the jth excited neutral expressed in units of particles per unit volume per unit time. The case that has been studied most extensively from a theoretical point of view is that of single-step ionization and excitation by electron impact, i.e., $s = \nu_i n_e$, and $S_j = \nu_j n_e$, where ν_i and ν_j are the ionization and excitation frequencies. The production rates S_i and the S_j are given by the standard formula

$$S = \int_0^\infty n_a Q(v) v f_e(v)\, d^3v = \nu n_e = a n_a n_e, \tag{24}$$

where $Q(v)$ is the cross section for the process. The axial field E_z and the electron temperature θ_e are constants in the simplest case, while n_e, n_p, and E_r are the functions of radius only. Needless to say, however, the solution must be obtained by numerical means because of the nonlinearity of the equations. Consequently, further approximations are usually made. For example, in the limit of a collision-dominated plasma with approximate charge neutrality, the inertia terms can be dropped and the ion density can be set equal to the electron density. Then Eqs. (21) and (22) can be combined to give the following equation for the charged particle density n:

$$\frac{1}{r}\frac{d}{dr}r\frac{dn}{dr} + \frac{\nu_i}{D_a}n = 0. \tag{25}$$

This equation was first derived and studied by Schottky (1924), in connection with studies of the positive column. The solution is, of course,

$$n(r) = n_0 J_0(kr), \tag{26}$$

where J_0 is the zero-order Bessel function of the first kind and $k^2 = v_i/D_a$. Schottky reasoned that the edge of the column $r = R$ was defined by the condition that $n(R) = 0$. For this to be true, Eq. (26) shows that kR must be equal to 2.405, which is the first zero of J_0. This is an important relation, because it defines the electron temperature required to maintain the column in a steady state. In other words, for the column to be maintained, the electron temperature must satisfy the relation

$$v_i(\theta_e)/D_a(\theta_e) = (2.405/R)^2. \tag{27}$$

Five years after Schottky's article, Tonks and Langmuir (1929) published a comprehensive article dealing with the positive column in the free-fall limit. This limit occurs at low neutral density, where the ions fall freely to the wall under the influence of the radial space-charge field. They did not use the moment equations given above. Instead, they used what has come to be known as the kinetic approach, in which the particle energy is treated as a whole rather than dividing it up into random energy and directed energy. They also derived an equation that determined the electron temperature required to maintain the column in a steady state:

$$(m_p/e\theta_e)^{1/2} v_i(\theta_e) = 1.092/R. \tag{28}$$

Some years later, Self (1963) published numerical solutions to the moment equations for the complete range of gas pressures from the free-fall limit to the collision-dominated limit. At the edge of the column $r = R$, Self relied on the result due to Persson (1962), who showed that the retention of the inertia terms in the moment equations results in a plasma–sheath boundary at the point where the ion drift velocity equals the isothermal sound speed, which in this case is $(e\theta_e/m_p)^{1/2}$. Self's results are identical to those of Schottky in the collision-dominated limit, and close to those of Tonks and Langmuir in the free-fall limit. For example, Self found that the electron temperature in the free-fall limit must satisfy the relation

$$(m_p/e\theta_e)^{1/2} v_i(\theta_e) = 1.109/R, \tag{29}$$

which is quite close to the result of Tonks–Langmuir cited above.

An analytical result similar to that of Self can be obtained by making use of the fact that the ion drift velocity is small compared to the isothermal sound speed except near the edge of the column. In the ambipolar limit $n_e = n_p$, the radial momentum balance equation is

$$(1 + 2v_i/v_{pm})nu = -(e\theta_e/m_p v_{pm})(1 - m_p u^2/e\theta_e)\, dn/dr. \tag{30}$$

By ignoring the inertia term $m_p u^2/e\theta_e$ in this equation, it can be combined with the continuity equation to give an equation quite similar to that of

Schottky:

$$\frac{1}{r}\frac{d}{dr}r\frac{dn}{dr} + \left[\left(1 + \frac{2\nu_{\mathrm{i}}}{\nu_{\mathrm{pm}}}\right)\frac{\nu_{\mathrm{i}}}{D_{\mathrm{a}}}\right]n = 0. \tag{31}$$

The solution of this equation is the same as the solution of Eq. (25), but the parameter k is now defined by the relation $k^2 = (1 + 2\nu_{\mathrm{i}}/\nu_{\mathrm{pm}})\nu/D_{\mathrm{a}}$. The value of k is determined by the condition

$$u = (e\theta_{\mathrm{e}}/m_{\mathrm{p}})^{1/2}$$

at the boundary. The indicial equation for k is found from Eqs. (26) and (30):

$$-\left(n\bigg/\frac{dn}{dr}\right)_R = \frac{J_0(kR)}{kJ_1(kR)} = \frac{(e\theta_{\mathrm{e}}/m_{\mathrm{p}})^{1/2}}{\nu_{\mathrm{pm}} + 2\nu_{\mathrm{i}}} \Rightarrow \frac{J_0(kR)}{J_1(kR)} = \left(\frac{\nu_{\mathrm{i}}}{\nu_{\mathrm{pm}} + 2\nu_{\mathrm{i}}}\right)^{1/2}. \tag{32}$$

In Self's notation, this equation becomes

$$As_{\mathrm{b}} = \frac{yJ_0(y)}{J_1(y)}, \tag{33}$$

where y is defined by the relations

$$y = \left(\frac{1 + A}{A}\right)^{1/2} As_{\mathrm{b}}, \qquad A = \frac{\nu_{\mathrm{i}}}{\nu_{\mathrm{pm}} + \nu_{\mathrm{i}}}.$$

The degree of approximation embodied in Eq. (33) is shown in Fig. 13 which shows Self's numerical results for As_{b} vs. A, and the approximate result

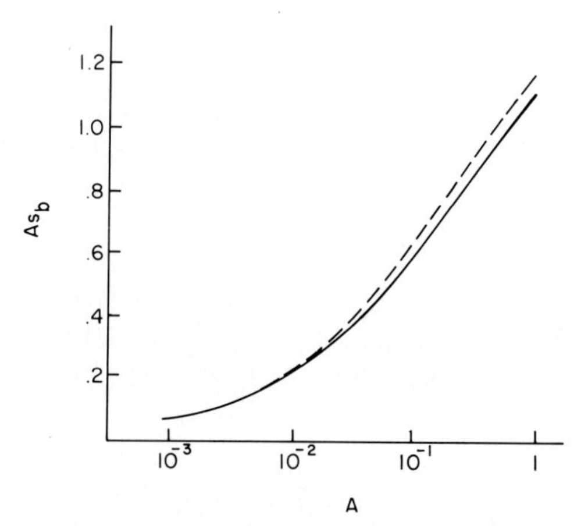

Fig. 13. Comparison of approximate (---) and exact (—) results for characteristic value of cylindrical positive column [in notation of Self (1963)].

given by Eq. (33). Equation (32) gives the Schottky result $y = 2.405$ in the limit $A \ll 1$, and gives $s_b = 1.19$ for the free-fall result in the limit $A = 1$. The value of 1.19 for s_b is about 7% higher than Self's result $s_b = 1.109$.

The indicial equation for the electron temperature, which is embodied in Eqs. (27), (28), and (32), is probably the single most important equation in the theory of the positive column. Once the electron temperature is known, all the other parameters of the column can be closely estimated. Therefore, it seems appropriate to elaborate on this maintenance equation for the positive column. For simplicity, consider the Schottky limit where Eq. (32) is reduced to

$$J_0(kR) = 0.$$

It has already been shown that this condition leads to the relation

$$\frac{v_i(\theta_e)}{D_a(\theta_e)} = \left(\frac{2.405}{R}\right)^2, \tag{34}$$

where v_i is defined by Eq. (24). It is customary to approximate the ionization cross section $Q_i(v)$ by the relation

$$Q_i(v) = C\left(\frac{mv^2}{2e} - V_i\right), \tag{35}$$

where C and the ionization potential V_i are constants that depend on the type of gas. When the electron distribution is Maxwellian, the integration indicated in Eq. (35) leads to the following expression for the ionization frequency:

$$v_i = n_a C(eV_i + 2kT_e)\bar{v} \exp(-eV_i/kT_e), \tag{36}$$

where $\bar{v} = (8kT_e/\pi m_e)^{1/2}$ is the mean speed of electrons with a Maxwellian velocity distribution at temperature T_e. To a very good approximation, the ambipolar diffusion coefficient is given by the expression

$$D_a = \theta_e \mu_0(760/p), \tag{37}$$

where μ_0 is the ion mobility at 273°K and 760 Torr. Consequently, Eqs. (34), (36), and (37) can be combined to give a universal curve relating electron temperature to physical properties of the gas and geometrical properties of the column:

$$\frac{\exp(V_i/\theta_e)}{(V_i/\theta_e)^{1/2} + 2(\theta_e/V_i)^{1/2}} = c^2R^2P^2, \tag{38}$$

where

$$c^2 = \frac{n_0 C}{\mu_0} \frac{(8eV_i/\pi m_e)^{1/2}}{760(2.405)^2} = 1.52 \times 10^4 \frac{n_0 C V_i^{1/2}}{\mu_0}$$

in which $n_0 = 3.53 \times 10^{16}$ cm^{-3} is gas density at 1 Torr and 273°K. Values of the constant c given by Von Engel and Steenbeck (1934) have been quoted in the literature for many years. It has been shown that Eq. (38) is in qualitative agreement with experiment. After the electron temperature has been determined, the axial electron drift velocity can be calculated by means of Eq. (22). If the current I is measured, then the electron density $n(0)$ at the center of the column can be found from the relation

$$I = E_z \int_0^R e\mu_e(r) n_e(r) \cdot 2\pi r \, dr. \tag{39}$$

When this is done, however, sometimes it is found that the results differ from measurements in one way or another. For example, Eq. (22) predicts that the electric field E is independent of the current I, whereas measurement usually shows that E decreases with increasing I, as shown in Fig. 14. The negative voltage–current curve, as it is called, of the low-pressure positive column has intrigued researchers and stymied engineers for many years. The re-

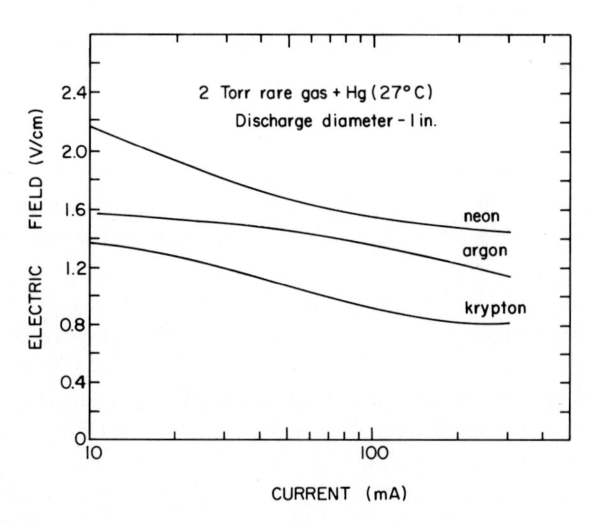

Fig. 14. Voltage–current curves measured in mercury–rare gas mixtures. (Courtesy of J. M. Anderson, private communication.)

searchers are interested because it is not explained by the simple Schottky theory, and engineers are concerned because an additional circuit element, called a ballast, is required to stabilize the discharge against runaway current.

Why does the positive column have a voltage–current curve with negative slope? In the literature can be found three possible reasons: the transition from free to ambipolar diffusion, gas heating, and multistep ionization. These processes are discussed in turn below. (The reader is cautioned that the transition from the dark discharge to the normal glow discharge, which is sometimes referred to as the subnormal glow, has a negative voltage–current curve that is a cathode-related phenomenon, not a property of the positive column.)

The transition from free to ambipolar diffusion in the mobility-controlled positive column was investigated theoretically by Allis and Rose (1954). By solving the standard diffusion equations for electrons and ions, coupled with Poisson's equation, they showed that the transition begins when the Debye length becomes shorter than about ten times the diffusion length of the column, and is nearly complete when the Debye length is a factor of ten or so less than the diffusion length. The negative voltage–current curve in the transition region is related directly to the loss of electrons from the positive column. In the free diffusion limit, the electrons diffuse to the wall of the discharge at a rate governed by the electron diffusion coefficient, whereas in the ambipolar limit, the electron loss rate is governed by the ampipolar diffusion coefficient, which is 10–1000 times smaller than the free electron diffusion coefficient. Consequently, the electric field must be significantly higher in the free diffusion regime than in the ambipolar diffusion regime to account for the higher loss rate. It is doubtful if this phenomenon has ever been observed in a steady-state discharge because the positive column is controlled by ambipolar diffusion for values of current density down to about 10^{-6} A cm^{-2}. Consequently, it is unlikely that this process is responsible for the negative voltage–current curve of the positive column for current densities larger than 10^{-6} A cm^{-2}.

At higher current densities where there is significant energy transfer from electrons to neutrals via elastic collisions, the neutral temperature in the center of the column becomes higher than at the wall. This causes the neutral density to be lower in the center than at the wall. In turn, this causes the electron mobility to be higher in the center than near the wall. It is possible for this effect to be so pronounced that the electric field decreases with increasing current. This effect was studied theoretically by Ecker and Zoller (1964) by adding energy balance equations for the neutral gas and the electron gas to the Schottky model of the positive column. Their results seem to be in qualitative agreement with experiment, and their theoretical model has formed the basis of subsequent models of the positive column. A result of their model that has not been verified unambiguously is a parabolic distribution of electron temperature. Most low-pressure positive columns are thought to have a fairly constant radial distribution of electron temperature. This is a crucial point because Ecker and Zoller found that the curvature of the electron density on the discharge axis becomes positive at high current, and this result must be due to their assumption that the electron temperature is proportional to the gas temperature, which becomes parabolic at high current. It is likely that inclusion of radiation losses in the energy balance equation would flatten the radial electron temperature profile because radiative processes depend exponentially on electron temperature.

Another mechanism that produces a negative voltage–current curve in the positive column is multistep ionization, i.e., electron impact ionization of an excited atom that itself was produced by electron impact. Examination of the energy balance equation (23) shows that the electron density cancels out of each term, giving the result that the electric field varies monotonically with the electron temperature. How does the electron temperature vary with current? In the Schottky analysis given above, it was found that the electron temperature did not vary at all with current, as a result of the indicial equation for the temperature. This indicial equation expresses a balance between loss and gain of charged particles. For single step ionization, this equation can be written

$$(D_a/\Lambda^2)N = \nu_i N, \tag{40}$$

where $\Lambda = R/2.405$ is called the diffusion length for an infinite cylinder, D_a is the ambipolar diffusion coefficient, and $N = 2\pi\int_0^R n(r)r\,dr$. This equation says that the rate at which particles are lost to the walls by diffusion is just balanced by the rate at which they are produced by single-step ionization, and that the electron temperature is independent of current. However, suppose there are other mechanisms of producing charged particles. For example, suppose the dominant source of charged particles is two-step ionization, caused by electrons ionizing excited neutrals, which themselves were produced by electron excitation. Then the source term in the continuity equations would be proportional to n_e^2, and the indicial equation for the electron temperature would be

$$(D_a/\Lambda^2)N = bN^2(\langle n^2/N^2\rangle), \tag{41}$$

where $\langle n^2\rangle = 2\pi\int_0^R n^2(r)r\,dr$. Based on the assumptions that the ratio $\langle n^2\rangle/N^2$ varies slowly if at all with current density, Eq. (41) shows that the electron temperature decreases with increasing current because b depends exponentially on temperature, whereas D_a depends linearly on electron temperature. The mathematical details of this theory have been presented by Spenke (1950) and others (Pahl, 1957; Ingold, 1970). They show that the maintenance condition for the positive column changes from $D_a/v_i\Lambda^2 = 1$ to

$$D_a/v_i\Lambda^2 \approx 1 + 0.675p/(1 + 0.675q) \tag{42}$$

as the cominant source of ions changes from one- to two-step ionization, where p is a measure of the amount of two-step ionization, and q is a measure of the amount of collisional deexcitation of the excited states that are being ionized. Consequently, a negative voltage–current curve can be caused by two-step ionization. A logical question at this point is, Can there be a glow glow discharge positive column with a positive voltage–current curve? By extension of the analysis given above, it is reasonable to expect a positive voltage–current curve when the rate of production of charge particles is independent of electron density, as in the case of ionization produced by fission fragments or some other source external to the discharge tube.

To illustrate qualitatively the relative influence of gas heating and two-step ionization on the course of the voltage–current curve of a low-pressure positive column, calculations based on a simple model that includes these two effects have been made for a discharge in helium. The discharge tube was assumed to have a radius $R = 1.2$ cm. The number of helium atoms per unit length was held fixed at $N = 1.46 \times 10^{18}$ cm^{-1}, and the wall temperature T_w was held fixed at 300°K, which corresponds to a gas pressure of 10 Torr at 300°K. The equations used in the calculations are as follows:

Conservation of electron (ion) density:

$$(D_a/\Lambda^2)n = (A_{gp}n_g - b_{pg}n^2)n + (A_{px}n_x - b_{px}n^2)n. \tag{43}$$

Conservation of electron momentum:

$$I = \pi R^2 e\mu_e nE. \tag{44}$$

Conservation of electron energy:

$$e\mu_e nE^2 = 3\frac{m_e}{m_a}v_{em}e(\theta_e - \theta_g)n + eV_i\frac{D_a n}{\Lambda^2} + eV_x\frac{n_x}{\tau_x}. \tag{45}$$

Conservation of excited atom density:

$$\frac{D_x}{\Lambda^2}n_x = (A_{gx}N_g - A_{xg}n_x)n - \frac{n_x}{\tau_x} - (A_{xp}n_x - b_{px}n^2)n. \tag{46}$$

Conservation of gas energy:

$$\frac{\lambda_a}{\Lambda^2}(T_g - T_w) = 3\frac{m_e}{m_a}v_{em}(\theta_e - \theta_g)n. \tag{47}$$

Conservation of nuclei:

$$\pi R^2(n_g + n + n_x) = N. \quad (48)$$

To simplify the calculations, spatial resolution has been sacrificed by approximating second derivative terms in the usual way:

$$D_a \nabla^2 n \rightarrow -(D_a/\Lambda^2)n.$$

This approximation gives good qualitative results because it is accurate in the low-current limit and the term that it represents is negligible in the high-current limit where electron–ion recombination is more important than diffusion.

The other symbols in these equations are defined as follows:

n_g = ground state atom density
n_x = excited state atom density
n = electron (ion) density
a_{gi} = ionization coefficient from ground state
a_{xi} = ionization coefficient from excited state
a_{gx} = excitation coefficient from ground state
b_{ig} = three-body electron–ion recombination coefficient to ground state
b_{ix} = three-body electron–ion recombination coefficient to excited state
a_{xg} = deexcitation coefficient from excited state by electron impact
τ_x^{-1} = deexcitation coefficient from excited state by photon emission
D_a = ambipolar diffusion coefficient
D_x = excited-atom diffusion coefficient
T_e = electron temperature
T_g = gas temperature
λ_g = thermal conductivity of gas
m_e = electron mass
m_a = atomic mass of gas
V_i = ionization potential of ground state atom
V_x = excitation potential of ground state atom
μ_e = electron mobility

For the calculations, the excitation and ionization coefficients were computed according to Eq. (24), and the corresponding coefficients for the respective reverse processes were computed on the basis of detailed balancing. For example, a_{xg} was computed from the relation

$$a_{gx}/a_{xg} = (w_x/w_g)\exp(-eV_x/kT_e), \quad (49)$$

where w is statistical weight and k Boltzmann's constant. The electron mobility was computed according to the relation

$$\mu_e = \frac{4}{3\sqrt{\pi}} \frac{e}{m_e n_a Q_{ea} V_0} [1 - d + d^2 e^d E_1(d)], \tag{50}$$

where $v_0 = (2e\theta_e/m_e)^{1/2}$, $Q_{ea} = 6 \times 10^{-16}$ cm^2 is the electron–atom momentum transfer collision cross section, and d is the ratio of the mean free path for electron–atom elastic collisions to that for electron–ion elastic collisions. The expression for d that is used here is given by

$$d = \frac{0.57}{4} \frac{\ln \Lambda}{Q_{ea} V_0^4} \left(\frac{e^2}{\varepsilon m_e}\right)^2 \frac{n}{n_a},$$

where ln Λ is the so-called coulomb logarithm. Equation (50) for the electron mobility reduces to the standard expressions in the limits of a weakly ionized gas ($d \ll 1$) and a fully ionized gas ($d \gg 1$). Equations (13)–(48) were solved by iteration to give the voltage–current curves shown in Fig. 15. Values of Q_{gi}, Q_{gx}, Q_{xi}, D_a, λ_g, D_x, τ_x, μ_e, V_i, and V_x pertaining to helium were used in the calculations to simulate the low-pressure helium positive column. The curves in Fig. 15 show that while both gas heating and two-step ionization play a role in causing the negative voltage–current curve, it appears that gas heating has the major influence and two-step ionization plays only a minor role. The experimental points indicated in Fig. 15 were obtained

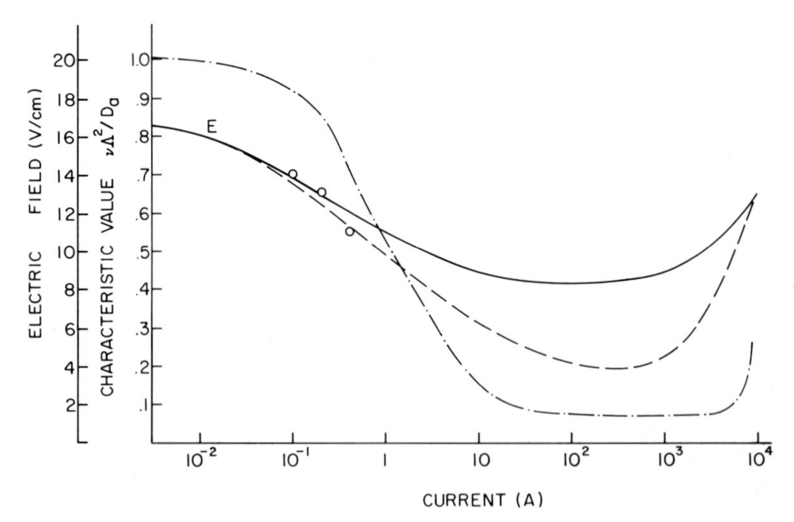

Fig. 15. Electric field E and characteristic value $\nu_i\Lambda/D_a$ vs. current calculated for helium positive column at 10 Torr and 0.12 m radius (theoretical). —, one-step ionization only: ---, one-step and two-step ionization.

from Goluborskii *et al.* (1968). The qualitative agreement between experiment and theory shows the usefulness of the rather simple model outline above.

The curves in Fig. 15 were extended to very high currents, even though it is realized that such a high power dissipation would require an unusual heat removal mechanism to keep the wall temperature at 300°K. The point to be emphasized here is that the voltage–current curve eventually becomes positive at very high currents, where the condition of a fully ionized plasma is approached. The corresponding variation of the temperatures of electrons and gas atoms is shown in Fig. 16. Note that the gas temperature approaches the electron temperature at high current. This condition is sometimes referred to as local thermodynamic equilibrium (LTE), which is characterized by nearly equal electron and atom temperatures and by excited state populations that are in temperature equilibrium with the ground state. A discharge in this condition is also referred to as a recombination controlled discharge, as opposed to one that is diffusion controlled.

It should be kept in mind that the foregoing discussion of the low-pressure helium positive column is intended to convey a qualitative picture of discharge behavior, and not a quantitative one. For example, it has been shown that the use of a Maxwellian energy distribution for the electrons together with only one-step ionization of helium gives results similar to those obtained with a more realistic energy distribution and two-step ionization (Mewe, 1970). By "more realistic energy distribution" is meant a distribution that is Maxwellian at low energy, and a depleted Maxwellian, or a Maxwellian at a lower temperature, in the high-energy tail. Other calculations based on a model with two electron temperatures, one for the "bulk" electrons and one for the "tail" electrons, in a low-pressure argon

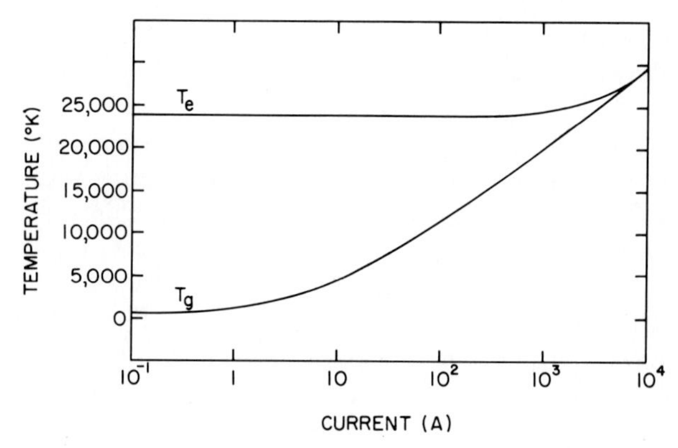

Fig. 16. Electron temperature and gas temperature vs. current for helium positive column at 10 Torr and 0.12 m radius (theoretical).

positive column show that deviations from the Maxwell distributions are significant for currents less than 0.5 A (Vriens, 1973).

In his comprehensive review article on the glow discharge, Francis (1956) makes the point that the major stumbling block in the way of understanding fully the characteristics of the positive column is inability to measure and to calculate accurately terms in the energy balance equation. Since that time, however, an exhaustive experimental study of the low-pressure Ar–Hg positive column has been made (Verweij, 1961; Koedam and Kruithof, 1962; Koedam *et al.*, 1963).

The positive column of the low-pressure Ar–Hg discharge is characterized by the following description. Argon at a pressure of a few Torr and mercury at a pressure of several mTorr are contained in a discharge tube a few centimeters in diameter. When a steady-state current density of a few mA/cm^2 passes through the discharge, the following physical phenomena take place. Electrons are accelerated by the applied field and give up the energy thus acquired in elastic and inelastic collisions with argon and mercury atoms. Because the density of argon atoms is 10^2–10^3 times greater than the density of mercury atoms, and because the average electron energy is about 1 eV, inelastic collisions with argon atoms are thought to be negligible, and elastic collisions with mercury atoms play a relative minor role in the discharge. Thus elastic collisions account for about 20% of the energy input to the electrons, inelastic collisions account for about 70%, and the remaining 10% is unaccounted for, as we shall see below. By far the great majority of inelastic collisions result in excited mercury atoms, which are deexcited by photon emission (90%) and by superelastic collisions with electrons (10%). Nevertheless, about 70% of the electrical energy input is converted to radiant energy by this discharge, which accounts for its usefulness as a light source.

The radiant energy contribution to the energy balance for the discharge described above has been carefully measured by Koedam and Kruithof (1962). They measured all the energy radiated in the mercury lines between 185 and 1711 nm. By coupling this information with measurements of the electric field, electron density, and electron temperature, and with careful calculations of the elastic loss based on the measurements of Kenty *et al.* (1951), they arrived at the energy balance picture presented in Fig. 17. These curves show that a combination of careful measurements and calculations gives a total energy output for the positive column that is about 10% less than the measured energy input. Possible explanations for this 10% discrepancy include undetected radiation below 185 nm or above 1711 nm, absorption of radiation in the wall of the discharge tube, and incorrect estimation of elastic loss. The discrepancy is probably not due to incorrect measurement of the 185 and 254 nm radiation, because these were verified

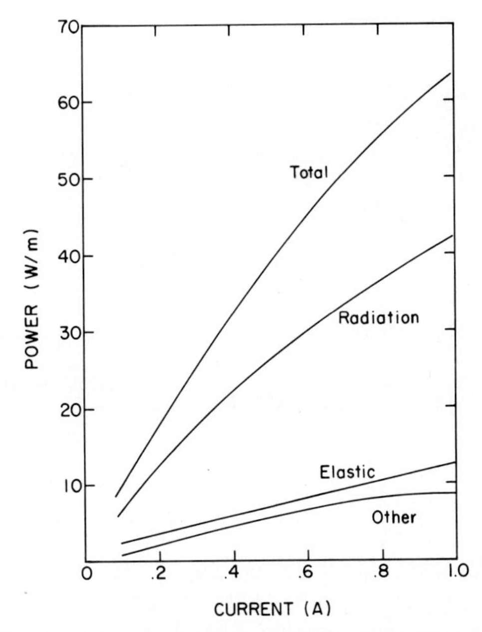

Fig. 17. Energy balance vs. current measured for Hg–Ar positive column at 3 Torr argon and 0.006 Torr mercury.

for identical conditions by Kenty *et al.* In spite of this discrepancy in the energy balance, this discharge appears to be one of the most thoroughly and accurately characterized low-pressure positive columns in the literature. For this reason it is worthwhile to look at some of its other properties.

The effect of variable argon pressure on this discharge is shown in Fig. 18 in the form of curves of the fractions of the power input that are radiated, lost in elastic collisions, and the remainder. These curves show that the elastic loss is relatively unimportant when the pressure is a few tenths of a Torr. In this region of pressure, the dominant energy losses are radiation and ion losses to the wall. As the pressure is increased, however, the elastic loss becomes larger than the ion loss to the wall, while the radiation slowly decreases. These curves were generated from the data given in Koedam *et al.* (1963) for a 0.4 A discharge in 3.6 cm diameter tube with a mercury pressure of about 6 μm. The main reason for this decrease in radiation with increasing argon pressure is decreasing electron temperature, as shown in Fig. 19. This curve indicates that the presence of argon has very little effect on the discharge below 0.1 Torr.

Several workers (Waymouth and Bitter, 1956; Cayless, 1962; Drop and Polman, 1972) have presented theoretical models that purport to be good

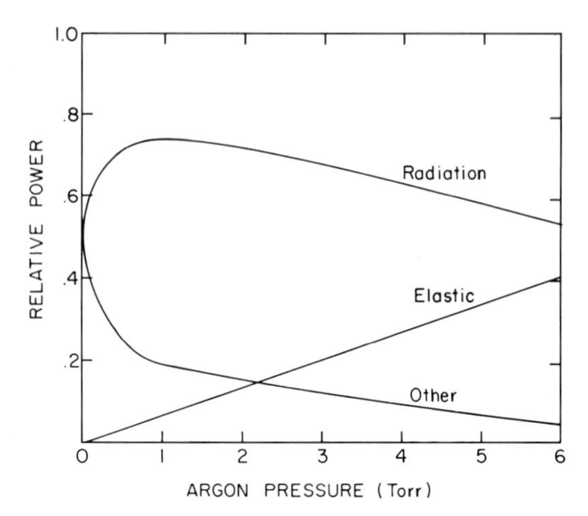

Fig. 18. Energy balance vs. argon pressure measured for Ar–Hg positive column at 0.4 A and 0.006 Torr mercury.

agreement with the data presented in Figs. 17–19. However, some of the important physical processes, such as gas heating and coulomb collisions, have been neglected in these models. Therefore, it seems inappropriate to try to compare these theories with the measurements. For this reason, calculations based on a model that includes most of the physical processes

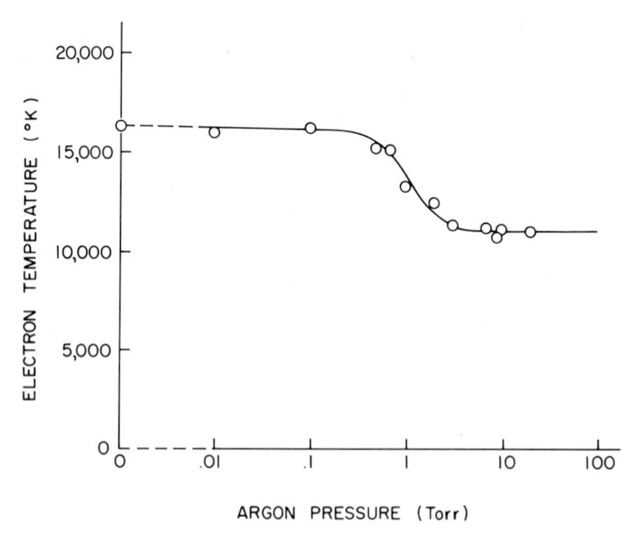

Fig. 19. Electron temperature vs. argon pressure measured for Hg–Ar column at 0.4 A and 0.006 Torr mercury.

thought to be important have bern made. This model is similar to the previous models cited above in that two-step ionization of the mercury is assumed to be the most important source of ionization, but it differs in the following ways:

(1) The triplet states of mercury (two metastable states and one radiating state) have been lumped together in one radiating state.
(2) Gas heating has been taken into account.
(3) Coulomb collisions have been taken into account in calculating the mobility of the electrons.

The equations used in the calculations are listed below:

Conservation of electron (ion) density:

$$\frac{1}{r}\frac{d}{dr} r D_{\mathrm{a}} \frac{dn}{dr} = \alpha_{\mathrm{gp}} n_{\mathrm{g}} n + \alpha_{\mathrm{rp}} n_{\mathrm{r}} n. \tag{51}$$

Conservation of electron momentum:

$$e\Gamma_{\mathrm{e}} = e\mu_{\mathrm{e}} n E. \tag{52}$$

Conservation of electron energy:

$$e\Gamma_{\mathrm{e}} E = 3\frac{m_{\mathrm{e}}}{m_{\mathrm{a}}} \nu_{\mathrm{em}} e(\theta_{\mathrm{e}} - \theta_{\mathrm{g}})n + \sum_{j=\mathrm{i,r,x,v}} eV_j(\alpha_{\mathrm{g}j} n_{\mathrm{g}} - \alpha_{j\mathrm{g}} n_j)n. \tag{53}$$

Conservation of primary radiating state density:

$$(\alpha_{\mathrm{gr}} n_{\mathrm{g}} - \alpha_{\mathrm{rg}} n_{\mathrm{r}})n - n_{\mathrm{r}}/\tau_{\mathrm{r}} - \alpha_{\mathrm{rv}} n_{\mathrm{v}} n = 0. \tag{54}$$

Conservation of secondary radiating state density:

$$(\alpha_{\mathrm{gx}} n_{\mathrm{g}} - \alpha_{\mathrm{xg}})n - n_{\mathrm{x}}/\tau_{\mathrm{x}} = 0. \tag{55}$$

Conservation of visible radiating state density:

$$(\alpha_{\mathrm{rv}} n_{\mathrm{r}} - \alpha_{\mathrm{vr}} n_{\mathrm{v}})n - n_{\mathrm{v}}/\tau_{\mathrm{v}} = 0. \tag{56}$$

Conservation of gas energy:

$$\frac{1}{r}\frac{d}{dr} r\lambda_{\mathrm{g}} \frac{dT_{\mathrm{g}}}{dr} = -3\frac{m_{\mathrm{e}}}{m_{\mathrm{a}}} \nu_{\mathrm{em}} k(T_{\mathrm{e}} - T_{\mathrm{g}})n. \tag{57}$$

These equations were solved numerically for the radial dependence of T_{e}, T_{g}, and the densities. The symbols in these equations are defined below, along with their numerical values where appropriate:

n_g = ground state atom density
n_r = excited-atom density, 254 nm radiation
n_v = excited-atom density, visible radiation
n_x = excited-atom density, 185 nm radiation
τ_r = effective lifetime of r state, 2.5×10^{-5} sec
τ_v = effective lifetime of v state
τ_x = effective lifetime of x state, 1.4×10^{-7} sec

The cross sections for the inelastic collisions were assumed to be of the form given by Eq. (35), with the values of C (in cm^2/V) listed below:

$$C_{gi} = 0.4 \times 10^{-16}$$
$$C_{gr} = 1.7 \times 10^{-16}$$
$$C_{gx} = 1.1 \times 10^{-16}$$
$$C_{ri} = 4.3 \times 10^{-16}$$
$$C_{rv} = 5.8 \times 10^{-16}$$

The rest of the symbols have the same meaning as previously. The values of C given above were chosen to give measured values of the respective radiation outputs at the measured electron temperature and current. Comparison of calculated values with the measured values of 254 nm radiation and total radiation is shown in Fig. 20. The calculated results for the variation of electric field with current are shown in Fig. 21, along with measured values. The upper curve labeled $Q_{ri} = 0$ in this figure was calculated in order to

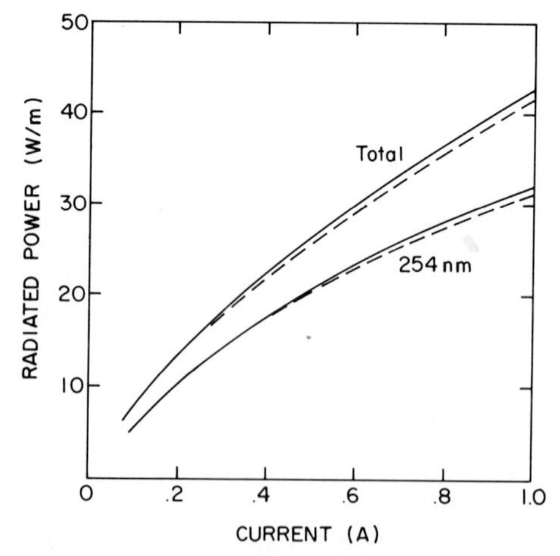

Dig. 20. Radiation vs. current measured for Hg–Ar positive column at 3 Torr argon and 0.006 Torr mercury.

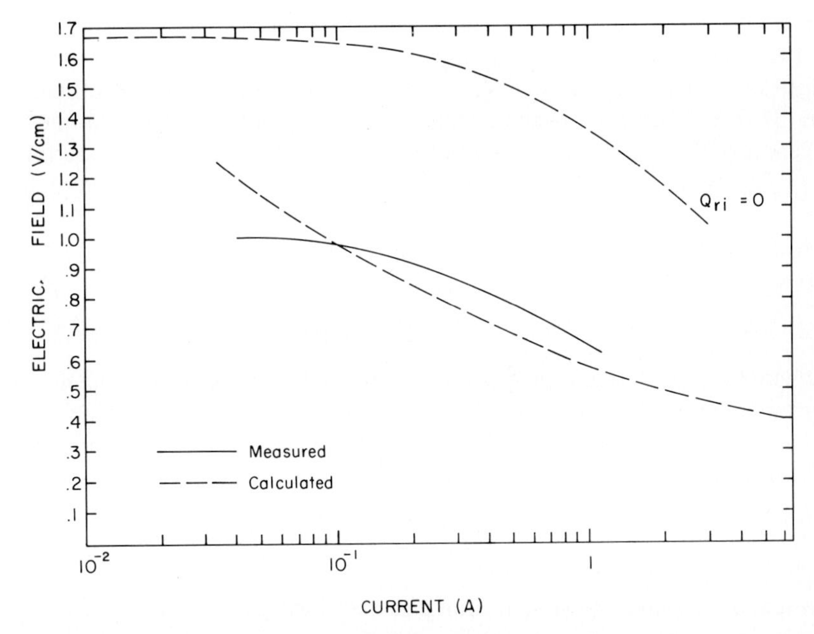

Fig. 21. Comparison of calculated (---) and measured (—) electric field vs. current for Hg–Ar positive column at 3 Torr argon and 0.006 Torr mercury. The dashed curve labeled $Q_{ri} = 0$ is for one-step ionization of mercury, the other for both one- and two-step ionization of mercury.

show the difference between one- and two-step ionization. According to these calculations, the difference here is larger than in the case of the helium discharge described earlier. The most interesting feature of the curves in this figure, however, is the shape—the curve for the measured electric field and that calculated for one-step ionization have negative curvature, while that calculated for two-step ionization has positive curvature. Furthermore, the calculated curve for two-step ionization gives a value for the electric field that is much higher than the measured value at low current. The reason for this is that the calculated curve for two-step ionization must approach the calculated curve for one-step ionization as the electron density, or the current, gets small. On the other hand, the measured curve approaches the value of 1.0 V cm^{-1}, which is smaller than the 1.7 V cm^{-1} based on one-step ionization of the mercury by electron impact. Consequently, it appears that the two-step ionization model for the low-pressure Ar–Hg positive column is inadequate. Furthermore, it appears that ionization in this discharge results from a one- rather than a two-step process.

A one-step process that has not been considered is the Ar–Hg Penning effect (Kruithof and Penning, 1936). Penning ionization occurs when an excited atom collides with an unexcited atom of another species that has a lower ionization potential and ionizes it. For example, the argon atom has four excited states near 11.6 V, whereas the ionization potential of mercury is 10.43 V. Therefore, excited argon atoms can ionize mercury atoms. The Penning effect has been known for many years, but the cross section for the Ar–Hg effect has not been measured directly. However, an estimate of $3 \times 10^{-15}\ \mathrm{cm}^2$ has been obtained from afterglow measurements (Phelps and Molnar, 1953). The rate of ionization of mercury by the Penning effect is given by the relation

$$\nu_{\mathrm{P}} = \bar{v} Q_{\mathrm{P}} n_{\mathrm{m}} n_{\mathrm{Hg}}/n, \tag{58}$$

where $\bar{v}$ is the relative velocity of the argon metastable atom and the mercury ground state atom, $Q_{\mathrm{P}} = 3 \times 10^{-15}\ \mathrm{cm}^2$ the Penning cross section, n_{m} the density of argon metastables, and n_{Hg} the density of mercury atoms in the ground state. When diffusion of the metastables can be neglected, the argon metastable density is determined by the rate equation

$$(A_{\mathrm{gm}} n_{\mathrm{Ar}} - A_{\mathrm{mg}} n_{\mathrm{m}})n - n_{\mathrm{m}}/\tau_{\mathrm{m}} - \nu_{\mathrm{P}} n = 0, \tag{59}$$

where the a_{ij} have the same meaning as previously, and $\tau_{\mathrm{m}} = 1.3 \times 10^{-5}$ sec is the effective lifetime of the argon excited atoms. When this equation is solved for n_{m}, the result is

$$n_{\mathrm{m}} = \frac{A_{\mathrm{gm}} n_{\mathrm{Ar}} n}{1/\tau_{\mathrm{m}} + \bar{v} Q_{\mathrm{P}} n_{\mathrm{Hg}} + A_{\mathrm{mg}} n}. \tag{60}$$

When this expression for n_{m} is substituted in Eq. (58), the final result for the Penning ionization frequency is

$$\nu_{\mathrm{P}} = \frac{\bar{v} Q_{\mathrm{P}} n_{\mathrm{Hg}} A_{\mathrm{gm}} n_{\mathrm{Ar}}}{1/\tau_{\mathrm{m}} + \bar{v} Q_{\mathrm{P}} n_{\mathrm{Hg}} + A_{\mathrm{mg}} n}. \tag{61}$$

To evaluate this expression, a value for the excitation rate of argon must be known. The cross section for this process has been measured (Lloyd *et al.*, 1972). When the measured cross section is expressed in the form of Eq. (35), the value of C turns out to be about $5 \times 10^{-18}\ \mathrm{cm}^2\ \mathrm{V}^{-1}$. Consequently, when a_{gm} is evaluated according to Eq. (24) and ν_{P} according to Eq. (61) for an electron temperature of 11,600°K, the result is $\nu_{\mathrm{P}} = 980\ \mathrm{sec}^{-1}$. On the other hand, by Eq. (27) the total ionization rate required to maintain the positive column is

$$\nu_{\mathrm{i}} = D_{\mathrm{a}}/\Lambda^2 = 812 \quad \mathrm{sec}^{-1},$$

where the mobility of the mercury ion in argon has been taken to be 1370 $\mathrm{cm}^2\ \mathrm{V}^{-1}\ \mathrm{sec}^{-1}$ at 1 Torr and 300°K. Therefore, it appears likely that main source of ionization in this discharge is Penning ionization, and not two-step ionization.

IV. ANODE REGION

The purpose of the anode is to transfer current from the glow discharge to the external circuit. When there is no positive column, the anode is usually in the Faraday dark space. In this case, the anode fall of potential can be very small, or even negative, because ions and electrons diffuse together to the anode from the negative glow in such a way that charge neutrality is maintained. When there is a positive column, however, ions do not diffuse to the anode, and charge neutrality cannot be maintained in the anode region. Consequently, when the anode current density is greater than about 10^{-7} A cm^{-2}, there is an electric field at the anode caused by electron space charge. In addition, the electrons in the anode fall region must supply by ionization enough ions to account for the ion current that flows out of the cathode end of the positive column. Therefore, the electron energy must increase in the anode fall region to supply this additional ionization.

The magnitude of the anode fall is usually on the order of the excitation or ionization potential of the gas. Measurements by Druyvesteyn (1937) have shown that the anode fall at the end of a positive column in a neon discharge is slightly less than the excitation potential. This result means that the requisite ionization probably comes from excited atoms being ionized by electrons in the anode fall region.

Regardless of the specific ionization mechanism, the following picture of the anode fall region appears to be valid qualitatively:

(1) The anode fall voltage V_a (cf. Fig. 3) is on the order of the ionization potential of the gas because electrons must be accelerated to an energy high enough to supply ion current for the positive column.

(2) The anode fall thickness d_a is determined primarily by space-charge requirements, and secondarily by the ionization requirement.

In the collision-dominated positive column, the ion current density is related to the electron current density by the equation

$$\Gamma_p = -(D_p/D_e)\Gamma_e. \tag{62}$$

Therefore, the ion current density is much smaller than the electron current density, because $D_p \ll D_e$. This means that the ionization requirement in the anode fall can be satisfied by a very thin ionization layer right next to the anode, where the electron energy is just slightly above the ionization potential of the gas. The ionization requirement is that the thickness δ of the ionization layer be sufficient to provide the ion current given by Eq. (62), i.e.,

$$\Gamma_p = -\int_0^\delta \alpha \Gamma_e \, dx = -(D_p/D_e)\Gamma_e.$$

Because the electron current density is nearly constant, this equation gives the following relation for the thickness of the ionization layer:

$$\int_0^\delta \alpha \, dx = D_p/D_e. \tag{63}$$

By way of illustration, let a monoenergetic electron distribution be assumed in the ionization layer, and let the gas be helium. Then the ionization coefficient α is given by Eq. (17), and the variation of electron energy in the ionization layer is governed by Eq. (18). When the ionization cross section for helium is taken to be

$$Q(\xi) = C(\xi - V_i),$$

where $C = 1.64 \times 10^{-18}\ \text{cm}^2\ \text{V}^{-1}$ (Kieffer, 1969, then Eq. (18) can be solved to give the following relation for the electron energy in the ionization layer, which is assumed to begin at $x = 0$:

$$\xi(x) = \frac{V_i[1 + a \tanh(ax/\lambda_i)]}{1 + \tanh(ax/\lambda_i)}, \tag{64}$$

where $p\lambda_i = p/n_a C V_i = 0.7$ cm-Torr for helium, and the parameter a is defined by the relation $a = (\lambda_i E/V_i + 1)^{1/2}$. This solution for $\xi(x)$ is based on the condition that $\xi(0) = V_i$, i.e., the ionization layer cannot form unless the electron energy has reached the ionization potential. The combination of Eqs. (17), (63), and (64) then gives the following relation for the thickness δ of the ionization layer:

$$\frac{D_p}{D_e} = \frac{a-1}{2a}\left[\frac{a}{\lambda_i} + \frac{1}{2}(e^{-2a\delta/\lambda_i} - 1)\right] \approx \frac{1}{2}a(a-1)\left(\frac{\delta}{\lambda_i}\right)^2, \qquad \frac{a\lambda}{\lambda_i} \ll 1. \tag{65}$$

When the parameters for helium are substituted into Eq. (65), it is found that $p\delta$ is about 10^{-1} Torr-cm, which is much less than pd_a, as shown below.

By assuming that the ionization layer is much less than the total thickness of the anode fall region, a simple model for the anode fall behavior can be constructed from the moment equations. For simplicity, the inertia terms will be neglected, which means that the following analysis is valid only when d_a is much greater than the electron and ion mean free paths. In this situation, the equations that describe the behavior of electrons and ions in the anode fall region are as follows:

Conservation of electron and ion density:

$$\frac{d\Gamma_e}{dx} = \frac{d\Gamma_p}{dx} = 0. \tag{66}$$

Conservation of electron and ion momentum:

$$\frac{d}{dx}(n_e\theta_e) - n_e\frac{dV}{dx} = -\frac{\Gamma_e}{\mu_e}, \qquad \frac{d}{dx}(n_p\theta_p) + n_p\frac{dV}{dx} = -\frac{\Gamma_p}{\mu_p}. \tag{67}$$

Conservation of electron energy:

$$\frac{d}{dx}\left(\frac{3}{2}\theta_e\Gamma_e\right) - \Gamma_e\frac{dV}{dx} - 0. \tag{68}$$

Poisson's equation:

$$\frac{d^2V}{dx^2} = \frac{e}{\varepsilon}(n_e - n_p). \tag{69}$$

This form of the energy balance equation expresses the assumption that there are no energy losses in the anode fall, and the assumption that the energy distribution of the electrons is a delta function in speed, i.e., a spherical shell in velocity space (Ingold, 1969). These equations can be combined to give

$$\theta_p\frac{d}{dx}(n_p\theta_p) - \theta_e\frac{d}{dx}(\theta_e n_e) + \left[\frac{\varepsilon}{2e}\left(\frac{dV}{dx}\right)^2 - \frac{\Gamma_e}{\mu_e}x + (n_e\theta_e)_0 + (n_p\theta_p)_0 - \frac{\varepsilon}{2e}\left(\frac{dV}{dx}\right)_0^2\right]\frac{dV}{dx} - \frac{\theta_e\Gamma_e}{\mu_e} + \frac{\theta_p\Gamma_p}{\mu_p} = 0, \tag{70}$$

where the subscript 0 means "evaluated at $x = 0$," which in this case denotes the entrance to the anode fall region. A useful approximate solution to this equation can be obtained when the gradient terms are ignored and θ_i/θ_e is neglected in comparison with unity. Then by introducing the variable changes

$$y = V/\theta_{e0}, \qquad N_e = eN_e/\varepsilon\theta_{e0}a^{2/3}, \qquad N_i = eN_p/\varepsilon\theta_{e0}a^{2/3},$$
$$z = a^{1/3}x, \qquad w = dy/dz, \qquad a = j_e/\varepsilon\theta_e\theta_{e0}{}^2.$$

Equation (70) can be expressed in the dimensionless form

$$0 = \tfrac{1}{2}W^3 + (N_0 - \tfrac{1}{2}W_0{}^2 - z)W - (1 + \tfrac{2}{3}y), \tag{71}$$

where $N_0 = N_e(0)$ and $W_0 = W(0)$.* At $z = 0$, this equation is reduced to $N_0 W_0 = 1$, which defines the values of N and W in the positive column. In terms of real variables, the condition $N_0 W_0 = 1$ becomes

$$j_e = en_e \mu_e E_0.$$

The solution to Eq. (71) can be found in parametric form by the following procedure. Solving for z and differentiating with respect to W gives

$$\frac{dz}{dW} = \frac{3}{5}\left(W + \frac{1 + \frac{2}{3}Y}{W^2}\right)$$

whence

$$\frac{dY}{dW} = W\frac{dz}{dW} = \frac{3}{5}\left(W^2 + \frac{1 + \frac{2}{3}Y}{W}\right).$$

Solving for Y gives

$$Y = \frac{3}{13} W_0{}^3(t^3 - t^{2/5}) + \frac{3}{2}(t^{2/5} - 1) \tag{72}$$

and Eq. (71) gives

$$z = \frac{1}{2} W_0{}^2(t^2 - 1) + \frac{1}{W_0}\left(1 - \frac{1 + \frac{2}{3}Y}{t}\right), \tag{73}$$

where $t = W/W_0$. These two parametric equations are equivalent to the high-pressure space-charge equation

$$V(x) = \left[\frac{8}{9}\frac{J_e/p^2}{\varepsilon\mu_e p}(px)^3\right]^{1/2}$$

in the limit of very large values of t. For values of t near unity, however, the potential distribution is different from the high-pressure space-charge equation. An example calculated for a discharge in helium at 10 Torr and a current density of 1.6×10^{-3} A cm^{-2} [Druyvesteyn's (1937) experimental conditions] is shown in Fig. 22. The solid curve is the potential distribution given by Eqs. (72) and (73) when $E_0/p = 1$ V cm^{-1} Torr^{-1}, $\theta_{e0} = 2$ eV, and $\varepsilon\mu_e p^3/J_e = 7.75 \times 10^{-9}$ V^{-2} cm^3-Torr^3. The dashed curve represents the distribution given by the high-pressure space charge equation for the same values of current density and anode fall thickness. The difference between

* When $\theta_p = \theta_e = \text{const}$, Eq. (70) can be put in the form

$$W''' = \tfrac{1}{2}W^3 + (N_{e0} + N_{i0} - \tfrac{1}{2}W_0{}^2 - z)W - 1$$

which has been studied for a few special cases (Painleve, 1902; Davis, 1960; Blue *et al.*, 1962).

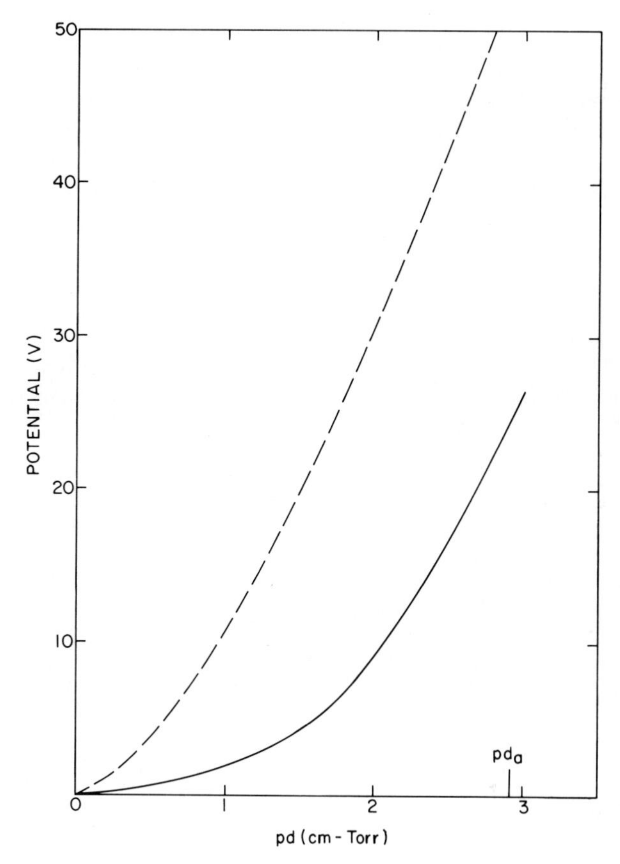

Fig. 22. Calculated potential distribution in anode fall with (—) and without (---) positive ions.

the two curves in Fig. 22 is due to the effect of ions on the high-pressure space-charge-limited potential distribution. This model predicts a value for the anode fall thickness pd_a of about 3 cm-Torr, which is much greater than the value of 0.1 found earlier for the thickness of the ionization layer. The model also predicts a potential distribution in the anode fall region that is approximately parabolic, in qualitative agreement with measurement (Chaundy, 1954).

REFERENCES

Allis, W. P., and Rose, D. J. (1954). *Phys. Rev.* **93,** 84.

Blue, E., Ingold, J. H., and Ozeroff, W. J. (1962). *J. Advan. Energy Conv.* **2,** 395.

Cayless, M. A. (1962). *Proc. Internat. Conf. Phenomena Ionized Gases Vth, Munich.*

Chapman, S., and Cowling, T. G. (1960). "The Mathematical Theory of Nonuniform Gases." Cambridge Univ. Press, Cambridge.
Chaundy, C. F. J. (1954). *Brit. J. Appl. Phys.* **5,** 255.
Cobine, J. D. (1958). "Gaseous Conductors." Dover, New York.
Davis, H. T. (1960). "Introduction to Nonlinear Differential and Integral Equations." U.S.A. E. C.
Drop, P. C., and Polman, J. (1972). *J. Phys. D.: Appl. Phys.* **5,** 562.
Druyvesteyn, M. J. (1937). *Physica* **4,** 669.
Druyvesteyn, M. J. (1938). *Physica* **5,** 975.
Druyvesteyn, M. J., and Penning, F. M. (1940). *Rev. Mod. Phys.* **12,** 87.
Ecker, G., and Zoller, O. (1964). *Phys. Fluids* **7,** 1996.
Francis, G. (1956). The glow discharge at low pressure, *in* "Handbuch der Physik," Vol. XXII. Springer, Berlin.
Goluborskii, Y. B., Kagan, Y. M., and Lyagushchenko, R. I. (1968). *Opt. Spect.* **24,** 79.
Gove, P. B., ed. (1961). "Webster's Third New International Dictionary." Merriam, Springfield, Illinois.
Guntherschulze, A. (1930). *Z. Phys.* **49,** 358; **59,** 433.
Ingold, J. H. (1969). *J. Appl. Phys.* **40,** 55, 63.
Ingold, J. H. (1970). *Z. Phys.* **233,** 89.
Kenty, C., Easley, M. A., and Barnes, B. T. (1951). *J. Appl. Phys.* **22,** 1006.
Kieffer, L. J. (1969). *Atomic Data* **1,** 19.
Koedam, M., and Kruithof, A. A. (1962). *Physica* **28,** 80.
Koedam, M., Kruithof, A. A., and Riemens, J. (1963). *Physica* **29,** 565.
Kozlov, N. P., and Khvesyuk, V. I. (1971). *Sov. Phys. Tech. Phys.* **15,** 1303.
Kruithof, A. A., and Penning, F. M. (1936). *Physica* **3,** 515.
Lloyd, C. R., Teubner, P. F. O., Weigold, E., and Hood, S. T. (1972). *J. Phys. B: Atom. Mol. Phys.* **5,** L44; **5,** 1712.
Mewe, R. (1970). *Physica* **47,** 373; **47,** 398.
Pahl, M. (1957). *Z. Naturforsch.* **12a,** 632.
Painleve, P. (1902). *Acta Math.* **25,** 1.
Persson, K. B. (1962). *Phys. Fluids* **5,** 1625.
Phelps, A. V., and Molnar, J. (1953). *Phys. Rev.* **89,** 1202.
Schottky, W. (1924). *Z. Phys.* **25,** 635.
Self, S. A. (1963). *Phys. Fluids* **6,** 1762.
Spenke, E. (1950). *Z. Phys.* **127,** 221.
Tonks, L., and Langmuir, I. (1929). *Phys. Rev.* **34,** 876.
Townsend, J. S. (1915). "Electricity in Gases." Univ. Press, Oxford.
Verweij, W. (1961). Philips Research Rep. Suppl. No. 2.
von Engel, A. (1965). "Ionized Gases," 2nd ed., p. 217. Clarendon, Oxford.
von Engel, A., and Steenbeck, M. (1934). "Elektrische Gasentladungen." Springer, Berlin.
Vriens, L. (1973). *J. Appl. Phys.* **44,** 3980.
Ward, A. (1962). *J. Appl. Phys.* **33,** 2789.
Waymouth, J. F., and Bitter, F. (1956). *J. Appl. Phys.* **27,** 122.
Weston, G. F. (1968). "Cold Cathode Glow Discharge Tubes." Arrowsmith, Bristol.

Part 2.2 Ionization Waves in Glow Discharges

ALAN GARSCADDEN
HIGH POWER BRANCH
AIR FORCE AERO PROPULSION LABORATORY
WRIGHT-PATTERSON AFB, OHIO

I. INTRODUCTION

In direct current discharges self-excited oscillations may occur over a very wide range of frequencies and discharge parameters (Table I; see also Rutscher and Wojaczek, 1964). Oscillations at microwave and ultrahigh frequencies (10^{10}–10^{8} Hz) excited by an injected electron beam or by the primary cathodic electrons at low pressures are explained as electron plasma oscillations (Briggs, 1964). Under similar conditions, oscillations or fluctuations at lower frequencies (10^{7}–10^{5} Hz) are usually ascribed to ion plasma oscillations. In the gas discharge positive column plasma, two main classes of instabilities are responsible for most of the fluctuations observed in the frequency range below 1 MHz. At low gas pressures (pR less than approximately 0.05 Torr-cm), the oscillations are found to be of the ion–acoustic

ISBN 0-12-349701-9

TABLE I

SPECTRUM OF OSCILLATIONS IN GLOW DISCHARGES

Approximate frequency range (Hz)	Wave type	Occurrence	Parameter relations	References
0	Standing striation	Molecular gases; boundary conditions important in rare gases	$E\lambda$ approximately equals excitation potential	Boyd and Twiddy (1959), Laska and Exner (1968)
10–10^5	Relaxation oscillation I (circuit)	Circuit dependent (low ballast R, distributed C)	No phase change along column: large amplitude	Wojaczek and Rutscher (1963)
10^3–10^5	Moving striations (MS)	Wide range in most gases ambipolar diffusion conditions	Modulation of n_e and T_{er}; dispersive wave phase changes along column	Pekarek (1971), Neodospasov (1963), Oleson and Cooper (1968); see text
10^4–10^5	Relaxation oscillation II (plasma boundary)	Enhanced by limited electrode area; often coupled to MS	Current dependent: nonpropagating	Cooper (1964), Rademacher and Wojaczek (1958a)
10^3–10^6	Ion acoustic	Low-pressure discharge	Modulation of n_e velocity, $\omega/k = (\gamma_e kT_e/m_i)^{1/2}$ little dispersion	Alexeff and Jones (1966), Swain and Brown (1971)
10^6–10^8	Ion plasma	Excited by electron beam–plasma interaction	Frequency, $f_{pi} = 9000(m_e n_e/m_i)^{1/2} q_i$	Spitzer (1962) Emeleus (1964)
10^8–10^{10}	Electron plasma		Frequency, $f_{p'} = 9000 n_e^{1/2}$	Briggs (1964)
Microwave noise	Incoherent radiation	All discharges "radiation temperature"	$kT = -m_e \int_0 Q(v) f(v) v^5\, dv / \int_0 Q(v)(Df/dv) v^4\, dv$	Bekefi (1966), Bekefi *et al.* (1961)

wave type. This wave shows up as a modulation of the ion density and it has approximately constant phase velocity. At higher pressure there is a transition to a striation wave. The term striation arose from the marked periodic changes in light intensity that are easily observed, often stationary or slowly moving in molecular-gas positive columns. These striations can be stationary relative to the discharge electrodes, move toward the cathode (positive striations), or move toward the anode (negative striations). A large number of subcategories are now distinguished (Pekarek, 1971). Despite their different physical natures, these two classes of instability often lead to waves with similar frequencies and wavelengths. Boundary or sheath-related fluctuations may also contribute to this spectral range, directly or by intermodulation with the other oscillations (Wojaczek and Rutscher, 1963; Rutscher and Wojaczek, 1964). However, there are clear distinctions in their physical properties and these become apparent in the dispersion of the waves (Pekarek, 1971). Our interest in the oscillations occurring in glow discharges will primarily concern the moving and standing striations.

Although the positive column of a rare gas glow discharge usually appears homogeneous and uniform, often upon examination by time-resolved techniques it is found to contain alternate bright and dark layers moving along the axis of the tube. These striations often involve large (20%) modulations of the electron energy and electron density. The light intensity modulation may approach 100%. The moving striations are then easily observed on a rotating mirror whose axis is parallel to the axis of the discharge. The plasma potential oscillation amplitude is approximately the ionization potential of the gas. The phase shifts between these plasma parameters are functions of the gases and excitation conditions. These aspects will be discussed later. In the more common low-pressure, low-current rate gas discharge, it has been found that the electron energy variations in a positive striation lead the electron density variation by about 90°. However, an important result found by Sicha and Drouet (1968) for the fast type *r* striations in neon is that the luminosity was approximately 180° out of phase with the "electron temperature." The moving striations have been made to appear stationary by flowing the gas with a velocity equal and opposite to the phase velocity (Gentle, 1966). Self-excited stationary or standing striations in nonflowing gas discharges are most prominent in the molecular gases or in contaminated rare gases. Standing striations are also often observed at the cathodic end of the positive column, or they may be induced locally by probes or external boundary conditions, especially those close to the characteristic wavelength. Moving and standing striations have been observed to coexist in many discharges (Cooper *et al.*, 1958).

The striated glow discharge occurs over such a wide range of discharge parameters that Pekarek (1971) has introduced the term ionization waves, emphasizing their basic nature. Much of the theory and controlled experiments have been on the rare gases (Pekarek, 1971; Oleson and Cooper, 1968). However, motivated by laser applications, significant progress also has been made recently in the understanding of instabilities in the molecular gases, including those discharges supported by external ionization (Nighan *et al.*, 1973; Nighan and Wiegand, 1974a; Bailey *et al.*, 1974; Dougherty *et al.*, 1976).

II. RANGES OF OCCURRENCE

In Fig. 1, general mappings of the pressure–current parameter range of existence of spontaneously excited waves in neon, argon, krypton, and xenon are given (Pfau *et al.*, 1969). At high current densities, a constricted–striated discharge occurs. The helium discharge data are less extensive. However, the parameter mapping shown in Fig. 2 would indicate that it is similar to the other gases. Pupp (1932) showed that moving striations in pure rate gases at constant pressures do not occur above a certain critical current (now referred to as the Pupp limit). Data from later experiments (Rademacher and Wojaczek, 1958b) confirmed and extended Pupp's data. The critical current can be represented by

$$I_u = C/p^\gamma, \tag{1}$$

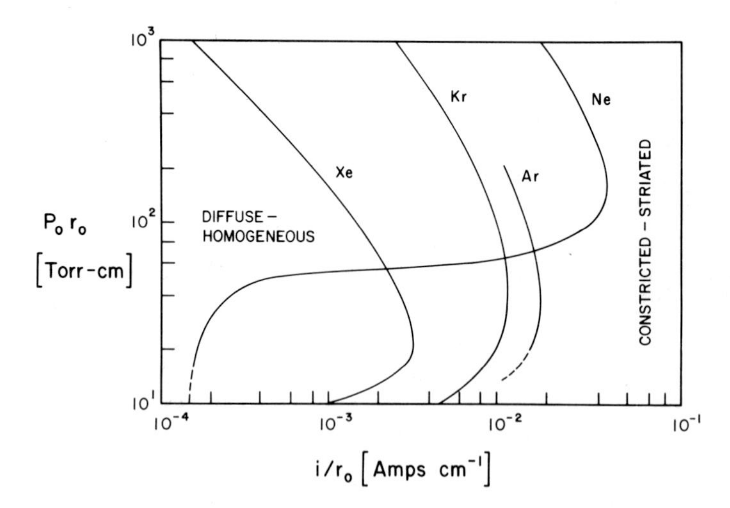

Fig. 1. Existence regions of column forms in neon, argon, krypton, and xenon for a medium pressure discharge.

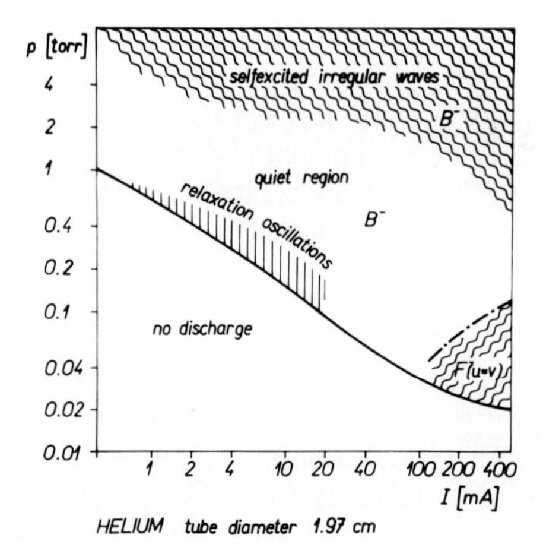

Fig. 2. Existence regions of oscillations in a low-pressure helium glow discharge. The notation follows Pekarek (1971): F, forward wave (phase and group velocities in the same direction); B, backward waves (phase and group velocities in opposite directions); +, direction of the applied electric field; −, direction opposite to the electric field.

where p is the pressure measured in Torr and the mean values for C and γ for the rare gas discharges, in tubes of diameter between 1.5 and 5.5 cm, are given in Table II. The low-current transition, homogeneous-to-striated, has been parameterized by Rutscher and Wojaczek (1964). Wellenstein and Robertson (1972) report for the low-current transition a particularly strong dependence of the critical current in helium on the product pR. They found that for I_1 in mA, R in cm, and p in Torr,

$$I_1 = \mathrm{A}/(pR)^5,$$

where A has a value between 160 and 30, depending on the point of observation as it is changed from the cathode to anode end of the column. These empirical relations are useful in deciding the design of light sources. Also,

TABLE II

COEFFICIENTS FOR DETERMINATION OF I_u (PUPP LIMIT) BY EQ. (1) IN RARE GAS DISCHARGES[a]

	He	Ne	Ar	Kr	Xe
C	12	7	2.2	1.4	1.1
γ	0.93	1	0.8	0.5	0.5

[a] From Rutscher and Wojaczek (1964).

they give approximate identification of where the ionization waves may be studied under small signal conditions.

The theoretical analysis of ionization waves near the Pupp limit is probably the most tractable because one can assume that the bulk of the election energy distribution is Maxwellian. Recently, a theoretical relationship for the upper limiting current has been derived by Alanakyan and Mikhalev (1975), who obtain

$$(I_u p)^2 = AI_u p(E/p) + B(p\Lambda)^2, \tag{2}$$

where $\Lambda = (D_a \tau)^{1/2}$, is the diffusion length, τ is the time constant for the ions to diffuse to the tube wall, A and B are constants depending on the gas and the electron temperature (which is itself a weak function of pressure, current, and radius of the discharge tube). At low pressures

$$I_u p = \mathrm{A}E/p, \tag{3}$$

while at high pressures

$$I_u = B^{1/2}\Lambda. \tag{4}$$

These relations give fair agreement with experiment (Rademacher and Wojaczek, 1958b; Rutscher and Wojaczek, 1964) and are useful for identifying the usually unstable regions. The variety of striations that have now been identified is much more complex. As an example, a more extensive diagram for neon is given in Fig. 3 (after Pfau *et al.*, 1969), where the data points are from discharges in tubes of diameter from 0.3 to 7 cm. These

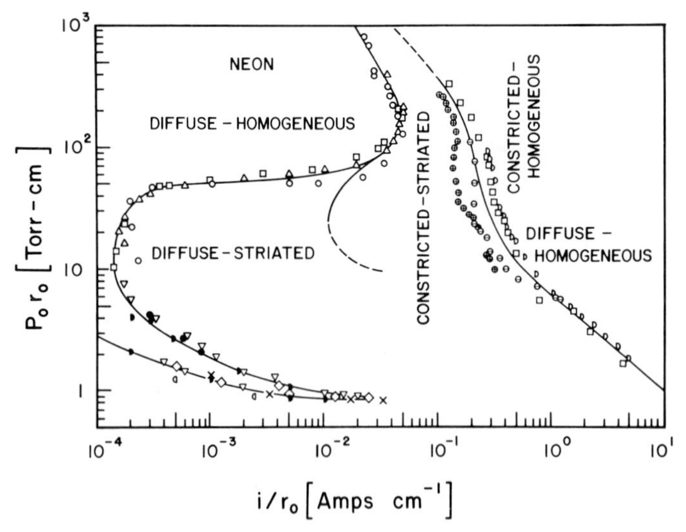

Fig. 3. Existence boundaries of moving striations in neon (after Pfau *et al.*, 1969).

moving striations in neon are further distinguished at low currents by their velocities and their $E\lambda$ product (Novak, 1960) (where E is the axial field and λ the (preferred) wavelength of the striation). The range and occurrence of the different types are shown in Figs. 4 and 5 (after Pfau and Rutscher, 1968). In neon there are two slow types (called p and s') and two fast types (r and s). $E\lambda$ is 9.2 V for the p-type, 12.7 V for r, and 19.5 V for s and s'. Sicha and Drouet (1968) have reported that $E\lambda$ is only approximately constant for each wave type and that the values may vary by up to 25% from Novak's (1960) values, depending on the tube diameter. It is thought that the exact value depends on the ratio of direct to secondary ionization rates with E decreasing with increasing secondary ionization rate. At higher discharge currents, above the Pupp limit only one type of wave can be stimulated and $E\lambda$ is not a constant (Rayment, 1974).

The occurrence of striations in the molecular gases is even more diversified and as illustration we show the existence regions for hydrogen (Stirand and Laska, 1967) (Fig. 6) and for nitrogen (Venzke, 1971; Laska and Exner, 1971) (Fig. 7). While these regions are very sensitive to gas purity, there is general agreement on the existence parameters. Further data and discussions of these experimental results up to 1968 are given in the review articles of Olson and Cooper (1968), Pekarek (1971). Nedospasov (1968), and Rutscher and Wojaczek (1964). More recent studies will therefore be emphasized here. Moving striations also occur in rare gas mixtures including typical helium–neon laser gas mixtures (Wallard and Woods, 1974; Baseau *et al.*, 1976). There is an additional contribution to the amplitude

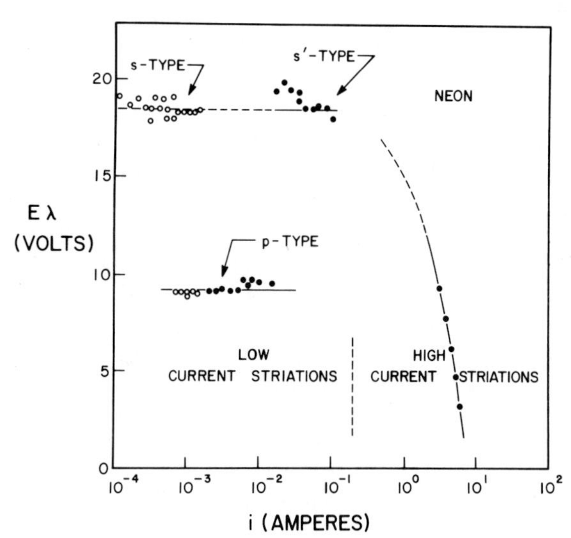

Fig. 4. Novak relations for neon identifying different types of ionization waves in the one gas.

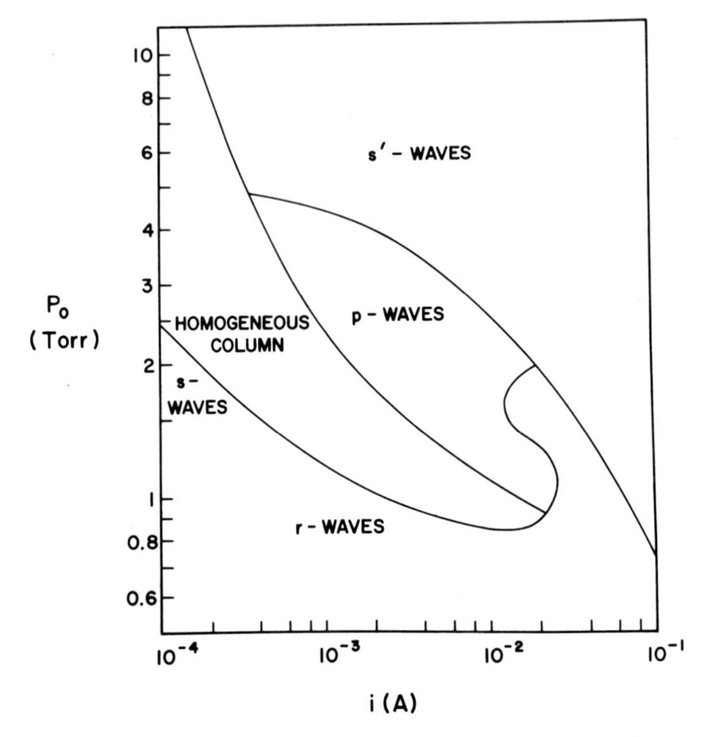

Fig. 5. Existence regions for the different types of ionization waves in neon.

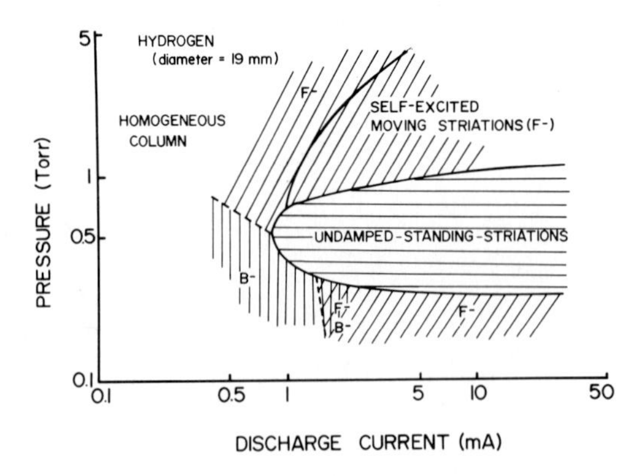

Fig. 6. Existence regions for the different types of ionization waves in hydrogen.

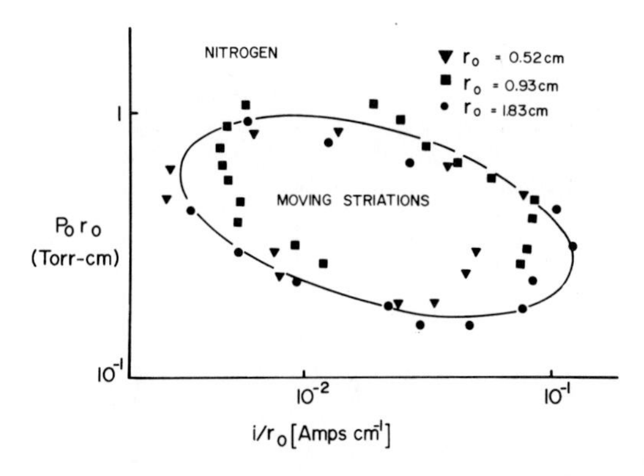

Fig. 7. Existence diagram for moving striations in nitrogen.

and frequency noises of the laser output. The actual coupling mechanisms include modulation of the excitation and refractive index changes.

Moving striations are not found in the alkali metal–vapor discharges and this gave impetus to an analysis (Robertson, 1957) linking metastable states with self-excited striations. There has been only one report (Foulds, 1956), later disputed (Crawford and Pagels, 1953), on moving striations in pure mercury vapor. Moving striations do occur readily in mercury vapor–rare gas mixtures, such as that in a fluorescent lamp or a germicidal lamp. The latter is a very convenient, and economical discharge for demonstrating the properties of these waves (Lee *et al.*, 1966). A simple arrangement using a pulse disturbance and a real or electronic equivalent rotating mirror gives an easy measure of the wave velocity and dispersion (Stirand *et al.*, 1966).

An important discovery by Pfau and Rutscher (1965) was the observation that forward-wave ionization waves in the axial direction occurred in association with the onset of constriction in the rare gases. This occurs at higher pressures and currents (Fig. 1). The comparison of the properties of these waves gave added support to the ionization wave model of Pekarek (Garscadden and Lee, 1966; Garscadden *et al.*, 1969). There is a report (Kiselevskii and Suzdalov, 1974) that radial ionization waves may also occur, but a similar detailed study has not been made.

The low-frequency waves and instabilities in a magnetoplasma have been studied by Duncan and Forrest, 1971) in the gas pressure range below 1 Torr and at magnetic fields up to several hundred gauss. At these pressures the waves are normally damped; however, the damping was shown to decrease with increasing pressure or increasing axial magnetic fields. Ion-

TABLE III

WAVE TYPES OBSERVED IN NEON-POSITIVE COLUMNS

Notation	Characteristic phase velocity	$E\lambda$ (V)
p	slow	9.2
s'	slow	19.5
r	fast	12.7
s	fast	19.5
High current	slow	Not constant

acoustic and drift waves were also observed. These instabilities were distinguishable by means of their dispersion and damping and by a careful comparison (Duncan and Forrest, 1971) with the predictions of a nonisothermal, but hydrodynamic, perturbed positive column model.

The occurrence of ionization waves in a transverse field such as in a nonequilibrium MHD generator (Neodospasov, 1968) above a critical magnetic field is well known and various comparisons have been made with theory, principally using the two-temperature model of Kerrebrock (1964).

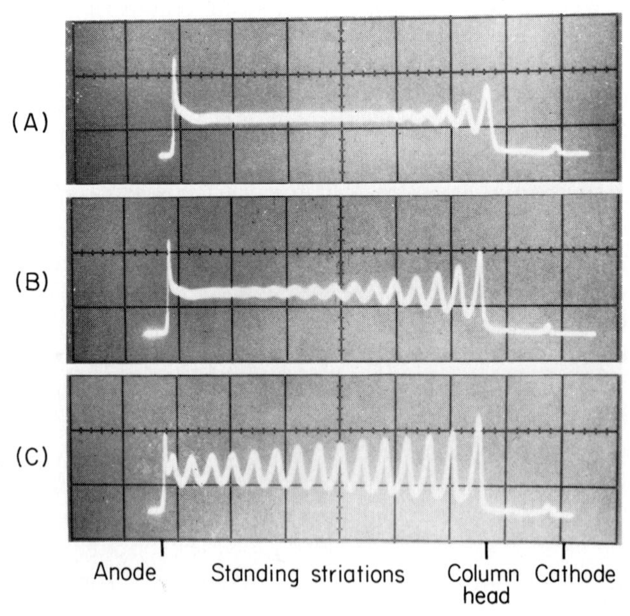

Fig. 8. Photomultiplier recordings of standing striations in a discharge in nitrogen, current 10 mA, diameter 48 mm. (A)–(F): 0.6–0.1 Torr at 0.1 Torr intervals. Anode to cathode spacing 39.4 cm.

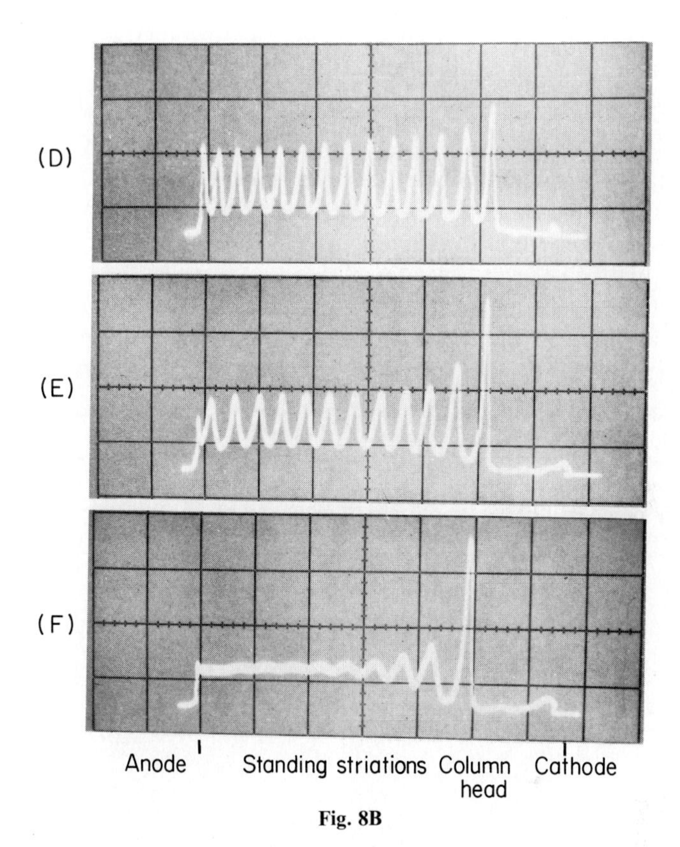

Fig. 8B

Standing striations also occur in discharges at very low currents (1 mA/cm^2). These layers can occur in the Townsend discharge or when a well-defined plasma has not yet been established. They are known as the Holst and Oosterhius layers. The reasons why it is not always so easy to observe them is that the wavelength is quite small (of the order of mm rather than the cm of moving striations) and also they are curved if the applied field is inhomogeneous. These layers have been studied in detail by Holscher (1967). Table III summarizes the moving striations observed in neon positive columns.

III. METHODS OF OBSERVATION AND MEASUREMENTS

A. Optical Measurements

The standing striations are easily observed visually; however, for quantitative measurements and detection of the small-amplitude striations, photomultiplier detection with slit aperturing of the discharge is appropriate. Examples at different pressures and constant current are shown in Fig. 8 for standing striations in nitrogen. The following characteristics are noted for diffusion-controlled conditions:

(1) The spatial oscillations are damped toward the anode.

(2) The attenuation increases as the gas pressure is increased.

(3) The wavelength of the standing striations does not vary rapidly with pressure.

(4) At lower pressures (in this example, below 0.15 Torr) the standing striations are rapidly attenuated, again in the anode direction. The "initial perturbation" at the head of the positive column is larger at lower pressures.

The moving striations are also easily measured using a photodiode or photomultiplier, with the signal output to an oscillograph or to a spectrum analyzer covering the frequency bandwidth of interest. Two photomultipliers can be used to measure the phase velocity and wavelength of the

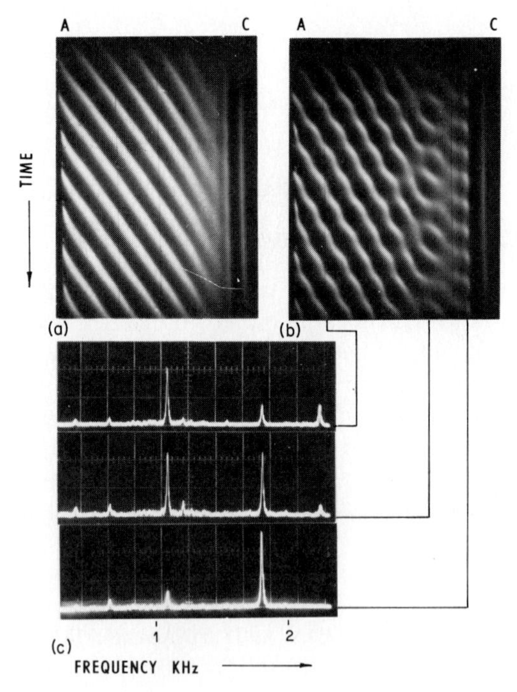

Fig. 9. (a) Self-excited ionization waves in mercury–argon discharge wave frequency 1105 Hz. (b) Same discharge, cathode modulation applied at 1785 Hz. (c) Spectrum analyzer record of light output at the points marked for the conditions of (b).

striations very accurately using conventional electronics. Correlation computers are useful if randomlike signals are present. More rapid surveys are possible using a rotating mirror whose axis is parallel to the discharge tube axis. Examples of such rotating mirror records or "smear photographs" are shown in Fig. 9a,b. The record of Fig. 9a shows the self-excited ionization waves in a mercury–argon discharge. The frequency of these waves was 1105 Hz and it was very stable and reproducible. The striations appear to peel off the anode and then to propagate at a uniform velocity in the direction anode to cathode, becoming attenuated at the cathode end of the column. For the record of Fig. 9b a modulation was applied at the cathode at a frequency of 1785 Hz. The space–time record shows that to satisfy the boundary condition a splitting of the self-excited ionization waves occurs at the cathode end of the column. The spectrum analyzer records of a photomultiplier observing at the points marked are shown in Fig. 9c. The discharge manages to satisfy the original frequency at the anode end of the column and also responds to the applied frequency at the cathode. In between both frequencies coexist as the velocity of the waves is periodically modulated by a disturbance propagating in the direction cathode to anode. The velocity of this disturbance is approximately equal to the striation velocity (i.e., the waves intersect at about the same angle with respect to the vertical axis).

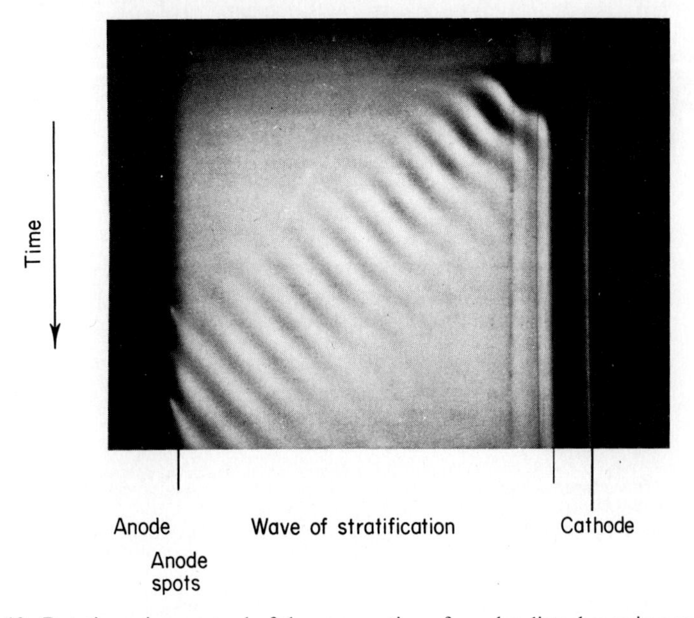

Fig. 10. Rotating-mirror record of the propagation of a pulse disturbance in a mercury–argon discharge just above the upper current limit for self-excited waves (100 mA). Time increases downward. Cathode on right-hand side.

Such observations give the clue to the nature of the wave. This is shown more dramatically in the record of Fig. 10. In this case the mercury–argon discharge was run at a current of 100 mA, which is just above the Pupp current for self-excited waves. A pulse disturbance was applied at the cathode and its subsequent propagation recorded. The disturbance propagates toward the anode and temporarily excites the ionization waves within the pulse. Thus, we are observing the dispersion of a convective backward wave. The disturbance or group velocity is in one direction while the ionization wave or phase velocity is in the opposite direction. This is the Pekarek "wave of stratification."

It will be noted that the ionization waves thus excited in the anode region frequency-pull and magnify the oscillation amplitude of a previously existing oscillating anode spot. The mechanical rotating mirror is still very useful for single-shot records. However, if the transient responses of the discharge are reproducible or if statistical information is acceptable, an electronic equivalent, due to Stirand *et al.* (1966), using a sampling technique is more convenient. The experimental requirements are illustrated in Fig. 11, where the record will be given in the left oscilloscope (Garscadden and

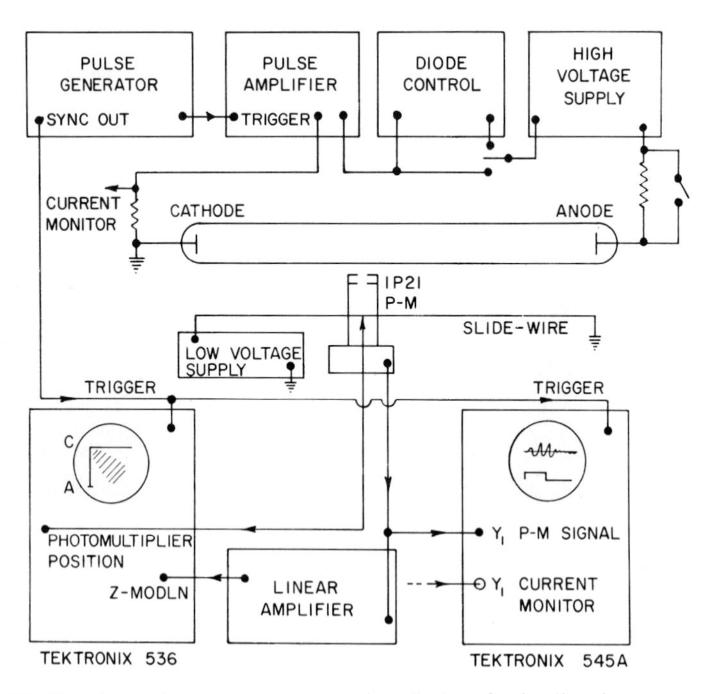

Fig. 11. Experimental arrangement: space–time display of pulse disturbance propagation in low-pressure positive columns.

Lee, 1966). The position of the photomultiplier determines the Y position. The photomultiplier response for a each Y position is amplified and used to modulate the beam intensity of the display. A suitable time-base is used for the X axis and it is triggered by a reference signal from the pulse generator. A slow uniform movement of the photomultiplier then gives a record such as that shown in Fig. 12. This is for a low-pressure nitrogen discharge. It is noted that the ionization waves excited within the disturbance propagate in the same direction as the disturbance pulse. This is therefore a "forward wave" response since the group and phase velocities are in the same direction. This diagnostic has been described because it is such a powerful yet inexpensive method, and using appropriate transducers, applicable to the individual parameters of the oscillation. If the discharge is driven by an external voltage controlled oscillator, a number of derivations of this circuit are possible to give a rapid display of the wave dispersion, i.e., the variation of the phase velocity and the variation of the wave amplification/attenuation vs. the driving frequency can be easily recorded.

A second optical survey method is to use an image converter (Lee *et al.*, 1966). This gives a two-dimensional record of the ionization waves. Modern image tubes have good time resolution and sensitivity so that often single exposures are possible, even at low currents. Then a variable time delay between a reference pulse and the opening of the shutter gives a time sequence of the discharge instability. An example of such a recording is given in Fig. 13 for a 50 mA pulsed-on mercury–argon discharge. If the discharge luminosity is weak, the reproducibility of the ionization waves often permits

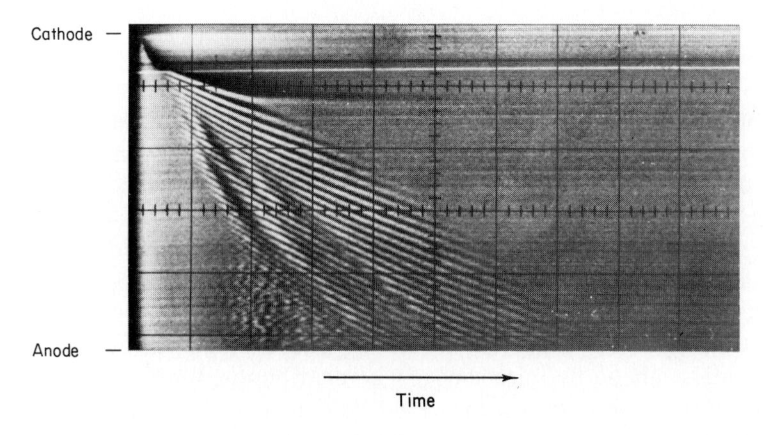

Fig. 12. Space–time display of ionization waves excited in a current-modulated nitrogen discharge: pressure 0.83 torr, current modulated from sustaining value of 5 to 28 mA, pulse width 5 msec. Time increases left to right: 200 μsec/division; cathode at top of photograph.

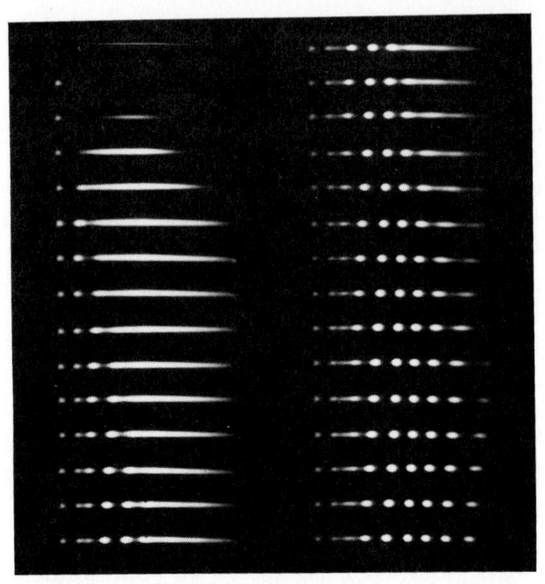

Fig. 13. Image converter records of the initiation of a glow discharge in mercury–argon. Each frame shows the discharge, cathode on the left-hand-side. Exposures 1 μsec, final discharge current 50 mA.

multiple samples at each delay time and subsequent integration. In this record it is again apparent that the formation of the waves is cathode to anode, while their phase velocity is anode to cathode.

B. Probe Measurements

Initially, Langmuir probe measurements were made by taking data at one bias voltage, recording the current as a function of time, and then repeating at a new bias voltage. The data were then reassembled and analyzed. Subsequently, the advent of fast sample and hold amplifiers permitted more rapid acquisition of a Langmuir probe curve for a specified phase of the striation. The modulated light signal was used to provide phase reference. However, the "locking" of the ionization wave frequency to an external source has been very important in reliable signal averaging and consequently information about the electron energy distribution function (EEDF) at higher energies. Methods using either computer analysis of such data or elegant time-resolved second-derivative methods have thus been established. An example of the former is shown in Fig. 14. The importance of these data is that they show that the temporal response of the EEDF to the changing

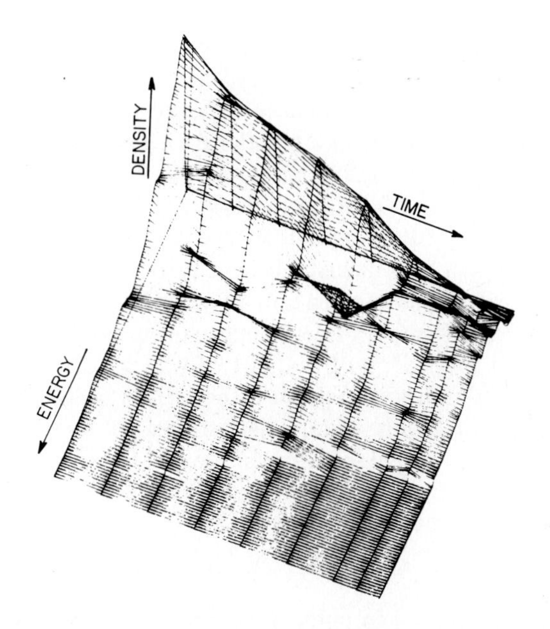

Fig. 14. Time-resolved electron energy distributions for 1.43 kHz, $E\lambda = 21$ V, ionization wave in neon. Pressure 0.8 Torr, radius 24 mm, current 40 mA. Z-axis, probability; X axis, energy; Y axis, time or phase (each main division is 36°) (Weber and Garscadden, 1970).

electric field is different at different electron energies, especially near excitation and ionization threshold energies. Consequently the moments of the distribution function such as the electron density, ionization rate, and average electron energy will have different phase relationships with the time-varying electric field. These phase relationships may be expected to change with the average value of the electric field. The most detailed time-resolved EEDF measurements on ionization waves are probably those of Rayment and Twiddy (1969), and Rayment (1974). Although often maligned, the time-resolved probe diagnostic has been invaluable in unraveling the non-Maxwellian EEDFs of the different types of ionization waves and giving impetus to theory. These measurements have been made in neon discharges *at low currents* on the p, r, and s waves. It was clearly shown that the different wave types are associated with spatial resonances of the electron gas. That is, the data reveal that in all of the low-current ionization waves, there is a spatial shift of the direct ionization rate maximum with respect to the electric field maximum. As a result, more new electrons are produced on the cathode side of the field maximum than on the anode side and the maximum moves toward the cathode. Further, Rayment has shown that this phase shift is associated with the arrival of a perturbation, or bump, in the EEDF, close to the ionization potential. This is suggested by Rayment as a physical explanation for the distinctive $E\lambda$ characteristics. It appears that in the p wave the bump is accelerated from a low energy around 1 eV to contribute to the direct ionization rate in two wavelengths. The s wave appears similar except that only one wavelength is required. The r wave is again similar but it appears to develop, not from very low energy but from about 13 eV. (It has been suggested that the original perturbation is caused by collisions of the second kind with a high metastable density, or that the inelastic collision cross section is near a minimum here, in an excited gas.) The p_2 wave appears to combine features of the p and r wave mechanisms in that the perturbation again grows from low energy, through two wavelengths. However, in the middle of the distribution it appears to be more rapidly accelerated than would be calculated from the electric field strength.

Finally, in the wave at high currents, the EEDF measured by Rayment is of constant form with a small sinusoidal variation in the mean electron energy. This can be represented in the form

$$F(V,t) = F_0(V)\{1 + cV[\cos(ft - 30^\circ) - 1]\},$$

where c is small compared to unity (Rayment finds $c \approx 0.015$) and f is the frequency. This expression shows that the average electron energy is reduced when the wave is present. There are no bumps in the EEDF. Calculations show that there are insufficient fast electrons to account for the ionization rate needed to balance losses. Therefore, Rayment suggests that this wave mechanism corresponds with the dependence of the ionization rate on the electron density (at constant electron temperature) due to cumulative ionization. These results and the achievement of very good agreement between theory and experiment by Wojaczek (1971) suggest that in the range of current and pressure above the Pupp limit, the understanding of ionization waves in a rare gas is satisfactory.

C. Microwave Methods

The electron temperature and the electron density in the striations has been measured with microwave diagnostics by many authors. Gentle (1966), with flowing gas to avoid time-resolved measurements, used an X-band bridge for electron temperature measurements and a microwave interferometer to measure the relative electron density changes. For the backward-wave ionization waves in argon he found that the phase difference between maxima of "electron temperature" and density was always close to 75°.

More extensive measurements with higher spatial resolution using a re-entrant toroidal cavity have been made by Sicha *et al.* (1967).

D. Identification of Different Varieties of Ionization Waves

The complexity of ionization wave types at low currents led Rother (1959) to suggest that two or more maxima existed on the amplification curve associated with a simple dispersion curve. The approaches of Pekarek (1971), later elaborated by Ruzicka and Rohlena (1973), showed that the low-current dispersion curves have to be modified for both the real and imaginary parts. In his theory of the successive production of ionization waves, Pekarek distinguishes two different chains of processes, illustrated in Fig. 15. Both are initiated by the additional axial space charge electric field caused by an assumed perturbation. This electric field causes, via the electron gas, a change in the ionization in the neighboring region (toward the anode) and creates another space charge but of opposite polarity. It was proposed that, when the additional ionization occurs through stepwise ionization of a metastable state, this results in the slow wave or *p* wave of ionization; however, if the ions are created by direct ionization of ground state atoms, the fast wave or *r* wave results. The difference in relaxation times is accounted for, since the metastable atoms decay by thermal diffusion and quenching while the atomic ions decay by the faster process of ambipolar diffusion. These concepts are quantitatively true for the *r* wave but the *p* wave relaxation time is about an order of magnitude shorter than the calculated metastable state lifetime (Rutscher, 1962). Therefore, experiments were made on the characteristics of the ionization waves at different gas temperatures by Behnke (1966) (at 298° and 623°K) and by Masek and Perina (1969) (at 290° and 77°K). At the same gas density, lowering the gas temperature causes the metastable lifetime to increase but the lifetime of the atomic ions will decrease. Masek and Perina found that the *p* and *r* waves displayed the

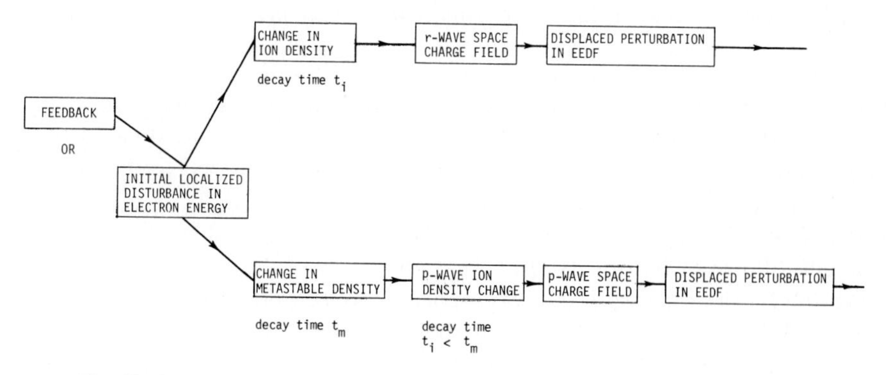

Fig. 15. Sequence illustrating direct (fast) and slow (metastable) ionization waves.

frequency dependences required in the model. Other experiments were performed on the *p* waves by Pfau and Rutscher (1964) (adding hydrogen to quench the metastables) and by Pekarek (1958) (illuminating the discharge with light from another discharge in the same gas) to show the importance of the metastable states on the slowest wave. It is interesting to note that although both the electric field E and the wavelength λ both changed, the Novak (1960) relation $E\lambda$ = const remained the same at the different temperatures. Measurements on the *s* wave, although less accurate, showed that this wave followed the *r* wave characteristics except in one feature. At higher pressures the *s* wave frequency increases with increasing discharge current like the *p* wave, whereas the *r* wave frequency decreases with increasing current.

IV. ANALYSES

A. Rare Gases

The work on low-frequency wave propagation in rare gas discharges at low pressures has led to a fairly complete description of ion acoustic waves (Ewald *et al.*, 1969). Recent work on pulsed ion wave excitation showed that nonlinear effects could be studied in a convenient time regime. When account is taken of the damping due to ion-neutral collisions at pressures less than 100 μm and also the steady-state drifts of charged particles, good agreement is obtained between theory and experiment. If the electron dynamics are analyzed to higher order, new types of propagation are predicted in an axially magnetized positive column. As the gas pressure is increased, the ion-acoustic waves become strongly damped and the predicted dominant form of the wave propagation becomes the electron drift waves. Again, rather good agreement has been obtained between theory and experiment for the propagation of the helical electron drift modes (Self *et al.*, 1969; 1969; Duncan *et al.*, 1969). These can be amplified leading to the instabilities observed in low magnetic fields due to the destabilizing effects of the axial drift and the diagmagnetic drift (caused by the interaction of the radial density gradient and the magnetic field). However, in the case of the $m = 0$ symmetric mode, which is driven only by the axial drift, the predicted damping is much greater and the predicted slight dispersion of this model is very much different from the experimental observations on rare gas ionization waves. In the rare gases the experimental ionization wave dispersion at low pressures approximates a hyperbolic relation ωk = const. Therefore, an approach other than the drift or density wave model is required. The group of Pekarek at Prague has had the longest dedication to the study of ionization waves and consequently some of the most interesting advances on developing a theory. The first model (Pekarek and Krejci, 1962) was based on four linearized equations for the ions and electrons:

$$\frac{\partial n_p}{\partial t} = D_p \frac{\partial^2 n_p}{\partial z^2} - \mu_p N_{e0} \frac{\partial e}{\partial z} - \mu_p E_0 \frac{\partial n_p}{\partial z} - \frac{1}{\tau_a}(n_p - n_e) + Z_\theta' N_{e0}\theta, \quad (5)$$

$$\frac{\partial n_e}{\partial t} = D_e \frac{\partial^2 n_e}{\partial z^2} + \mu_e N_{e0} \frac{\partial e}{\partial z} + \mu_e E_0 \frac{\partial n_e}{\partial z} + Z_\theta' N_{e0}\theta, \quad (6)$$

$$0 = \frac{\partial e}{\partial z} - 4\pi q(n_p - n_e), \quad (7)$$

$$0 = \frac{\partial \theta}{\partial z} - a_1\theta + b_1 e, \quad (8)$$

where

$$n_j = N_j - N_{j0}, \qquad |n_j| \ll N_j, \qquad j = \mathrm{p, e},$$
$$\theta = \Theta - \Theta_0, \qquad |\theta| \ll \Theta, \qquad \Theta = 3kT_e/2,$$
$$\bar{e} = \bar{E} - \bar{E}_0, \qquad |e| \ll E.$$

These are the small-amplitude deviations in ion and electron densities, in electron temperature, and in electric field, respectively; $\tau_a = \Lambda^2/D_a$ is the diffusion time under ambipolar conditions, where D_a is the ambipolar diffusion coefficient; $Z_\theta' = (\partial z/d\Theta)_{T_e = T_{e0}}$, where Z is the ionization frequency and T_{e0} the steady-state electron temperature; a_1^{-1} is the electron temperature relaxation length (Granowski, 1955), and b_1 is a constant approximately equal to $3q/2$. In this model the electron and ion inertia have been omitted from the continuity equations. Charge neutrality is not assumed, and thus Eq. (7) is obtained from application of Poisson's equation. The fourth equation expresses the fact that the perturbation in electron temperature is not exactly proportional to the perturbation of the electric field but is also influenced by the potential difference through which the electrons passed beforehand (this phase shift of θ with respect to e is very important to provide amplification in this model). The electron pressure has been neglected. The frequencies of interest permit neglect of the time delay between the electric field and the electron temperature.

Equation (5) is further simplified by neglecting the axial ambipolar diffusion term compared to the drift term. Equation (4) is simplified by assuming that the electron concentration fluctuation is given by

$$\partial n_e/\partial t = Z_\Theta' N_{e0}\theta$$

so that Eq. (6) becomes

$$0 = D_e \frac{\partial^2 n_e}{\partial z^2} + \mu_e E_0 \frac{\partial n_e}{\partial z} + \mu_e N_{e0} \frac{\partial e}{\partial z}. \tag{9}$$

A solution of the simplified coupled equations gives a dispersive wave with the backward properties. Pekarek also derived an analysis for the response of the plasma to an initial perturbation of the form

$$n_p(z,t = 0) = (\Lambda^2 \pi)^{-1/2}[\exp(-z^2/\Lambda^2)]. \tag{10}$$

If we substitute (9) into (7) we obtain an expression for the subsequent perturbation in electron concentration

$$n_e \approx n_p + \frac{2 l_D{}^2}{l_1} \frac{\partial n_p}{\partial z} + l_D{}^2 \frac{\partial^2 n_p}{\partial z^2}, \tag{11}$$

where

$$l_1 = 2D_e/\mu_e E_0, \tag{12}$$

and the Debye length

$$l_D = (kT_e/4\pi q^2 N_{e0})^{1/2}. \tag{13}$$

Then from Poisson's equation and Eq. (8), the perturbation in electric field is

$$e = -4\pi q l_D{}^2 \left(\frac{2}{l_1} \frac{\partial n_p}{\partial z} + \frac{\partial^2 n_p}{\partial z^2} \right). \tag{14}$$

Then

$$\frac{\partial n_p}{\partial t} \approx D_a \frac{\partial^2 n_p}{\partial z^2} + Z_\theta' N_{e0} \theta, \tag{15}$$

and finally, using Eq. (8), an equation with only one dependent variable is obtained:

$$\frac{\partial n_p}{\partial t} = D_a \frac{\partial^2 n_p}{\partial z^2} + Z_\theta' b_1 \theta n_p - Z_\theta' b_1 (\theta + E_0) \int_z^\infty e^{a_1(z-\xi)} n(\xi)\, d\xi. \tag{16}$$

Several methods have been used to solve Eq. (16) including numerical evaluation (Pekarek, 1963), Fourier series expansion (Lee *et al.*, 1966), and Laplace transform calculus (Garscadden *et al.*, 1969). The dispersion relation that is obtained is

$$\Omega^* - \frac{K}{K^2 + 1} - i\left(\alpha K^2 + \frac{1}{K^2 + 1}\right) = 0 \tag{17}$$

and is illustrated in Fig. 16. Here the pseudo-frequency

$$\Omega^* = \Omega + i\beta, \tag{18}$$

where the discharge parameter $\beta = k/a$, and

$$\Omega = \omega[\tfrac{3}{2}Z_\theta'(\theta + E_0/a)]^{-1} \tag{19}$$

is given as a function of a normalized wavenumber $K = k/a$, where $k = 2\pi/\lambda$ and is the wavelength of the ionization wave. The variable parameter for the imaginary part of Ω^* is another discharge parameter

$$\alpha = \frac{Da^2}{\tfrac{3}{2}Z_\theta'(\theta + E_0/a)}. \tag{20}$$

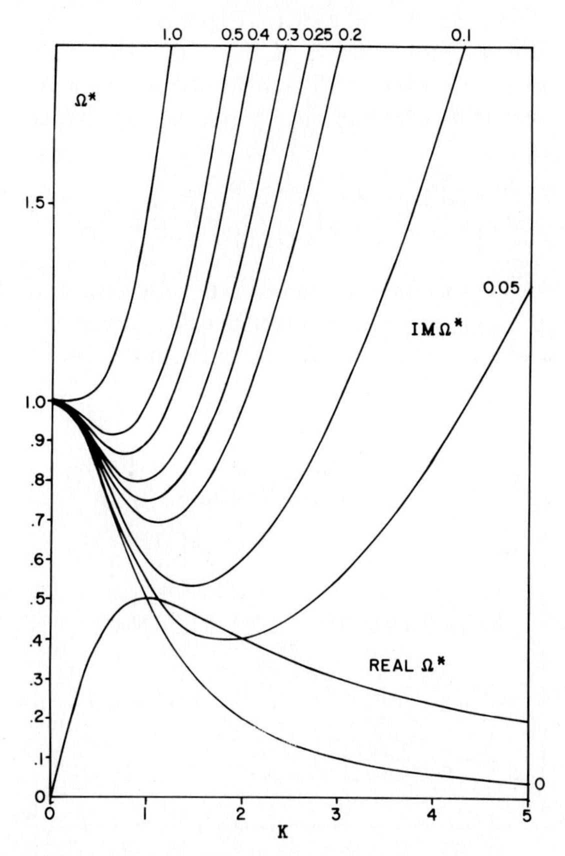

Fig. 16. Normalized dispersion diagram for ionization waves in rare gases showing the pseudofrequency $\Omega^* = \Omega - i\beta$ as a function of real wavenumber K for several values of the discharge parameter α.

Imaginary $\Omega < 0$ corresponds to the unstable half-plane so that the unstable wavenumbers are those where $\beta > \operatorname{Im}\Omega^*$. The slope of $\operatorname{Re}\Omega^*$ determines the group velocity. Thus if $K < 1$, the ionization wave will be a forward wave in the anode to cathode direction, while if $K > 1$, the wave will become a backward wave in the cathode to anode direction. It will be noted that depending on whether α is less than or greater than 0.25 determines whether the ionization wave instability in rare gases will be a forward wave or a backward wave, respectively. Physically interpreted, the condition $K < 1$ means that we are considering a wavelength shorter than the electron temperature relaxation length. Perhaps it is for this reason that these forward waves have only been seen in association with constricted discharges (Pfau and Rutscher, 1965). The stability analysis of this dispersion relation using the criteria of Derfler (1967) showed that there can exist stable, convectively unstable, and absolutely unstable waves. Expressed in the parameter space $\beta - \alpha$, it was shown (Garscadden *et al.*, 1969) that the existence regions of these wave types are as given in Fig. 17. Modifications and improvements have been made in this analysis, e.g., by writing the change in ionization rate as

$$b_1\left[\left(\frac{\partial Z}{\partial \theta}\right)_0 + \frac{1}{\tau_a{}^2}\left(\frac{\partial \tau_a}{\partial \theta}\right)_0\right] + \left(\frac{\partial Z}{\partial n_e}\right)_0 N_{e0}, \tag{21}$$

thus allowing for the variation of the loss rate with respect to electron temperature change, and the effect of electron density changes on the ionization rate. The latter term might be important due to cumulative processes. How-

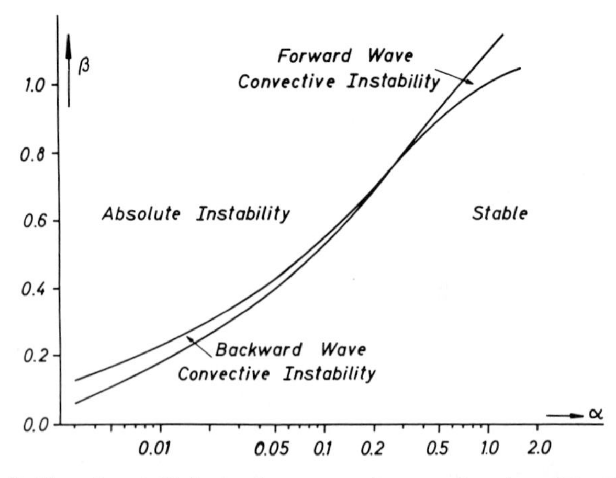

Fig. 17. Stable and unstable ionization wave regions as a function of the discharge parameters α and β.

ever, the basic nature of the instabilities remains unchanged and the model provides a start for unraveling the different experimental results. Thus, in a convectively unstable discharge, the occurrence and properties of ionization waves will be very much influenced by the boundary conditions and other oscillations in the discharge. On the other hand, in the absolutely unstable discharge, the boundary conditions will not influence the existence of the unstability although the properties may be influenced by the requirement that certain wavenumber eigenvalues are more likely (Garscadden and Bletzinger, 1970).

The most thorough analysis for ionization waves near the Pupp limit has been performed by Wojaczek (1971). He used a complete expression for the electron energy equation, including thermal conductivity and diffusion due to the electron temperature gradients. Wojaczek also included the effects of inelastic collisions on the high-energy tail of the EEDF (which was otherwise Maxwellian) in his calculation of the ionization frequency. Impressive agreement of his calculated dispersion was obtained in the comparison with Venske's (1970) experiments in low-pressure argon discharges. Duncan and Forrest (1971) subsequently extended the electron drift fluid model by permitting electron temperature variations. Swain and Brown (1971) later gave an extension of this treatment and combined some of the earlier approach of Gentle by treating two models.

In their first model, the moment equations are used to treat both the electron and ion motion. This gives a dispersion relation quadratic in frequency; thus in general, two solutions for $\omega(k)$ are obtained, one corresponding to the ion-acoustic branch and one corresponding to the ionization or striation branch. In the appropriate limit, the latter corresponds to Pekarek's formulation. Pekarek's original theory of ionization wave neglects the ion inertia term in the momentum equation and thus he has a diffusion type of equation (appropriate at higher pressures) rather than a wave type of equation. The treatment of Swain and Brown is valuable in that it gives a dispersion curve that crosses the wavenumber axis, and thus includes standing striations, whereas Pekarek originally had to propose a phenomenological curve to explain the experimental results. A minor point is that the comment by Swain and Brown, that standing striations are not included in Pekarek's equation, is not completely correct. As has been emphasized by Rognlien and Self (1972), the dispersion solution to be sought depends critically on the experiment performed. Certainly, the more usual experiment is where a given frequency is applied to the discharge and the wavenumbers; or wavelengths that are excited are measured. However, if there is a selection of some wavenumbers due to the finite length or particular geometry of the discharge, then one should solve the dispersion relation for $k(w)$, the complex wavenumber as a function of frequency. It was demonstrated by Lee and Garscadden (1972) that Pekarek's equation under these conditions leads to standing striation solutions.

Swain and Brown's second method recognizes that the equation relating the ionization frequency to the electron temperature is only of value if the EEDF is Maxwellian. Pekarek and his colleagues had emphasized this point also in their earlier papers by only making experimental comparisons for conditions at high currents near the Pupp limit. Thus, the approaches using the hydrodynamic equations are valuable for a qualitative description of the ionization waves.

However, a Boltzmann approach must be used for the electron dynamics at low currents. The most outstanding work is a series of papers by the Prague group, including Rohlena *et al.* (1972), Ruzicka and Rohlena (1973), and Perina *et al.* (1975). The Boltzmann equation expanded in the angular dependence in velocity space was solved numerically to obtain the perturbed distribution function for a periodic field. The perturbed EEDF was then integrated to obtain the rate coefficients in the atomic ion and metastable continuity equations. From these balance equations the dispersion relations for both fast and slow waves have been obtained. This procedure has been accomplished for neon (Rohlena *et al.*, 1972) and for helium (Perina *et al.*, 1975). The theory gives six possible spatial resonances for neon. It appears that four wave varieties (three connected with the electron gas resonances and one of hydrodynamic origin) of the six have been observed experimentally in neon (Table IV). The term, electron gas resonances, means that at certain wavelengths the perturbed electron density deviates very much from the value predicted by hydrodynamic theory. A physical interpretation might be as follows. The direct variety of ionization wave exists because it is more probable that an electron will reach the ionization threshold in a steeply changing electric field than in a gradual one. When one allows

TABLE IV

TIME- AND SPACE-SELECTED WAVE VARIETIES FOR NEON AT LOW VALUES OF DISCHARGE CURRENT

Space selection	First resonance, $E = 19$ V	Hydrodynamic maximum, $E = 13$ V	Second resonance, $E = 9.5$ V
Time scale selection	s^1 Variety	r^1 Variety	p-Variety
Slow wave (metastable guided)	Observed	Not observed	Observed
Fast wave (ion guided)	s Variety observed	r-Variety observed	p^1 Variety not observed

for the presence of other inelastic thresholds and the presence of metastable excited states it is found that there are certain preferred wavelengths favoring the ionization process. These special wavelengths are given approximately by the relation (Ruzicka and Rohlena, 1973)

$$\lambda = U_a(1 + p)/qE, \qquad p = \tfrac{2}{3}m/M(n_g Q_m U_a/E)^2, \tag{22}$$

where m, M are the electron and ion masses, U_a the first inelastic potential, E the electric field, n_g the neutral density, Q_m the momentum transfer cross section, and $p \ll 1$ the ratio of the elastic and inelastic energy losses. The space resonances strongly modify the dispersion curves of the ionization waves. Figure 18 is an example of the dispersion and amplification calculated from this theory for a given set of parameters. At the two resonant wavenumbers sharp maxima appear on the amplification curve accompanied by inflections on the dispersion curve. The resonant wavenumbers correspond to those observed for the s^1 and p varieties of the slow wave. The correct values of $E\lambda$ are also obtained. Another important result of this theory is that the modifications in the dispersion curves permit interaction and modulation of the ionization waves, which satisfy the energy and momentum relationships. Previously with the ωk = const dispersion relation, this was not possible.

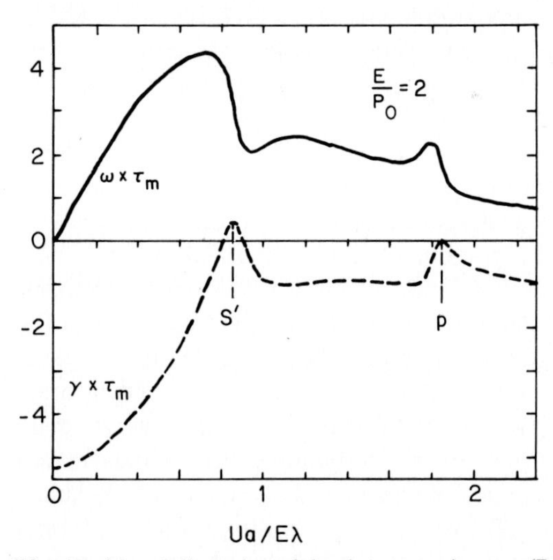

Fig. 18. Amplification (γ) and dispersion of the slow waves in neon (Rohlena *et al.*, 1900), $I/R = 1.27$ mA/cm, $P_0^R = 3$ cm Torr, $n_m/n_g = 3.16 \times 10^{-6}$ (degree of excitation of the metastable), $\tau_m/\tau_{miv} = 0.29$ (ratio of total and wall metastable lifetimes). Two distinct s^1 and p resonances appear separated by a weak hydrodynamic maximum.

B. Molecular Gases

Advances in the understanding of ionization waves in molecular gases have been much slower, because the phenomena were nearly always studied under nonlinear conditions. Also, the discharges were usually sealed so that there was no control on the species of ions or complex reactions that can occur with a mixture of dissociated products and even a small amount of impurities. Recently, due to laser applications, many discharges have been studied under open-cycle conditions. It was not appreciated earlier how important a role negative ions would play in these gases or gas mixtures. However, until recently, the influence of large vibrational inelastic cross sections on the EEDF was not known, and consequently the various rate constants associated with ionization, attachment, etc., could not be calculated. It is now known that the EEDF in nitrogen, carbon monoxide, hydrogen, etc., and in mixtures thereof deviates markedly from a Maxwellian distribution function. The electron kinetic rates therefore must be calculated from a solution of the collisional Boltzmann equation. On the other hand, once these requirements are satisfied, the subsequent calculations are simpler than in the rare gases. This is because the existence of large inelastic cross sections at most energies ensures that the different portions of the EEDF are closely coupled with respect to temporal or spatial changes of the electric field. Thus we do not obtain spatial resonances of the electron gas, and the hydrodynamic approach using coupled rate equations can be used. The number of species equations is increased by the inclusion of the negative ions and excited states (this approach is almost a return to the methods used by Robertson (1952) and Reece-Roth (1967) in attempts to describe the ionization waves in the rate gases).

A completely generalized approach along these lines including, among others, ionization, attachment, detachment, vibrational, rotational, thermal, and electronic excited state excitation, has been accomplished by Haas (1973). Some of the resulting modes couple together. However, Haas, using the large differences in characteristic times, demonstrates that the resulting tenth-order dispersion equation (!) can be coupled into a set of relatively independent dispersion relations of third order or less. Haas developed the linearized equations for hydrodynamic instabilities in a five-component plasma where the species considered are neutral molecules (n), metastable molecules (n_m), negative ions (n_n), positive ions, and electrons. The first three conservation equations for number density, momentum, and energy density are established for each species. Using the separation of time scales (Allis, 1976) it is possible to obtain the following ionization instability criterion for a recombination-dominated plasma *without negative ions:*

$$\mathrm{Re}(i\omega) = nk_{\mathrm{i}}\left(\frac{-2\cos^2\phi}{\hat{\nu}_{\mathrm{u}}'}\hat{k}_{\mathrm{i}} - 1\right) > 0, \tag{23}$$

where the caret indicates the logarithmic derivative of a rate constant with respect to electron temperature,

$$\hat{k} = \frac{T_{\mathrm{e}}}{k}\frac{\partial k}{\partial T_{\mathrm{e}}} = \frac{\partial \ln k}{\partial \ln T_{\mathrm{e}}}. \tag{24}$$

ϕ is the angle between the unperturbed current density (and electric field) vector and the wave propagation vector k, and

$$\hat{\nu}_{\mathrm{u}}' = 1 + \hat{\nu}_{\mathrm{u}} - \hat{\nu}_{\mathrm{m}}\cos 2\phi, \tag{25}$$

where $\hat{\nu}_{\mathrm{u}}$ is the total electron energy exchange collision frequency and ν_{m} the electron momentum-transfer collision frequency. k_{i} is the ionization rate coefficient, which has been calculated from the cross section and the EEDF. Although $\hat{\nu}_{\mathrm{u}}'$ depends on the spatial orientation of the disturbance wave, calculations show that $\hat{\nu}_{\mathrm{u}}'$ is almost always positive. Since $\hat{k}_{\mathrm{i}} \gg 0$, Eq. (23) cannot be satisfied for instability. The physical reason for this result is that when direct ionization and recombination dominate electron production and loss, an increase in electron temperature is accompanied by an increase in electron density. Haas showed the quasi-steady relationship between the wave perturbation amplitudes $T_{\mathrm{e}}(k)$ and $n_{\mathrm{e}}(k)$:

$$\frac{T_{\mathrm{e}}(k)}{T_{\mathrm{e}0}} = -\frac{2\cos^2\phi}{\nu_{\mathrm{a}}'}\frac{n_{\mathrm{e}}(k)}{N_{\mathrm{e}0}}. \tag{26}$$

Thus these quantities are originally out of phase and the local perturbation does not reenforce the condition. Therefore this plasma condition is stable. The criterion for an ionization instability *when negative ions are present* was also derived as

$$\left[\left(\frac{-2\cos^2\phi}{\nu_{\mathrm{u}}'}\right)\left(1 - \frac{k_{\mathrm{a}}\hat{k}_{\mathrm{a}}}{k_{\mathrm{i}}\hat{k}_{\mathrm{i}}}\right)Nk_{\mathrm{i}}\hat{k}_{\mathrm{i}}\right] - \left(\frac{n_{\mathrm{e}}}{n_{\mathrm{p}}}n_{\mathrm{p}}k_{\mathrm{r}}^{\mathrm{e}} + \frac{n_{\mathrm{n}}}{n_{\mathrm{e}}}Nk_{\mathrm{a}} + \frac{n_{\mathrm{n}}}{n_{\mathrm{p}}}n_{\mathrm{p}}k_{\mathrm{r}}^{\mathrm{i}} + \frac{n_{\mathrm{e}}}{n_{\mathrm{n}}}Nk_{\mathrm{a}} + \frac{N}{n_{\mathrm{e}}}S\right) > 0, \tag{27}$$

where $k_{\mathrm{r}}^{\mathrm{e}}$ and $k_{\mathrm{r}}^{\mathrm{i}}$ are the electron and ion recombination rates, k_{a} and k_{d} the electron attachment and detachment rates, and S an external ionization source, if present.

A necessary condition for ionization instability is

$$k_a \hat{k}_a / k_i \hat{k}_i = \delta k_a / \delta k_i > 1. \tag{28}$$

Physically this requires that the electron attachment coefficient be an increasing function of electron temperature and that its response to an electron temperature change be larger than the response in ionization rate. If this occurs, then an increase in electron temperature can result in a decrease in the local electron density, thus reenforcing the preferred quasi-steady condition. Therefore an attachment–ionization instability can develop if this term is large enough to compensate the various damping terms in the second bracket of Eq. (27). It is also of interest to note that the maximum instability will occur for $\phi = 0$, i.e., in the unperturbed applied field direction. In a series of detailed papers Nighan and Wiegand (1974a,b; Nighan *et al.*, 1973) have examined the application of Haas' stability criteria to gas laser mixture plasmas (also including thermal instability, which we have not considered here). These calculations and comparisons have proved very valuable. In laser applications, they have proved the advantages of establishing a reliable and self-consistent data bank so that the rate coefficients can be estimated with reasonable accuracy and the stability criteria used to define an accessible parameter range of stable discharge operation. These criteria are usually applied in the limit $k \rightarrow 0$, i.e., the long-wavelength limit; also the applications primarily have been to high-pressure recombination-dominated laser-type discharges. This approach has been described in detail recently (Nighan and Wiegand, 1974a) and so we shall adopt a different procedure, which is closer to the development previously described for the rare gases. This follows earlier work by Bailey *et al.* (1974), which has been presented but not published in detail.

C. Electronegative Gases

At low pressures the steady-state positive column with negative ions present has been studied by Thompson (1959). He calculated the diffusion coefficients* for a three-component plasma (n_e, n_n, n_p) assuming particle continuity and charge neutrality. A summary of his results is given in Fig. 19. In a cylindrical discharge with sufficient negative ions present, the ion density

* There is some reservation on the general applicability of these results. Oskam (1958) has derived a particular solution for the situation when the profiles are similar ($\nabla n/n$ is the same for all species). Sabadil (1973) has derived numerical results for the oxygen discharge; these show that the effective diffusion coefficients are complicated functions of density ratios and radial position. Under these circumstances the term diffusion coefficient loses its definition.

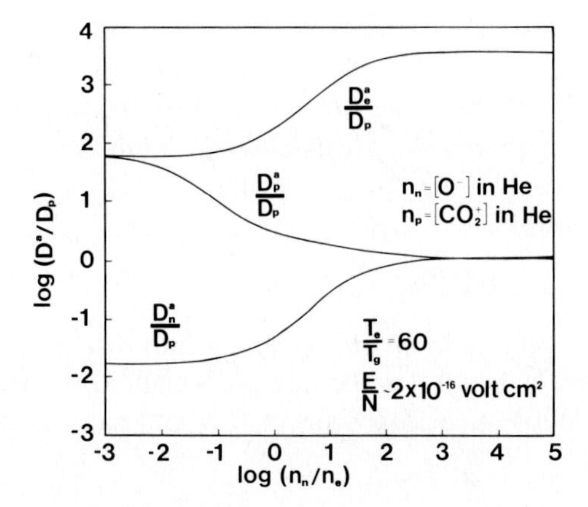

Fig. 19. Dependence of ambipolar diffusion coefficients on the ratio of negative ion density to electron density in a three-component plasma (Thompson, 1959).

profiles are close to a zero-order Bessel function; however, the electron density has a rather flat profile except near the boundaries. It is noted that the diffusion coefficients are fairly insensitive to E/N, when the ratio T_e/T_g is large (as assumed here); however, they are sensitive to the relative concentration of negative ions. The decrease in the positive ion diffusion coefficient D_p is due to the reduced ambipolar field. The increase in the electron diffusion coefficient is partially compensated by the increase in the characteristic length of the electron density gradient.

For the subsequent analyses, the parameters of the discharge that are functions of E/N were calculated using integrals from the non-Maxwellian distribution function derived from solutions of the collisional Boltzmann equation. Because there are large collisional frequencies for both elastic and inelastic collisions, the EEDF is assumed to follow any electric field changes instantaneously. The coupled equations considered are

$$\frac{\partial N_e}{\partial t} = N_e N k_i - \frac{D_e{}^a N_e}{\Lambda_e{}^2} + N_n N k_d - N_e N k_a - k_r{}^e N_e N_p, \tag{29}$$

$$\frac{\partial N_n}{\partial t} = N_e N k_a - \frac{D_n{}^a N_n}{\Lambda_n{}^2} - N_n N k_d - k_r{}^i N_n N_p, \tag{30}$$

$$\frac{\partial N_p}{\partial t} = N_e N k_i - \frac{D_p{}^a N_p}{\Lambda_p{}^2} - k_r{}^i N_n N_p - k_r{}^e N_e N_p. \tag{31}$$

These are the equations of species conservation; we then include the discharge current

$$I = \int_0^R qN_e(r)\mu_e(E/N)E2\pi r\,dr \tag{32}$$

and gas temperature

$$\bar{\nabla}\cdot(\lambda\nabla T_g) + f(E/N)qN_e(r)\mu_e E^2 = 0, \tag{33}$$

where $f(E/N)$ is the fractional energy going into discharge heating for the particular gas mixture and λ is the mixture-weighted thermal conductivity. This leaves the following as functions of E/N and gas mixture: $D_p{}^a$, $D_n{}^a$, k^i, k^a, k^d, k_r, e.

1. Steady-State Solution

Equations (29)–(31) have been solved for initial conditions of interest for low-pressure laser discharges. An example of the results is shown in Fig. 20. To understand these results, we start at the left-hand side when N_n/N_e is small. The equations take an especially simple form if the recombination term is small compared to ionization and attachment:

$$N_n/N_e \simeq k_a/k_d, \qquad N_p/N_e \simeq Nk_i\Lambda_p{}^2/D_p{}^a, \tag{34}$$

$$(Nk_i\Lambda_p{}^2/D_p{}^a) - (k_a/k_d) = 1. \tag{35}$$

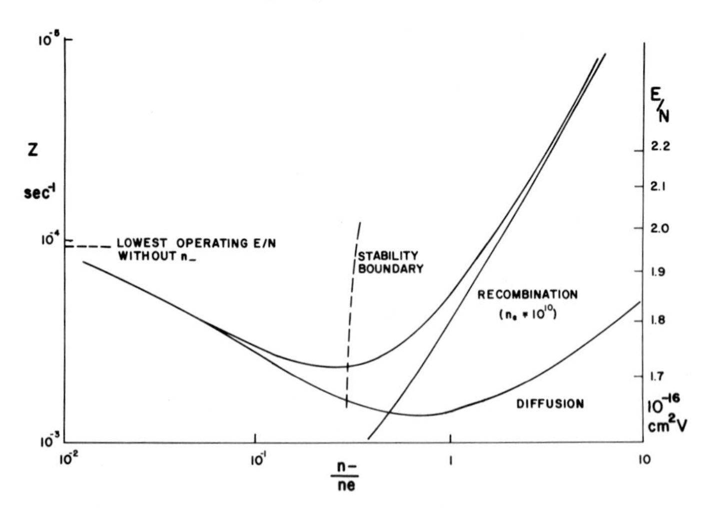

Fig. 20. E/N, ionization rate, diffusion loss, and recombination loss in a low-pressure laser mix discharge as functions of the negative density to electron density ratio. Positive ion loss at 10 Torr; $R = 1$ cm; $CO_2:N_2:He$ ratio 5:10:85.

As the relative negative ion concentration is increased, the positive ion loss due to diffusion decreases at first due to the decrease in the diffusion coefficient. However, it becomes an increasing function of N_n/N_e due to the increase in ion concentration compared to the electron concentration (ion diffusion loss is proportional to N_p, whereas ion production is proportional to N_e). The full equations must then be used. The ion–ion recombination loss eventually becomes dominant, for this example at about $N_n/N_e \simeq 0.6$.

Under many circuit conditions the discharge will tend to operate close to the minimum of the total loss curve, provided the discharge is stable over the whole parameter space.

2. Linearized Perturbation Analysis

To determine the parameter range over which the discharge is stable, the equations were examined in the usual manner as to the stability of solutions to small amplitude perturbations. For this analysis the gas temperature was held constant at the steady-state value. The following condition for stability for long wavelengths was obtained:

$$k_i(\hat{k}_i - 1) - k_a(\hat{k}_a - 1) + k_d + (D_p{}^a/\Lambda_p{}^2) + k_r{}^i(N_{p0} + N_{n0}) + k_r{}^e(N_{p0} + N_{e0}) > 0. \tag{36}$$

and if $k_r{}^i N_p \ll k_d$, then

$$N_n/N_e \lesssim (\hat{k}_a - 1)^{-1} \tag{37}$$

is the upper limit on the ratio N_n/N_e in a stable discharge. Typically this will occur when $N_n \simeq N_e$. The stability boundary is shown in Fig. 20 as the dashed line intersecting close to the minimum of the ion loss curve, where the carets now denote the logarithmic derivatives of the ionization and attachment rates with respect to E/N (which we found to be a more accessible function). As $\hat{k}_i > 1$ in a self-sustaining discharge, the only quantity that can give rise to instability is $\hat{k}_a$. Equation (36) shows that a large attachment rate that increases rapidly with E/N is destabilizing, while ionization, detachment, diffusion, and recombination are all stabilizing. The instability is now known as the attachment instability. It has been derived from different approaches by Haas (1973) and by Douglas-Hamilton and Mani (1973). The present type of analysis has been extremely useful in comparing the effects of contaminants or additives on molecular gas laser mixture discharges (Bletzinger *et al.*, 1975).

If we assume that current continuity is maintained, $\bar{\nabla} \cdot J_e = 0$, then it is possible to derive the phase relationships of the parameters of the wave:

$$\frac{\tilde{E}}{E_0} = \frac{-\cos^2\theta - iKl_0\cos\theta}{1 + \hat{\mu}_e\cos^2\theta + iKl_0T_e\cos\theta}\frac{N_e}{N_{e0}}, \tag{38}$$

where $l_0 = T_e/E$, and θ is the angle between the wave vector and the field. For wavenumber $K = 0$,

$$\frac{\tilde{E}}{E_0} = -\frac{\cos^2\theta}{1 + \hat{\mu}_e\cos^2\theta}\frac{N_e}{N_{e0}},$$

where μ_e is the logarithmic variation in electron mobility with E/N. This may be compared with the analogous expression for the rare gases at high currents (Ruzicka and Rohlena, 1972)

$$\tilde{E}/E_0 = -(1 + iKl_0)\tilde{N}_e/N_{e0}, \tag{39}$$

which shows that as $K \to 0$ the electric field and the electron density oscillate out of phase, in both cases.

3. Numerical Comparisons

A numerical calculation of the coupled equations confirms these phase relationships. The results of the time-dependent calculation for a typical laser mix are given in Fig. 21. The oscillations in this figure describe the moving striations in the positive column. The properties of the computer-simulated waves with respect to waveform, frequency, and damping agree well with those observed experimentally in molecular gases with negative ions. Oxygen is a gas with a large $\hat{k}_a$ due to dissociative attachment. It has been added to the gases to induce instability. The electron density and field fluctuations are locked out of phase, satisfying current continuity. A local increase in electron density results in a decrease in the electric field. In the absence of negative ions, the resulting decrease in ionization rate would bring the electron density back down if $\hat{k}_i > 1$. In the presence of negative ions, the detachment process provides another source of electrons. The negative ion density closely follows the increase in electric field because of the large attachment cross sections in these gases. The maximum in the negative ion concentration is controlled by the detachment process, eventually contributing significantly to the electron density. The rate of decrease of the negative ion density when the electric field starts decreasing is determined by the amount of detaching gas. If we add CO to an oscillating discharge, the increased detachment rate steepens the decrease in negative ion density. The minimum negative ion density of the cycle is actually controlled by the

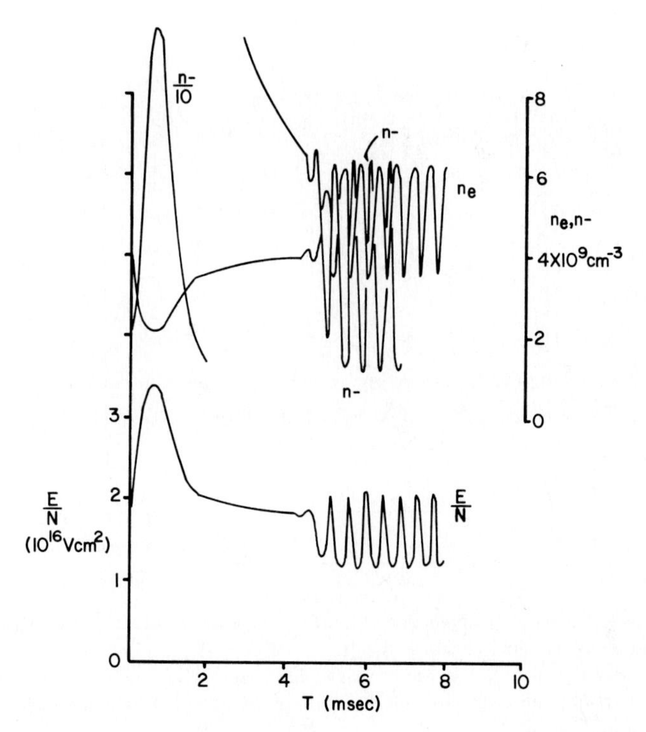

Fig. 21. Results of time-dependent calculations of growth of instability in laser mix of Fig. 20 to which oxygen has been added ($O_2:CO_2:N_2:He$, 2:5:10:83). The upper curves show the variations in N_n and N_e and the lower curves shows the associated changes in E/N. $I = 10$ mA, $R_p = 10$ Torr cm, $T_w = 300$°K, $T_g \sim 350$°K.

diffusion and recombination losses of the electron density at its peak, which then causes the electric field to reestablish control to satisfy current continuity.

Figures 22 and 23 show the effects that take place in a computer experiment by adding CO to an unstable discharge. Figure 22 displays the ionization cycle of an ionization–attachment–detachment instability* cycle as a function of the ratio N_n/N_e. Superimposed on the steady-state positive-ion loss curve. The stability boundary is located at $N_n/N_e = 0.2$ for the condi-

* Although it is a long description, the term ionization–attachment–detachment (IAD) waves is most appropriate. In the numerical calculations, it was found that a small amount of detaching gas was necessary to show the oscillations (in this case, it was CO provided by dissociation of the CO_2). In order to remove the time delay, before CO was produced in the calculation, an amount of CO was added to the original mix. The dispersion characteristics were not influenced by the parameterization of a small amount of CO. These results were also observed in the laboratory experiments.

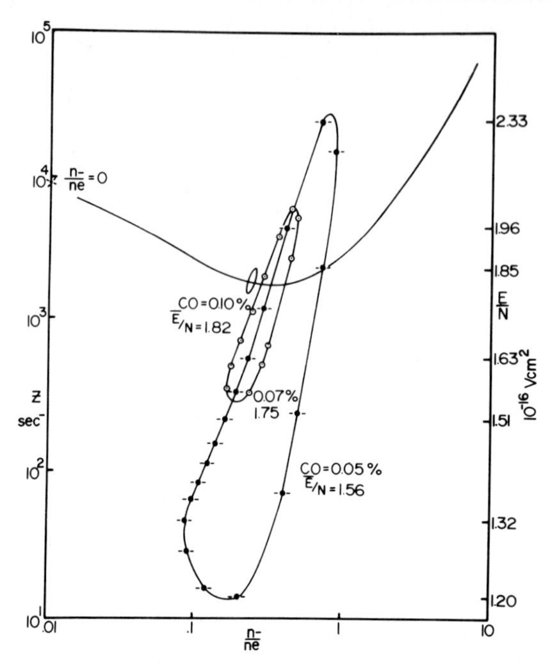

Fig. 22. The effect of detachment on the striation cycle of an $O_2:CO_2:N_2:He$ (2:5:10:83) plasma. Z ($\equiv k_i N_e$) is the ionization rate (sec^{-1}), E/N is the ratio of electric field to neutral number density (V-cm²), E/N is the cycle averaged value. Data points are spaced at a time interval of 10 μsec in order to show the variation of the period of oscillation with detachment rate (amount of CO added).

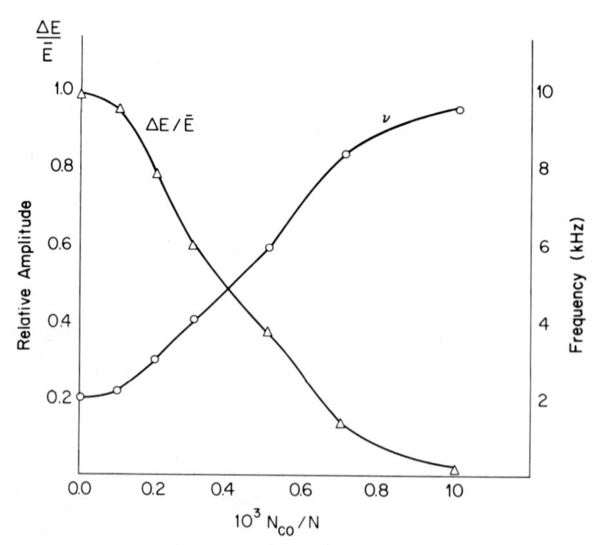

Fig. 23. Variation of instability amplitude and frequency with detachment rate (amount of added CO).

tions chosen. For $N_n/N_e > 0.2$ (which, as shown, can be controlled by the addition of CO) instability occurs. The amplitude and the frequency of the oscillation are strongly influenced by the ratio N_n/N_e (Fig. 23). The cycle averaged value $\langle E/N \rangle$ decreases with decreasing detachment and is always lower than the steady-state value. Thus the impedance and average power dissipated by the discharge are less for the striated column than for a uniform column. The cycle time decreases (i.e., the ionization wave frequency increases) as detachment increases and approaches the steady-state condition uniformly.

Laboratory experiments were performed by adding oxygen to a laser mixture and then comparing the relative amounts of oxygen and carbon monoxide that were required to produce instability or reestablish stability, respectively. Figure 24 shows the results for a discharge that was originally stable. The stability envelopes as functions of discharge current occur because the discharge in the slowly flowing gas mixture creates its own pattern of dissociated products (Tannen *et al.*, 1974; Wiegand and Nighan, 1973). By monitoring the instability, it is thereby possible to measure the relative ratios of different attachers or detachers to cause instability or stability, respectively, or in conjunction with mass-spectrometer diagnostics, to measure their effective rates under discharge conditions of vibrational excitation.

Similar experiments have been performed on the ionization waves in oxygen discharges (Nighan and Wiegand, 1974a; Dettmer and Garscadden,

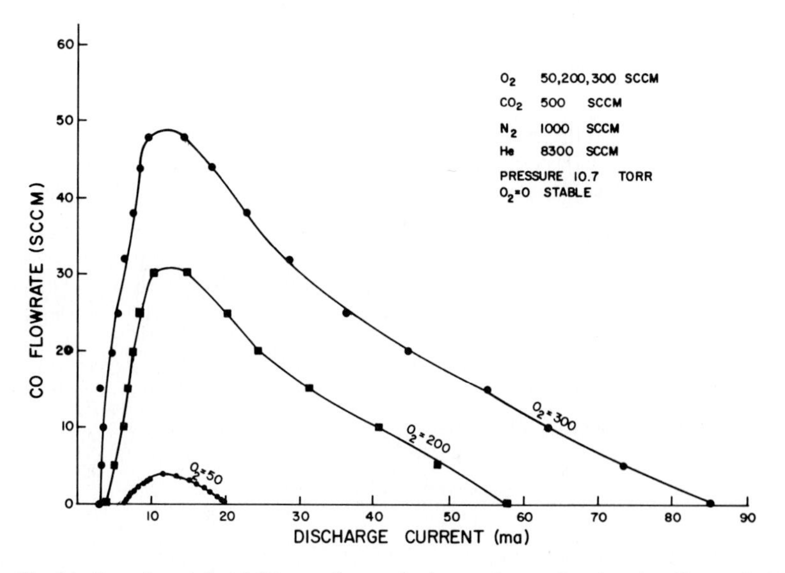

Fig. 24. Experimental stability envelopes of a laser mixture showing the effects of adding amounts of O_2 (attaching) and CO (detaching) gases. The discharge was originally stable.

1974). It was found that the instabilities could be controlled or suppressed by addition of a detaching gas. The abrupt transition from the low-field, T form to the high-field, H form was determined by the amount of added detacher, thus suggesting that the self-switching that occurs is due to the discharge generation of a sufficient amount of detaching species, e.g., a candidate is O_2 $(a^1\Delta_g)$. The Gunn instability of Sabadil (1973) thus appears to be an example of the ionization–attachment instability in a "pure" gas.

4. Dispersion of Ionization Waves in Molecular Gases

Long *et al.* (1975; Long, 1975) calculated the dispersion of the ionization waves in molecular gases using the above model. Using a computer code developed by Bailey and Long for the solution of the Boltzmann equation they took into account the non-Maxwellian EEDFs and their effects on the ionization rates and transport properties of the discharge. The results are difficult to normalize because the rates and coefficients are sensitive functions of E/N and the specific gas mixture. In a molecular gas like nitrogen without stable negative ions, one obtains a wave that is a forward wave propagating

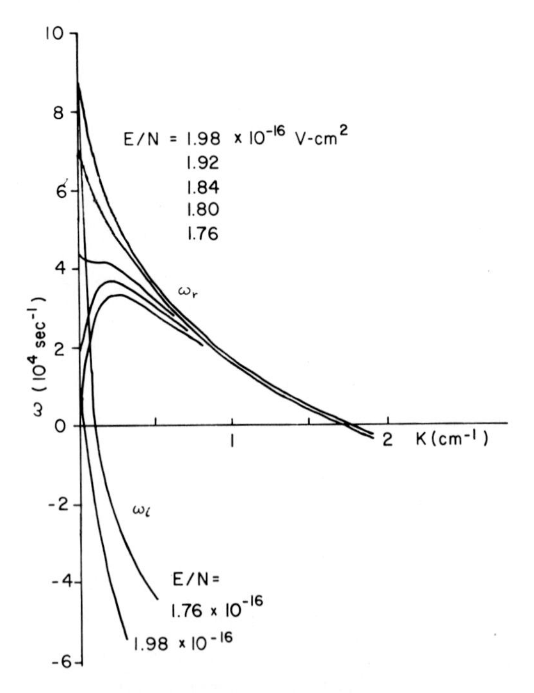

Fig. 25. Calculation dispersion of anode to cathode waves in a laser mixture. Pressure, 10 Torr; O_2:CO_2:N_2:He, 2:5:10:83 with 0.02% CO added.

cathode to anode. The wave is normally damped and is only slightly dispersive. However, the phase velocity is predicted to be sensitive to the value of E/N and therefore should be affected by gas temperature and pressure. The phase velocity is calculated to be typically 10^4 cm sec^{-1} at $E/N = 1.5 \times 10^{-15}$ V cm^2, and 2.5×10^4 cm sec^{-1} at $E/N = 2 \times 10^{-15}$ V cm^2. These conclusions and values are in qualitative agreement with some experimental measurements Garscadden and Bletzinger, 1968) (Fig. 12). The variety of waves observed by Laska and Exner (1971) is not obtained from the analysis. However, other wave types would be expected by the inclusion of vibrational and electronic excited states (metastable) along the lines indicated by Haas.

In the presence of negative ions, Long *et al.* (1975) obtained a wave similar to the above variety and the ionization–attachment wave. The latter can have amplifying solutions. Results for the laser mixture discussed in the previous section are given in Figs. 25 and 26. The ionization wave, Fig. 25, is predicted to have a phase velocity in the anode to cathode direction at short wavelengths and to be a backward wave (normally damped). However, it will, below a certain lower E/N value, change over to become a forward wave at long wavelengths. The analysis predicts that the wave may be amplified under these conditions. We think that this was experimentally

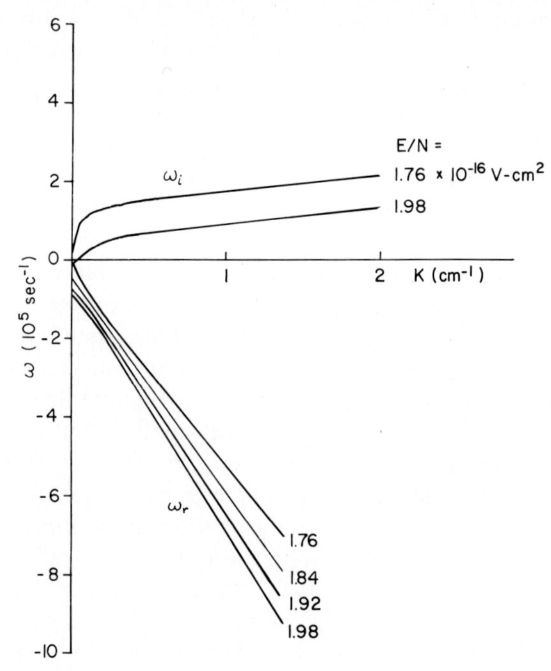

Fig. 26. Calculated dispersion of cathode to anode waves in a laser mixture. Conditions as in Fig. 25.

observed at the low-current edge of the stability envelopes shown in Fig. 24. The ionization–attachment waves have dispersion as in Fig. 26. They have cathode to anode phase velocities and are predicted to have broadband amplification and relatively little dispersion. These cathode to anode waves show some dispersion at low wavenumbers, where the damping is most.

V. SUMMARY AND CONCLUSIONS

The intent of the preceding outline of the present knowledge of ionization waves in glow discharges was to show that substantial progress has been made in the abilities to describe and predict the instability. These abilities are now much more quantitative and less empirical than just a few years ago. However, the ability to analyze the ionization waves requires a rather detailed knowledge of excitation and ionization cross sections and a fair degree of sophistication in mathematical computation and computer modeling. The information required to describe in detail ionization waves in the rare gases is mostly available. It is less certain that the data for many molecular gases permit quantitative comparisons. These gases can involve nonequilibrium vibrational distributions as well as nonequilibrium EEDFs, nonequilibrium dissociation, and a complicated interplay between these effects, energy deposition in many channels, and the subsequent relaxation into translational energy. All in all, the discharge in the self-sustained mode is very demanding to analyze and only in special situations does it give easily accessible quantitative information on collisional processes. If one attempts nonlinear modeling to describe the limit cycle and saturation mechanisms of standing striations, there are several choices, and there is no assurance that the proper effects have been included. At present, the maximum return from a study of the ionization waves comes not from an increase in our knowledge of atomic and molecular processes but rather from the fact that the discharges can be used with full awareness of their properties. In some applications the ionization wave occurrence is acceptable. In other situations, it may be totally unacceptable. In laser applications at high pressures it appears to trigger some thermal instabilities. The use of the instability itself in a glow discharge has, with some exceptions (Grabec, 1971; Sato, 1973) been overlooked as a test apparatus to study nonlinear wavegrowth and interactions. Large-amplitude ionization waves have sufficient harmonic content, yet in a long positive column, the amplitude and phase velocity of the fully developed wave do not change. It has been shown (Garscadden and Bletzinger, 1970) that ionization waves exhibit amplitude dispersion that could balance the frequency dispersion. However, an analysis treating the saturated amplitude wave as a "soliton" has not been accomplished.

It should be possible to use Whitham's (1965) criteria to examine amplitude dispersion. There is some evidence (Garscadden *et al.*, 1965) that pulse splitting occurs in a nonlinear wave about the wavenumber $k = 1.7$ (reciprocal of the electron temperature relaxation length), which is predicted (Whitham, 1965) from dispersion relation Eq. (17).

With the advent of discharges using external ionization sources, so that it is possible to control the discharge and choose any desired E/N value, some of the difficulties of using ionization wave characteristics as a diagnostic disappear. As the technology of electron beam devices and other methods matures, the technique will become available to a greater number of groups. It is reasonable to expect that attachment, detachment, and other rates will be measured for complex gas mixtures, and as functions of E/N, vibrational temperature, and pressure. Some interesting studies have already been done along these lines (Douglas-Hamilton, 1973) and eventually they may permit cross sections rather than rates to become available.

REFERENCES

Alanakyan, Y. R., and Mikhalev, L. A. (1975). *Sov. Phys.–Tech. Phys.* **20,** 5.

Alexeff, I., and Jones, W. D. (1966). *Phys. Fluids* **9,** 1871.

Allis, W. P. (1976). *Physica* **82C,** 43.

Bailey, W. F., Long, W. H., Pond, D. R., and Garscadden, A. (1974). *IEEE Plasma Sci. Conf., Knoxville, Tennessee, May 1974.*

Basaev, A. B., Molchanow, M. I., and Yaroshenko, N. G. (1976). *Sov. Phys.–Tech. Phys.* **20,** 1215.

Behnke, J. (1963). Physiker-Tagung (Leipzig 1966) Vortrag Diplomarbeit, Physikalische Institut der Ernst-Moritz-Arndt-Universitat, Greifswald, Germany.

Bekefi, G. (1966). "Radiation Processes in Plasmas." Wiley, New York.

Bekefi, G., Hirschfield, J. L., and Brown, S. C. (1961). *Phys. Fluids* **4,** 173.

Boyd, R. L. F., and Twiddy, N. D. (1959). *Proc. Roy. Soc.* **A250,** 53.

Bletzinger, P., LaBorde, D. A., Bailey, W. F., Long, W. H., Jr., Tannen, P. D., and Garscadden, A. (1975). *IEEE J. Quantum Electron.* **QE-11,** 317.

Briggs, R. J. (1964). "Electron Stream Interaction with Plasmas." MIT Press, Cambridge, Massachusetts.

Cooper, A. W. (1964). *J. Appl. Phys.* **35,** 2877.

Cooper, A. W. M., Coulter, J. R. M., and Emeleus, K. G. (1958). *Nature* **181,** 1326.

Crawford, F. W., and Pagels, H. R. (1963). *J. Opt. Soc. Am.* **53,** 734.

Derfler, H. (1967). *Phys. Lett.* **24A,** 763.

Dettmer, J. W., and Garscadden, A. (1974). Paper MB-5, *Abstr. 27th Annu. Gaseous Electron. Conf., Houston, Texas, 1974.*

Dougherty, J. D., Mangano, J. A., and Jacob, J. H. (1976). *Appl. Phys. Lett.* **28,** 581.

Douglas-Hamilton, D. H. (1973). *J. Chem. Phys.* **58,** 4820.

Douglas-Hamilton, D. H., and Mani, S. A. (1973). *Appl. Phys. Lett.* **23,** 508.

Duncan, A. J., and Forrest, J. R. (1971). *Phys. Fluids* **16,** 1973.

Duncan, A. J., Forrest, J. R., Crawford, F. W., and Self, S. A. (1969). *Phys. Fluids* **12,** 2607.

Emeleus, E. G. (1964). *Adv. Electron. Electron Phys.,* **20,** 590.

Ewald, H. N., Crawford, F. W., and Self, S. A. (1969). *Phys. Fluids* **12,** 303.

Foulds, K. H. W. (1956). *J. Electron. Control* **2,** 270.
Garscadden, A., and Bletzinger, P. (1968). *Phys. Lett.* **27A,** 203.
Garscadden, A., and Bletzinger, P. (1970). *Proc. Am. Inst. Phys. Conf. Feedback and Dynamic Control of Plasmas* (T. Chu and H. W. Hendel, eds.). AIP, New York.
Garscadden, A., and Lee, D. A. (1966). *Int. J. Elect. Control* **20,** 567.
Garscadden, A., Bletzinger, P., and Simonen, T. C. (1969). *Phys. Fluids* **12,** 1833.
Gentle, K. W. (1966). *Phys. Fluids* **9,** 2203.
Grabec, I. (1971). *Beitrage Plasmaphys.* **11,** 285.
Granowski, W. L. (1955). "Der Electrische Strom im Gas," Chap. 8, Akademie-Verlag. Berlin, Germany.
Haas, R. A. (1973). *Phys. Rev.* **8,** 122.
Holscher, J. G. A. (1967). *Physica* **35,** 129.
Kerrebrock, J. L. (1964). *AIAA J.* **2,** 1072.
Kiselevskii, L. I., and Suzdalov, I. I. (1974). *Sov. Phys.–Tech. Phys.* **18,** 1318.
Laska, L., and Exner, V. L. (1968). *Czech. J. Phys.* **B18,** 1472.
Laska, L., and Exner, V. L. (1971). *Czech. J. Phys.* **B21,** 126.
Lee, D. A., and Garscadden, A. (1972). *Phys. Fluids* **15,** 1827.
Lee, D. A., Bletzinger, P., and Garscadden, A. (1966). *J. Appl. Phys.* **37,** 377.
Long, W. H., Jr., (1975). Gen. B. A. Schreiver Award Paper, *USAF Sci. Eng. Symp. 1975* (unpublished).
Long, W. H., Jr., Bailey, W. F., and Garscadden, A. (1975). Paper B4, *28th Annu. Gaseous Electron. Conf., Rolla, Missouri.*
Masek, K., and Perina, V. (1969). *Czech. J. Phys.* **B19,** 956.
Neodospasov, A. V. (1968). *Usp. Fiz. Nauk* **94,** 439.
Nighan, W. L. (1976). *In* "The Principles of Laser Plasmas" (G. Bekefi, ed.). Wiley, New York.
Nighan, W. L., and Wiegand, W. J. (1974a). *Phys. Rev.* **10,** 922.
Nighan, W. L., and Wiegand, W. J. (1974b). *Appl. Phys. Lett.* **25,** 633.
Nighan, W. L., Wiegand, W. J., and Haas, R. A. (1973). *Appl. Phys. Lett.* **22,** 579.
Novak, M. (1960). *Czech. J. Phys.* **10,** 954.
Oleson, N. L., and Cooper, A. W. (1968). *Adv. Electron. Electron Phys.* **24,** 463.
Oskam, H. J. (1958). *Philips Res. Rep.* **13,** 335.
Pekarek, L. (1958). *Czech. J. Phys.* **8,** 742.
Pekarek, L. (1963). *Proc. Int. Conf. Ionization 6th,* "Phenomena in Gases" (T. Hubert and E. Cremieu-Alcan, eds.), Vol. 2, p. 133. Bureau des Editions, Centre d' Etudes Nuclearies de Saclay, Paris.
Pekarek, L. (1971). *Proc. Int. Conf. Phenomena in Ionized Gases 10th* (P. A. Davenport, ed.).
Pekarek, L., and Krejci, V. (1962). *Czech. J. Phys.* **B12,** 296.
Pfau, S., and Rutscher, A. (1964). *Beitrage Plasmaphys.* **4,** 41.
Pfau, S., and Rutscher, A. (1965). *Proc. Int. Conf. Phenomena in Ionized Gases 7th, Belgrade.*
Pfau, S., and Rutscher, A. *Proc. Czech. Conf. Electron. Vacuum Phys. 4th, Prague,* 217.
Pfau, S., Rutscher, A., and Wojaczek, K. (1969). *Beitrage Plasmaphys.* **9,** 333.
Perina, V., Rohlena, K., and Ruzicka, T. (1975). *Czech. J. Phys.* **B25,** 660.
Pupp, W. (1932). *Phys. Z.* **33,** 844.
Rademacher, K., and Wojaczek, K. (1958a). *Ann. Phys.* (*Leipzig*) **7,** 47.
Rademacher, K., and Wojaczek, K. (1958b). *Ann. Phys.* **2,** 57.
Rayment, S. W. (1974). *J. Phys. D* **1,** 871.
Rayment, S. W., and Twiddy, N. D. (1969). *Brit. J. Appl. Phys.* (*J. Phys. D*), *Ser. 2* **2,** 1747.
Reece-Roth, J. R. (1967). *Phys. Fluids* **10,** 2712.
Robertson, H. S. (1957). *Phys. Rev.* **105,** 368.
Rognlien, T. D., and Self, S. A. (1972). *J. Plasma Phys.* **7,** 13.

Rohlena, K., Ruzicka, T., and Pekarek, L. (1973). *Phys. Lett.* **40A,** 239.
Rother, H. (1959). *Ann. Phys.* **7,** 373.
Rutscher, A. (1962). *Czech. J. Phys.* **A12,** 521.
Rutscher, A., and Wojacek, K. (1964). *Beitrage Plasmaphys.* **4,** 41.
Ruzicka, T., and Rohlena, K. (1972). *Czech. J. Phys.* **B22,** 906.
Ruzicka, T., and Rohlena, K. (1973). *Int. Conf. Ionization Phenomena in Gases 11th, Prague* **61.**
Sabadil, H. (1973). *Beitrage Plasmaphys.* **13,** 235.
Sato, M. (1973). *Beitrage Plasmaphys.* **13,** 11.
Self, S. A., Crawford, F. W., and Ewald, H. N. (1969). *Phys. Fluids* **12,** 316.
Sicha, M., and Drouet, M. G. (1968). *Canad. J. Phys.* **46,** 1175, 2491.
Sicha, M., Pilan, J., Gajudsek, J., Novak, J., Fuchs, V., and Lukac, P. (1967). *Czech. J. Phys.* **B17,** 48.
Spitzer, L. Jr. (1962). "Physics of Fully Ionized Gases," 2nd ed., Sec. 3.2, p. 55. Wiley (Interscience), New York.
Stirand, O., and Laska, L. (1967). *Czech. J. Phys.* **B17,**
Stirand, D., Krejci, V., and Laska, L. (1966). *Rev. Sci. Instrum.* **37,** 1481.
Swain, D. W., and Brown, S. C. (1971). *Phys. Fluids* **14,** 1383.
Tannen, P. D., Bletzinger, P., and Garscadden, A. (1974). *IEEE J. Quantum Electron.* **QE-10,** 6.
Thompson, J. B. (1959). *Proc. Phys. Soc.* **73,** 818.
Venzke, D. (1970). *Beitrage Plasmaphys.* **10,** 441.
Venzke, D. (1971). *Beitrage Plasmaphys.* **11,** 141.
Wallard, A. J., and Woods, P. T. (1974). *J. Phys. E.* **7,** 209, and references therein.
Weber, R. F., and Garscadden, A. (1970). Aerospace Res. Lab. Rep. ARL 70-0079, AD #710597.
Wellenstein, H. F., and Robertson, W. W. (1972). *J. Appl. Phys.* **43,** 4823.
Wiegand, W. J., and Nighan, W. L. (1973). *Appl. Phys. Lett.* **22,** 583.
Whitham, G. B. (1965). *Proc. Roy. Soc.* **A283,** 238.
Wojaczek, K. (1971). *Beitrage Plasmaphys.* **11,** 335.
Wojaczek, K., and Rutscher, A. (1963). *Beitrage Plasmaphys.* **3,** 217.

Part 2.3 Nonuniformities in Glow Discharges: Electrophoresis

L. M. CHANIN

DEPARTMENT OF ELECTRICAL ENGINEERING
UNIVERSITY OF MINNESOTA
MINNEAPOLIS, MINNESOTA

I. TERMINOLOGY

In 1893 Baly reported, "When an electric current is passed through a rarefied mixture of two gases, one is separated from the other and appears in the negative glow" (Baly, 1895). The term that physicists commonly use to describe this type of partial separation of gases in a mixture is cataphoresis. Cataphoresis can be regarded as the establishment of differences in the partial pressures of the gas constituents in the anode and cathode regions. The segregation occurs not only in an axial or longitudinal sense but also radially.

The general term electrophoresis has been used for the past 50 years to refer to the differences in the total gas pressure between the anode and cathode ends of a dc discharge. Usually, the term has been applied to de-

ISBN 0-12-349701-9

scribe the resulting pressure difference for single gases. In this section the term electrophoresis will be used to refer to phenomena other than thermal effects, which result in the establishment of pressure gradients. Electrophoresis involves radial as well as longitudinal effects. Clearly by subjecting a gas mixture to a discharge one can produce not only pressure gradient or electrophoretic effects, but also the segregation of the gas component or cataphoretic effects. Unfortunately, while these terms have been used in the literature for many years, considerable, and perhaps understandable, confusion still seems to exist concerning their use.

A. Introduction

The first report of an axial pressure gradient believed to have been produced by electrophoresis appeared in 1892 (Warberg, 1892). In 1904 Stark reported larger pressures at the cathode than at the anode for dc discharges in air. Later observations by Wehnelt and Franck (1910) were consistent with Stark's observations. Measurements in argon by Skaupy (1917) indicated that in contrast to the earlier reports, the pressure at the anode was greater than the cathode pressure. Subsequent results reported by Hamburger (1917) from studies on argon and nitrogen were in agreement with Skaupy's observations. In 1922 Rüttenauer undertook a series of extensive observations of pressure gradients in various noble gases. Rüttenauer's results were compared with theory by both Langmuir (1923) and Druyvesteyn (1935). Fair agreement was found; however, detailed comparisons suggested basic difficulties. Measurements of pressure differentials were reported in the 1930s by Kenty (1938) and Klarfeld and Poletaev (1939). Later, measurements by Cairns and Emeléus (1958) showed that the gradient Δp increased to a maximum with decreasing pressure and then subsequently decreased.

Considerable progress was made in understanding electrophoresis in the 1960s as a consequence of both experimental and theoretical investigations. Theories were introduced by Leiby (1966), Leiby and Oskam (1967), and also by Chester (1968). Experimental results were reported by Leiby and Rogers (1967, 1969), Bergman and Chanin (1968, 1969), Chester (1968), Bridges and Halsted (1967), Bridges *et al.* (1966), Bergman (1968), and Tombers (1970). Many of these recent reports contain extensive references to the earlier investigations.

The first part of this survey primarily considers recent theoretical advances. Examples are then given of some of the more recent pressure gradient measurements. Finally, the effect of electrophoresis on gas laser performance is considered and examples are given of retrograde or cathode-directed electrophoresis.

II. THEORY

The first generally accepted explanation was proposed by Langmuir in 1923. According to Langmuir there could be no resultant force acting on the gas as long as the charged particles remained in the gas. Any imbalance, however, in the rates at which the electron gas and the ion gas transferred axial momentum to the walls of the discharge tube could result in a net transfer of axial momentum to the neutral gas. Since the axial ion drift momentum is larger than the electron drift momentum the greater ion momentum loss at the tube walls could result in a net anode-directed force on the gas, in the vicinity of the walls. In 1935 Druyvesteyn proposed a modified version of Langmuir's explanation, which involved two additional effects that would result in cathode-directed forces. These included the transport of gas as positive ions to the cathode. Also, due to the negative space charge sheath at the walls of the tube, the ion concentration would be slightly greater than the electrons. Hence, the ion gas would receive more axial momentum per second from the electric field than the electron gas. The anode-directed force, which was believed to be important only within a few ion mean free paths from the wall, would set up an anode-directed motion of the neutral gas, which would be resisted by viscous damping at the tube walls. The gas flow would be only slightly offset by the cathode-directed motion of atoms represented by the ion current and would result in a buildup of the anode–cathode pressure difference until the net gas flow vanished. Based on Langmuir's and Druyvesteyn's interpretation, for a discharge tube closed at both ends, and operating under steady-state conditions, in the region of the walls of the discharge tube the gas would move toward the anode while in the center region of the tube the gas would return toward the cathode. Measurements by Kenty (1967) on the gas flow direction at the center of a dc discharge tube that was closed at both ends, however, conflicted with the direction predicted by theory. Instead of a cathode-directed flow along the axis of the tube, the flow was found to be anode directed. Calculations of the magnitude of the pressure gradient were found to consistently underestimate the magnitude of the effect. In addition, the Druyvesteyn theory at low pressures predicted for certain gases that the cathode pressure would be greater than the anode pressure—a result that was contrary to experimental observations.

Recently, significant progress has been made in establishing an improved theoretical understanding of electrophoresis. In 1966 Leiby postulated a volume force exerted on the gas particles in the direction of the anode as a consequence of the existence of charged-particle density gradients in the plasma. Its magnitude was related to the difference between the momentum invested per second in speeding up newly created electrons and ions to their steady-state drift velocities. The resulting force was an order of magnitude larger than the Langmuir and Druyvesteyn forces and appeared to explain part of the experimental observations. In 1967 Leiby and Oskam obtained a more general expression for the force acting on the neutral gas particles as a consequence of charged–neutral interactions in a dc discharge. These authors showed that Langmuir's and Druyvesteyn's explanations were incomplete and also that the force predicted by Leiby is a limiting case of a more general form of the force equation.

The conservation of mass and momentum equations were applied to a steady-state plasma consisting of singly charged positive ions of one type, electrons, and one species of neutral particle. A spatial-dependent force $\bar{F}_d$ called the driving volume force on the neutral particles was expressed in the form

$$\bar{F}_d = q(n_p - n_e)\bar{E} - \bar{F}_{CMD}, \tag{1}$$

where $\bar{F}_{CMD}$ is the charged-particle momentum divergence force, which is related to the sum of the net electron and ion momentum flows out of (into) the volume. The first term on the right-hand side of Eq. (1) is the space charge imbalance volume force discussed by Druyvesteyn. By using assumptions appropriate to the positive column in which the motion of the charged particles is determined by collisions with neutral particles, an expression was obtained for the charged-particle momentum divergence force in the positive column,

$$F_{CMD_z}(r) = -m_i \frac{\mu_p}{\mu_e} \frac{J}{q} \frac{D_p}{\Lambda^2} \frac{p_0}{P_0} \left(1 + \frac{T_e}{T_i}\right), \tag{2}$$

where D_p is the ion diffusion coefficient, T the Maxwellian temperature of the charged particles, J the current density, p_0 the reduced gas pressure, and Λ the characteristic diffusion length of the discharge tube. One of the assumptions used to obtain the form for $\bar{F}_{CMD}$ shown in Eq. (2) is that the disappearance of electrons and ions from the plasma is governed by ambipolar diffusion to the wall of the discharge tube and also that the ionization process is linear. The axial force given by Eq. (2) has a maximum value at the center of the discharge tube and is a minimum at the wall for an electron density distribution given by the fundamental diffusion mode. Neglecting the influence of the space charge imbalance force introduced by Druyvesteyn, this will result for a closed system in a neutral gas flow pattern opposite to that proposed by Langmuir and Druyvesteyn. The force is zero at the tube wall since the charged particle density is zero there. When the tube wall is removed from the active discharge region, the force will be zero when the net rate of production of the charged particles is zero. Outside the active region, since the electron–ion recombination rate will be larger than the rate of electron–ion production, the charged particle momentum divergence force reverses sign and is oppositely directed to the force inside the active region. If the walls of the plasma container are sufficiently far removed from the active discharge region such that no charged-particle momentum is dissipated at the wall, no net charged-particle momentum is dissipated at the wall, no net charged-particle momentum is dissipated at the wall and the charged-particle momentum force integrated over the wall and the charged-particle momentum force integrated over the plasma volume is zero. Here local neutral gas flow will occur. Hence the existence of a F_{CMD} does not depend on the presence of a discharge tube wall. The existence of a net force, however, requires that the opposite momenta dissipated by the electrons and ions at the discharge container walls are not equal. To relate Eq. (1) to experimental results, Oskam (1969) derived an expression for $\partial p/\partial z$ in terms of F_{d}. Equation (3) gives the predicted dependence of the pressure gradient due to the axial component of the volume force $F_{\mathrm{vz}}(0)$ for a closed dc discharge (i.e., no net gas flow) when it is assumed that the gas flow is laminar and the neutral gas drift velocity $v_{\mathrm{nz}}(r)$ is independent of z and its magnitude is much smaller than the ion drift velocity.

$$\frac{\partial p}{\partial z} = -\frac{16}{j_{01}^3} J_1(j_{01}) F_{\mathrm{vz}}(0). \tag{3}$$

Here j_{01} is the first root of the zeroth-order Bessel function and $J_1(j_{01})$ is the first-order Bessel function evaluated at j_{01}. From Eq. (3), $\partial p/\partial z = \Delta p/L$, where Δp is the anode–cathode pressure difference and L is the tube length. Equation (2) predicts that Δp will increase linearly with tube length L and and current I and will vary with tube radius as R^{-4}. Druyvesteyn's expression for Δp also varies linearly with L and I; however, an R^{-5} dependence is predicted.

In 1968 Chester published an article that considered gas pumping in discharge tubes. This report contained extensions to Druyvesteyn's theory attributed to Halsted (Bridges and Halsted, 1967; Bridges *et al.*, 1966). The major contribution, however, involved an analysis of a gas pumping force that acts throughout the discharge volume rather than confined to regions close to the wall, as in Langmuir's and Druyvesteyn's analyses. Chester regarded this force as resulting from the unequal radial distances traveled by ions and electrons between collisions with gas atoms. In the positive column ions and electrons move radially outward with a common average velocity due to the combined effects of diffusion and the radial field. When a charged particle collides with a neutral gas atom, delivering to it axial momentum that it has gained from the axial electric field, it has traveled radially outward a distance of about $v_{dr}\tau$ since its last collision. Here, v_{dr} is the radial drift velocity of the charged particle and τ is its mean free time (which may be controlled either by its radial drift velocity or by its thermal velocity). For the case of higher pressures, $v_{dr}\tau \ll R$. Then the axial momentum delivered to atoms at radius r is characteristic of the charged-particle density at radius $r - v_{dr}\tau$. Since v_{dr} must be nearly the same for ions and electrons (because their wall currents are equal and their densities are nearly equal at any radius) and since usually $\tau_e \ll \tau_p$, the effect on the transfer of axial momentum to the gas is the same as a radial charge separation, and so can result in a net force on the gas. This type of differential momentum transfer produces an axial force on the gas distributed throughout the discharge volume.

Druyvesteyn's analysis may be expressed in terms of the volume of gas pumped past position z per second toward the anode:

$$V(z) = V_1(z) + V_2(z) - \left(\frac{2\pi}{16\eta}\right) R^4 \frac{\partial}{\partial z} [n(z)kT_g(z)]. \tag{4}$$

In Eq. (4) η is the viscosity, $n(z)$ the neutral-atom number density, and T_g the kinetic neutral-atom temperature. $V_1(z)$ is the force arising from the lack of charge neutrality; $V_2(z)$ is due to forces exerted within a few ion mean free paths of the tube wall and is usually much larger than $V_1(z)$. For Chester's analysis, $V(z)$ is given by

$$V(z) = V_{1s}(z) + V_{2s}(z) + V_{3s}(z) - \left(\frac{2\pi}{16\eta}\right) R^4 \frac{\partial}{\partial z} [n(z)kT_g(z)]. \tag{5}$$

The subscript s on the various V terms represents attempts to correct for the presence of ion sheaths at the walls of the discharge tube. These corrections are greatest at high pressures, where the ion mean free path $\lambda_p \ll R$. $V_{1s}(z)$ represents the small volume force arising from the lack of charge neutrality. $V_{2s}(z)$ represents gas pumping due to forces exerted close to the discharge tube walls. The term $V_{3s}(z)$ in the high-pressure limit is the same as that obtained by Leiby and Oskam. This term is nonnegative and always results in a flow of neutral gas toward the anode. Usually $V_{1s}(z)$ is much smaller than $V_{2s}(z) + V_{3s}(z)$, and the ion sheath can be shown to effect essentially only the term $V_2(z)$. In calculating the volume force, corrections were considered for the contributions due to "newly" ionized atoms, which give much less momentum to the gas on the average than "old" ions. Corrections to the volume force were also considered when charge exchange collisions occurred between the ions and atoms. Modified equations were given in this case for parameters such as S_p, the distance traveled by an ion since its last collision.

Chester's (1968) analysis, which was compared with measurements of the pressure difference between the cathode and anode bottles of a standard CW argon ion laser tube, included a term that accounted for a gas return path external to the discharge. That comparison is considered in Section III,B.

III. RECENT RESULTS

This section considers examples of recent measurements of electrophoretic pressure gradients. Results are given for low current density discharges ≤ 1 A/cm^2 and also for current densities as large as 120 A/cm^2. The fundamental dependence of Δp on various discharge parameters is also discussed. Finally, the effect on Δp of the presence of striations, constriction, and changing ion species is considered.

A. Low Current Density Discharges

Many measurements of pressure gradients caused by electrophoretically induced gas flows due to low current density (≤ 1 A/cm^2) dc discharges have been reported within the past ten years (Leiby and Rogers, 1967, 1969; Bergman and Chanin, 1968, 1969; Bergman, 1968; Tombers, 1970). Results reported by Leiby and Rogers (1969) were obtained using a racetrack type of discharge tube. The length of the plasma in the 6 mm bore of the racetrack was 53 cm. The perimeter of the racetrack was 150 cm. The pressure difference generated by the flow of room temperature gas in the nondischarge portion of the racetrack was measured by a MKS Baratron capacitance manometer, which had an accuracy of ± 0.03 μm. System volumes were symmetrically arranged about the active discharge to avoid differential volume effects. This symmetry was checked by reversing the direction of the discharge current (each electrode contained an oxide cathode and a molybdenum anode, and so either could act as the discharge cathode or anode). No noticeable change in the measured pressure differentials was observed. A 0.5 liter expansion volume served to limit system pressure increases (caused by gas heating in the active discharge region) to less than 3%. Thermocouples were used to measure discharge tube wall temperatures. Two floating probes were used to measure electric fields. A waveguide was used to measure microwave conductivities of the plasmas that were used for electron density determinations.

Measured values of pressure differentials Δp for helium are shown in Fig. 1. The reproducibility is good, except at high pressures, where the dis-

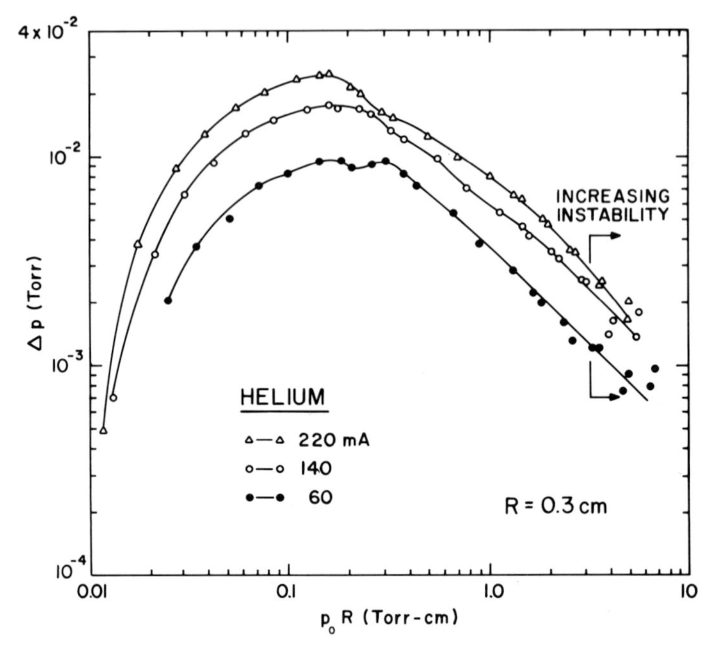

Fig. 1. Pressure differential measured between the midpoints of the racetrack tube vs. p_0R for dc discharges in helium. Evenly spaced points represent one experimental run. Remaining points represent the results of a second run undertaken to determine the degree of reproducibility of the data (Leiby and Rogers, 1969).

charges tended to be somewhat unstable at the anode. The values of Δp can be used to compute the corresponding gas flow velocities in the nondischarge portion of the tube. Theoretically predicted values for the gas flow velocities were calculated using the high-pressure approximation to the Leiby–Oskam theory. The plasmas investigated were ambipolar diffusion controlled over approximately 95% of the pressure ranges where stable discharges were established. The calculated gas flow velocities involved using electron temperatures and densities obtained from microwave measurements, electric field values in the positive column obtained from probe measurements, discharge gas temperatures that were assumed equal to measured discharge tube wall temperatures, and published ion mobility values. For helium and also neon the measured and calculated gas flow velocities were in agreement to within factors of 2.5 or better. For the 220 mA discharge, the maximum gas flow velocity in helium was 31.7 cm/sec.

Figure 2 shows measurements of the pressure differential Δp for krypton. For argon, krypton, and xenon the agreement between calculated and measured velocities is not as good as for helium and neon. These calculations

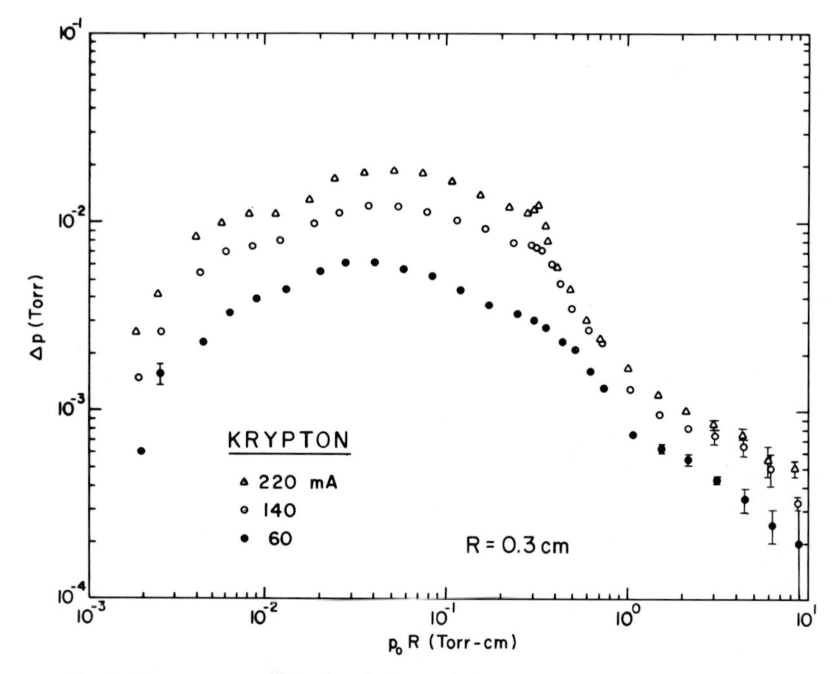

Fig. 2. Measurements of Δp for discharges in krypton (Leiby and Rogers, 1969).

used the simple steady-state approximation for the Leiby–Oskam theory. Discharges in these gases, however, become increasingly constricted as the gas pressure is increased; moreover, in certain pressure ranges, striations were present. If one uses the modifications to the theory proposed by Leiby (1971) to account for the effects of striations and also changes in gas temperatures due to constriction (Section III,D), the agreement between calculated and measured flow velocities is significantly improved.

Figure 3 shows Δp data obtained from studies on xenon. At high pressures, oscillations in the values of Δp were observed. Xenon exhibited stronger constriction effects than krypton. Taking constriction and striations into account, the theoretically predicted flow velocities at the higher pressures were in appropriate agreement with the observed velocities. As in the case of krypton as the pressure is reduced and the constriction effects diminish, the agreement between the steady-state theory and experiment improves.

Figure 4 shows a comparison of Δp measurements in helium reported by various investigators (Bergman and Chanin, 1969). Data shown in Fig. 4, obtained from previous investigations, were scaled assuming that $\Delta p \propto IL/R^4$ as predicted by the Leiby–Oskam (1967) and Chester (1968) theories.

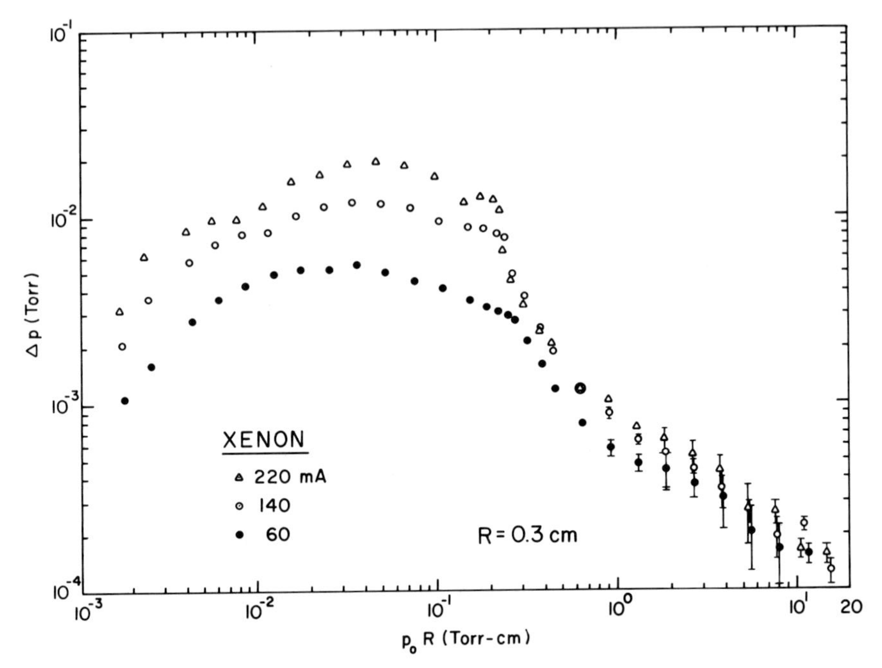

Fig. 3. Measurements of Δp for discharges in xenon (Leiby and Rogers, 1969).

The tube radii used by various investigators are indicated in the figure. It is interesting to note that the maximum values of Δp occur at approximately the same pR value in all reported studies. This may be related to the occurrence of the maximum in Δp for a given gas at a particular electron temperature.

B. High Current Density Discharges

Chester has reported measurements of the pressure difference between the anode and cathode bottles of a standard cw argon ion laser tube (1968). The quartz discharge capillary was 30 cm long with a 1.25 mm inner radius and 1-mm-thick walls. The cathode and anode bottles were each roughly 20 cm long by 5 cm in diameter. Gas flowed through most of the volume of the electrode bottles before entering the gas return.

A capacitance manometer sensing head was connected between the electrode bottles, and separate thermocouple gauges were connected to each bottle. The tube was connected to the vacuum station at the cathode end, thus providing a large cold volume that served to hold the pressure constant at the cathode end. The absolute magnitude of the pressure differences

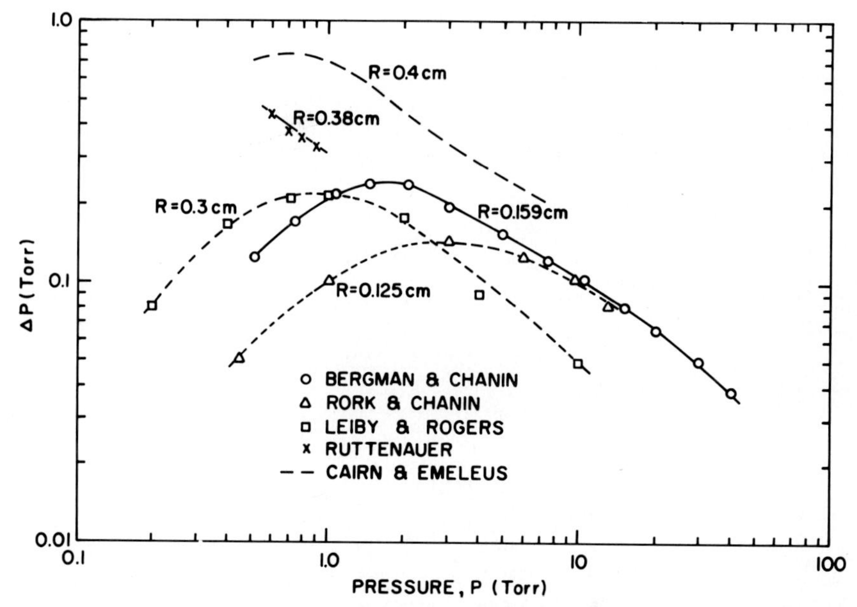

Fig. 4. Comparison of the results in helium at 100 mA with those obtained by previous investigators. Results were scaled to $R = 0.159$ cm, $L = 50$ cm, $I = 100$ mA, assuming that $\Delta p \propto IL/R^4$ (Bergman and Chanin, 1969).

measured involved a maximum error of 4.7%; however, their relative values were expected to be more accurate (1–2%). Experimental results show the presence of a discontinuity in the I–V characteristics. This occurs at a particular value of discharge current that varied with tube radius and filling pressure. Simultaneously, discontinuities appear in many other discharge properties, including the pressure differential Δp and the intensity of light emission in the spectral lines. The axial electric field and the electron temperature are strongly dependent on current below the discontinuity. Since some discharge parameters were measured with confidence only above the discharge discontinuity, comparisons of gas pumping experiments with theory were made only beyond this region. This occurred at current densities of a few tenths of amperes per square millimeter in the pR range from 1 to 5 Torr mm.

Results in Fig. 5 show the observed pressure dependence. The curves are quite regular except for the distortion in the 1 A curve caused by the presence of the discharge discontinuity (and to a small extent in the 2 A curve). If this distortion is neglected, the pressure dependence is almost the same at all currents, being roughly independent of p below ~ 1.2 Torr and varying as p^{-1} above that pressure. If one uses a simplified form of the theory that assumes the gas pumping is entirely due to the volume force term V_3,

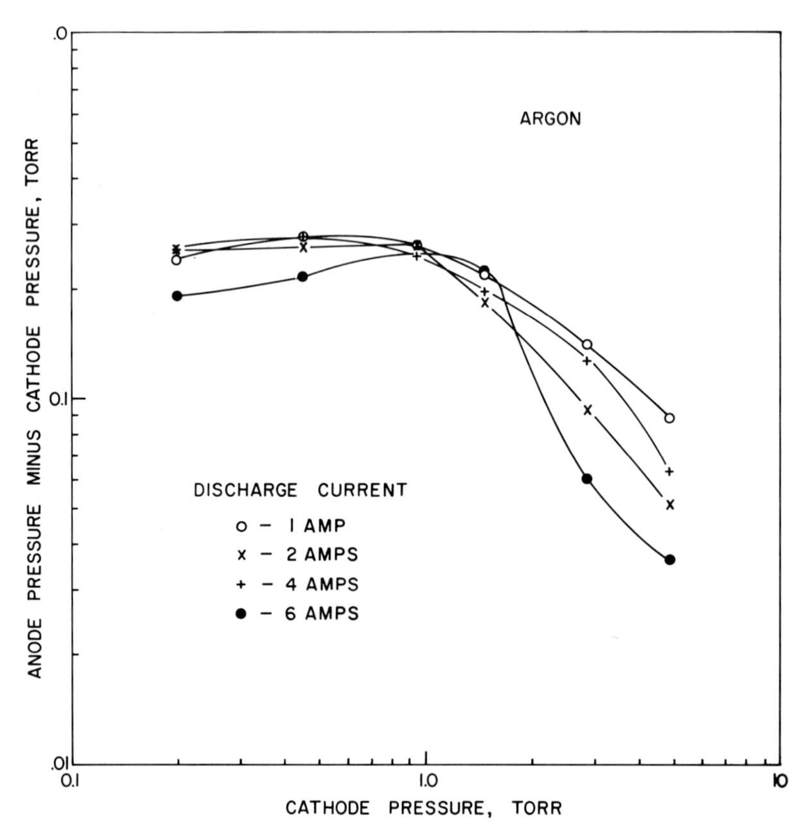

Fig. 5. Experimentally measured pressure difference as a function of cathode pressure (Chester, 1968, p. 184).

neglecting the force terms used by Druyvesteyn, the predicted pressure difference curves closely approximate the data shown in Fig. 5. Chester has noted that this similarity may be deceptive, due to compensating terms in Eq. (5). According to Chester, since the theory involves the assumption of viscous flow, the theory may fail at values of pR below ~ 1 Torr mm and should be fully applicable only at the highest pressures used. Despite this qualification the agreement is good regarding the functional form and magnitude. The current dependence predictions are also in a good agreement at the higher pressures. While no attempt was made to calculate the magnitude of the predicted pumping at low pressures, with certain changes in various parameters the following pressure and current dependence is predicted:

$$\Delta p \propto I T_g^{-1/2} p^0 \qquad (6)$$

where p^0 is the pressure in the gas return path. Equation (6) includes only the ion transport term and the V_3 volume force term. The V_2 wall force term can be neglected at low pressures. The agreement in functional form with the low-pressure experimental data in Fig. 5 is satisfactory.

C. Dependence of Δp on p, R, I

Measurements of the functional dependence of the pressure gradient Δp on various parameters such as gas pressure, tube radius, and discharge current were first undertaken by Rüttenauer. With the subsequent introduction of various theories it has become of interest to compare experimental observations with theoretical predictions. Within the past ten years several investigators have performed extensive measurements of Δp. These include results obtained by Leiby and Rogers (1969), Chester (1968), Bridges and Halsted (1967), Bergman and Chanin (1969), and Tombers and Chanin (1970). Unfortunately many of these results were obtained under differing conditions so that a comparison of various results may not be very meaningful. Data obtained by Leiby and Rogers, Bergman and Chanin, and Tombers and Chanin were, for comparable discharge current densities, usually ≤ 1 A/cm^2. By contrast, Chester's results refer to current densities of 20–122 A/cm^2 and Halsted's data relate to densities of 70–350 A/cm^2. Clearly the plasma excitation processes may differ for these conditions. For example, ion production processes and ion species may be electron density and energy dependent. In addition, under high-density conditions, striation, constriction, and instability phenomena are more likely to be present than at low densities. Moreover, the necessary wall cooling in such discharges could enhance temperature and density gradients in the plasma. Finally, the use of different discharge geometries may also account for differences in measured Δp values.

With the considerations noted above, most recent measurements made at low current densities (Bergman and Chanin, 1969; Bergman, 1968; Leiby and Rogers, 1969; Tombers, 1970; Tombers and Chanin, 1970) suggest a functional dependence for Δp of the form

$$\Delta p/L \propto I/pR^4. \tag{7}$$

Moreover, results reported by Chester (1968) at higher current densities are in accord with Eq. (7). The high current density data of Halsted (Bridges and Halsted, 1967; Bridges *et al.*, 1966) indicate that at high currents and low pressures the volume force term becomes increasingly less important. Accurate comparisons of these results with theory are difficult due to the lack of data such as the dependence of gas temperatures on discharge current. The theory proposed by Leiby and Oskam (1967) and also Chester's analysis, predicts at high pressures the dependence given in Eq. (7). Chester has reexamined the results reported by Rüttenauer (1922), who reported that $\Delta p/L \propto I/pR^4$ for helium, neon, and argon. Such an evaluation is complicated by the lack of auxiliary data such as those relating to electron temperatures and axial fields.

In addition to a knowledge of the dependence of Δp on the parameters discussed above, a reasonable estimate of the magnitude of Δp involves the use of values for the electron and ion mobilities as well as the gas and electron temperatures. Reliable values for the electron temperature are in general difficult to obtain since the temperature is a function of pressure, tube radius, and current density. Both theory and experiment indicate that in the vicinity of the Δp maximum, the electron temperature begins to rise rapidly as the pressure is decreased. Another effect involving the electron temperature should be noted. In the vicinity of or below the maximum, the anode pressure is noticeably greater than the cathode pressure. Since the electron temperature is a function of pressure, an electron temperature gradient would be expected.

Theoretical estimates also involve decisions concerning the dominant ion species, so that appropriate ion mobility values can be assigned. In this respect, ion sampling data such as those obtained by Fitzwilson and Chanin (1973) provide reasonable guides. Considerations such as laminar flow conditions and Reynolds number estimates relating to the Leiby–Oskam theory have been discussed by Bergman and Chanin (1969). Chester (1968) has discussed viscous-flow assumptions, low-energy ion mean free paths, etc., relating to his theory.

D. Striation (Ionization Wave) Effects

The simple steady-state approximation to the Leiby–Oskam theory consistently underestimates the magnitudes of the pressure gradients observed in the heavier noble gases. Leiby observed that correlations existed between the magnitude of constriction and striation phenomena in the discharge, and the magnitude of the discrepancy between experimental results and values calculated using steady-state theory. The use of discharge tube wall temperatures for estimates of gas temperatures can result in serious overestimates of gas densities in constricted discharges. Part of the lack of agreement between experimental predictions based on the steady-state theory for the Ramsauer gases (argon, krypton, and xenon) is believed due to errors in the assumed gas temperatures. To take into account the time-dependent phenomena associated with striations, Leiby (1971) has considered a more sophisticated approximation to the Leiby–Oskam theory, which takes into account temporal variations of the plasma parameters. In the presence of striations, electric fields, electron densities, and temperatures are all strongly modulated. Since it is the product of these quantities that determines the magnitude of the electrophoretically induced gas flow, the time average of this product could be considerably different than the product of their individual time averages. To include the effects of time-varying phenomena, the Leiby–Oskam volume force was rederived retaining partial derivatives with respect to time. It was assumed that the partial derivative of the electron drift momentum with respect to time is negligible compared to the partial derivative of the ion drift momentum. To facilitate analysis the time variation of plasma parameters $E(t,z)$, $n_e(t,z)$, $T_e(t,z)$ associated with striations is approximated by

$$X = \langle X \rangle [1 - \alpha_x \cos(\omega t + \delta_x + \beta z)] \tag{8}$$

Here δ is an arbitrary phase angle, α a modulation degree, ω the striation radian frequency, and $\langle X \rangle$ the time-averaged value of the quantity X. Since measured gas flows reflect the time average of the volume force F_d, the force per unit volume is averaged over the striation oscillation period. According to Leiby (1971), Garscadden indicates that the electron temperature and the electric field tend to be in phase. For such cases, the ratio of the forces predicted by the time-dependent and steady-state theories has the simple form

$$\langle F_{dz}(t) \rangle / F_{dz}(\text{steady-state}) \approx 1 + \tfrac{1}{2}[\alpha_E \alpha_T + \alpha_n(\alpha_E + \alpha_T)\cos\delta_{ET}]. \tag{9}$$

Calculations of the ratio given in Eq. (9) for representative modulation degrees and reported lag angles indicate that for strongly striated plasmas, an error of a factor of 2 to 2.5 in predicted gas flow velocities could easily result if time-dependent phenomena are not taken into account. This is roughly the difference between the observed gas flow velocities in the Ramsauer gases and the velocities predicted by the steady-state approximation to the volume force theory.

According to Leiby a comparison of electron momentum transfer rates (as inferred from measured electron mobilities) with those calculated using published cross sections and measured electron temperatures shows them to be in good agreement when striations are absent. In the presence of striations, however, observed momentum transfer rates are up to 20 times larger than the calculated rates. Leiby suggests that this excess electron momentum transfer rate may constitute the force driving the observed striations.

E. Ion Species

Measurements of electrophoretic pressure gradients indicate that while steady-state theories agree well at high pressures for helium and neon, discrepancies occur for the heavier gases (argon, krypton and xenon). It has been suggested that these discrepancies may be due, in part, to a lack of knowledge of the relative ion concentrations and the variation of these concentrations with changing discharge pressure and current (Bergman and Chanin, 1969). Although it is generally believed that in noble gas discharges the molecular ion dominates at sufficiently high pressures, no systematic study of the molecular to atomic ion concentrations exists for the noble gases in moderately high current density capillary discharges. Consequently, a study was undertaken to measure the ratio of the molecular to the atomic ion density for small diameter discharges in argon, krypton, and xenon (Fitzwilson and Chanin, 1973). Current densities varied from 0.28 to 0.75 $Å/cm^2$.

The experimental method consisted of permitting a small number of the ions that reach the discharge tube wall via ambipolar diffusion to pass through a small sampling orifice. The effusing ions were then mass analyzed. A multicomponent continuity equation was used to develop a theoretical model for the ion density ratio within the plasma. Magnitudes of the several rate constants required to fit this model to the data were determined and compared to previous measurements. The effect of two-step ionization and excitation processes on the theoretical model was also discussed. This investigation has provided information regarding several of the assumptions employed in the development of the theoretical expression for the electrophoretic pressure gradient (Fitzwilson and Chanin, 1973). Specifically, large-amplitude moving striations were observed in all three gases. As noted in Section III,D the incorporation of striation effects substantially increases the agreement between experiment and theory. The ion ratio results also suggest the importance of electron–ion recombination processes for intermediate- and high-pressure discharges in capillary tubing for argon, krypton, and xenon. It is not clear exactly how the incorporation of this ion loss process would affect the predicted pressure gradient but it is believed that it would be somewhat reduced.

The ion ratio data also indicate that the majority ion changes from atomic to molecular in the intermediate pressure range 1–20 Torr. A simple p_0^{-1} pressure dependence is predicted from the model if only one type of ion is present. However, the experimental evidence, for example, in argon, is that the pressure gradient decreases more slowly than p_0^{-1} for pressures between 1 and 10 Torr (Bergman and Chanin, 1969). This effect could be partially due to the increase in molecular ion density in this pressure range. The maximum change that could be expected is a factor of 3 based on the relative masses and mobilities of the atomic and molecular ions.

IV. ELECTROPHORETIC EFFECTS

This section discusses undesirable effects that electrophoresis may have on gas laser performance. Methods of minimizing electrophoretic effects and their relative efficiencies are also considered. Finally, examples of negative pressure gradients or retrograde electrophoresis are discussed.

A. Gas Lasers

Gordon and Labuda (1964) were among the first to report the influence of electrophoresis on gas laser operation. According to these investigators, for gas ion lasers operating in small bore tubing at discharge currents of several amperes, pressure differences between the anode and cathode in excess of 10:1 can be established in less than one minute of discharge operation. Since the optimum pressure range for laser operation is narrowly defined relative to the range of pressures existing in the discharge tube, gas pumping effects deteriorate or extinguish laser action shortly after turn-on. To minimize both electrophoretic as well as cataphoretic effects in gas mixtures these authors suggested the use of a bypass or a connecting tube between the anode and cathode. As a result, high gas flow connecting tubes have become common practice in gas laser technology.

Tombers and Chanin (1970) have reported measurements of the effect of a bypass tube on pressure gradients in dc discharges in helium and xenon and in a 10% mixture of neon in helium. The pressure sensor used was a capacitance manometer in conjunction with a precision voltmeter for increased accuracy. Small floating tungsten probes were used to measure the electric field in the positive column of the discharge. A series of thermocouples was used to measure the wall temperature at various positions on the outside of the discharge tube and the bypass tubes. The leads for sampling the pressure were located so as to ensure that Δp measurements were performed on the positive column.

Figure 6 shows examples of measurements of the pressure gradient Δp as a function of the normalized discharge gas pressure ($p_0 = 273p/T$) for helium. Three curves are given for each current: $(\Delta p)_c$, the gradient for the closed system (no bypass); $(\Delta p)_b$, the gradient measured with the bypass open; and $(\Delta p)_{b\,calc}$, the gradient calculated using Eq. (10) with the bypass open. The values of $(\Delta p)_{b\,calc}$ were calculated using the experimental values

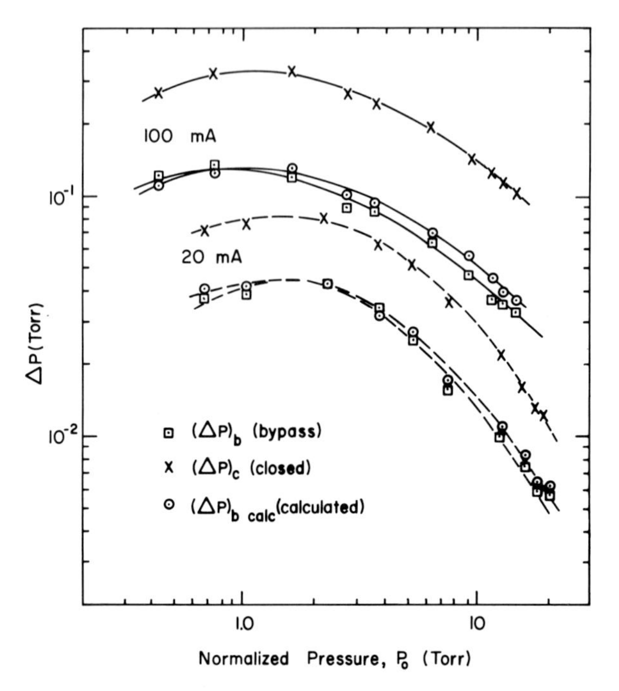

Fig. 6. Measurements in helium for 20 and 100 mA discharges. $(\Delta p)_b$, $(\Delta p)_c$ values were obtained with the bypass open and closed, respectively. Values of $(\Delta p)_{b\,calc}$ were calculated, and involved using values of $(\Delta p)_c$ and wall temperature measurements (Tombers and Chanin, 1970).

of $(\Delta p)_c$ and what were believed to be appropriate values for the various parameters:

$$(\Delta p)_c/(\Delta p)_b = \left\{1 + (R_2/R_1)^4[L_2/L_1 + L_3/L_1(R_2/R_3)^4]^{-1} \times \frac{\eta_1}{\eta_2}\left(\frac{T_1}{T_2}\right)\right\}. \quad (10)$$

Here η refers to the gas viscosity, L and R to the tube length and radius, and the subscripts 1, 2, and 3, respectively, to the discharge and bypass tubes. The above relation was obtained from the steady-state condition that the mass flow through both tube sections was equal and opposite. Equation (10) is a modification of a relation derived by Leiby and Rogers (1967). In Fig. 6 a comparison of the 20 and 100 mA curves indicates that at high pressures, Δp varies approximately linearly with current and inversely with p_0. This dependence is predicted by the Leiby–Oskam theory. From Fig. 6 it is evident that for a given discharge current, experimental values of the gradient using the bypass $(\Delta p)_b$ are in good agreement with $(\Delta p)_{b\,calc}$ values calculated using Eq. (10).

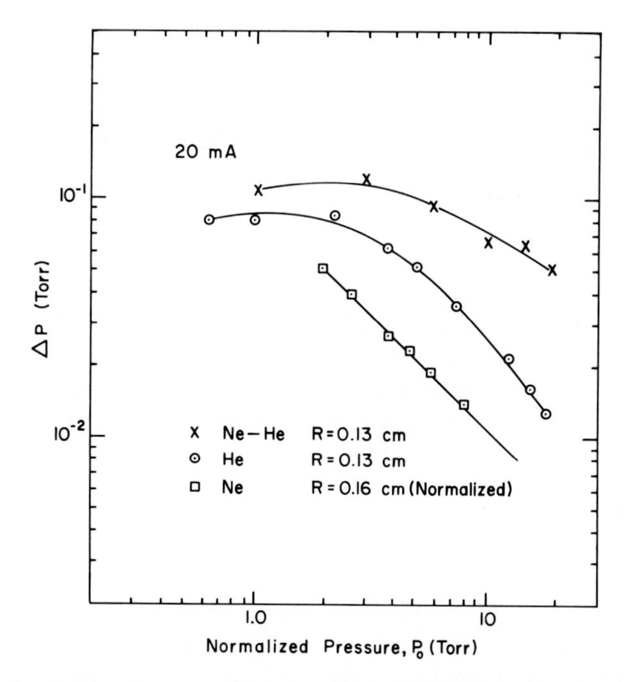

Fig. 7. Comparison of pressure difference data for 20 mA discharge in helium, neon, and 10% neon in helium. The neon data were obtained using a discharge tube with a radius of 0.16 cm and were normalized to 0.13 cm assuming an R^{-4} dependence (Tombers and Chanin, 1970).

A comparison of results for helium, neon, and a 10% neon in helium mixture is shown in Fig. 7. The data refer to measurements without a bypass and for a 20 mA discharge. Neon data were obtained using a tube radius $R = 0.16$ cm; however, the measurements were normalized to $R = 0.13$ cm assuming an R^{-4} dependence for Δp. It will be noted from Fig. 7 that values for the mixture lie well above the pressure differences in either of the pure gases. The larger values measured in the mixture may be explained since theory predicts that, at high pressures, the pressure gradient will be proportional to the square of the ion mobility, the ion mass, and the electron temperature. Assuming Ne^+ is the dominant ion, the pressure difference for the mixture will be larger than for pure neon due to the larger mobility of Ne^+ in the mixture.* The gradient would also be larger if the electron

* Estimates based on measurements by Gaur and Chanin (1969) in He–Ne mixtures indicate that Ne^+ would be the dominant ion. It should be noted, however, that the same general conclusions would apply if $Ne_2{}^+$ were the major ion.

temperature were greater in the mixture. The pressure difference would of course be greater in the mixture than in helium, due to the larger ion mass.

Differences in the relative effectiveness of the bypass for various gases may be accounted for on the basis of different gas temperatures. In practical applications, to achieve a reduction (≈ 100) in the pressure gradient for comparable discharge and bypass lengths, the bypass radius should be at least three times larger than the discharge tube.

Laser perturbation experiments have been conducted by Freiberg and Weaver (1965) on xenon laser discharges. Lasing was found to result in electron energy and density changes. The dominant feature of the discharge was that all the parameters displayed a considerable axial variation. This was attributed to electrophoresis, which caused an anode-directed pressure gradient. This produced an electron density gradient between the cathode and anode, which in some circumstances could give the anomalous result that lasing depopulated the lower level. This was thought to be due to the fact that, although the discharge was capable of lasing, the increased electron density had changed the local excited-state populations near the anode so that this portion of the discharge was absorbing on the laser transition.

Halsted has reported results of studies concerning gas pumping effects in gaseous ion lasers. Measurements were made in argon, krypton, and xenon in 1.8, 2, and 3 mm bore tubes for discharge currents from 1 to 30 A with pressures from ~ 0.1 to 3.0 Torr and for axial magnetic fields as high as 1.7 kG. Halsted reports that the pressure differential decreased with increasing discharge current. This decrease was attributed to gas heating effects, which reduced the neutral density and hence increased the gas viscosity. Unfortunately, results were not expressed in terms of the reduced pressure p_0 nor were gas temperatures related to discharge currents, pressures, and tube geometry. The results were compared to a modified version of Druyvesteyn's original formulation. The variations of the observed and predicted pumping rates with current density were in approximate agreement. Experimental and theoretical curves were, however, laterally displaced due to uncertainties in the neutral gas densities. It is apparent that in the region of low gas pressures additional experimental and theoretical studies are necessary to clarify the dominant pumping mechanisms.

Many reports have appeared in recent years concerning the detection of electrophoretic effects in various types of devices that utilize gas lasers (Bloom, 1966; Garscadden, 1967). For example, Podgorski and Aronowitz (1968) have investigated "Langmuir flow effects" in a laser gyro. To achieve the high accuracy associated with a practical inertial sensor of this type, various sources of error were analyzed. One of the major effects involved the flow of gain atoms due to electrophoresis.

B. "Retrograde" Electrophoresis

The earliest observations of the pressure differential between the anode and cathode ends of a dc discharge reported that the cathode pressure exceeded the pressure at the anode. These measurements were performed at low pressures ($p < 0.1$ Torr) in air and nitrogen. Few subsequent measurements of pressure gradients have been reported in the literature on molecular gases. Unpublished results by Rork and Chanin (1967) on nitrogen and oxygen and hydrogen, however, yielded normal anode-directed gradients. Measurements were made at current densities <1 A/cm^2 and for gas pressures as low as 0.2 Torr. The first quantitative studies, which were performed by Rüttenauer on noble gases, in contrast to the early measurements reported larger anode than cathode pressures. With several exceptions (Section IV,C and measurements on magnetically confined ion gas lasers, which will be considered later in this section), most subsequent studies of pressure gradients have reported an anode-directed gradient. Unfortunately some misunderstanding still seems to exist concerning the conditions under which a reversal in the gradient might occur (Cairns and Emeléus, 1958; Francis, 1956). It is easily inferred from some literature that a change-over from an anode- to a cathode-directed gradient will occur as the gas pressure is reduced to <0.1 Torr. This conclusion contrasts to experimental observations obtained in the early 1930s by Kenty (1938) and Klarfeld and Poletaev (1939). These authors reported anode-directed gradients for gas pressures as low as 10^{-3} Torr for current densities from $\simeq 0.1$ to $\simeq 1.0$ A/cm^2. Part of the misunderstanding may also be related to the improper use of Druyvesteyn's theory, which seriously underestimates the volume forces and consequently predict at low current densities a cathode-directed gradient. With the exceptions of some high current density gas lasers such as described below and mentioned in the succeeding section, the present understanding of pressure gradients is that they are normally anode directed.

Ahmed and Faith (1966) have reported the observation that the application of an external B_z field at high discharge currents results in a reversal of the pressure differential. Experiments were conducted on a dc, magnetically confined argon laser discharge in an 8 mm bore, 40 cm long quartz tube situated in a 1500 G axial magnetic field. The results show that at high currents, ion drift effects cause the gas pump-out to reverse from the more familiar pumping toward the anode to pumping toward the cathode. This causes the pressure to drop considerably along the tube toward the anode section, with important consequences for laser operation. Measurements were made over two ranges of discharge pressure. CW measurements were made with the laser discharge tube filled to initial pressures from 0.11 Torr to 0.34 Torr. Lower pressure operation (0.02 Torr to 0.05 Torr) was investigated in a quasi-CW pulse mode. At higher pressures, it was found that threshold laser action started only some time after discharge initiation (1.4 sec for a 25 A discharge and 0.11 Torr initial fill) when, due to ion gas pump-out toward the cathode, pressure in the narrow bore region near the anode bulb had dropped considerably (typically to 0.04 Torr).

Figure 8 shows results obtained with an initial filling of the tube to 0.11 Torr pressure. The 25 A current was the lowest at which laser action could be obtained for this initial pressure filling. Laser oscillations start 1.4 sec after the discharge is switched on. During this period, the anode region pressure drops to 0.05 Torr while the cathode region pressure rises to 0.14 Torr. The discharge current and voltage remain practically constant throughout the duration of the discharge. Garscadden (1967) has speculated that the reduction of diffusion losses due to the B field results in the dominance of the ion drift current term. He also suggests that the possible onset of anomalous diffusion effects may influence the pressure gradients (Powers, 1966).

Results reported by Bridges and Halsted (1967) indicate a reduction of the pressure gradient with increasing axial magnetic fields. It should also be noted that Halsted observed a normal electrophoretic behavior, i.e., the anode pressure exceeded the cathode pressure. The current densities, gas pressures, and magnetic field values were, however, comparable to those used by Ahmed and Faith. More theoretical and experimental studies are needed to improve the understanding of electrophoresis when axial magnetic fields are applied to dc discharges.

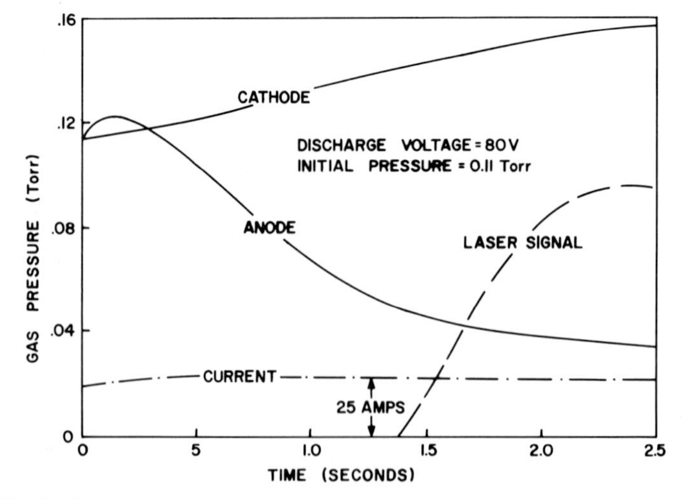

Fig. 8. Gas pressure, current, and laser output vs. time (Ahmed and Faith, 1966).

C. Nonuniform Positive Columns

Robinson *et al.* (1970) have observed that, for a mercury discharge tube that has a spatially periodic cross section, the normal increase of pressure at the anode can under certain conditions be replaced by a rarefaction. These authors indicate that this negative pressure gradient has a profound effect on the current-carrying capacity of such positive columns and could possibly be the cause of current limitation in many practical devices. Mercury vapor densities were determined by measuring the absorption of the resonance line radiation (2537 Å) using a modulated mercury lamp and a phase-sensitive detector. Vapor density measurements along the periodically varying diameter discharge tune were compared with similar measurements performed on a uniform diameter discharge tube. Measurements were made at discharge currents of 1, 2, and 4 A for several cathode chamber pressures. For the periodic discharge tube at 10^{-3} Torr and a 4 A discharge current, a pressure reduction from the cathode toward the anode was observed that was more pronounced at 5×10^{-4} Torr and 2 A (in contrast to the large positive vapor density gradient that was observed for the uniform diameter tube). The authors suggest that the wall corrugations in the periodic tube might lead to a reduction in the effective electron mobility because they give rise to a net transfer of axial anodeward electron momentum to the wall. This may imply a higher mean axial field, larger plasma density, and increased ratio of ion to electron axial flow. If the vapor density gradient along the column depends on this ratio, this might lead to an anode rarefaction in contrast to the results obtained using the uniform tube.

Robinson *et al.* suggest that their tentative explanation should receive further experimental investigation. Since many gas discharge devices often contain a series of grids or electrodes, each constituting a form of constriction on the cross section of the positive column, this type of effect may be of considerable practical importance. Obviously the effect may involve complicating factors such as axial temperature gradients and striation phenomena.

REFERENCES

Ahmed, S. A., and Faith, J. T. Jr. (1966). *Proc. IEEE* **54,** 1470.
Baly, E. C. C. (1895). *Phil. Mag.* **35,** 200.
Bergman, R. S. (1968). Ph.D. Thesis. Univ. of Minnesota, Minneapolis.
Bergman, R. S., and Chanin, L. M. (1968). *Bull. Am. Phys. Soc.* **13,** 211.
Bergman, R. S., and Chanin, L. M. (1969). *Phys. Fluids* **12,** 2348.
Bloom, A. L. (1966). *Appl. Opt.* **5,** 1500.
Bridges, W. B., and Halsted, A. S. (1967). Hughes Research Laboratories Tech. Rep. Nu AFAL-TR-67-89, pp. 171–209.
Bridges, W. B., Clark, P. O., and Halsted, A. S. (1966). *J. Quant. Electron.* **QE-2** (3B-1), XIX.

Cairns, R. B., and Emeléus, K. G. (1958). *Proc. Phys. Soc.* (*London*) **71,** 694.
Chester, A. N. (1968). *Phys. Rev.* **169,** 172, 184.
Druyvesteyn, M. J. (1935). *Physica* **2,** 255.
Fitzwilson, R. L., and Chanin, L. M. (1973). *J. Appl. Phys.* **44,** 5337.
Francis, G. (1956). *In* "Handbuch der Physik" (S. Flügge, ed.), Vol. 22. Springer-Verlag, Berlin.
Freiberg, R. J., and Weaver, L. A. (1965). *J. Appl. Phys.* **38,** 250.
Garscadden, A. (1967). Invited Paper, *Int. Conf. Phenomena in Ionized Gases, 8th, Vienna, Austria.*
Gaur, J. P., and Chanin, L. M. (1969). *J. Appl. Phys.* **40,** 256.
Gordon, E. I., and Labuda, E. F. (1964). *Bell Syst. Tech. J.* **43,** 827.
Hamburger, L. (1917). *Proc. Amsterdam Accd.* **25,** 1045.
Kenty, C. (1938). *J. Appl. Phys.* **9,** 765.
Kenty, C. (1967). *J. Appl. Phys.* **38,** 4517.
Klarfeld, B., and Poletaev, I. (1939). *Compt. Rend.* (*Dokl.*) **23**(5), 460.
Langmuir, I. (1923). *J. Frank. Inst.* **196,** 751.
Leiby, C. C. Jr. (1966). *Proc. Annu. Conf. Phys. Electron. 26th* (R. E. Stickney, ed.). MIT Press, Cambridge, Massachusetts.
Leiby, C. C. Jr. (1971). *Annu. Gaseous Electron. Conf. 24th, Gainesville, Florida, October.*
Leiby, C. C. Jr., and Oskam, H. J. (1967). *Phys. Fluids* **10,** 1992.
Leiby, C. C., Jr., and Rogers, C. C. *Proc. Int. Conf. Phenomena in Ionized Gases 8th.* Springer-Verlag, Berlin.
Leiby, C. C. Jr., and Rogers, C. C. (1969). *Annu. Gaseous Electron. Conf. 22nd, Gatlinburg, Tennessee, October.*
Oskam, H. J. (1969). *Phys. Fluids* **12,** 2449.
Podgorski, T. J., and Aronowitz, F. (1968). *IEEE J. Quant. Electron.* **QE-4**(1), 11.
Powers, E. J. (1966). *Proc. IEEE* **54,** 804.
Robinson, T. J., Sandahl, S., and Torvén, S. (1970). *J. Phys. D, Appl. Phys.* **3,** 69.
Rork, G. D., and Chanin, L. M. (1967). Unpublished results.
Rüttenauer, A. (1922). *Z. Phys.* **10,** 269.
Skaupy, F. (1917). *Ver. Deut. Phys. Ges.* **19,** 264.
Stark, J. (1904). Boltzmann Festschrift, 399.
Tombers, R. B. (1970). Ph.D. Thesis, Univ. of Minnesota, Minneapolis.
Tombers, R. B., and Chanin, L. M. (1970). *J. Appl. Phys.* **41,** 2433.
Warburg, E. (1892). *Wied. Ann.* **45,** 1.
Wehnelt, A., and Franck, J. (1910). *Ver. Deut. Phys. Ges.* **12,** 444.

Part 2.4 Nonuniformities in Glow Discharges: Cataphoresis

L. M. CHANIN
DEPARTMENT OF ELECTRICAL ENGINEERING
UNIVERSITY OF MINNESOTA
MINNEAPOLIS, MINNESOTA

I. INTRODUCTION

The major purpose of Sections II and IV will be to consider some of the more recent theoretical and experimental results. Later sections will discuss some recent applications. No attempt will be made to provide a comprehensive survey of results obtained prior to 1960. The earliest literature on

ISBN 0-12-349701-9

cataphoresis has been summarized by Lehmann (1898). Later, systematic experimental studies by Skaupy (1916) and Skaupy and Bobek (1925) were summarized by Mierdel (1929). Several experimental contributions in the 1930s by investigators such as Vygodskii and Klarfeld (1933) and Penning (1934) appeared in advance of Druyvesteyn's (1935) important theoretical interpretation.

Considerable interest was shown in the 1950s in cataphoresis primarily due interest in gas purification techniques. Loeb in 1958 summarized results and current understanding of the effect. Major experimental contributions in this decade were made by Riesz and Dieke (1954) and Matveeva (1959). In more recent years the publications by Schmeltekopf (1964), Freudenthal (1967a), Shair and Remer (1968), and Gaur and Chanin (1969) have involved extensive references to earlier investigations.

A. Terminology

As noted in Section I, Part 2.3, the terms cataphoresis and electrophoresis are sometimes used to refer to the same phenomenon. One probable reason for this is that the term electrophoresis, which is derived from Greek, means "borne by electricity," and hence is a general term. In this regard cataphoresis should represent a special case of electrophoresis. The term cataphoresis is used in this section in the same general sense as it is used in most English physics literature. Specifically, the term will be used to refer to the partial segregation of gas components that occurs when a gas mixture is subjected to an electrical discharge.

II. THEORY

The first attempt to explain cataphoresis theoretically was made by Druyvesteyn (1935). For noble gas–magnesium mixtures, he reported that the metal vapor concentration was enhanced in the cathode region of the discharge. By equating the cathode-directed metal vapor positive-ion current to the back diffusion current of the metal atoms, for $n_p = n_e$ he obtained

$$n(z) = n(0) - A(\mu_+ T_e^{1/2} I/\lambda_e D R^2)z, \tag{1}$$

where n, n_p, and n_e are the concentrations of the magnesium atoms, the ions, and the electrons, respectively. A is a constant, μ_+ the mobility of Mg^+, D the diffusion coefficient of the magnesium atoms, T_e the electron temperature, λ_e the mean free path of the electrons, z the axial distance from the cathode along the positive column, and I the discharge current.

When $n_p \ll n_e$,

$$n(z) = n(0) \exp[-Bn_e\beta S/(D + 0.22n_e\beta S^2)]z, \tag{2}$$

where B is a constant, $S = 0.26R^2E/T_e$ with R the tube radius, E the electric field, β a function of T_e, and $n(0)$ the value of n at $z = 0$. Druyvesteyn suggested that normally Eqs. (1) and (2) would be valid for small and large values of z, respectively.

The above equations were obtained assuming steady-state conditions and that the magnesium concentration was independent of radial position. Assuming that the concentration was independent of distance from the cathode, Druyvesteyn obtained equations for the radial dependence of n. For $n_p = n_e$,

$$n(r) = n_w - n_e(0)(2\mu_+/3D)T_eJ_0(2.4r/R) \tag{3}$$

assuming $n_w - n \ll n$. Here J_0 is the zeroth-order Bessel function and n_w refers to the concentration of the metal vapor at the discharge tube wall. When $n_p \ll n_e$ and n_e is independent of r,

$$n(r) = n(0)J_0\left[I\frac{(\beta n_e)^{1/2}}{D}r\right]. \tag{4}$$

To examine the predicted functional dependence of cataphoresis on discharge parameters such as gas pressure and current, it is convenient to re-express Eq. (1) in the form

$$n(z) = n(0) - A'\left[\frac{P_C T_e^{1/2}}{T_g^2 R^2} PI\right] z, \tag{5}$$

where A' is a constant, P_C the electron collision probability, T_g the gas temperature, and p the gas pressure. From Eq. (5) it is evident that measured values of dn/dz should show a linear dependence on discharge current and gas pressure.

While significant progress was made in understanding cataphoresis following the introduction of Druyvesteyn's theory, most of the investigations were experimental. Few further theoretical contributions to this topic appeared until the mid-1960s. At that time several articles were published that discussed various aspects of the phenomenon theoretically.

In 1967 Pekar (1967a) reported results of studies of the transverse (radial) separation of a mixture in a longitudinally uniform plane-symmetrical column of a discharge operating in the diffusion mode. Using simplifying assumptions that involved the neglect of viscosity, and temperature gradient effects, the diffusion equations for the electrons and the minor gas component were solved for two limiting cases. It was assumed that only the minor constituent concentration was excited and ionized in the positive column and that the electron temperature was constant over the discharge tube cross section. The diffusion equations were then solved using boundary conditions believed to be appropriate when (1) the minor gas was a metal vapor whose pressure at the walls was determined by the wall temperature, and (2) when the minor gas was an inert type of gas. A system of parametric equations was obtained for each of these conditions. The theory was used to calculate various parameters for a helium–mercury mixture using other simplifying assumptions relating to the ionization process. The theoretical analysis led to the conclusion that a decrease in the admixture concentration along the column would produce an increase in the transverse separation of the gas components and an increase in the electron temperature.

Pekar (1967b) also considered the longitudinal separation of a gas mixture under glow discharge conditions. An attempt was made to construct a theory for cataphoresis in a longitudinally inhomogeneous column, which allowed for a transverse or radial separation of the mixture. Expressions for the electron temperature, longitudinal potential gradient, minor constituent concentration of the electrons and neutral atoms, and the radiation intensity were obtained for the limiting cases of large and small concentrations. A number of simplifying assumptions and approximations were used to obtain solutions to the appropriate continuity equations. For large concentrations of the preferentially ionized gas component, assuming negligibly small ionization compared to the major constituent, a solution was obtained of the form

$$n(rz) = G \exp(-z/b)F(z), \tag{6}$$

where G and b are constants and $F(z)$ is a Mathieu function. In the transition region of the positive column where the ionization frequencies of both gas components are comparable, a number of approximations are necessary to describe the discharge characteristics. Here, under certain conditions,

$$n = \frac{\tilde{P}}{kT}\left(1 - \frac{3\gamma}{2\alpha + 2}\right), \tag{7}$$

where $\tilde{P}$ is a term involving the electron pressure and γ and α are terms involving parameters such as diffusion and ionization coefficients. The theoretical results were applied to a helium–mercury mixture. According to Pekar's analysis, the transition can occupy a large part of the column; also in the presence of strong radial effects spectroscopic measurements do not always allow reliable estimates to be made of the longitudinal admixture distribution.

Freudenthal (1967) was one of the first investigators to consider transient cataphoresis theoretically. In 1967 he developed a model for transient and also steady-state cataphoresis. The stationary-state analysis represented an extension of Druyvesteyn's theory and involved several simplifying assumptions, including the neglect of transport effects due to momentum transfer by electrons and ions. The solution for the axial steady-state distribution assuming nearly all ions are admixture ions n_{pa} is given by

$$n_{\mathrm{a}}(z) = n_{\mathrm{a}}(0) - cz, \tag{8}$$

where $n_{\mathrm{a}}(0)$ is the admixture density at the cathode $z = 0$, and $c = \mu_{\mathrm{pa}} J/\mu_{\mathrm{e}} Dq$, where J is the tube current density, and c is assumed independent of z. μ_{pa} and μ_{e} are, respectively, the mobilities of the admixture ions and the electrons. For discharges where the gas temperature and pressure are uniform, Eq. (8) yields a linear distribution for cataphoresis in agreement with Eq. (1). When the admixture ions represent a small fraction of the total ions, then

$$n_{\mathrm{a}}(z) = n_{\mathrm{a}}(0) \exp(-\mu_{\mathrm{pa}} E\phi/D)z, \tag{9}$$

where $\phi = n_{\mathrm{pa}}/n_{\mathrm{a}}$ is the ionization degree of the admixture. Clearly Eq. (9) is of the same form as Eq. (2). By normalizing $n_{\mathrm{a}}(z)$ to the average density of the admixture $n_{\mathrm{a}}{}^*$ so that

$$\int_0^L n_{\mathrm{a}}(z)\, dz = n_{\mathrm{a}}{}^* L, \tag{10}$$

where L is the discharge tube length, Eq. (9) can be reexpressed as

$$n_{\mathrm{a}}(z) = n_{\mathrm{a}}{}^* \frac{uL}{D} \frac{1}{1 - e^{-uL/D}} e^{-uz/D}, \tag{11}$$

where $u = \mu_{\mathrm{pa}} E\phi$. This normalization is valid when the cathode and anode are mounted at the ends of the discharge tube. Freudenthal also considered the influence of radial effects on the axial gas transport.

The time dependence of the axial segregation was obtained from a solution of the diffusion equation

$$\frac{\partial}{\partial t} n_{\mathrm{a}}(z,t) = D \frac{\partial^2}{\partial z^2} n_{\mathrm{a}}(z,t) - u \frac{\partial}{\partial z} n_{\mathrm{a}}(z,t), \tag{12}$$

where the second term on the right accounts for ion transport effects. Equation (12) was solved subject to the conditions that $n_{\mathrm{a}}(z,0) = n_{\mathrm{a}}{}^*$ for all z

and no gas losses at the ends of the discharge tube. Here

$$n_a(z,t) = n_a(z,\infty) + n_a{}^*B\exp\left(-\frac{u}{2D}z - \frac{u^2}{4D}t\right)$$
$$\times \sum_{n=1}^{\infty} C_n\left(\cos\frac{\pi n}{L}z - \frac{uL}{2\pi Dn}\sin\frac{\pi n}{L}z\right)\exp\left(-\frac{\pi^2 Dn^2}{L^2}\right). \quad (13)$$

In this equation B and C_n are functions of parameters such as L, D, and u. According to Freudenthal (1967a), calculated values of the slopes of the time dependence for a Ne + 0.2% Ar mixture using Eq. (13) were within 10% of the experimental results. The above analysis neglects the effect of end bulbs, which according to Shair and Remer (1968) can have a significant effect on the transient behavior.

In 1968 Shair and Remer proposed a theoretical model for cataphoretic gas separation in a glow discharge. This model also represented a refinement and extension of Druyvesteyn's original interpretation and was developed using the macroscopic continuity equations. The basic assumptions in the model are that after breakdown the level of ionization of the admixture (impurity) component and the axial electric field are constant. For these conditions the system involving rapid ionization–recombination reactions was regarded as equivalent to a system in which no reaction occurs but in which the effective ion mobility is a product of the true ion mobility and the fraction of impurity ionization. The influence of end bulbs commonly used in experiments was analyzed and found to affect the characteristic time required to reach steady state. As in previous analyses, several simplyfying assumptions were used; for example, only direct electron impact ionization processes were considered. The main loss of charged particles was assumed to be through ambipolar diffusion losses to the walls of the discharge tube. The radial diffusion terms were assumed to be large compared to the longitudinal terms. The ionization level of the impurity $\langle n_{pa}\rangle/(\langle n_{pa}\rangle + \langle n_a\rangle)$ was assumed to be constant and independent of z and time in the discharge. The brackets refer to radially averaged quantities. A general cataphoretic equation was obtained of the form

$$\frac{\partial\psi}{\partial\tau} = \frac{\partial^2\psi}{\partial\eta^2} + \alpha\frac{\partial\psi}{\partial\eta}. \quad (14)$$

In this equation

$$\psi = \langle n_a\rangle + \langle n_{pa}\rangle, \qquad \tau = \frac{tD}{L^2}, \qquad \eta = \frac{z}{L}, \qquad \alpha = \left(\frac{\langle n_{pa}\rangle}{\langle n_a\rangle + \langle n_{pa}\rangle}\right)\mu\frac{EL}{D}.$$

Shair and Remer solved Eq. (14) by taking into account the size of the electrode volumes. The boundary conditions assumed were that, for $\eta = 0$,

$$\frac{\partial \psi}{\partial \eta} + \alpha \psi = \delta \frac{\partial \psi}{\partial \tau}$$

and, for $\eta = 1$,

$$\frac{\partial \psi}{\partial \eta} + \alpha \psi = -\varepsilon \frac{\partial \psi}{\partial \tau}. \tag{15}$$

Here

$$\delta = \frac{\text{volume of the cathode bulb}}{\text{volume of the discharge tube}}, \qquad \varepsilon = \frac{\text{volume of the anode bulb}}{\text{volume of the discharge tube}}.$$

The initial condition is $\psi = 1$ for all η. The solution of Eq. (14) involves a result similar in form to Eq. (13). Differences exist, however, as a consequence of considering the effect of the end bulbs in Shair and Remer's analysis. These authors also considered the collapse of the steady-state profile when the discharge was extinguished. A quantitative comparison was made between the model and the data obtained from rare gas mixtures and mixtures of hydrogen and deuterium reported by Beckey *et al.* (1953) and by Matveeva (1959).

Oskam (1968) used the continuity equation and an approximation to the momentum conservation equation to derive the equations that had been used as the basic equations for analysis of experimental results. The approximation involved the assumptions that (a) the drift energy of the particles is small compared to the random energy, (b) the random-energy distribution is Maxwellian with a temperature that is independent of spatial coordinates, and (c) viscosity effects could be neglected. The analysis involved the derivation of an equation analogous to the one-dimensional equation used by Freudenthal in his discussion of time-dependent segregation as well as the steady-state equation. The validity of the assumptions and of ion production and loss processes for a given mixture was also considered. In an earlier publication the pressure dependence of cataphoresis for helium–neon mixtures was considered by Oskam (1963) and explained on the basis of the Ne^+ formation reaction, which involved $He_2{}^+$.

Gaur and Chanin (1969) proposed an analysis of cataphoretic effects for helium–neon mixtures. One of the purposes of this description was to attempt to incorporate specific ion production and loss processes for a given gas mixture into the cataphoretic theory. The analysis was based on the use of the continuity equations for the ith particle, i.e.,

$$0 = \bar{\nabla} \cdot \bar{\Gamma}(i) - S(i) + L(i), \tag{16}$$

where $\bar{\Gamma}(i)$, $S(i)$, and $L(i)$ are, respectively, the particle current density, the net volume production, and the loss rate of the ith type of particle. For small concentrations of neon in helium at high pressures, Ne^+ ions are produced primarily by the reaction

$$He_2^+ + Ne \xrightarrow{k_1} Ne^+ + 2He.$$

The production of Ne^+ by other processes is assumed to be small. Apart from ambipolar diffusion loss and possible conversion to Ne_2^+, which may be neglected for small neon concentrations, Ne^+ is primarily lost due to the reaction

$$Ne^+ + 2He \xrightarrow{k_2} HeNe^+ + He$$

k_1 and k_2 refer, respectively, to the rate coefficients for the production and loss of Ne^+. In equilibrium, since the production rate of one of the particles of interest equals the loss rate of another particle, the addition of equations appropriate to Ne, Ne^+, $HeNe^+$, and He_2^+ yields

$$\bar{\nabla}\cdot[\bar{\Gamma}(Ne) + \bar{\Gamma}(Ne^+) + \bar{\Gamma}(HeNe^+) + 2\bar{\Gamma}(Ne_2^+)] = 0. \tag{17}$$

Integrating (17) over the tube volume and assuming that Ne^+, $HeNe^+$, and Ne_2^+ have the same dependence on r and that Ne^+ is the dominant ion yields

$$\left\langle\left(\frac{\partial}{\partial z}\right)[\Gamma_z(Ne) + \Gamma_z(Ne^+)]\right\rangle_{av} = 0, \tag{18}$$

where the averaging is performed over the tube cross section. Using the continuity equation for Ne^+ and assuming that the dependencies f of Ne^+ and Ne in the z direction are identical($f' \equiv \partial f/\partial z$) one obtains

$$-\left(\frac{2.4}{R}\right)^2 D_a + \mu E\left(\frac{f'}{f}\right) + \frac{k_1[He_2^+]}{\phi(0,z_0)} - k_2[He]^2 = 0. \tag{19}$$

In Eq. (19) ϕ is approximately equal to the degree of ionization ρ at the center of the tube $[\rho = \phi/(1 + \phi)]$. One can also show that f'/f may be expressed in the form

$$f'/f = -(\mu E/D_0)\phi(0,z_0)\bar{R}. \tag{20}$$

This equation was obtained by assuming that D_0, the diffusion coefficient of neon in helium, and E are independent of r and that viscosity effects are negligible. Using Eqs. (19) and (20) yields

$$\left[\frac{(\mu E)^2}{D_0}\right][\bar{R}\phi(0,z_0)]^2 + \left\{\left(\frac{2.4}{R}\right)^2 D_a + k_2[He]^2\right\} \times [\bar{R}\phi(0,z_0)] - k_1\bar{R}[He_2^+(0,z_0)] = 0. \tag{21}$$

From Eq. (21) it is apparent that if parameters such as μ, E, D_0, and the rate rate constants k_1 and k_2 are known, the averaged degree of ionization of neon may be calculated. The solution of Eq. (20) for a fixed position z_0 in the positive column is given by

$$f(z - z_0) = \exp[-(\mu E/D_0)\phi(0,z_0)\bar{R}(z - z_0)]. \tag{22}$$

Equation (22) is of the same form obtained by Druyvesteyn, i.e., Eq. (2). For large concentrations of neon in helium the value of the exponent is small; hence the neon atom concentration would vary approximately linearly with distance from the cathode as predicted by Druyvesteyn. From Eq. (22) when E, D_0, and μ are known, $\bar{\phi}$ can be estimated from measurements of the radial ion current I_r due to Ne^+ arriving at an area ΔA on the discharge tube wall. The latter quantity is given by

$$I_r = -\Delta A q D_a \bar{\nabla}_r[Ne^+(r,z)]\big|_{r=r_0}. \tag{23}$$

A comparison of calculated values of ϕ from using Eq. (21) and measured values using Eq. (22) resulted in agreement within a factor of four.

In a later article Gaur and Chanin (1970) extended their helium–neon cataphoresis analysis to helium–argon mixtures. For this Penning mixture, measurements of the axial distribution of Ar^+ and values of the rate coefficient for the reaction

$$He^m + Ar \rightarrow Ar^+ + He + e$$

enabled estimates to be made of the helium metastable particle densities He^m. Results obtained in this manner were found to be in agreement with values obtained using other experimental techniques.

It is apparent that the incorporation of ion production and loss processes into a theoretical description of cataphoresis is in general complicated. Nevertheless, it is evident that the inclusion of production and loss processes is of major importance if one wishes for a given gas mixture to account for the cataphoretic dependence on common discharge parameters.

III. EXPERIMENTAL TECHNIQUES

The various experimental methods that have been used to investigate cataphoretic segregation are briefly described in the following section.

A. Radiation Sampling

The first investigation of cataphoresis, using optical techniques, was the original report by Baly (1893). Since then the majority of the experimental studies have utilized this diagnostic method. The optical system used by Riesz and Dieke (1954) used a beam splitter and two monochromators for the direct recording of the ratio of two spectral lines. Light at any point along the axis of the discharge tube could be sampled using an optical bench system for the axial displacements. Variations of this technique have been used by Schmeltekopf (1964) and others.

When radiation measurements are used to detect cataphoretic action, care must be used to sample only from the region of the positive column. Matveeva (1959) noted that the determinations of changes in gas composition based on variations in spectral emission along the discharge column have limitations as far as quantitative information is concerned. This follows since the intensity of spectral lines depends not only on the concentration of a given gas but also on the conditions of excitation that in the presence of separation differ in various parts of the positive column. In this regard Schmeltekopf (1964) has shown that for helium–neon mixtures the extreme conditions of excitation near the cathode render spectral intensity interpretations in this region difficult if not misleading.

B. Gas Sample Mass Analysis

Within the past 20 years a number of investigators have used mass analysis of gas samples that were withdrawn from the discharge tube as a method of studying cataphoretic behavior. Becky *et al.* (1953) were the first investigators to use this technique. The method has subsequently been used by Schmeltekopf (1964) and Freudenthal (1966, 1967a,b). Typically, in such measurements gas samples are withdrawn at various axial distances from the cathode to determine the axial or longitudinal dependence of the admixture gas component. An example of this type of system is that used by Freudenthal shown in Fig. 1. The distribution of the gas along the positive column of the discharge (tube length 129 cm, inner diameter 2.4 cm) was determined by sampling gas at several places in the positive column (Freudenthal, 1967b). Three molecular gas leaks were mounted in the wall of the tube and were connected to a mass spectrometer. By reversing the discharge for nonsymmetrical positions of the leaks, the composition of the gas could be determined at six different positions in the positive column. The total gas pressure was measured using a capacitance manometer. The vacuum system was evacuated to pressures $\simeq 5 \times 10^{-7}$ Torr prior to the initiation of measurements. By sampling through gas leaks as indicated in Fig. 1 rather than collecting samples in volumes for subsequent analysis as in some of the earlier studies of this type, Freudenthal was also able to measure mutual diffusion coefficients for neon–argon mixtures. Such measurements in-

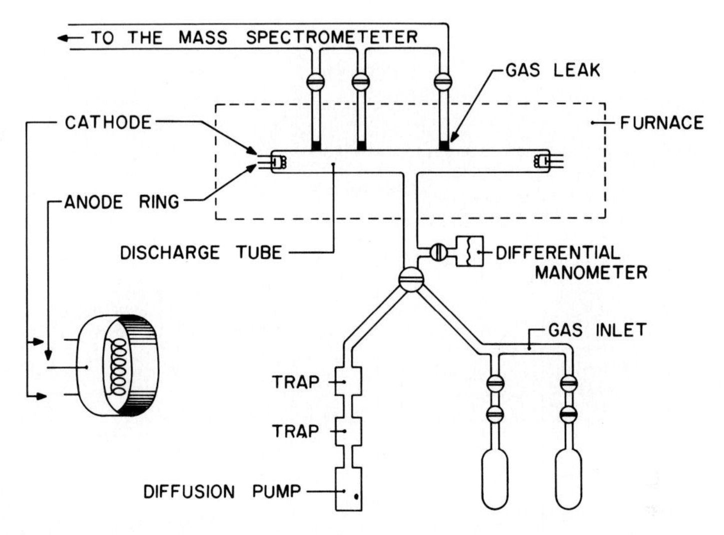

Fig. 1. Schematic diagram of the experimental arrangement for measuring cataphoresis in mixtures of gases. The actual electrode geometry is represented in the inset (Freudenthal, 1967b).

volved time-dependent mass analysis determinations following the termination of the discharge.

C. Ionic Sampling

Ion production and loss processes are of importance for a detailed understanding of cataphoresis for a particular gas mixture. One experimental method that permits direct analysis of a sample of positive ions from the positive column of the discharge is the procedure used by Gaur and Chanin (1969). The apparatus permits direct sampling of the ion species at a desired position relative to the cathode of the discharge. Versions of this system enable measurements to be made of the neutral constituents present in the discharge, as well as the radiation emitted from the plasma (Gaur and Chanin, 1969; Kuehn and Chanin, 1972). Figure 2 shows a schematic drawing of a version of the system that was used for ionic and radiation sampling studies. Most of the system was mounted on a standard high-vacuum gas-handling system. The part of the discharge tube containing the sampling orifice projected into an evacuated housing that contained the quadrupole mass spectrometer and ion multiplier. The tube consisted of a 100 cm length of precision bore 2.22 cm i.d. Pyrex tubing. The cathode was shielded and movable relative to the sampling orifice, thus enabling measurements to be

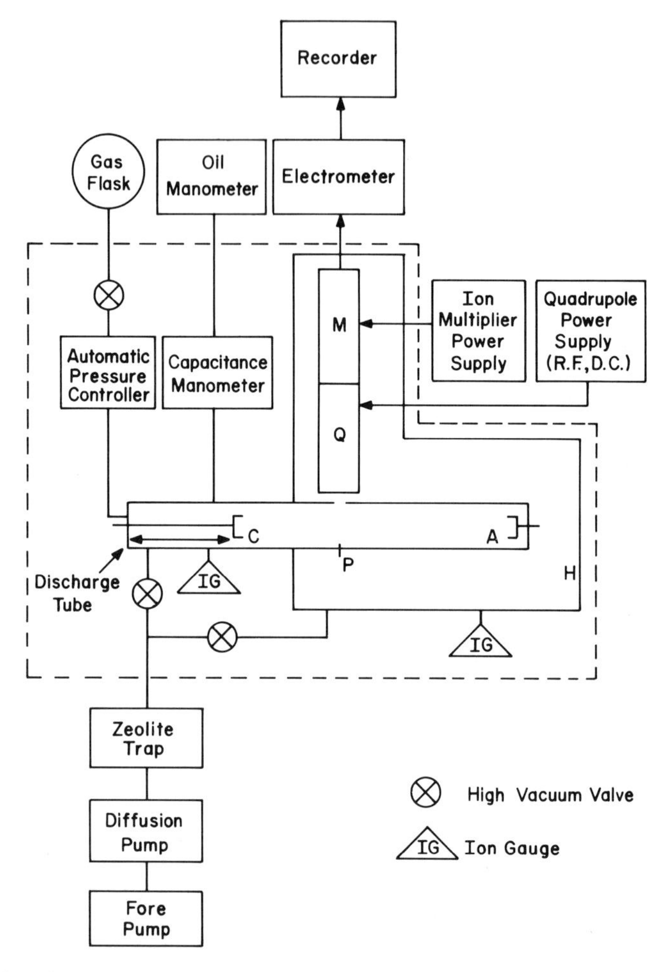

Fig. 2. Schematic drawing of the experimental apparatus. The portion enclosed by the dashed line was subjected to high-temperature bakeout (Gaur and Chanin, 1970). The cathode *C* was movable. *P* represents a small probe located opposite to the ion extraction orifice.

made of the ions effusing from the positive column at various distances from the cathode. Values of the electric field in the positive column could be obtained by measuring the potential drop across the discharge tube as a function of electrode separation, at a fixed current. The sampling orifice was conically shaped with the smallest diameter 35 μm. A metallic shield around the discharge tube in the orifice region served as a radiation shield and a reference electrode. The discharge was maintained by two separately regulated power supplies connected in series with the common terminal grounded. Measurements were made under pressure and discharge current conditions such that striation phenomena were absent. Data were obtained sufficiently far removed from the cathode so that the measurements were representative of the positive column.

D. Thermal Conductivity

The use of a tungsten filament lamp as a gauge for cataphoretic gas analysis has been reported by Remer and Shair (1969, 1971). The method of analysis of the gas composition involves thermal conductivity probe measurements at the anode end of the discharge tube. Mixtures involving high thermal conductivity gases, such as He or H_2, are best suited for this method. Results obtained using this technique have apparently not been extensively compared with earlier spectroscopic measurements. The method clearly offers a very simple way of determining the cataphoretic dependence of certain gas mixtures on some discharge parameters. The information obtained, however, is qualitative and does not provide any appreciable insight into the basic cataphoretic processes.

IV. RECENT RESULTS

In this section examples are given of results obtained using various types of gas mixtures. The greatest experimental interest has centered on steady-state conditions, especially for noble-gas mixtures. Reasons for this relate to concern for noble-gas purification methods. Also discharges in these mixtures are better understood and hence the results are more amenable to interpretation.

A. Noble-Gas Mixtures

Figure 3 shows an example of data obtained for commercial helium gas samples by Gaur and Chanin (1969). The ion signals of Ne^+ and $HeNe^+$ are shown as a function of sampling distance from the cathode for various gas pressures and discharge currents. The existence of the $HeNe^+$ ion was postulated by Oskam (1957) and was independently verified using mass analysis in the same year by Pahl and Wiemer (1957). For the conditions shown in Fig. 3, $HeNe^+$ is primarily formed by the reaction

$$Ne^+ + 2He \rightarrow NeHe^+ + He. \tag{24}$$

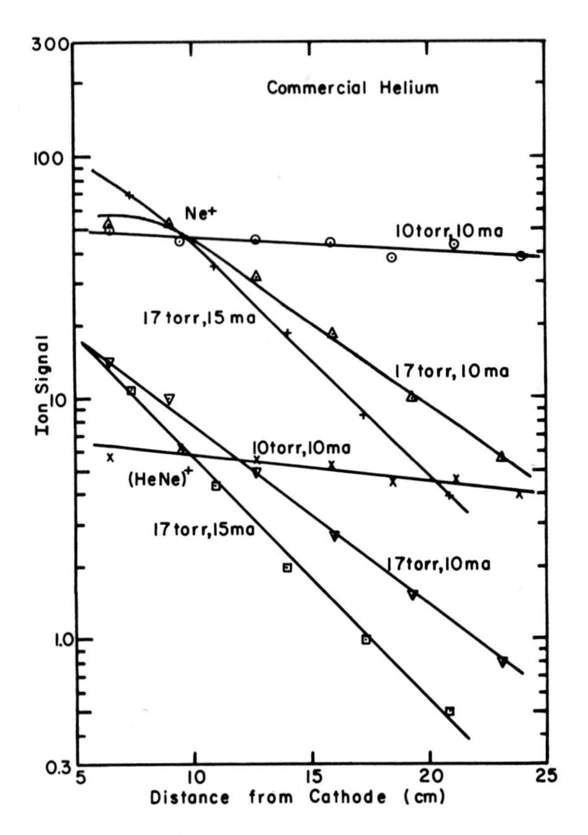

Fig. 3. Relative signals of Ne^+ and $HeNe^+$ as a function of distance from the cathode for commercial helium gas samples (Gaur and Chanin, 1969).

Assuming that the primary production and loss processes for $HeNe^+$ are the above reaction and ambipolar diffusion, respectively, measured values of $[HeNe^+]/[Ne^+]$ enable estimates to be made of the reaction rate constant of (24). Values obtained in this manner are within a factor of three of results obtained using afterglow techniques (Veatch and Oskam, 1970).

Figure 4 shows examples of results obtained by Gaur and Chanin (1970) using ion sampling techniques for helium containing 0.05% argon at 3 Torr. These data demonstrate the well-known current dependence of cataphoresis, which is predicted by theory and has been confirmed for a number of gas mixtures.

Figure 5 shows the major ion signals observed at various distances from the cathode in helium containing 0.05% argon for 15 Torr and 25 mA. A similar functional dependence of these ions was observed at other pressures and currents. The Ne^+ signals are believed to result from traces of neon

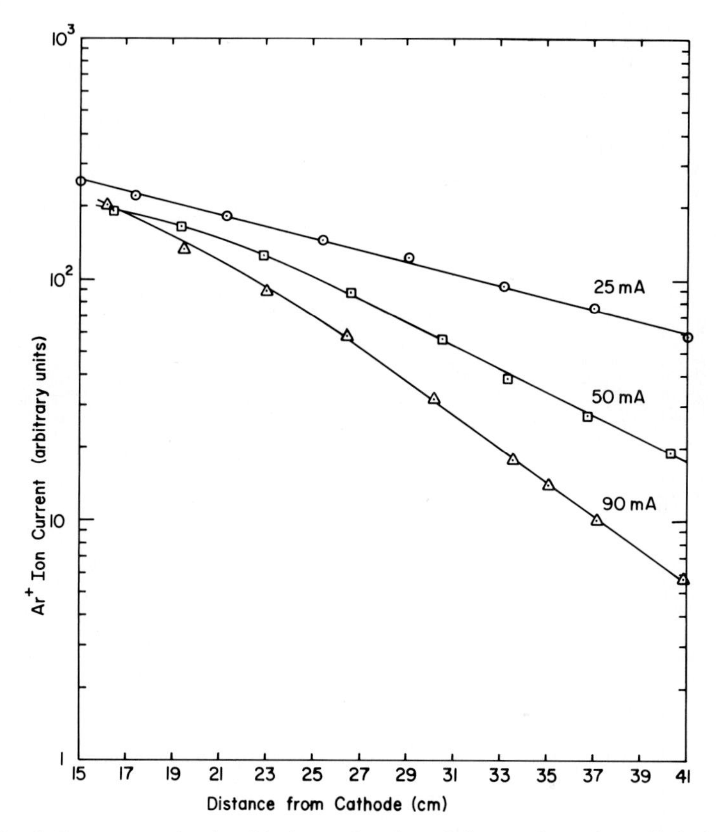

Fig. 4. Ion current signals of Ar^+ as a function of distance from the cathode for 0.05% argon in helium at 3 Torr. Data are shown for various discharge currents (Gaur and Chanin, 1970).

present in the gas samples. The quoted neon concentration in the research grade helium gas samples that were used was less than 10 ppm. The absence of $Ne_2{}^+$ and $NeAr^+$ ions also suggests very low concentrations of neon. It will be noted from Fig. 5 that at distances far from the cathode (>25 cm), the slopes of the Ar^+ and Ne^+ signals are not influenced by the relative magnitudes of the Ne^+ or Ar^+, respectively. This suggests that, for small concentrations of neon and argon, the two segregation mechanisms act independently. This result would be expected since the production of Ar^+ and Ne^+ are by He^m and $He_2{}^+$, respectively, i.e.,

$$He^m + Ar \rightarrow Ar^+ + He + e, \tag{25}$$

$$He_2{}^+ + Ne \rightarrow Ne^+ + 2He. \tag{26}$$

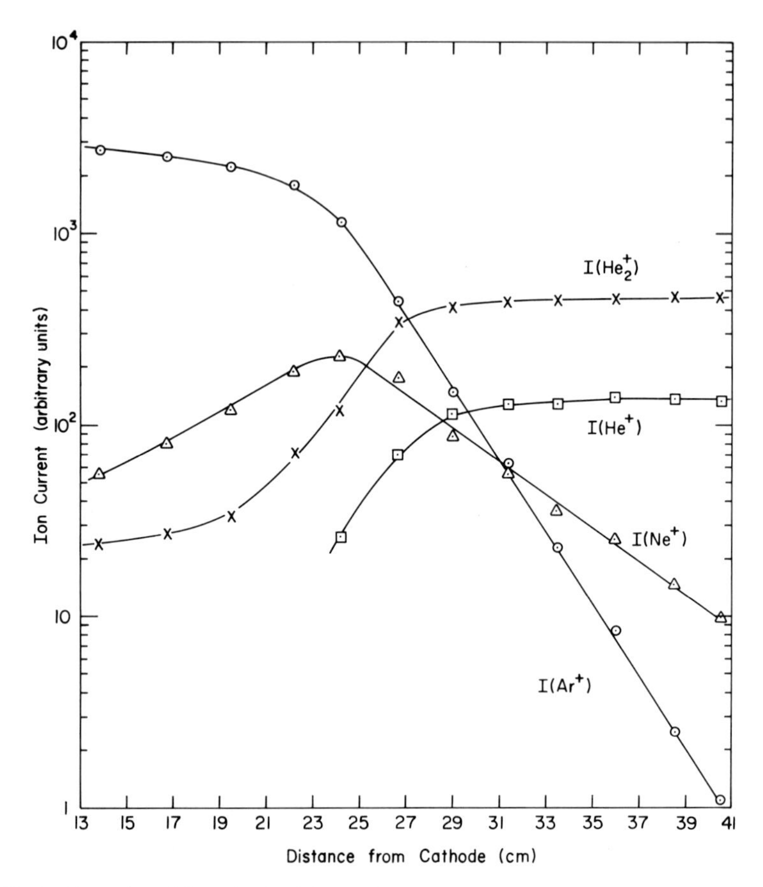

Fig. 5. Examples of data obtained for 0.05% argon in helium at 15 Torr and 25 mA (Gaur and Chanin, 1970).

It is interesting to note from Fig. 5 that at distances far from the cathode (>27 cm), where the $He_2{}^+$ ion density is independent of the axial position, the Ne^+ ion signals exhibit a normal cataphoresis behavior. However, when the $He_2{}^+$ ion current shows an abnormal decrease comparatively close to the cathode, the Ne^+ ion signals decrease as the cathode is approached. This result is consistent with the belief that the Ne^+ ion is produced by reaction (26).

Riesz and Dieke (1954) first reported results on the current and pressure dependencies of heavy noble-gas mixtures such as argon–krypton and krypton–xenon. In 1969 Bhattacharya reported results of studies to determine if commercially available krypton, which generally contains 25–100 ppm of xenon, could be cataphoretically purified. Ionic mass-spectrometric analyses were made using a quadrupole mass spectrometer. The plasma was produced in a cylindrical Pyrex glass tube by a high-frequency (110 MHz) pulse of 10 msec duration. The ions effused through a small hole of $\simeq 50\ \mu m$ diameter in a flange at one end of the tube into the quadrupole mass spectrometer. Kr^+, $Kr_2{}^+$, Xe^+, $Xe_2{}^+$, and $(XeKr)^+$ appeared as the major ion species. Measurements of the time dependencies of the number density of the ions in the afterglow indicated that $Kr_2{}^+$ ions were rapidly destroyed by the formation of Xe^+ ions, which in turn were quickly converted into $Xe_2{}^+$ ions. When the gas sample was subjected to a discharge (20–40 Torr, 40–60 mA) subsequent mass spectra clearly showed the absence of Xe^+, $Xe_2{}^+$, and $(XeKr)^+$, indicating that xenon impurities were removed by cataphoresis. The major production of Xe^+ and $Xe_2{}^+$ was believed to occur by the reactions

$$Kr_2{}^+ + Xe \rightarrow Xe^+ + 2Kr,$$

$$Xe^+ + Xe + \begin{Bmatrix} Kr \\ Xe \end{Bmatrix} \rightarrow Xe_2{}^+ + \begin{Bmatrix} Kr \\ Xe \end{Bmatrix},$$

$$(XeKr)^+ + Xe + Kr \rightarrow Xe_2{}^+ + 2Kr.$$

Hackam (1973) has recently studied cataphoresis for 2.2% argon in helium. The discharge was scanned spectroscopically to investigate the segregation dependence on gas pressure and discharge current. The ratio of argon to helium was found to increase with increasing distance from the anode. Furthermore, the quality of cataphoresis improves with increasing pressure and current in accordance with theoretical predictions. The effect of the temperature of the walls on the quality of cataphoresis was found to decrease with increasing temperature in agreement with theory and with Schmeltekopf's experimental study in helium–neon mixtures.

Hackam (private communication) also investigated a mixture of 0.1% argon in neon. The quality of cataphoresis in this mixture is found to improve with increasing gas pressure and the discharge current until a saturation is reached—a result in agreement with previous investigations. The degree of ionization of argon was found to increase with gas pressure for a fixed discharge current. A similar dependence was reported by Gaur and Chanin (1970) for 0.05% argon in helium.

B. Helium–Molecular-Gas Mixtures

When gas handling systems are used for gaseous electronics studies in noble gases, molecular components are commonly the major gas impurities. Many of the very early cataphoresis studies were performed on molecular gases; however, relatively little quantitative information is available concerning such mixtures. It is obvious that ion transport in molecular gases can be strongly influenced by effects such as dissociation and atom–atom recombination. Another effect that can significantly alter cataphoretic processes in such mixtures is gas clean-up at the cathode of the discharge. Removal of small concentrations of the readily ionized gas component can easily occur through gas burial in sputtered deposits in the cathode region of the discharge. For gas purification purposes this is desirable; for other types of studies, such effects could prove to be highly undesirable.

Figure 6 shows an example of the results obtained by Tombers *et al.* (1971) for the major ion signals in 0'05% N_2 in helium. The rapid decrease

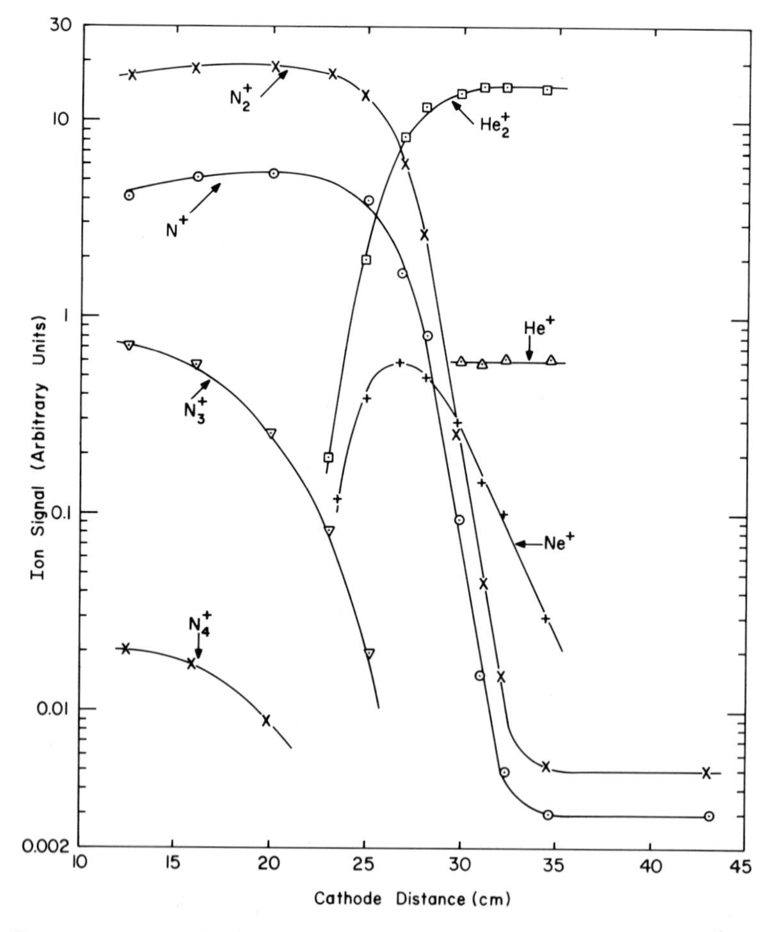

Fig. 6. The variation of the major ion signals as a function of distance from the cathode of 0.05% N_2 in helium. Measurements were made at 30 Torr, a discharge current of 15 mA, and using a shielded planar gold cathode (Tombers *et al.*, 1971).

of the He_2^+ signals close to the cathode, the sharp increase in the N_2^+ and N^+ signals, and the axial independence of N_2^+ and N^+ at large distances from the cathode shown in Fig. 6 were typical of the data obtained using a shielded gold cathode (Gaur and Chanin, 1968). This type of cathode minimized gas clean-up effects and enabled steady-state measurements to be made at low N_2 concentrations. The axial variations of the various nitrogen ions emphasizes the relative complexity of the ionization processes. The presence of Ne^+ is believed due to neon impurities present in the helium sample. The production processes for the major nitrogen ions, N_2^+, and N^+, are believed to be

$$He_2^+ + N_2 \rightarrow N_2^+ + 2He, \tag{27}$$

$$He^m + N_2 \rightarrow N_2^+ + He + e, \tag{28}$$

$$He^+ + N_2 \rightarrow N_2^+ + He, \tag{29}$$

$$He^+ + N_2 \rightarrow N^+ + N + He, \tag{30}$$

where He^m refers to metastable energy states. From Fig. 6 it is apparent that N_3^+ and N_4^+ are relatively important only close to the cathode. This result is understandable, since these ions are secondary ions, and hence formation will only be appreciable in those regions of the discharge where the neutral nitrogen concentration becomes relatively great. The formation processes for N_3^+ and N_4^+ are believed to be

$$N^+ + N_2 + He \rightarrow N_3^+ + He,$$

$$N_2^+ + N_2 + He \rightarrow N_4^+ + He.$$

The decrease of the impurity ion Ne^+ close to the cathode may be explained by the observation that He_2^+ also decreases in this region, since the probable formation reaction is $He_2^+ + Ne \rightarrow Ne^+ + 2He$. Ne^+ and He_2^+ decrease close to the cathode due to a decrease in the average electron energy as a consequence of the increased nitrogen concentration. Since the ionization potential of neon is greater than N_2, one can understand the increase of Ne^+ as the cathode is approached, before the signals of N^+ and N_2^+ increase. From a series of measurements such as shown in Fig. 6 results were obtained that indicated a normal cataphoretic dependence on gas pressure and discharge current, i.e., the slope of the axial dependence of ions such as N_2^+ increased with current and pressure before saturating at large values.

For common types of cathodes, gas clean-up effects were readily observed for small N_2 concentrations. Figure 7 shows an example of data relating to the time dependence of the N_2^+ signals for various types of cathodes. Data are given for both shielded and unshielded hollow molybdenum cathodes and also for gold and aluminum cathodes (Gaur and Chanin, 1968). In general the effect of gas clean-up decreases with increasing pressure due to

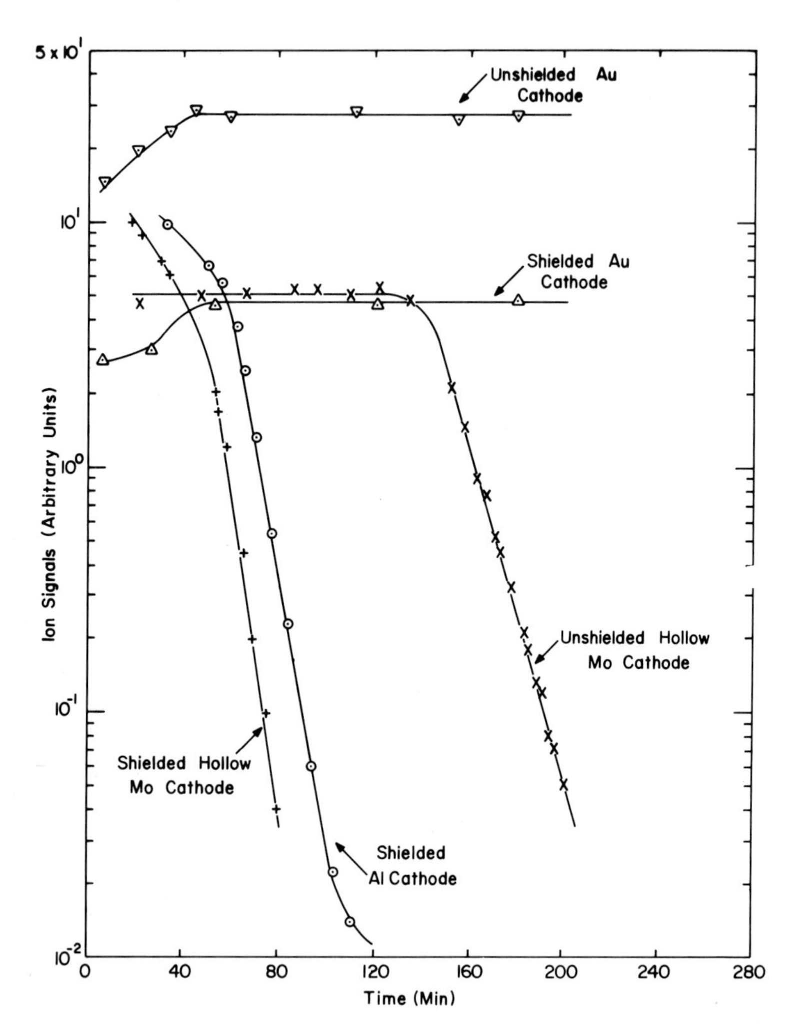

Fig. 7. Comparison of N_2^+ ion signals as a function of time following the initiation of the discharge for various types of cathodes. The aluminum and gold cathodes were planar cylindrical electrodes. Measurements were made with 0.28% nitrogen in helium, 16 Torr, 50 mA, sampling distance 25 cm (Tombers *et al.*, 1971).

the reduction in mean free path of the sputtered atoms. Gas removal due to clean-up is believed to be primarily associated with gas burial in sputtered cathode deposits. The net removal rate for a given type of cathode material will be determined not only by the sputtering rate of the material but also by the rate of desorption from the sputtered deposits. In general, materials with high sputtering yields tend to readily desorb buried gases. This accounts for the lack of clean-up observed at low pressures $\simeq 10$ Torr using gold cathodes and that clean-up was observed at $\simeq 20$ Torr using aluminum cathodes. For the application of cataphoresis to helium purification, gas removal in the cathode region greatly increases the overall efficiency. When removal is observed using the molybdenum and aluminum cathodes, evidence indicates that it is of a permanent nature. Under normal discharge operating conditions, extinguishing the discharge does not result in a return to the relative concentrations prior to the discharge. In this respect the separation of the gas constituents is considerably different from the effect observed in helium–neon or helium–argon mixtures, where extinguishing the discharge results in a return to the same relative concentrations before the discharge. Extrapolation of these results to gas mixtures containing other molecular gases is difficult. The results, however, suggest that due to gas-pumping effects, considerable care must be used in gas discharge studies, especially those involving small concentrations of a molecular gas.

Sanctorum (1975) has recently conducted mass analysis measurements of axial cataphoresis in mixtures of 1.00 and 0.20% nitrogen in neon. Great care was taken to avoid errors due to gas removal processes. The experimental results indicate that in a mixture of neon with high relative nitrogen percentages the normal cataphoretic segregation occurs. The dependence of the segregation on reduced gas pressure and discharge current is similar to the dependence in neon–noble-gas mixtures and appears to be consistent with theory.

Remer and Shair (1971) reported results for $He–N_2$, $He–O_2$, He–CO, and $He–CO_2$ mixtures. As indicated earlier, the experimental method is primarily a technique for qualitative studies rather than for quantitative investigations. Their results, however, indicate that cataphoretic segregation improves with increasing pressure and discharge current, and with decreasing initial composition of the minority component. This type of behavior is in accord with that observed using other experimental techniques.

C. Noble-Gas–Metal Vapor Mixtures

The first reports of cataphoresis in noble-gas–metal vapor mixtures were by Penning (1934). Recently, investigations have been conducted on mixtures containing mercury and also on helium–cadmium systems. Xenon–mercury results are discussed in Section V.

Measurements of cataphoresis using ion and radiation sampling methods for neon–mercury mixtures, have been reported by Tombers and Chanin (1973). Figure 8 shows an example of measurements made with traces of xenon present in neon–mercury mixtures. The mercury percentage was

0.01%; the xenon percentage was less accurately known but was estimated to be ≈0.01%. The axial variations of Hg^+ and Xe^+ have a dependence on discharge current comparable to that reported for noble-gas mixtures. It should be noted, however, that the axial independence for Hg^+ at 30 and 60 mA far from the cathode is somewhat unusual, and is analogous to that previously reported for He–N_2 mixtures (Tombers *et al.*, 1971). A similar effect was not observed for Xe^+. If the interpretation previously applied to account for this behavior in the He–N_2 mixtures is correct, one concludes that the Hg^+ variation at large cathode distances is in part determined by wall effects. From the axial dependence of Hg^+ such as shown in Fig. 8, one can calculate the average degree of ionization of mercury using the analysis described in Section II. For 10, 30, and 60 mA, $\bar{\phi} = 1.6 \times 10^{-3}$, 2.0×10^{-3}, and 2.8×10^{-3}, respectively. The general increase of $\bar{\phi}$ with

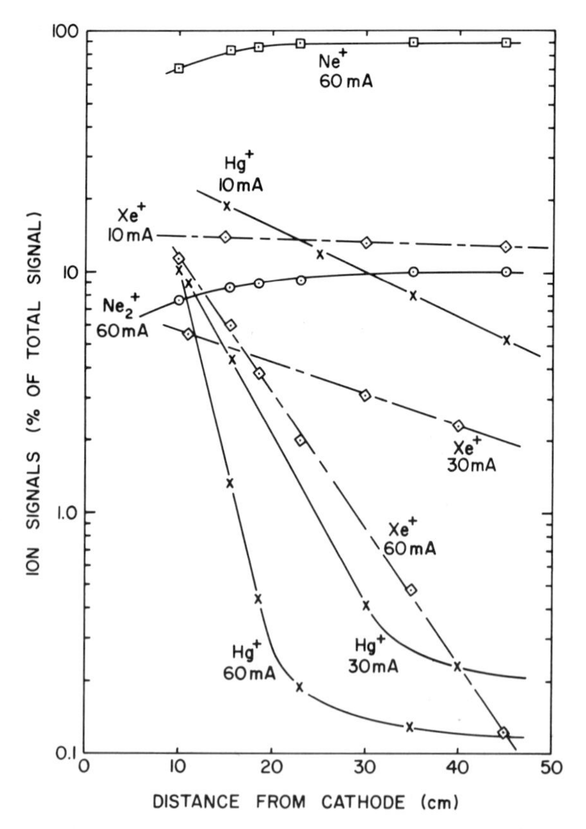

Fig. 8. Ion current signals at $p_0 = 2.0$ for 0.01% mercury and ≃0.01% xenon in neon. The signals for various discharge currents are shown vs. distance from the cathode. Measurements at 2 Torr (Tombers and Chanin, 1973).

current is similar to that previously reported for helium–neon and helium–argon mixtures. For comparable discharge conditions, the values of $\bar{\phi}$ for the neon–mercury mixtures are significantly greater than those for helium–argon (Gaur and Chanin, 1970). Since the magnitudes of $\bar{\phi}$ are a measure of the ionization efficiency, one concludes that very efficient ionization processes exist for mercury in these mixtures. Cumulative ionization and/or radiation trapping mechanisms may be responsible.

From Fig. 8 it is apparent that for a given discharge current the slope of the axial variation of Hg^+ is larger than for Xe^+. This result is probably due to the fact that Hg^+ can be produced by neon as well as by xenon particles. The xenon–mercury mixture is not a Penning mixture; ionization of mercury atoms, however, could occur through the charge exchange reaction $Xe^+ + Hg \rightarrow Hg^+ + Xe$. The larger values of the slopes of the axial variations of Hg^+ relative to Xe^+ indicate that the degrees of ionization are greater for mercury than xenon.

In 1969 Sosnowski reported results of cataphoretic studies for a helium–cadmium laser discharge tube. The objective of this investigation was to determine how the axial cadmium density distribution is affected by a discharge in a small-bore tube containing cadmium in helium. The discharge tube had an electrode at either end and a single cadmium source near the center. The tube was constructed of 4 mm bore pyrex, and utilized commercially available hollow cathodes as electrodes. Separate ovens were provided for the cadmium side arm and the discharge tube so that their temperatures could be individually controlled.

The side light intensity of the cadmium 4799 Å spectral line was measured as a function of distance along the discharge tube. The line intensity is proportional to the density of atoms in the upper excited state of the line. Assuming a Boltzmann distribution, the upper state of the line is proportional to the cadmium density. In measuring the side light as a function of distance along the tube, a monochromator was drawn atuomatically along the length of the tube and the output of the attached photomultiplier tube recorded directly on an $x - y$ chart recorder.

As the discharge tube was operated, deposits formed on the inside walls and altered the transmission constant from point to point along the length. The true relative cadmium intensity vs. length was obtained using calibration procedures. Due to preferential ionization, the discharge causes the cadmium vapor to flow toward the cathode, the flow being greater at high currents. At currents greater than 20 mA almost no cadmium is found a few centimeters from the source in the discharge region between it and the anode. In contrast to this, large amounts of cadmium are found in the discharge region between the source and the cathode. It is concluded that an approximately uniform density of metal vapor may be maintained in a helium–cadmium laser by placing a single cadmium source near the anode and passing currents greater than 50 mA through the tube. The experimental results were compared with a simplified model of the discharge using a theoretical analysis involving conservation equations similar to those used by Freudenthal (1967a) and Shair and Remer (1968). The assumptions involved the neglect of radial cataphoretic and electrophoretic effects. Moreover, it was assumed that the fractional ionization of cadmium was independent of z and that the axial electric field was constant. Measured values of the percentage ionization of cadmium varied from 0.02 to 0.7%, depending primarily on discharge current. For currents of 20 mA and greater, agreement between experimental and theoretical values in the source cathode region becomes relatively poor. The theory, however, as an aid to the qualitative understanding of cataphoretic processes, does explain the gross features of the experimental observations.

Investigators such as Springer and Barnes (1968) have also observed the effects of cataphoresis in cadmium–noble-gas discharges. In studies of the UV power radiated in low-pressure neon and argon discharges that contained small amounts of cadmium, these authors reported significant differences in the power radiated between ac and dc operation, a result that was attributed to cataphoretic segregation.

D. Radial Effects

In 1963 Cayless reported results of a theoretical study of the positive column of mercury–noble-gas discharges. A theory of the uniform positive column was developed for dc fluorescent lamp-type discharges in mercury and noble-gas mixtures. The theory involved considerations of multistage excitation and ionization processes, available cross-sectional data, and the imprisonment of the resonance radiation. Distributions of the ions, unexcited and excited atoms, and the temperature throughout the discharge were calculated and used to evaluate discharge characteristics, using digital computer methods. The general form of the theory applied to the radial distributions of unexcited mercury atom densities predicted appreciable depletion at high discharge currents. Mercury depletion is believed to arise partly from the temperature increase and primarily from the ionic pumping process. The comparatively rapid ambipolar diffusion of ions to the walls produces a concentration gradient of neutral mercury atoms diffusing back to the center. The result is a general broadening of the discharge as the current is increased. The majority of the mercury excitation eventually occurs near the walls since the effect is more pronounced at high noble-gas and low mercury pressures.

Radial distributions for argon–cesium mixtures have been extensively studied in recent years (Bleekrode and van der Large, 1969; Waszink and Polman, 1969). Bleekrode and van der Laarse have investigated, using absorption spectroscopy, the radial and axial distributions of cesium ground-state ($6\ ^2S_{1/2}$) atoms for low-pressure argon–cesium discharges. A discharge tube (length 55 cm, diameter 3 cm) filled with 5 Torr argon was provided with vacuum chambers at both ends of the tube so that the discharge could be viewed along the axis. Light from a 100 W dc mercury discharge lamp was passed through a system of lenses and diaphragms to obtain a narrow beam. This beam had a diameter of about 0.4 cm over the

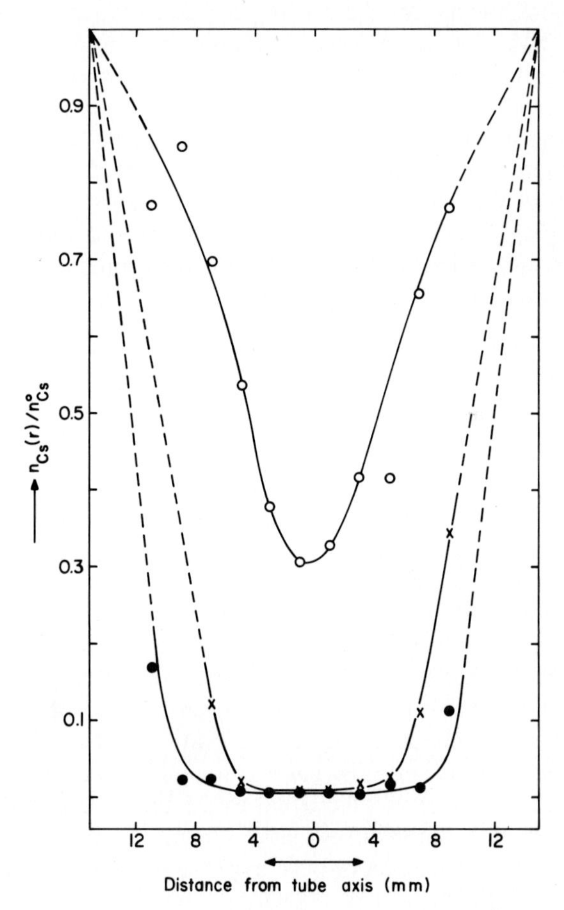

Fig. 9. Measured radial concentration distributions of cesium ($6^2S_{1/2}$) in a cesium–argon low-pressure dc discharge at 378°K. ○, 0.500 A; ×, 0.85 A; ●, 0.95 A (Bleekrode and van der Laarse, 1969).

length of the tube. The shape of the concentration profile could be measured over a distance of 1 cm on both sides of the tube axis. Measurements were made at 0.2 cm intervals.

Concentration profiles of ground-state cesium, indicated by $n_{Cs}(r)$, were determined at various values of the wall temperature, that is, for various values of n_{Cs}. In Fig. 9 a number of profiles are shown measured at 378°K. The results clearly demonstrate the increasing effect of the ground-state depletion with increasing current density. This is believed to be caused by the relatively rapid ambipolar diffusion of cesium ions to the wall where recombination takes place, and the slow back-diffusion of neutral atoms into the discharge. Measurements were also made to determine the cesium atom distribution along the axis of the discharge. Within experimental accuracy, the cesium ($6\,^2S_{1/2}$) concentration was found to be constant over the whole length of the positive column. The observed radial distributions were in part attributed to ionic pumping effects, i.e., radial cataphoresis.

Recently van Tongeren (1974) has developed a model for the positive column of low-pressure dc discharges in argon–cesium mixtures. The model includes radial depletion of cesium ground-state atoms and also involves radiation trapping. Experimental results obtained over a wide range of discharge currents and cesium vapor densities are in reasonable agreement with the theoretical predictions.

E. Time Dependence

Unfortunately, few measurements of the time dependence have been made under sufficiently well controlled conditions to permit a reliable comparison with theoretical predictions. Matveeva (1959) was one of the first investigators to measure the variation in the concentration of the readily ionized constituent following the initiation of the discharge. For 9% argon in helium measurements were also made of the variation following the cessation of the discharge. According to Shair and Remer (1968) whose theoretical analysis considered transient cataphoresis and incorporated the effect of the influence of electrode end bulbs, to obtain agreement between Matveeva's results and theory it was necessary to assume an effective volume for each electrode volume equal to 2.5 times that of the reported value. It is apparent that more investigations involving comparisons between experiment and theory are necessary to improve the understanding of transient cataphoretic effects.

F. Alternating Current Cataphoresis

Alternating current effects were first noted by Kenty (1967) in 1932 when measurements were made on traces of argon in 2 Torr neon. It was observed that there was a marked segregation of the argon toward the middle of the tube. Kenty (1958) later reported the observation of gas segregation effects when ac excitation is applied to mercury–noble-gas mixtures. According to Kenty there can be a segregation of the trace gas of lower ionization potential either toward the center of the tube, as in argon–mercury, neon–mercury, and to some extent krypton–mercury, or to the ends, as in helium–mercury. With xenon–mercury the effect was observed to go either way, depending on the current.

Figure 10 shows an example of initial measurements of ac cataphoresis effects obtained by G. E. Sery and L. M. Chanin (private communication). Fiber optics were used to scan the length of a 70 cm, 2.22 cm diameter discharge tube; data accumulation involved using a multichannel analyzer. Data shown in Fig. 10 are for a 0.1% argon in neon mixture at 5 Torr, 40 Hz, and 20 mA current. The upper curve in Fig. 10 shows the ratio of the argon 4200 Å line intensity to the neon 5852 Å line intensity. It is evident that an enhancement occurs close to the electrodes and in general is slightly greater in the region between the electrodes. A detailed understanding of

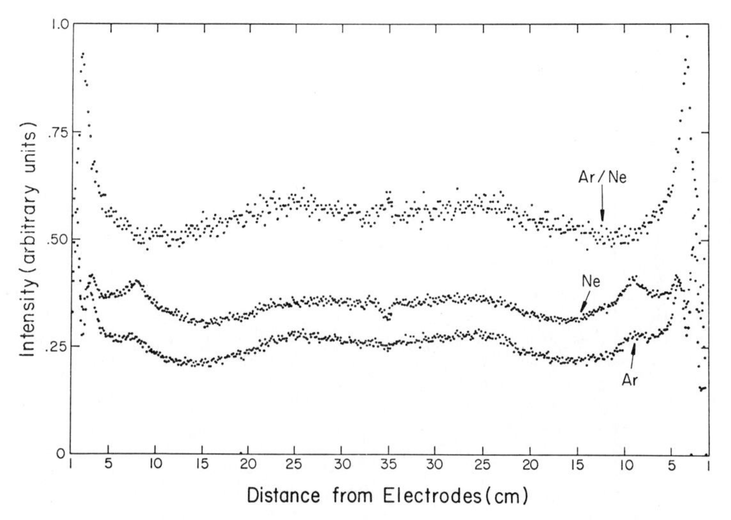

Fig. 10. The ac cataphoresis observed in 0.1% argon in neon, at 40 Hz, 5 Torr, and 20 mA (G. E. Sery and L. M. Channin, private communication).

this behavior is not available at the present time. Clearly, additional quantitative results are needed to achieve better understanding of ac cataphoresis phenomena.

V. RETROGRADE CATAPHORESIS

In 1958 Kenty reported measurements of the rate of transport of mercury in the positive columns of mercury–noble-gas mixtures. Since the ionization potential of mercury is less than those of the noble gases, mercury should be preferentially ionized in these binary mixtures and hence carried in the cathode direction. For dc discharges in xenon–mercury at currents >1 A,

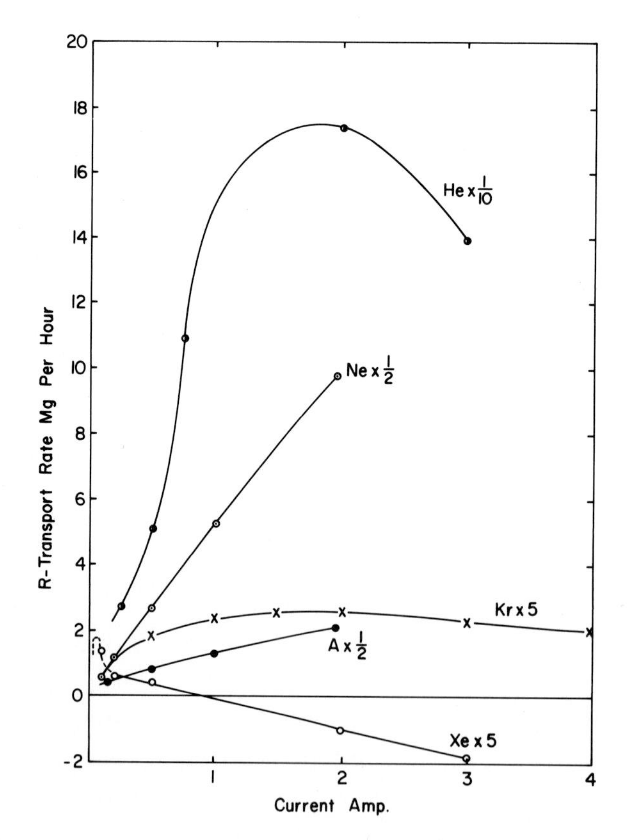

Fig. 11. The rate of liquid mercury transfer in five gases as a function of current (note scale changes on curves). Gas pressure 3 Torr (Kenty, 1968).

Kenty observed that the mercury was carried toward the anode rather than the cathode end of the discharge tube. This effect was termed retrograde cataphoresis (Loeb, 1958).

Values of the rates of cataphoretic transfer measured for a 3.4 cm diameter discharge tube are shown in Fig. 11. Most significant is the transfer of mercury in xenon toward the anode above 1 A. According to Kenty, this retrograde effect is due to selective pressure of the electrons on the neutral mercury in their flow toward the anode. This selective pressure is a consequence of (a) the large elastic collision cross section of mercury for electrons, (b) the small cross section of xenon for electrons of about 1 eV energy (Ramsauer effect), and (c) the fact that above 1 A, more and more of the ions of the discharge are xenon ions, as demonstrated by the progressive appearance of xenon lines in the discharge, as the current is increased. From Fig. 11 it is clear that krypton–mercury shows some of the effects present in xenon–mercury. It seems possible that at still higher currents the flow of mercury in krypton–mercury might also be retrograde. The behavior of helium–mercury at high currents was considered to be possibly due to radial cataphoretic effects.

R. B. Tombers and L. M. Chanin (private communication) have studied cataphoretic effects for xenon–mercury mixtures. Ion and radiation sampling methods were used to measure the axial variations of the mercury ions and the mercury line at 5461 Å. Figure 12 shows a comparison of some results obtained in xenon–mercury mixtures with data obtained earlier from neon–mercury mixtures using the same apparatus (Tombers and Chanin, 1973). It is apparent for neon–mercury that the axial variations for Hg^+ and Hg* (5461 Å) are almost identical. Measurements using other lines such as Hg* (4385 Å) were in excellent agreement with the (5461 Å) data. The neon–mercury data show the normal exponential decrease with distance from the cathode typical of small admixture concentrations. In contrast to the neon–mercury data, the xenon–mercury results indicate a very slight decrease in signal amplitude with distance from the cathode. This dependence is distinctly different from the exponential behavior observed for neon–mercury mixtures, which is usually observed for small concentrations of the readily ionized component in other gas mixtures. A series of measurements were made of the changes that occurred in the axial variation of the Hg* (5461 Å) intensities in xenon–mercury following the initiation of a discharge in a baked discharge tube. The comparatively slow changes in the variations that occurred over a period of several days may possibly be associated with surface migration of the mercury from its source behind the cathode.

The above results seem to be consistent with Kenty's (1967) earlier observations of an abnormal transport of mercury in xenon–mercury mixtures. A direct comparison of results with Kenty's is not very meaningful

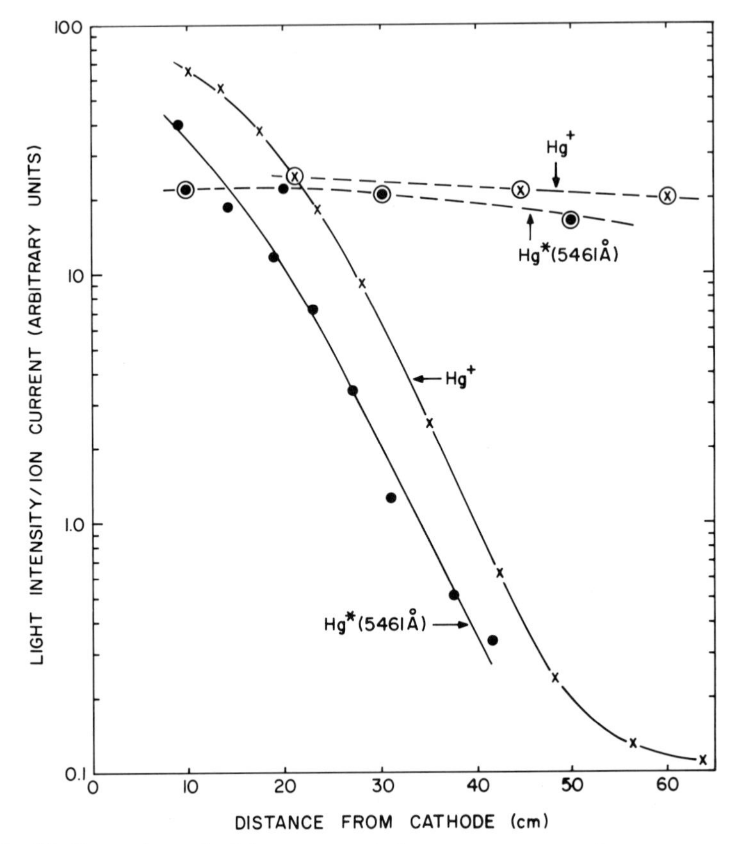

Fig. 12. Comparison of axial measurements in xenon–mercury and neon–mercury mixtures. Measurements at 2.0 Torr, 10 mA; —, 0.01% neon in mercury; ---, 0.017 xenon in mercury (R. B. Tombers and L. M. Chanin, private communication).

since the rate of transport of mercury was not measured in Tombers and Chanin's investigations. Moreover, the current densities used were much smaller. The results of both investigations, however, indicate complex transport phenomena in discharges created in xenon–mercury mixtures. The effect may be most pronounced for xenon due to its low thermal conductivity and the consequent tendency of the discharge to become constricted. Side radiation measurements showed some degree of constriction even at the low pressures investigated (≤ 10 Torr).

R. S. Bergman (private communication) has shown that if a simple model is used for the discharge, one may account for Kenty's observed rate of mercury transport in xenon. The model is based on Kenty's explanation that the anode-directed flow of mercury in a high-current xenon–mercury discharge may be related to a preferential pressure of the electrons on the neutral mercury. Bergman's analysis involves using the momentum conservation equations for Xe, Hg, Xe^+, Hg^+, and the electrons as well as several simplifying assumptions such as the dominance of Xe^+ relative to Hg^+, the assumption that the mercury concentration is axially independent, and the neglect of other effects such as those due to electrophoretic forces.

It is apparent that the term "retrograde cataphoresis," which was used to refer to an abnormal rate of transport of mercury, is somewhat of a misnomer in the sense that it may refer to a preferential electron pressure on neutral gases with subsequently differing rates of transport rather than preferential ionization and subsequent segregation processes. Finally it should be noted that in addition to complications of "normal" cataphoretic processes arising from electrophoresis, evidence suggests that for certain mixtures that tend to be constricted, surface diffusion effects may also be of importance.

VI. APPLICATIONS

One of the earliest reports on the applications of cataphoresis involved using the effect to prevent magnesium attack of glass windows that were used in magnesium radiation sources (Druyvesteyn, 1935). The most common recent application of cataphoresis has been its use in noble-gas purification methods. Within the past few years, cataphoresis has also been used in noble-gas–metal vapor lasers as a method of ensuring an appropriate lasing mixture. These applications are considered in the following sections.

A. Gas Purification

The purification of gas samples using cataphoresis was first considered by Emeleus and Sayers (1952) and later by Riesz and Dieke (1954). As a result of these investigations and others by Mittlestadt and Oskam (1961), Schmeltekopf (1962). Freudenthal (1967a), and Shair and Remer (1968) the technique of removal of minor concentrations of impurities from noble-gas samples using cataphoretic segregation has become routine in many laboratories.

The method, which was primarily developed by Riesz and Dieke (1954), consists in operating a glow discharge in a long cylindrical tube connected to a volume containing the gas to be purified. Through cataphoretic segregation the minor gas component will be enhanced in the cathode region of the discharge if its ionization probability is larger than the major gas constituent. The gas sample at the anode end of the discharge will consequently be comparatively free of the admixture constituent. Riesz and Dieke have also noted that the same procedure used for noble-gas purification can also be used to increase the sensitivity of spectroscopic analysis of the gas. This is achieved by using cataphoresis to enhance the "normal" concentration of the minor gas component in the cathode region. By using calibration procedures of spectroscopic analysis on samples of known proportions, Riesz and Dieke indicate that for mixtures of neon in helium, one part of neon in 2×10^8 parts of helium could easily be detected.

Schmeltekopf (1962) investigated the functional dependence of cataphoresis on various discharge parameters for helium–neon mixtures. He concluded that cataphoresis improved with increasing pressure except at high pressures, >40 Torr, where the quality of cataphoresis improved only slowly with pressure. The term "quality" of cataphoresis was defined as the distance between the anode and the point at which the neon radiation was visible. The slope of the neon concentration was observed to increase with current at low currents up to $\simeq 20$ mA, and then became independent of current. The region from 20 to 100 mA where the slope was essentially current independent was contrary to the predictions of Druyvesteyn's theory. In this region Schmeltekopf suggested an empirical formula for estimating the effectiveness of cataphoresis for 1% neon in helium:

$$n(z) = n(0) - 3 \times 10^{18}(pz/T_g^{\,2}) - 4 \times 10^{18}(I/R^2).$$

The cataphoretic efficiency was observed to decrease with increasing temperature. An increase in tube diameter and a decrease in the length of the discharge tube also decreased the effectiveness of cataphoresis. Schmeltekopf concluded that cataphoresis could also be effective in flowing systems (3×10^4 cm^3/h).

The removal of traces of argon from neon by cataphoretic segregation was studied by Miller (1963). The sensitivity of the breakdown potentials in neon to traces of argon was used as an indicator of the degree of purity of the neon. In this investigation the limit of sensitivity for argon was 1:10 (Penning, 1934). For 125 mA discharges in 1.6 cm O.D. and 2 m long tubes, the original argon concentrations of 50 ppm and $2 \pm 1\%$ of argon in neon were reduced to ≤ 0.1 and ≤ 1 ppm, respectively. These results may be compared with Schmeltekopf's result that for 0.5% neon in helium, the resultant neon concentration at the anode was $<0.06\%$. Matveeva (1959) reported that for 3% argon in helium the concentration was reduced to $10^{-3}\%$ one hour later.

Matveeva, Freudenthal, and Shair and Remer have also discussed the dependence of steady-state cataphoresis on various discharge parameters. For gas purification purposes the tube current density J is quite important, since the segregation increases with increasing values of J. However, the temperature of the wall of the discharge tube is dependent on the current density and the segregation decreases with increasing temperature. Freudenthal has indicated that the maximum value of J is given by $J = C''/ER$, where C'' is a constant relating to the heat current density at the wall of the discharge tube.

To obtain maximum segregation, Freudenthal (1967a) suggested that, for a given gas mixture, an optimum value exists for Rp, the product of the tube radius and gas pressure. This value must be determined experimentally for a given gas mixture. For neon–argon mixtures the optimum Rp value is believed to be $\gtrsim 3$ Torr cm. Clearly, the product Rp will influence basic ionic production and loss processes that are in part determined by parameters such as the electron temperature. For a given gas mixture and a particular value of Rp for those mixtures such as helium–argon where the ionization of neon increases with pressure, it is advisable to use a small value for R and hence large p. The tube length is also an important parameter, since the segregation increases exponentially with the length of the discharge tube. Typically, cataphoresis tubes are coiled to minimize their overall length when used on gas handling systems. Clearly, the length of the tube will in part be determined by the power supply available since the discharge voltage increases with length. By using an extra volume attached to the cathode side of the discharge tube containing neon with small traces of argon for 6.76 Torr and a 0.50 A discharge, Freudenthal reported a reduction in the argon concentration on the anode side by a factor of 4000.

Riesz and Dieke (1954) noted that the time required to reach equilibrium was dependent on the anode volume. According to Shair and Remer (1968) a large end bulb at the cathode will have the effect of decreasing the steady-state concentration of the admixture near the anode region. Matveeva's measurements have shown that the time required to achieve steady-state conditions increases linearly with an increase in the impurity concentration. This may be a consequence of the possible increase in the percentage ionization of the impurity with decreasing concentration of this component.

Extrapolation of the results obtained by Tombers *et al.* (1971) for Ne–N_2 mixtures to other mixtures containing molecular gases as minor gas constituents is difficult. However, to enhance the removal of a molecular gas impurity that may occur by burial in sputtered cathode deposits, care should be taken to use a cathode material with a low desorption rate of the trapped gas. Since desorption rates are observed to decrease with decreasing temperature, the cathode design should allow relatively low-temperature cathode operation. The gas pressure should be restricted to low or intermediate pressures, so that gas-sputtering effects are not greatly reduced. Alternatively, operation at higher pressures would not prove detrimental, provided the cathode had a shield or was constructed so that it was located close to the walls of the discharge tube.

Gaur and Chanin (1970) have used the parameter $E\phi/T_g$ to discuss cataphoresis. This parameter determines the minor gas concentration at any point z in the positive column. Here $\phi = \langle n_{pa}\rangle/\langle n_a\rangle$. It should be observed that this quantity is analogous to the electrostatic Peclet number α used by Remer and Shair (1971) in discussing cataphoresis, where α is defined by

$$\alpha = \frac{\langle n_{pa}\rangle}{\langle n_a\rangle + \langle n_{pa}\rangle} \frac{\mu EL}{D}.$$

Muller and Tubbs (1963) have used cataphoretic segregation effects in neon–mercury discharges to estimate various discharge parameters. The method is based on the result that for these mixtures a sharp line of demarcation is observed between the region that primarily contains only the noble gas and the region containing the gas mixture. These authors have shown that while the boundary depends in a complex manner on basic collision processes, it can be used in a simple way to estimate parameters such as electron densities and electron and ion mobilities. The method is not applicable to mixtures of metal vapors and the heavier noble gases, where the ion current involves both gas components.

B. Gas Lasers

Recently laser transitions have been discovered in a number of metal vapors (Fowles and Hopkins, 1962; Silfast, 1968, 1969; Silfast and Szeto, 1970). While these transitions were interesting, the practical problem of maintaining a uniform vapor distribution in the discharge proved difficult until the development of a cataphoretic pumping scheme, which was independently developed by Sosnowski (1969), Goldsborough (1969), and Fendley *et al.* (1969). Sosnowski showed that when a single metal source such as cadmium was used in the bore region, a uniform distribution of cadmium was obtained between the source and the cathode when the discharge current was higher than a certain critical value (depending on the bore size). This uniformity is maintained only as long as the bore diameter remains constant and the bore temperature does not drop below the value where cadmium condensation occurs. The critical current in a 4 mm bore discharge in a helium–cadmium system is approximately 50 mA. At lower discharge currents it is necessary to add more sources along the discharge region to keep a more uniform density.

In using this method the bore region must be maintained at a higher temperature than the metal source to prevent the metal from condensing on the walls. This is usually achieved as a result of the heating effect of the discharge in the bore region. As the metal is cataphoretically pumped from the source to the cathode region it will condense along the tube walls near the end of the furnace as soon as a temperature is reached where the saturated vapor pressure is equal to the effective pressure of the metal in the bore. Thus it is necessary to expand the bore just before the condensing region so that the condensed metal will not obstruct the bore. As noted in Section IV,C, Sosnowski measured the side radiation from discharges formed in helium–cadmium mixtures. Such measurements enabled estimates of the cadmium density distributions within the tube. A simple theory based on cataphoretic considerations gave a qualitative explanation of the main features of the measured density distribution curves.

The cataphoresis effect in helium–cadmium vapor lasers has also been investigated by Mash *et al.* (1971). These authors used a scanning interferometer and a helium–cadmium dc discharge tube to obtain estimates of the velocity of Cd^+ and also the Cd^+ concentration.

Hernquist (1972) has described the construction and performance of helium–cadmium lasers that have a diffusion return path for cadmium recirculation back from the cathode to the anode end of the laser tube. The geometry, which utilizes Kovar honeycomb baffles in the diffusion return path to prevent electrical breakdown, is reminiscent of the use of bypass tubes in gas lasers. The recirculation geometry permits long-time operation and also results in low radiation noise.

It is apparent from the above discussion that cataphoretic effects can play a very significant role in the operation of certain types of gas lasers. Clearly, in certain instances they could also prove to be detrimental, for example, if the axial variation of the admixture were sufficiently pronounced so as to result in appreciable differences in excitation processes along the axis of the tube. In practice, the use of tubes that bypass the bore of the laser and connect the anode and cathode volumes tends to minimize not only electrophoretic but also cataphoretic effects (Gordon and Labuda, 1964).

Garscadden (1966) has reported on measurements of some of the consequences of cataphoresis in a helium–neon laser including its influence on moving striations and associated "noise" occurring in the discharge. Evidence of the influence of cataphoresis on moving striations was obtained from measurements of the noise frequency spectrum of the discharge side light as a function of time at various positions along the discharge tube.

Tombers and Chanin (1970) have reported measurements of the reduction of cataphoretic effects due to a bypass tube for a 10% mixture of neon in helium. Figure 13 shows examples of spectral measurements obtained at 26 Torr, for 50 and 100 mA discharges with the bypass open and closed. Data were obtained by sampling the radiation from the discharge at various distances from the cathode using a scanning spectrometer. Comparison of these curves shows the typical cataphoretic behavior, i.e., the efficiency increases with discharge current. The dashed vertical lines indicate the relative positions of the pressure sampling leads, or bypass connections. Sharp breaks were drawn in the curves for the bypass open, at the positions of the bypass connections. The effect of the bypass will be to increase the gas mixing and thus reduce the cataphoretic dependence. However, to the left of the bypass 11 cm from the cathode, the bypass is ineffective, and hence the neon concentration will increase as indicated. A similar argument holds for the region between the anode and the second connection, 61 cm from the cathode. The data shown in Fig. 13 clearly show the reduction in the cataphoresis effect due to the use of a bypass tube.

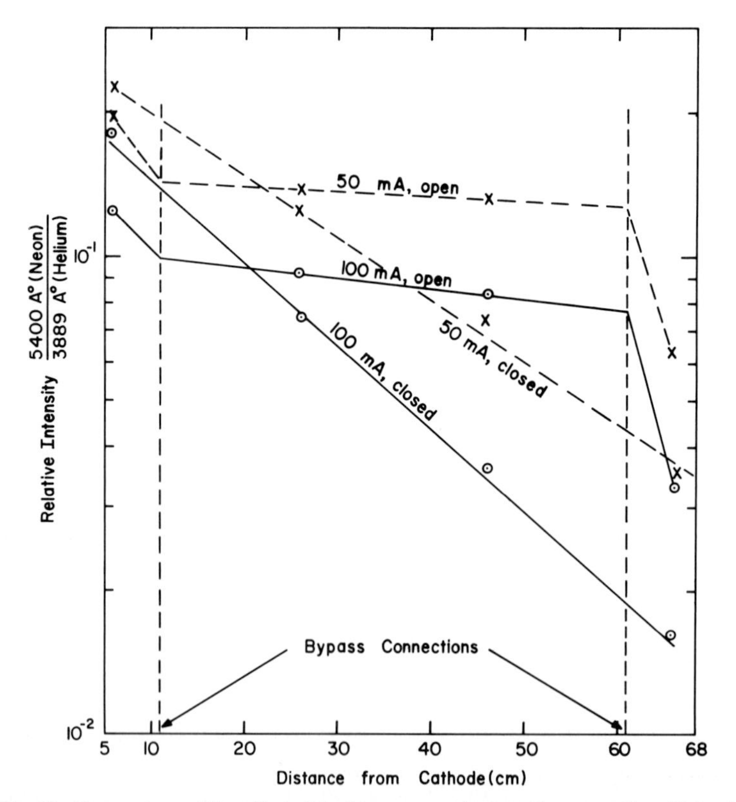

Fig. 13. Comparison of the effect of the bypass on cataphoretic segregation. Data obtained for 10% neon in helium at 26 Torr for 50 and 100 mA discharges. The dashed vertical lines indicate the positions of the bypass and pressure sampling leads (Tombers and Chanin, 1970).

Axial cataphoretic effects in continuously pulsed copper halide lasers have been observed to rapidly deteriorate the laser performance (Liu *et al.*, 1977). Dissociated copper ions are cataphoretically pumped toward the cathode by the applied electric field, causing nonuniform laser discharges and preferential copper condensation near the cathode. The problem of axial cataphoretic density gradients can be eliminated by altering the polarity of the applied voltage either mechanically or electronically on a time scale comparable to or faster than the characteristic cataphoresis time (Freudenthal, 1967a; Shair and Remer, 1968). This technique extends laser performance and operating lifetime.

The use of copper and other metal vapor halides for use as a lasant has received considerable attention within the past several years (Chen and Russell, 1975; Akirtava *et al.*, 1975; Subotinov *et al.*, 1975). Cataphoretic distributions in metal-halide–helium discharges have been measured recently using optical techniques to determine the longitudinal distribution of the sidelight emitted from the positive column (Cross *et al.*, 1976; Cross and Funk, 1977). Studies of $ZnCl_2$, $SnCl_2$, and $AlCl_3$ have shown that considerable concentrations of the metal atoms and ions can be obtained when the temperature of the bore is far too low to allow the use of the parent metal itself. These investigators concluded that when small diameter discharge tubes are used, a careful appraisal of the temperature of the bore must be made since, contrary to what might be expected, the cataphoresis is less effective at higher temperatures (Cross *et al.*, 1976).

C. Isotopic Enhancement

Isotope separation in the axial direction of a steady-state discharge was first reported many years ago (Beckey *et al.*, 1953: Groth and Harteck, 1939; Beckey and Dreeskamp, 1954; Beckey and Warneck, 1955). For a H_2–D_2 gas mixture, deuterium enrichment was reported at the cathode end of the discharge. This result may be understood since the masses of the diffusing neutral molecules and the atomic ions ($p \leq 1$ Torr) significantly differ. Moreover, since the thermal velocities of the atomic ions differ by a factor of $\sqrt{2}$, mass-dependent ion recombination may also contribute to the enhancement.

Freudenthal (1966) has examined the separation of ^{36}Ar and ^{40}Ar isotopes in a neon–argon mixture. No separation of the isotopes was observed in the axial direction of the discharge tube when the discharge was operated under steady-state conditions. Under non-steady-state conditions, however, a small separation effect was detected, which was reported to be in agreement with the predicted behavior.

Rogoff (1976) has noted that ambipolar diffusion can be an effective pumping mechanism for a selectively ionized isotopic species in a low-pressure gas mixture. For sufficient ion and electron densities, ambipolar space-charge fields can result in pumping of the species away from the ionization region. The ionization and subsequent radial cataphoresis can result in considerable depletion of the isotope near the axis and enrichment near the wall. Calculations have been made for atomic uranium of natural relative isotopic abundance using argon as a buffer gas (Rogoff, 1976, 1977). Assuming an external radiation source produces selective photoexcitation of ^{235}U leading to ionization, calculations indicate substantial depletion of ^{235}U near the axis with a corresponding increase in density near the walls.

REFERENCES

Akirtava, O. S., Dzhikiya, V. L., and Oleinik, Yu. M. (1975). *Sov. J. Quant. Electron.* **5,** 1001.

Baly, E. C. C. (1893). *Phil. Mag.* **35,** 200.

Battacharya, A. K. (1969). *Appl. Phys. Lett.* **15,** 362.

Beckey, H. D., and Dreeskamp, H. (1954). *Z. Naturf.* **8a,** 735.

Beckey, H. D., and Warneck, P. (1955). *Z. Naturf.* **10a,** 62.

Beckey, H. D., Groth, W. E., and Welge, K. H. (1953). *Z. Naturf.* **8a,** 556.

Bleekrode, R., and Laarse, J. W. v. d. (1969). *J. Appl. Phys.* **40,** 2401.

Cayless, M. A. (1963). *Brit. J. Appl. Phys.* **14,** 863.

Chen, C. J., and Russell, G. R. (1975). *Appl. Phys. Lett.* **26,** 505.

Cross, L. A., and Funk, L. L. (1977). *J. Appl. Phys.* **48,** 94.

Cross, L. A., Willis, R. K., Funk, L. L., and Gokay, M. C. (1976). *J. Appl. Phys.* **47,** 2395.

Druyvesteyn, M. J. (1935). *Physica* **2,** 255.

Emeléus, K. G., and Sayers, J. (1952). *Proc. Phys. Soc. Lond.* **A,** 219.

Fendley, J. R., Jr., Goroy, I., Hernquist, K. G., and Sun, C. (1969). *RCA Rev.* **30,** 422.

Fowles, G. R., and Hopkins, B. D. (1962). *IEEE J. Quant. Electron.* **3,** 319.

Freudenthal, J. (1966). Thesis, Utrecht.

Freudenthal, J. (1967a). *J. Appl. Phys.* **38,** 4818.

Freudenthal, J. (1967b). *Physica* **36,** 354.

Garscadden, A. (1966). *Appl. Phys. Lett.* **8,** 85.

Gaur, J. P., and Chanin, L. M. (1968). *Rev. Sci. Instrum.* **39,** 1948.

Gaur, J. P., and Chanin, L. M. (1969). *J. Appl. Phys.* **40,** 256.

Gaur, J. P., and Chanin. L. M. (1970). *J. Appl. Phys.* **41,** 106.

Goldsborough, J. P. (1969). *Appl. Phys. Lett.* **15,** 159.

Gordon, E. I., and Labuda, E. F. (1964). *Bell Syst. Tech. J.* **43,** 827.

Groth, W., and Harteck, F. (1939). *Naturwissenschaften* **27,** 390.

Hackam, R. (1973). *J. Appl. Phys.* **44,** 3113.

Hernquist, K. G. (1972). *IEEE J. Quant. Elect.* **QE-8,** 740.

Kenty, C. (1958). *Bull. Am. Phys. Soc.* **3,** 82.

Kenty, C. (1967). *J. Appl. Phys.* **38,** 4517.

Kuehn, D. G., and Chanin, L. M. (1972). *J. Appl. Phys.* **43,** 339.

Lehmann, O. (1898). *Elek. Lichterscheinungen*, p. 265.

Liu, C. S., Feldman, D. W., Pack, J. L., and Weaver, L. A. (1977). *J. Appl. Phys.* **48,** 194.

Loeb, L. B. (1958). *J. Appl. Phys.* **29,** 1369.

Mash, L. D., Rabkin, B. M., and Rybakov, B. V. (1971). *ZhETF Pis'ma Red.* **13,** (5) 240.
Matveeva, N. A. (1959). *Bull. Acad. Sci. USSR, Phys. Ser.* (English Transl.) **23,** 1009.
Mierdel, G. "Handbuch der Experimental Physik" Vol. XIII, 3, 519. Academische Verlagesellchaft, Leipzig, Germany, 1929.
Miller, H. C. (1963). *J. Appl. Phys.* **35,** 1745.
Mittelstadt, V. R., and Oskam, H. J. (1961). *Rev. Sci. Instr.* **32,** 1408.
Muller, W., and Tubbs, E. F. (1963). *J. Appl. Phys.* **34,** 969.
Oskam, H. J. (1957). Dissertation, University of Utrecht.
Oskam, H. J. (1963). *J. Appl. Phys.* **34,** 711.
Oskam, H. J. (1968). *Physica* **44,** 209.
Pahl, M., and Weimer, U. (1957). *Naturwissenchaften* **44,** 487.
Pekar, Yu A. (1967a). *Sov. Phys.* **11,** 1024.
Pekar, Yu A. (1967b). *Sov. Phys.* **12,** 800.
Penning, F. M. (1934). *Physica* **1,** 763.
Remer, D. J., and Shair, F. H. (1969). *Rev. Sci. Instr.* **40,** 968.
Remer, D. J., and Shair, F. H. (1971). *Chem. Eng. Prog. Symp. Ser.* (112) **67,** 60.
Riesz, R., and Dieke, G. (1954). *J. Appl. Phys.* **25,** 196.
Rogoff, G. L. (1976). *Bull. Am. Phys. Soc.* **21,** 173.
Rogoff, G. L. (1972). *U.S.–Japan Joint Seminar on the Glow Discharge and Its Fundamental Processes, Boulder, Colorado, July 1977.*
Sanctorum, C. (1975). *11th Int. Conf. on Phenomena in Ionized Gases, Eindhoven*, p. 70.
Schmeltekopf, A. L., Jr. (1962). Ph.D. Thesis University of Texas.
Schmeltekopf, A. L. Jr. (1964). *J. Appl. Phys.* **35,** 1712.
Shair, F. H., and Remer, D. S. (1968). *J. Appl. Phys.* **39,** 5762.
Silfast, W. T. (1968). *Appl. Phys. Lett.* **13,** 169.
Silfast, W. T. (1969). *Appl. Phys. Lett.* **15,** 23.
Silfast, W. T., and Szeto, L. H. (1970). *Appl. Optics* **9,** 1484.
Skaupy, F. (1916). *Verh. Deut. Phys. Ges.* **18,** 230.
Skaupy, F., and Bobek, F. (1925). *Z. Tech. Phys.* **6,** 284.
Sosnowski, J. P. (1969). *J. Appl. Phys.* **40,** 5138.
Springer, R. H., and Barnes, B. T. (1968). *J. Appl. Phys.* **39,** 3100.
Subotinov, N. V., Kalchev, S. D., and Telbizov, P. K. (1975). *Sov. J. Quant. Electron.* **5,** 1003.
Tombers, R. B., and Chanin, L. M. (1970). *J. Appl. Phys.* **41,** 2483.
Tombers, R. B., and Chanin, L. M. (1973). *J. Appl. Phys.* **44,** 3087.
Tombers, R. B., and Chanin, L. M. (Private communication).
Tombers, R. B., Gaur, J. P., and Chanin, L. M. (1971). *J. Appl. Phys.* **42,** 4855.
von Tongeren, H. (1974). *J. Appl. Phys.* **45,** 89.
Veatch, G., and Oskam, H. J. (1970). *Phys. Rev. A* **2,** 1442.
Vygodskii, Ya S., and Klarfeld, B. N. (1933). *Zh. Tekh. Fiz.* **3,** 610.
Waszink, J. H., and Polman, J. (1969). *J. Appl. Phys.* **40,** 2403.

Chapter 3

High Frequency and Microwave Discharges

A. D. MACDONALD and S. J. TETENBAUM
LOCKHEED PALO ALTO RESEARCH LABORATORY
PALO ALTO, CALIFORNIA

I. INTRODUCTION

Electrical discharges initiated by high frequency or microwave fields differ significantly from dc or low frequency discharges in a number of ways. These include the initiation of the discharge and the conditions required to keep it operating. In order to describe the manner in which the higher frequencies affect the discharge we will first consider the basic processes which go on in a gas subjected to an electric field. These processes, involving the interaction of the field, electrons, atoms or molecules, ions, and containing walls, determine the values of the experimental parameters by which we describe the phenomena. These considerations lead us directly to one of the

ISBN 0-12-349701-9

great difficulties of describing all gaseous electronic phenomena, namely the large number of independent variables which must be taken into account. The breakdown electric field is a function of the ionization potential of the gas, of the collision properties of the electrons in the gas, of the container dimensions, and of the frequency of the applied field. When we consider steady state discharges, at least one more variable, the electron density, must be considered and the situation is even more complicated. Fortunately, it is possible to make use of dimensional analysis to simplify the situation somewhat; the manner in which this is done will be described later.

A. Basic Phenomena

When a high frequency or microwave electric field is applied across a gas, charged particles in the volume are accelerated. Because of the difference in mass, electrons are accelerated much more than ions and energy transfer to these electrons is so much greater than to the ions, that for the most part, we can ignore the heavier particles. Electrons are almost always present in a given gas volume because of cosmic radiation; in the event there are no electrons in the gas during the time a field is applied, no transfer of energy takes place and there will be no discharge. When the direction of the field changes, the direction of the force on the electron changes and so the electron will oscillate within the volume of the gas, provided the walls of the container are sufficiently far apart. This is the characteristic that distinguishes high frequency or microwave discharges from low frequency or dc discharges. When frequencies are low, the reversal of direction of acceleration does not take place before the electron strikes the walls of the container. The electron impact is likely to cause the release of other electrons or impurity atoms from the walls, thus introducing complicating factors into the situation.

At high frequencies, if there were electrons only and no gas atoms, the electrons would oscillate out of phase with the field and no energy would be transferred. In the presence of the gas, electron–atom collisions occur, the electrons are accelerated by the field after each collision, sometimes losing energy and sometimes gaining energy, but on the average, a net transfer of energy from the field to the electrons will take place. While the electrons are gaining energy from the field, they also lose energy via collisions with atoms. If the electric field is large enough, an electron may gain sufficient energy to exceed the excitation energy level of the atom and, on a subsequent collision, the electron may excite the atom and thus lose most of its energy. This energy subsequently becomes radiation when the atom returns to its ground state. Thus, in addition to energy losses by elastic collisions, there are generally losses by inelastic collisions which must be taken into account. Since only a fraction of the electrons which have energy above the excitation energy will in fact excite atoms, some will acquire sufficient kinetic energy so that they may ionize the atoms. When this happens we have two electrons where before we had one. Thus we have a means whereby electrons are produced. At the same time this is happening, electrons are being lost by diffusion to the walls, recombination with positive ions, or by attachment to neutral atoms or molecules. The relative values of these production and loss rates determine the value of the electron concentration and this in turn determines the electrical behavior of the system. The production and loss rates are complicated functions of the kind of gas, the gas density, electric field magnitude and frequency, and the geometry of the container. We must now consider in a more detailed and quantitative fashion how these variables affect the properties of discharges.

The electrical breakdown of the gas, which initiates the discharge, will be considered first. The various processes which determine breakdown thresholds are also of primary importance in steady state discharges.

B. Proper Variables

The breakdown electric field is a function of several variables and in order to describe the phenomenon in an economical fashion, care must be taken in the combination of variables used. The breakdown field, E_b, may be written as a function of four other variables

$$E_b = E(u_i, l, \Lambda, \lambda), \tag{1}$$

where u_i is the ionization potential, l is the mean free path of the electron in the gas, Λ is the characteristic diffusion length of the container, and λ is the free space wavelength of the electric field. The mean free path is a measure of the gas density, which under normal conditions is directly proportional to pressure. The characteristic diffusion length is a measure of the size and shape of the container and is, for example, equal to L/π for infinite parallel plates a distance L apart. Use of these quantities as measures of the variables involved enable us to write Eq. (1) using only two dimensions, voltage and length. There is a theorem in dimensional analysis which enables one to determine how many nondimensional quantities need be used to completely describe relationships among a number of variables. This is generally known as Buckingham's Π-theorem; it states that the relationship between n dimensional quantities is equivalent to a relationship between $n - k$ nondimensional quantities, where k is the number of independent dimensions involved.

In Eq. (1) there are five dimensional variables and two independent dimensions so that application of the Π-theorem tells us that three independent dimensionless variables will form a functional relationship which is equivalent to Eq. (1). Groups of dimensionless quantities may be chosen in a variety of ways. For example, $E\Lambda/u_i$, λ/l, and Λ/l would be a useful set, but it would also be valid to choose another group such as $E\lambda/u_i$, El/u_i, and λ/Λ. The variation of one such dimensionless quantity in terms of the other two describes a complete solution of the problem. For historical reasons it is useful to use somewhat modified variables in describing microwave breakdown phenomena. The mean free path is inversely proportional to the gas pressure under normal conditions. Furthermore, for a given gas the ionization potential is a constant and so it is convenient to make up sets of variables such as $E\Lambda$, $p\lambda$, and $p\Lambda$ or $E\lambda$, E/p, and λ/Λ which correspond to the groups above. These variables, which are based on dimensional analysis, have been called proper variables, and perform the same function as the sets of dimensionless variables. They have proved extremely useful in the development of high frequency and microwave breakdown theory and will be used extensively in this chapter.

One can see immediately on considering the sets of variables the reason that high frequency breakdown is much more difficult to describe than dc breakdown. When wavelength is not a variable, the number of required dimensionless or proper variables is reduced to two so that an appropriately chosen two-dimensional plot suffices, whereas in the microwave case, a three-dimensional surface is required to present all the information.

II. BREAKDOWN

A. Diffusion Theory

A brief description of the different types of collisions electrons make with atoms and molecules will be useful in describing the diffusion theory of breakdown. At this point we will consider only those aspects of collisions needed in the development of breakdown theory; the reader interested in a detailed account of collision phenomena in ionized gases is referred to McDaniel (1964).

1. Electron–Atom Collisions

(a) *Elastic Collision* A collision is described as elastic when the electron simply bounces off the atom and there is no change in the internal state of the atom, though there may be exchange of kinetic energy. Collisions have been described by the use of a number of different concepts which are commonly used in gas discharge literature. If electrons and atoms are considered to be small spheres moving randomly in space, the path of an individual electron will be a series of straight lines forming a zigzag pattern. The individual distances between the encounters which change the direction of motion are the free paths and the mean free path, usually designated l, is the average of these free paths.

The concept of collision cross section is also based on the idea of electrons and atoms consisting of hard spheres. When an electron travels through a gas, the probability of a collision with gas atoms is proportional to the cross sectional area Q that the atoms present to the line of direction of motion of the electron. This concept has been extended so that the difficulties of the rigid sphere model are overcome and much experimental data is interpreted in terms of cross sections. Furthermore, cross sections have been defined for the angular distribution of electrons scattered from a beam. This differential scattering cross section is used in the derivation of a cross section for momentum transfer. The latter is important in gaseous electronics because in considering energy loss of electrons by collision, the momentum transfer cross section is a more convenient parameter to use than is the total cross section. The difference between the two is generally small although with some gases such as argon it may be quite significant. Detailed calculations and experiments are presented in Massey and Burhop (1952, 1969). Huxley and Crompton (1974) have collected a great deal of information on momentum transfer cross sections determined from swarm experiments. These are typically at the lower energies only, generally under 10 eV. The collision probability P_c is frequently used to describe electron–atom interactions in gaseous electronic phenomena. It is defined by the simple equation $pP_c l = 1$ where p is the gas pressure in Torr and l is the mean free path in centimeters. P_c is related to collision cross section Q (cm^2) by the factor 2.82×10^{-17} cm^3-Torr.

Perhaps the most useful description of the collision process uses the collision frequency, ν_c, the number of collisions per second between electrons and atoms. It is equal to the speed of the electron divided by the mean free path v/l and is therefore related to the collision probability and cross section in the following way:

$$\nu_c = vpP_c = 5.93 \times 10^7 u^{1/2} pP_c = 21.0 \times 10^7 u^{1/2} pQ,$$

where u is the electron energy in electron volts, p is the pressure in Torr, and Q is the cross section in square angstroms. We can also define a collision frequency for momentum transfer ν_m, which is related to the collision frequency ν_c in the same way that the cross section for momentum transfer is related to the total collision cross section. Figure 1 shows the collision cross sections for hydrogen and helium, which are unusual in that ν_c is independent of energy to a very considerable extent. Figure 2 shows similar curves for

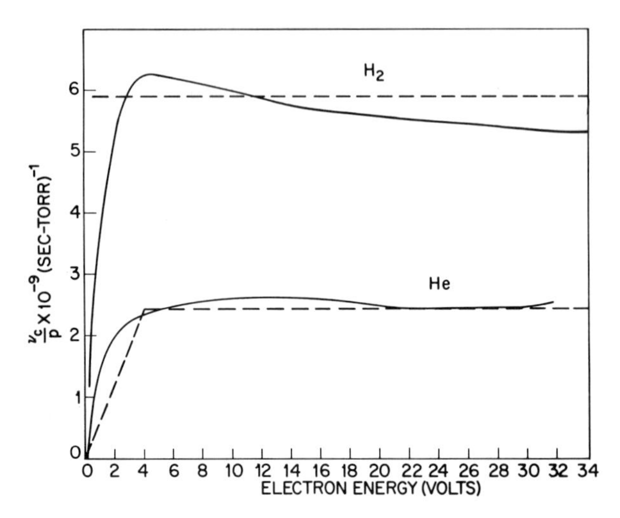

Fig. 1. Electron collision frequencies in hydrogen and helium.

Ne, Ar, Kr, and Xe. The data used in calculating these curves are basically those used by MacDonald (1966), modified in the low energy regions by the data of Crompton and his colleagues collected by Huxley and Crompton (1974).

(b) *Inelastic Collisions* An inelastic collision may occur if an electron has energy in excess of a threshold value characteristic of the gas before colliding

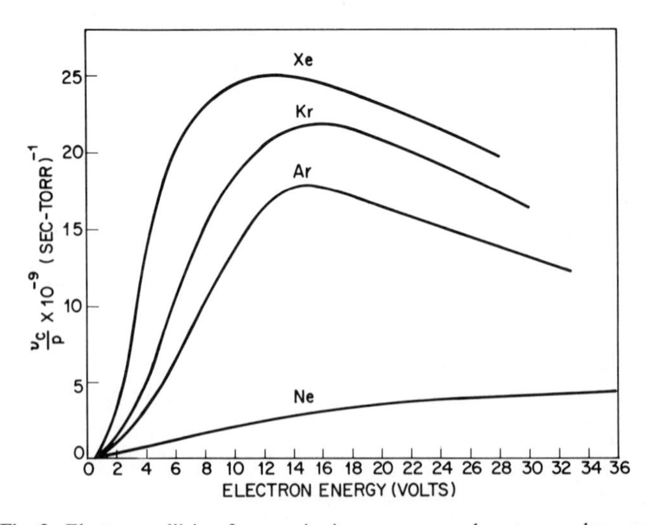

Fig. 2. Electron collision frequencies in neon, argon, krypton, and xenon.

with a gas atom. Some of the energy is spent in internal rearrangement of the atom which subsequently returns to the ground state either by radiating some energy or by losing an electron. The excited state of the atom generally exists an extremely short time, although there are some levels of excitation which are metastable and may last for times of the order of milliseconds. Inelastic collisions have been extensively treated in the literature (McDaniel 1964; Massey and Burhop 1952, 1969; Huxley and Crompton 1974), but for our present purposes the data shown in Figs. 3 and 4 will serve to illustrate the basic idea. Figure 3 shows the efficiency of excitation, i.e., the fraction of collisions which result in excitation, for hydrogen, argon, neon, and helium. The experiments are very difficult and the data presented here are those of Maier-Leibnitz (1935) for helium, neon, and argon, and Ramien (1931) for hydrogen. Similar data for ionization efficiencies are shown in Fig. 4. Ionization measurements are easier to do and much information has been collected by Brown (1959), McDaniel (1964), and others. The data shown for helium and neon are those of Fox (1961), for argon those of Fox (1961) and Bleakney (1930), and for hydrogen those of Tate and Smith (1932).

(c) *Diffusion* Diffusion is the process whereby a flow of particles takes place in a direction which reduces a gradient in either particle concentration or velocity. When we consider electron diffusion, the flow of electrons constitutes a current, and if we set the current density $\mathbf{\Gamma}$ equal to the gradient of the particle concentration n, multiplied by a constant D, we have defined the diffusion coefficient by the equation

$$\mathbf{\Gamma} = -\nabla(Dn).$$

The continuity equation for electrons is

$$(\partial n/\partial t) + \nabla \cdot \mathbf{\Gamma} - P = 0,$$

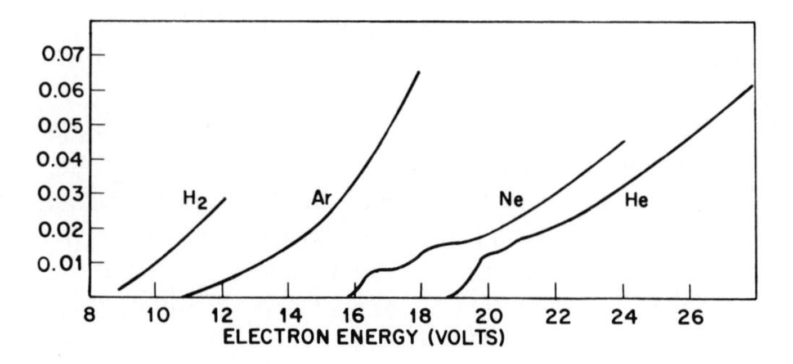

Fig. 3. Efficiency of excitation in hydrogen, argon, neon, and helium.

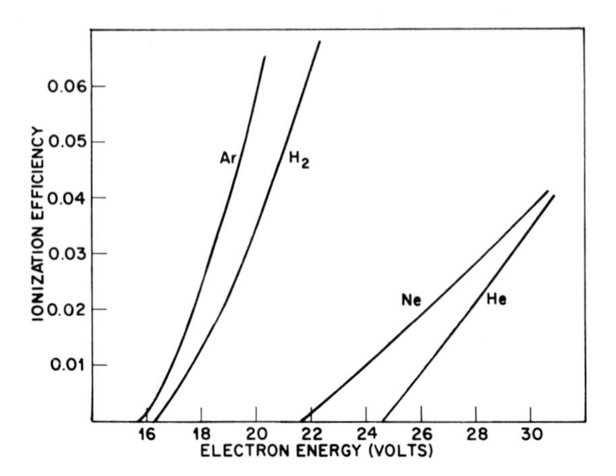

Fig. 4. Efficiency of ionization in hydrogen, argon, neon, and helium.

where P is a net production or loss rate. Generally this will be written $n\nu_i$ where ν_i is the net ionization rate per electron. Combining the above equations, we obtain

$$(\partial n/\partial t) = \nabla^2(Dn) + n\nu_i. \tag{2}$$

This equation has many solutions depending on the initial and boundary conditions, but one of the most useful can be written

$$n = n_0 \exp[(\nu_i - D/\Lambda^2)t], \tag{3}$$

where Λ is the characteristic diffusion length which is determined by the boundary conditions. For a right circular cylinder of length L and radius R, for example, it is given by

$$1/\Lambda^2 = (\pi/L)^2 + (2.405/R)^2.$$

(d) *Attachment* When an electron becomes attached to an ion in a microwave discharge, the net effect is the same as though it were lost from the region in which the field acts. This is so because the negative ion which replaces the electron is at least two thousand times as heavy and so will be accelerated so little during a cycle of the field that energy transfer to it will be negligible. The attachment rate is significant in oxygen and in air and so is of much importance in breakdown phenomena in the atmosphere. There has been considerable work done on attachment in oxygen and much less on attachment rates in air. (Burch and Geballe, 1957; Schulz, 1962; Craggs *et al.*, 1957; Buchel'nikova, 1959; Bradbury, 1933; Harrison and Geballe, 1953.)

(e) *Recombination* Electrons can disappear from the discharge by combining with a positive ion to form a neutral. This process of recombination has been studied extensively, but in most cases has relatively little influence on breakdown phenomena. When an electron and ion recombine, the resulting atom is in an excited state and the excess energy may then be radiated away. The rates for this process are very small. Another recombination process which is more likely to occur involves collisions between molecular ions and electrons which result in dissociation of the molecule, leaving one ground state atom and one excited atom to share the excess kinetic energy. This process has been studied by Biondi and Brown (1949), as well as by Rogers and Biondi (1964). The dissociative recombination rate is substantially greater than the radiative recombination rate, but it is negligible for the types of discharges we consider in this chapter.

2. Boltzmann Equation

A description of microwave breakdown phenomena is much simplified by assuming that each electron is moving with an average energy, but the real situation is of course much more complicated, and electrons and atoms are moving in random directions with speeds varying over very wide limits. Furthermore, the distribution of velocities may also vary in space and in time depending on the physical constraints on the systems. Any physical measurement of the ionized gas is an average of some kind and in order to compute any quantity we can use the kinetic theory of gases provided we have knowledge of the distribution of electrons in space and time. A detailed analysis of the Boltzmann equation for an ionized gas, including consideration of the simplifying assumptions particularly as they relate to breakdown, is available (MacDonald, 1966). In this chapter we will indicate how the analysis leads to calculations of breakdown fields for specific gases for which the assumptions are valid. The excellent agreement between theory and experiment is a verification that we understand the basic physical processes going on.

The Boltzmann equation is a phase space continuity equation for electrons and describes the variation of the electron distribution function $F(\mathbf{v}, \mathbf{s}, t)$, which is the number of electrons that have velocities close to $\mathbf{v}$, in a small volume of space at the end of a vector $\mathbf{s}$, at a time t. The equation can be written

$$C = (\partial F/\partial t) + \mathbf{v} \cdot \nabla F + \mathbf{a} \cdot \nabla_v F, \tag{4}$$

where C represents all changes in velocity or positions caused by collisions, $\partial F/\partial t$ is the local rate of change at the phase space point $\mathbf{s}, \mathbf{v}$; ∇ is the gradient in configuration space and ∇_v is the gradient in velocity space. In order to solve Eq. (4) it is convenient to expand F in spherical harmonics in space and in a Fourier series in time. In addition we must relate the collision term, which is also expanded in spherical harmonics, to changes in the distributions caused by both elastic and inelastic collisions. This procedure results in many equations for the various components of the distribution function, but if we assume rapid convergence of the series and take account of the physical significance of some of the terms, realizing also that most breakdown problems can be adequately approximated by steady-state solutions, we arrive at a set of four simultaneous equations for components of the distribution function:

$$\frac{m}{M}\frac{2}{u^{1/2}}\frac{\partial}{\partial u}(u^{3/2}\nu_{\mathrm{m}}F_0{}^0) - h\nu_{\mathrm{c}}F_0{}^0 = \frac{v}{3}\left[\mathbf{v}\cdot\mathbf{F}_1{}^0 + \frac{1}{2u}\frac{\partial}{\partial u}(u\mathbf{E}_{\mathrm{p}}\cdot\mathbf{F}_1{}^1)\right], \tag{5}$$

$$\nabla\cdot\mathbf{F}_1{}^1 = -\frac{1}{u}\frac{\partial}{\partial u}(u\mathbf{E}_{\mathrm{p}}\cdot\mathbf{F}_1{}^0), \tag{6}$$

$$-\nu_{\mathrm{m}}\mathbf{F}_1{}^0 = v\,\nabla F_0{}^0, \tag{7}$$

$$-\nu_{\mathrm{m}}\mathbf{F}_1{}^1 = j\omega\mathbf{F}_1{}^1 + v\mathbf{E}_{\mathrm{p}}\frac{\partial F_0{}^0}{\partial u}, \tag{8}$$

where $F_n{}^m$ is the component of the distribution which is nth order in space and mth order in time, m and M are the electron and atom masses, respectively, u is $mv^2/2e$, e is the electron charge, v the velocity, and $\mathbf{E}_{\mathrm{p}}$ is the peak value of the electric field of frequency $\omega/2\pi$. The variable u has the units of energy divided by charge, i.e., voltage, and it is frequently convenient to analyze discharge phenomena in terms of the voltage variable. If we can solve for $F_0{}^0$, the symmetrical term of the distribution function, we have all the information necessary to calculate any electrical property of the discharge. Equations (5)–(8) can be combined to solve for $F_0{}^0$ which is a function of both space and velocity (or voltage). However, the space variation in the resulting equation appears only in the Laplacian operator so that if we set $F_0{}^0(u, x, y, z) = f(u)g(x, y, z)$, we can separate out the space variation and simply replace ∇^2 by $1/\Lambda^2$, where Λ turns out to be the characteristic diffusion length discussed above. If in addition we replace $E_{\mathrm{p}}{}^2/2$ by E^2, then E will be the rms value of the electric field and the equation for the energy varying component of the spherically symmetric term of the distribution function f is

$$\frac{2e}{3m}\frac{E^2}{u^{1/2}}\frac{d}{du}\left(\frac{u^{3/2}\nu_{\mathrm{m}}}{\nu_{\mathrm{m}}{}^2 + \omega^2}\frac{df}{du}\right) + \frac{2m}{Mu^{1/2}}\frac{d}{du}(u^{3/2}\nu_{\mathrm{m}}f) = \left(h\nu_{\mathrm{c}} + \frac{2eu}{3m\nu_{\mathrm{m}}\Lambda^2}\right)f. \tag{9}$$

In this equation h is the efficiency of the inelastic collision process and includes the effect of excitation, ionization, attachment, and recombination. Specification of the electron energy variation of the collision frequencies ν_c and ν_m, and of the inelastic collision efficiencies makes possible a solution of Eq. (9), at least in principle. When the distribution function is obtained, the basic theory of gases enables us to calculate all of the electrical properties of the discharge.

One can think of Eq. (9) both in terms of particle balance and energy balance, though care must be used in such an interpretation. The first term represents the number of particles being raised to a given energy range per unit time, while the other three terms represent those lost from this range by elastic collisions, inelastic collision and by diffusion. The first term can also be thought of as a measure of the electrical energy transferred from the field to the electrons while the other three terms represent loss of energy by electrons because of elastic collisions, inelastic collisions, and diffusion from the region, respectively.

(a) *Collision Frequency Independent of Energy* It is clear from examination of Eq. (9) that in general a solution will be very complicated, but there is one special case that yields an analytic solution and also is applicable to some practical situations. This happens when the electron collision frequencies ν_c and ν_m are independent of energy. Reference to Fig. 1 shows that for both helium and hydrogen the collision frequencies are indeed very nearly independent of electron energies over a considerable energy range, in particular over that part of the energy range which is of most interest in breakdown problems. As an illustration of what one can do, we will sketch the method of solving Eq. (9), which enables us to calculate the electron distribution and predict breakdown fields for this restricted set of conditions. Details of this calculation as well as ways of applying the electron energy distribution functions to other cases can be found in MacDonald (1966).

It is convenient to define an effective electric field by the equation

$$E_e^2 = E^2/(1 + \omega^2/\nu_m^2). \tag{10}$$

If, in addition, the independent variable is converted to a dimensionless form by letting

$$w = \frac{3m}{M} \cdot \frac{m}{3} \cdot \frac{\nu_m^2}{E_e^2} u,$$

then Eq. (9) becomes

$$\frac{1}{w^{1/2}} \frac{d}{dw} \left[w^{3/2} \left(\frac{df}{dw} + f \right) \right] = f(\mu h + \eta^2 w), \tag{11}$$

where

$$\mu = \frac{M}{2m}\frac{\nu_c}{\nu_m} \quad \text{and} \quad \eta = \frac{e}{m}\cdot\frac{M}{3m}\cdot\frac{E_e\Lambda}{\nu_m{}^2\Lambda^2}.$$

Further manipulation of the variables by letting $\delta^2 = 1/(1 + 4\eta^2)$, a second transformation of the independent variable to $y = w/\delta$, and transformation of the dependent variable by defining $g = f \exp[(1 + \delta)y/2]$ lead to the equation

$$y\frac{d^2g}{dy^2} + \frac{dg}{dy}\left(\frac{3}{2} - y\right) - g(\alpha + \mu h\delta) = 0, \tag{12}$$

where $\alpha = \frac{3}{4}(1 - \delta)$.

The term h in Eq. (12) represents inelastic collisions. For electron energies below the lowest excitation level, h is zero, because attachment in both helium and hydrogen is negligible and for the low ionization levels recombination is also negligible. When h is zero, the equation is much simplified and can be solved in terms of the confluent hypergeometric function. When the electron energy is above the excitation level, the h terms must be accounted for, but there is a fortunate combination of circumstances that enables us to make appropriate calculations for the right combination of gases. Let us consider the gas mixture which has been called Heg, a mixture consisting mostly of helium with a small amount of mercury. The lowest excitation level of helium occurs at 19.8 V and is metastable, having a lifetime measured in milliseconds rather than in nanoseconds. During a millisecond an atom collides with other atoms thousands of times at the concentrations of interest, so that if there is one part of mercury per thousand of helium, a metastable helium atom is likely to have some encounters with mercury atoms during its lifetime. When a collision occurs between a metastable helium atom and a mercury atom there is a very high probability that some of the internal energy of the excited state will transfer to the mercury atom and ionize it. It is easy to adjust the relative concentrations of the two gases so that enough encounters between the atoms occur during a metastable lifetime to ensure that virtually every inelastic collision in the gas results in an ionization. The amount of mercury needed is so small that the mixture behaves like helium insofar as elastic collisions are concerned. This means that Heg is a gas with an electron collision frequency independent of energy, in which there are in effect no excitations, in which essentially all electrons reaching energies of 19.8 eV produce an ionization, and in which hardly any electrons have energies above 19.8 eV.

This set of circumstances allows us to solve Eq. (12) in closed form and to obtain a concise expression for breakdown fields. When h is zero, taking account of the transformation between the function g and the distribution function we obtain

$$f = A \exp[(\tfrac{3}{2}\alpha - 1)y][M(\alpha; \tfrac{3}{2}; y) + CW(\alpha; \tfrac{3}{2}; y)], \tag{13}$$

where $M(\alpha; \gamma; y)$ and $W(\alpha; \gamma; y)$ are the independent solutions of the confluent hypergeometric equation and A and C are constants to be determined by the boundary conditions. Because γ is not integral in this case, the second solution is expressible in terms of the M functions and is $y^{-1/2}M(\alpha - \tfrac{1}{2}; \tfrac{1}{2}; y)$. The distribution function goes to zero at a voltage slightly above 19.8 V, which for this gas is in effect the ionization voltage, so that if we designate the value of u at this point u_i, and the corresponding value of y as y_i, we can find the value of C immediately:

$$C = -y_i^{1/2} M(\alpha; \tfrac{3}{2}; y_i)/M(\alpha - \tfrac{1}{2}; \tfrac{1}{2}; y_i). \tag{14}$$

Equations (13) and (14) describe the distribution function completely and can now be used to calculate properties of the discharge.

Breakdown fields can be calculated by finding expressions for the ionization rate and the diffusion rate since breakdown occurs when the rate of production of electrons by ionization very slightly exceeds the loss rate caused by diffusion. In Eq. (3) the electron concentration is expressed in terms of these rates. The ionization rate is a very strong function of the electric field, often proportional to the 8th or 10th power of the electric field, and in addition, both the ionization rate and diffusion rate are numerically large so that when ν_i is slightly less than D/Λ^2 a small percentage change in ν_i, which can be caused by a very tiny change in the electric field, is sufficient to change the exponent very radically and increase the concentration many orders of magnitude. This is what happens at breakdown when the electron concentration can change typically from about 10^2 electrons/cm^3 to 10^8 or 10^9 electrons/cm^3. At the same time the gas starts to glow and becomes conducting. Therefore, it is reasonable to define the breakdown electric field as that for which $\nu_i = D/\Lambda^2$.

The ionization rate can be calculated for Heg by computing the number of inelastic collisions per electron per second, ν_i. The kinetic theory equation is

$$n\nu_i = 4\pi \frac{e}{m}\left(\frac{2e}{m}\right)^{1/2} \int_0^\infty h_i \nu_c u^{1/2} f \, du. \tag{15}$$

There appears at first sight to be a mathematical difficulty in Eq. (15) since h_i is zero below u_i and f is zero above u_i. However, we can let the range of integration be $u_i - \epsilon$ to infinity and let the distribution function vanish at $u_i + \epsilon$ instead of at u_i. Integrating by parts, using the appropriate boundary conditions, and then letting ϵ vanish, yields

$$n\nu_i = \frac{2\pi}{3}\left(\frac{2e}{m}\right)^{5/2} \frac{E_e^{\ 2}}{\nu_m} \left(u^{3/2} \frac{df}{du}\right)_{u_i}. \tag{16}$$

The diffusion coefficient is the average value of one third of the electron velocity squared divided by the collision frequency. This leads to the equation

$$nD = \frac{2\pi}{3}\left(\frac{2e}{m}\right)^{5/2} \int_0^{u_i} \frac{fu^{3/2}}{\nu_c}\, du. \tag{17}$$

After considerable mathematical manipulation that will not be reproduced here, it can be shown that the result of dividing Eq. (17) by Eq. (16) leads to

$$D/\Lambda^2 = \nu_i[M(\alpha; \tfrac{3}{2}; y_i)\exp(-\tfrac{3}{2}\alpha y_i) - 1]$$

or a breakdown condition which requires simply that

$$M(\alpha; \tfrac{3}{2}; y_i) = 2\exp(\tfrac{2}{3}\alpha y_i), \tag{18}$$

where

$$\alpha = \tfrac{3}{4}(1 - \delta), \tag{19}$$

$$\delta = [1 + [2.24(E_e/p)^2/(p\Lambda)^2]]^{-1/2}, \tag{20}$$

and

$$y_i = 1.34u_i/\delta(E_e/p)^2. \tag{21}$$

For Heg gas, $E_e^{\ 2} = E^2/[1 + (78.6/p\lambda)^2]$, with λ the free space wavelength of the electric field in centimeters, E the rms value of the field in volts per centimeter, p the pressure in Torr, and Λ the characteristic diffusion length in centimeters. The breakdown condition [Eq. (18)] can be further refined by making corrections for two assumptions implicit in the analysis. The collision frequency is not independent of energy over the whole range but is better approximated by a linear variation at the very low energies as indicated in Fig. 1. A calculation taking account of this leads to a correction in the calculated breakdown fields at the high pressures only. At the high pressures a larger fraction of the collisions takes place while the electron has a low energy. This correction does not exceed a few percent at the most (MacDonald and Brown, 1949a). At the lower pressures the average energy gained between collisions is higher and under some circumstances electrons will gain energies exceeding 19.8 V by a volt or two before producing ionizations. This also can be taken into account in a rather complicated calculation (MacDonald and Brown, 1949) which leads to some modification of the field calculated by means of Eq. (18) for the lower pressures only. Figure 5 compares the breakdown electric fields calculated by Eq. (18), with the

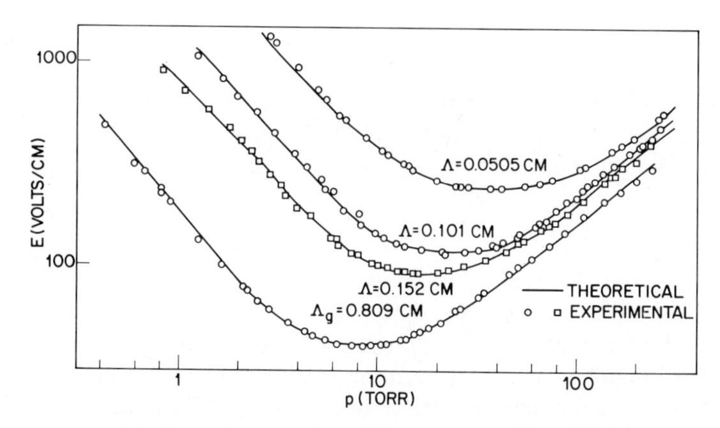

Fig. 5. Breakdown theory and experiment in Heg at 2.8 GHz.

corrections noted, with experiment. The breakdown fields are measured at 2.8 GHz in different sized resonant cavities to provide a range of values of the characteristic diffusion length and over pressures ranging from about 0.1 to 300 Torr. It is to be noted that there are no adjustable parameters in the theory and that theory and experiment agree within the small experimental error. This very striking agreement shows that the processes as described provide an accurate picture of the phenomena involved in microwave gas discharges. A similar calculation has been done for the other gases for which the electron collision frequency is independent of energy, namely hydrogen and pure helium. The detailed calculations and comparisons with theory will not be included here but can be found in MacDonald (1966).

(b) *Variable Collision Frequency* The calculations of breakdown field in Heg, helium, and hydrogen were much simplified by the energy independence of ν_c. For other gases this assumption is not valid and Eq. (9) is not solvable in closed form without further approximations. Neon gas has a mean free path which is almost independent of energy so the ν_c is proportional to $u^{1/2}$. Although this is probably the simplest of the cases in which the energy variation of ν_c must be taken into account, the analysis becomes very complicated as can be seen from the form that Eq. (9) takes:

$$\frac{2e}{3m}E^2\frac{d}{du}\left(\frac{ku^2}{k^2u+\omega^2}\frac{df}{du}\right)+\frac{2m}{M}\frac{d}{du}(u^2kf)=f\left(khu+\frac{2e}{3mk}\frac{u}{\Lambda^2}\right). \quad (22)$$

This has been solved for the special case corresponding to many cycles of the electric field per collision and that corresponding to many collisions per electric field oscillation, the low and high pressure cases, respectively. Breakdown fields were calculated for these cases and compared with experimental data obtained in neon at 2.8 GHz. The calculations are lengthy and complicated and will not be included here (MacDonald and Betts, 1952; MacDonald, 1966). Agreement with theory is reasonably good but the theoretical approximations to experimental collision data are not as good as in the case of Heg with the result that theory does not represent experiment so well.

3. Nonuniform Fields

Microwave fields typically have wavelengths of the same order of magnitude as the dimensions of the experimental apparatus so that the magnitude of the field varies within the containers used. For example, in a right circular cylinder resonant in the TM_{010} mode the electric field is uniform axially but varies from a maximum at the center of the cylinder to zero at the edges. In the analysis presented thus far it has been assumed that ionization and diffusion competed uniformly over the whole region. The reason that the analysis is valid is that we have in reality calculated and made measurements for those situations in which the height of the right circular cylinders used was much less than the radius, so that in essence all electron diffusion takes place to the parallel plate ends of the cylinder. In this case there is a region in the center of the resonant cavity in which the electric field is varying slowly enough, so that the effect of the radial walls is negligible and the theory is valid. The definition of the characteristic diffusion length for the right circular cylinder given earlier,

$$\frac{1}{\Lambda^2} = \left(\frac{\pi}{L}\right)^2 + \left(\frac{2.405}{R}\right)^2,$$

gives an indication of the dimensions for which the nonuniformity of the field is important. If the second term is large enough so that it significantly affects the calculations of Λ, then account must be taken of the nonuniform field. It is, in fact, a diffusion problem because when the length of the cavity is large enough so that diffusion to the side walls is important, the electrons formed at the center in the high field region drift out to where the field is lower and where breakdown will not occur. This electron loss from the central region then is equivalent to a loss from the whole container since the balance between diffusion and ionization is only important where the breakdown takes place. Thus the effective radius in the formula is smaller than the actual radius.

This diffusion problem has been solved for cylindrical cavities by Herlin and Brown (1948) and for spherical cavities by MacDonald and Brown (1950). These solutions have made it possible to extend significantly the range of values of the characteristic diffusion length from which breakdown calculations can be made with reasonable assurance of agreement with experiment.

4. Magnetic Fields

The introduction of a dc magnetic field changes the motion of the electrons because the acceleration term changes from $(e/m)\mathbf{E}$ to $(e/m)(\mathbf{E} + \mathbf{v} \times \mathbf{B})$, where $\mathbf{B}$ is the magnetic induction. The electrons move in spiral fashion changing the mode of diffusion differently in different directions so that diffusion is no longer isotropic. When the modified force term is put into Eqs. (4)–(8) the analysis is very complicated and leads to a modified expression for the effective electric field:

$$E_{eb}^2 = \frac{E^2 \nu_m{}^2}{2}\left[\frac{1}{\nu_m{}^2 + (\omega - \omega_b)^2} + \frac{1}{\nu_m{}^2 + (\omega + \omega_b)^2}\right], \tag{23}$$

where ω_b is equal to eB/m. Assuming that the collision frequency ν_m is independent of energy, the effect of the magnetic field on energy transfer is taken account of by replacing E_e by E_{eb} in all equations. Because the electron paths are changed by the magnetic forces, the diffusion rates are also changed. The analysis, which in general leads to a second-order tensor diffusion coefficient, will not be reproduced here. For the simple case of a magnetic field applied along the axis of a right circular cylinder, the characteristic diffusion length Λ can be replaced by Λ_b, where

$$\frac{1}{\Lambda_b{}^2} = \frac{1}{\Lambda_r{}^2}\left(\frac{\nu_m{}^2}{\nu_m{}^2 + \omega_b{}^2}\right) + \frac{1}{\Lambda_z{}^2}, \tag{24}$$

where Λ_r and Λ_z are, respectively, $R/2.405$ and L/π. Thus the diffusion in directions perpendicular to the magnetic field is reduced by an amount equivalent to increasing the dimension by a factor

$$\left(\frac{1 + \omega_b{}^2}{\nu_m{}^2}\right)^{1/2}$$

Lax *et al.* (1950) made measurements in Heg gas in S-band microwave fields and with the magnetic fields applied both transverse and parallel to the electric field. Figure 6 shows both experimental data and the theoretical prediction based on the breakdown as in Eq. (18) with the effective field and diffusion length modified as in Eqs. (23) and (24). The cyclotron resonance, the reduced breakdown field caused by the magnetic field, and the excellent agreement between theory and experiment are evident from the figure.

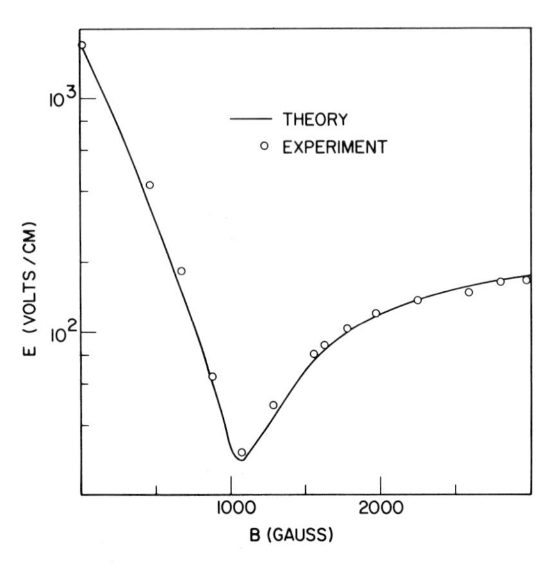

Fig. 6. Breakdown fields in Heg with applied magnetic field.

5. Laser Initiated Gas Breakdown

During the decade since the first published reports of the initiation of gas breakdown at optical frequencies (Meyerand and Haught, 1963), there has been much interest in the mechanisms involved. Some of the earlier accounts stressed the relatively large magnitude of the quanta (about 1.7 eV for a ruby laser) as compared with usual energy increments between collisions when microwave fields are applied. Furthermore, the very short pulse length, of the order of tens of nanoseconds, was considered to be an obstacle to applying the classical electromagnetic theory. However, it has been shown (MacDonald, 1966) that because of the very high frequency there are so many oscillations of the field during a pulse that the field is in effect a continuous wave. Furthermore, the quantum effect is not important in this process. Calculations made using the breakdown condition in this chapter have been compared with experiment in helium gas and show very good agreement. Figure 7 compares the data of Meyerand and Haught on optical breakdown in pure helium with calculations based on Eq. (18). Furthermore, application of the microwave breakdown theory to laser initiated breakdown experiments in other gases, where the approximation of an energy independent collision frequency is not as accurate, have led to satisfactory agreement between theory and experiment. It now appears that there is considerable evidence that the basic ideas of microwave breakdown can

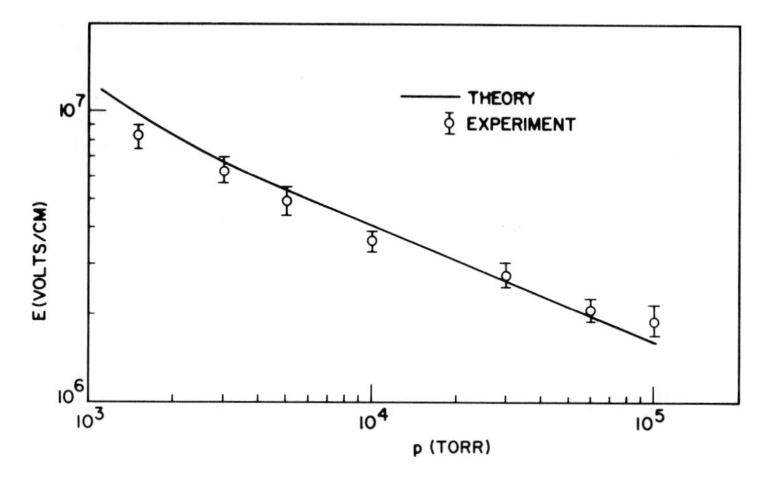

Fig. 7. Laser breakdown field in helium compared with theory.

be readily extended to explain laser initiated breakdown in gases (Gill and Dougal, 1965; MacDonald, 1966; Raizer, 1966).

B. Air Breakdown

1. Experimental Data

Helium and hydrogen have properties that make their study satisfying in terms of comparison between theory and experiment and also in terms of clarifying the basic processes that go on in a gas subject to applied microwave fields. Nonetheless, one of the major engineering and practical interests in microwave discharges is related to air and the atmospheric gases. For this reason we will consider microwave breakdown in the atmosphere in some detail, both in experiment and theory.

Typical experimental data are shown in Figs. 8 and 9. In Fig. 8 are shown continuous wave (cw) breakdown data for air, oxygen, and nitrogen at a frequency of 992 MHz in a resonant cavity with characteristic diffusion length of 1.51 cm. It is to be noted that the breakdown fields are much the same for the three gases with oxygen having the highest breakdown field and nitrogen the lowest, and with what little divergence there is being most pronounced at the higher pressures, where attachment is relatively more important. Figure 9 shows the breakdown in air in cavities of different characteristic diffusion length for a frequency of 9.4 GHz. Much experimental work has been done on breakdown in air; Table I summarizes some of the work

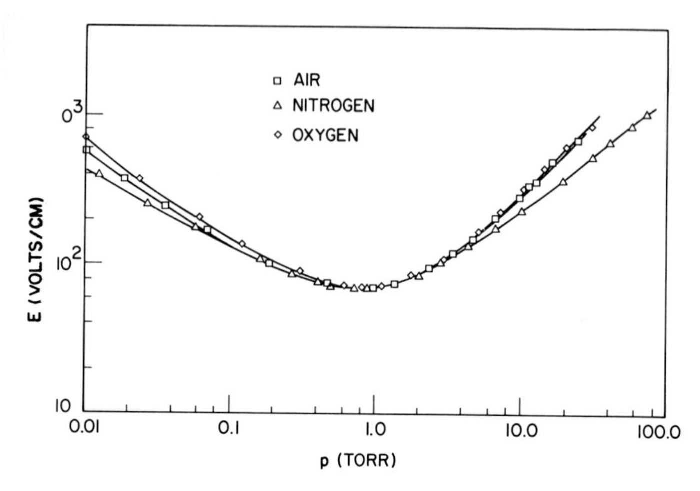

Fig. 8. Breakdown field in air, oxygen, and nitrogen at 0.992 GHz and $\Lambda = 1.51$ cm.

reported to date for uniform fields, including frequency, characteristic diffusion length, and pressure or altitude, for both cw and pulsed conditions. It is unsatisfactory to have to make measurements of breakdown under each set of conditions required and therefore a theoretical description which permits prediction is desirable. We will therefore describe in the next section a method of calculation which does provide fairly accurate prediction of breakdown under a wide range of conditions.

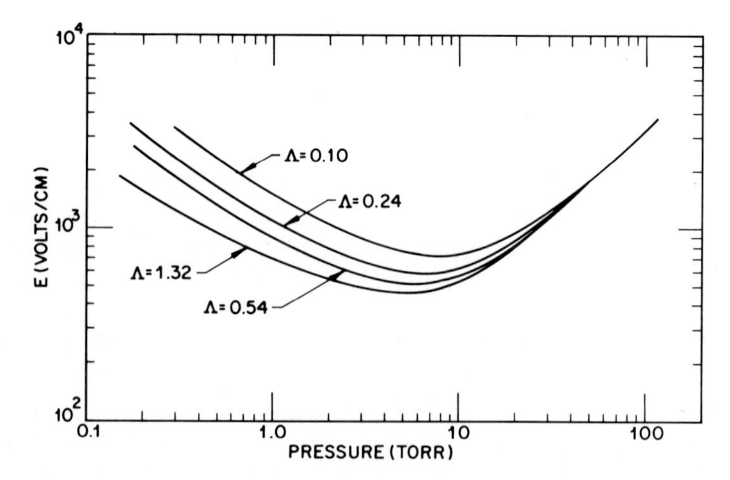

Fig. 9. Breakdown fields in air at 9.4 GHz for various characteristic diffusion lengths.

TABLE I

MICROWAVE BREAKDOWN IN AIR

f (GHz)	Λ (cm)	p (Torr)	Observer
0.20	0.02–0.03	50–1000	Pim (1949)
0.99	0.6–2.6	0.1–100	MacDonald *et al.* (1963)
2.8	0.21	7–100	Gould and Roberts (1956)
2.85	0.21	0.2–8.0	Rose and Brown (1955)
3.06	0.72	1–1000	Tetenbaum *et al.* (1971)
3.06	0.39	0.1–100	Bandel and MacDonald (1970)
3.13	0.02–0.1	0.1–60	Herlin and Brown (1948)
9.4	0.10–1.3	0.2–100	MacDonald *et al'* (1963)

2. Attachment-Diffusion Theory

Any description of microwave phenomena which is to represent the data in a reasonably economical way must make use of the effective field concept in some fashion. Unfortunately, the collision frequency for electrons in air varies with electron energy and the effective field equation cannot be used over a range of frequencies without introducing some error. It has been shown (MacDonald *et al.*, 1963) that if a value of ν_c equal to $5.3 \times 10^9 p$ is used, the experimental data on air breakdown over a large range of frequency and characteristic diffusion length form a good single line over a considerable fraction of an $E_e\Lambda - p\Lambda$ plot, but with some divergence at the lower values of $p\Lambda$. It therefore makes sense to use the effective field concept in this theory.

The electrons which are produced by the microwave field in air are lost by diffusion and attachment. The breakdown condition requires that losses equal gains so that

$$\nu_i = \nu_a + D/\Lambda^2, \tag{25}$$

where ν_a is the attachment rate per electron. Knowledge of D as a function of the experimental variables would enable us to obtain a set of data for the net ionization rate $\nu_i - \nu_a$ directly from experiment. D is not directly measurable when the electric field is high but a theoretical calculation can be made on the basis of some assumptions which prove reasonable. By definition, D is one third of the average value of the square of electron velocity divided by the electron collision frequency, $\langle v^2/3\nu_c \rangle$, so that we can set

$$D = \frac{2}{3\nu_c} \frac{e}{m} \int_0^\infty f u^{3/2}\, du \Big/ \int_0^\infty f u^{1/2}\, du, \tag{26}$$

assuming that ν_c is independent of energy u. The form of the distribution function chosen will make a difference in the magnitude of D but the energy variation ascribed to ν_c is relatively more important. If a Maxwellian distribution, i.e., $f \propto \exp(-3u/2u_0)$, is chosen,

$$D = 2.2 \times 10^5 u_0/p \text{ cm}^2\text{-sec}, \tag{27}$$

where u_0 is the average value of u in volts, and p is in Torr. u_0 can be related to E/p through the data of Crompton *et al.* (1953). MacDonald *et al.* (1963) used this to calculate the approximation

$$Dp = [29 + (0.9E_e/p)] \times 10^4 \text{ cm}^2\text{-Torr/sec}, \tag{28}$$

in which E_e is used instead of E, because it better represents the action of the field in enhancing diffusion.

3. Continuous Wave Breakdown

It proves convenient to regroup the quantities in Eq. (25) and to multiply by λ so that the breakdown condition becomes $(\nu_i - \nu_a)\lambda = D\lambda/\Lambda^2$. If we then introduce a new variable S, and express the effective field in terms of this variable, we can set

$$D(\lambda/\Lambda^2) = 10^4(\lambda/\Lambda)^2 S, \tag{29}$$

where

$$S = \frac{1}{p\lambda}\left[29 + \frac{0.9E\lambda}{((p\lambda)^2 + (35.6)^2)^{1/2}}\right].$$

S is plotted in Fig. 10 as a function of $p\lambda$ with $E\lambda$ as a parameter.

A representation of the ionization rate to use along with the diffusion variable in predicting breakdown can be obtained from the experimental $E-p$ curves of the type shown in Figs. 8 and 9. The high frequency ionization coefficient ζ defined as ν_i/DE^2, is first calculated (in this case $(\nu_i - \nu_a)/DE^2$). At breakdown, ζ is $1/(\Lambda E)^2$ because of the breakdown condition. Successive multiplication by E^2/p^2, $p\lambda$, and Dp leads to values of $(\nu_i - \nu_a)\lambda$. These are plotted in Fig. 11 as a function of $p\lambda$ with $E\lambda$ as a parameter. It is to be noted that the experimental breakdown data used in computing the curves of Fig. 11 include most of those referred to in Table I.

In Figs. 10 and 11 we have a set of curves for the net ionization rate multiplied by λ, and a set of curves for the diffusion loss rate multiplied by λ (except for a factor $10^4(\lambda/\Lambda)^2$ which typically is constant through a set of measurements). The second set of curves can be plotted on transparent paper and superposed on the first in such a way that the point $S = 1$ is coincident with the point $\nu\lambda = 10^4(\lambda/\Lambda)^2$. Then at each point where the curves with the

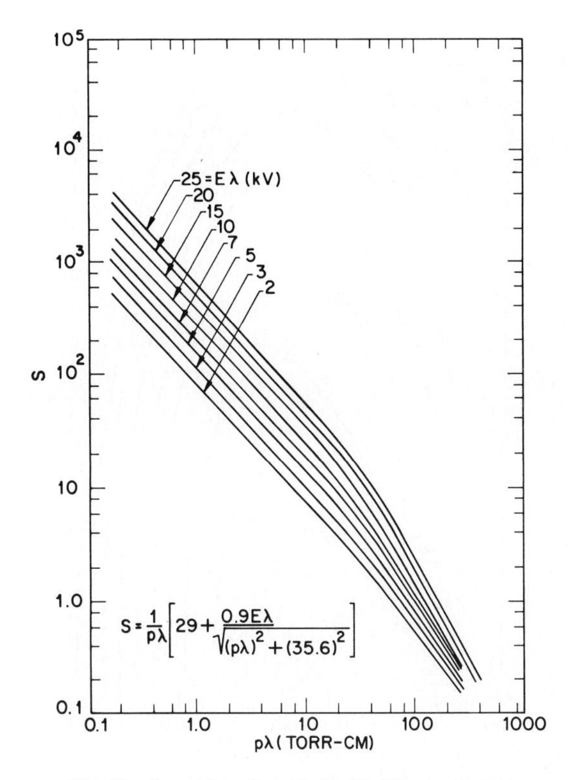

Fig. 10. S as a function of $p\lambda$ with $E\lambda$ as parameter.

same value of $E\lambda$ from the two plots intersect, the breakdown conditions are fulfilled, and therefore the value of $E\lambda$ for a particular $p\lambda$ is found. Since the wavelength and the characteristic diffusion length are known, the value of the breakdown electric field as a function of pressure can be obtained.

4. Pulsed Breakdown

Figure 11 can also be used to predict breakdown initiated by pulsed microwave power. The criteria by which we define breakdown fields are different when short pulses are used because it often happens that a pulsed source for which the peak electric field is well above cw breakdown level will not break down the gas. This is because of the time required for the electron concentration to build up from ambient to a level which starts the processes we associate with breakdown. During the buildup period the electron concentration can be expressed quite accurately by

$$n = n_0 \exp[(v_i - v_a - D/\Lambda^2)t]. \tag{30}$$

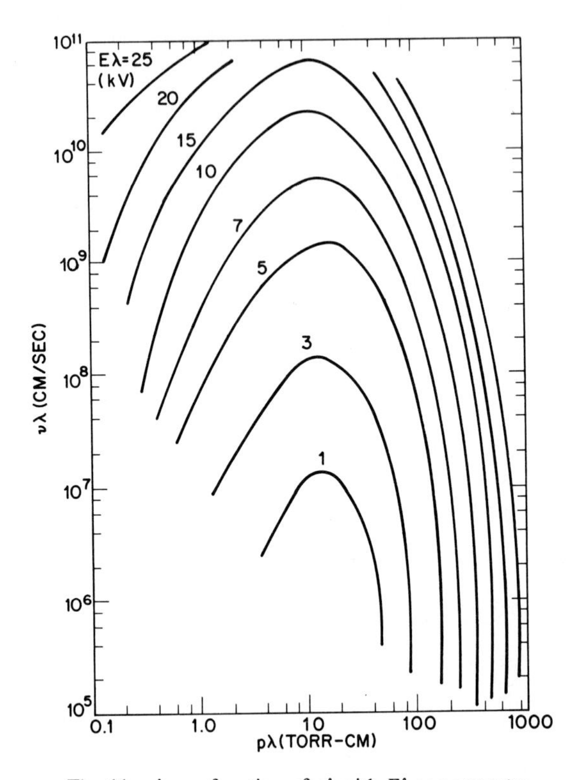

Fig. 11. $\nu\lambda$ as a function of $p\lambda$ with $E\lambda$ as parameter.

If the concentration reaches the plasma concentration n_p (the concentration for which plasma resonance occurs, approximately equal to $10^{12}/\lambda^2$, where λ is in centimeters), by the last 5% of the pulse, we can say that breakdown has occurred. Therefore simply replacing n by n_p and t by τ (the pulse length) in Eq. (30) gives a breakdown condition. This condition can be used quite simply by first finding D and thus D/Λ^2 by use of Eq. (28). (An estimate of breakdown field is needed, but iteration to a consistent value of the field is simple.) The calculation of $(\nu_i - \nu_a - D/\Lambda^2)t$ which fulfills the breakdown condition, along with the value of the pulse length and the calculated value of D/Λ^2, leads directly to a value of $(\nu_i - \nu_a)\lambda$ for breakdown. This can be used directly in Fig. 11 to yield a value of the breakdown field for the pulse length under consideration. This technique has been shown to work very well and has been successful in accurate prediction of pulsed breakdown fields over a considerable range of experimental variables (MacDonald, 1966).

Very good agreement between theory and experiment has been achieved by the use of Figs. 10 and 11 in the manner described above. Both pulsed and cw breakdown fields can be calculated for air for frequencies ranging from about 100 MHz to 100 GHz over a pressure range corresponding to an altitude variation from ground level to about 100 km, and for a very large range of variation of characteristic diffusion length.

C. Experimental Data

A wealth of experimental data on microwave breakdown fields has been reported. Just as in the case of theory, we can conveniently divide this data into three groups: the gases for which electron collision frequency is independent of energy, the atmospheric gases, and others. Having already dealt with the atmospheric gases, we now consider the other categories.

1. ν_c INDEPENDENT OF ENERGY

The three gases for which this is a good approximation are pure helium, Heg, and hydrogen. As pointed out earlier, the validity of the effective field concept makes it practical to present all the experimental data for each gas in a single curve. We have already considered Heg both in theory and experiment.

Figure 12 shows the experimental data for hydrogen, with $E_e\Lambda$ plotted as a function of $p\Lambda$. Figure 13 shows similar data for pure helium. The lines

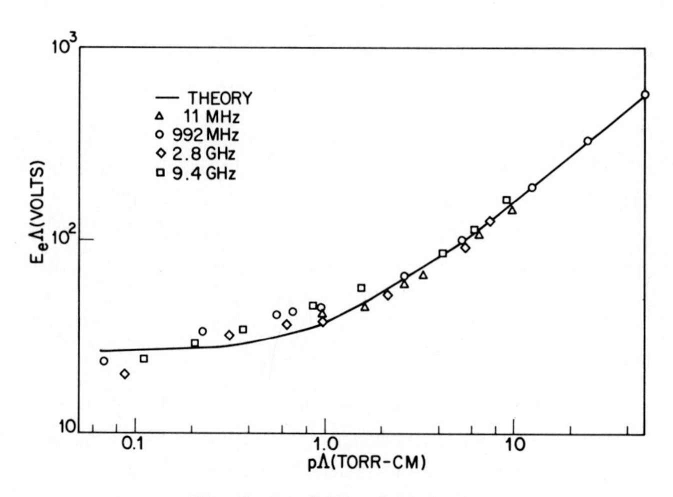

Fig. 12. Breakdown fields in hydrogen.

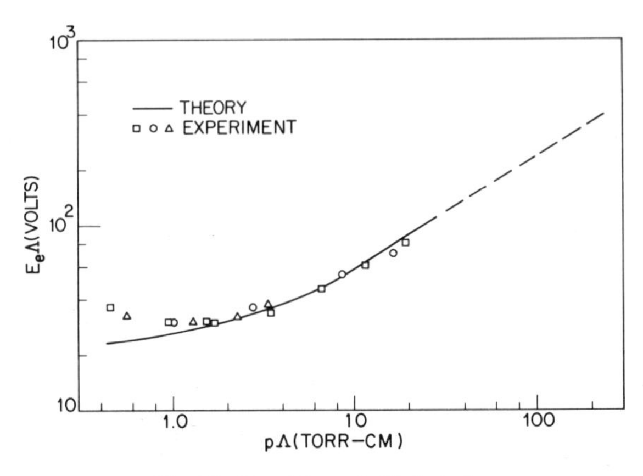

Fig. 13. Breakdown fields in helium.

give the results of theoretical analysis which shows quite good agreement (MacDonald and Brown, 1949b; Reder and Brown, 1954).

2. Other Gases

It is not practical in a limited space to present the experimental data for other gases because in general the variation of electron collision frequency with energy is such that the effective field concept does not work well, so the data for a given gas require many graphs to cover the ranges of frequency, characteristic diffusion length, and pressure. In Table II references are given to much of the experimental work which has been done on the noble gases, including the range of variation of the experimental variables. There have been theoretical analyses of some of those gases with some success in the case of neon (MacDonald and Betts, 1952).

Table III includes the ranges of variation of experimental parameters for a number of investigations of microwave breakdown in gases other than the noble gases, hydrogen, and air.

Tables II and III are indicative of the work done rather than inclusive. In general, they include the work reported for uniform fields for which characteristic diffusion lengths are given or readily calculated.

3. Antenna Breakdown

There has been much interest in the transmission of high power microwave signals from high flying missiles and from hypersonic vehicles re-entering the earth's atmosphere. The breakdown threshold levels in front of the antennas limit the amount of power that can be transmitted. In addition to the altitude variation of power level, one must consider for a hypersonic

TABLE II

MICROWAVE BREAKDOWN IN NOBLE GASES

Gas	f (GHz)	Λ (cm)	p (Torr)	Observer
He and Heg	0.99	0.63	0.3–100	MacDonald (1966)
	2.8	0.05–0.81	0.3–300	MacDonald and Brown (1949a)
	2.8	0.05–0.2	1–100	Reder and Brown (1954)
	10.	~0.3	10–100	Golant (1959)
Ne	0.99	0.63	0.02–100	MacDonald *et al.* (1963)
	2.8	0.1–0.2	1.0–300	MacDonald and Betts (1952)
	2.8	1.51	0.05–300	MacDonald and Matthews (1955)
	3.06	0.1–0.39	0.1–400	Bandel and MacDonald (1969)
	10.	~0.3	10–200	Golant (1959)
Ar	0.4–1.6	0.80	1–100	Anashkin (1968)
	0.99	0.63–1.51	0.01–100	MacDonald *et al.* (1963)
	2.8	0.05–0.15	0.05–200	MacDonald and Matthews (1956)
	2.95	0.135	4.0–100	Krasik *et al.* (1949)
	10.0	~0.3	1–50	Golant (1959)
Kr and Xe	2.8	0.10	0.05–100	Bradford *et al.* (1959)
	10.	~0.3	3–30	Golant (1959)

vehicle the effects of the shock wave and the ionization produced in the sheath surrounding the vehicle. Furthermore, the near field of the antenna presents additional problems because breakdown will likely occur at the high field points.

Taylor *et al.* (1971) have written a comprehensive review of antenna breakdown in which engineering aspects are emphasized and in which there is consideration of possible means of preventing breakdown. We will not deal with the subject in detail here but will simply illustrate the effect of the parameters involved by dealing with one example. The reader is referred to the Taylor *et al.* review for detailed treatment and a comprehensive list of references.

TABLE III

MICROWAVE BREAKDOWN IN OTHER GASES

Gas	f (GHz)	Λ (cm)	p (Torr)	Observer
CO	9.36	0.40	0.5–8.0	Mentzoni (1973)
NO	9.36	0.40	0.5–10.0	Mentzoni (1974)
H_2O	3.06	0.391	0.7–17.	Bandel and MacDonald (1970)
SF_6	3.06	0.1–0.72	0.01–200	Tetenbaum *et al.* (1973)
CO_2, CH_4	9.4	~0.5	0.3–30	Dawson and Lederman (1973)

Consider the problem of predicting breakdown fields in front of an antenna near the nose of a Mach 6 vehicle. In the shock wave region, the gas density increases so that at a given altitude breakdown takes place as though the altitude were lower than it actually is. But the electric field also exists outside the region of the shock wave so that when the breakdown field is lower in the ambient atmosphere, that is the level we calculate. This is illustrated in Fig. 14, which shows the breakdown field for a frequency of 3 GHz.

On the low altitude side of the minimum, the breakdown curve follows what one would calculate from zero speed, while on the high altitude side, where an increase in gas density makes breakdown easier, the curve follows the curve one would calculate for the maximum density in the shock wave. The additional curve shown in Fig. 14 is a correction for ambipolar diffusion, which takes account of ionization produced around the vehicle by the shock wave. The calculations we have outlined in this chapter have been based on the assumption that electron and ion concentrations were low enough so that there was no significant interaction. However, if the concentration is high enough, the Coulomb attraction between electrons and ions can couple their motions so that ion diffusion is speeded up and electron diffusion is slowed. This phenomenon is called ambipolar diffusion and the transition from free to ambipolar diffusion has been studied in detail by Allis and Rose (1954). On the basis of their work, we can estimate that the transition occurs in air at a concentration of approximately $10^{-4}\, n_p$, a value used in the calculations leading to the curves in Fig. 14. The ambi-

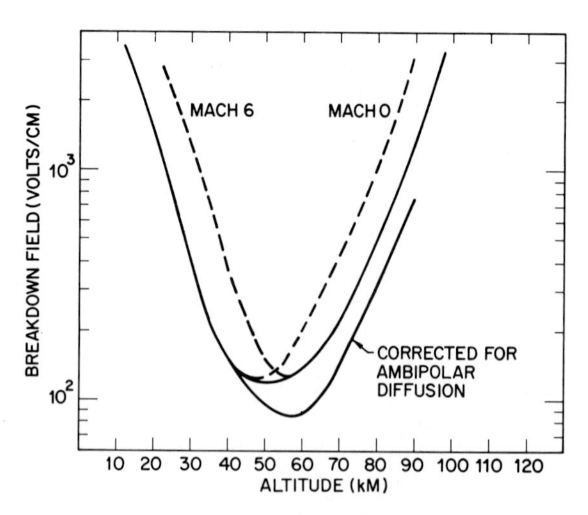

Fig. 14. Breakdown in air near a Mach 6 vehicle.

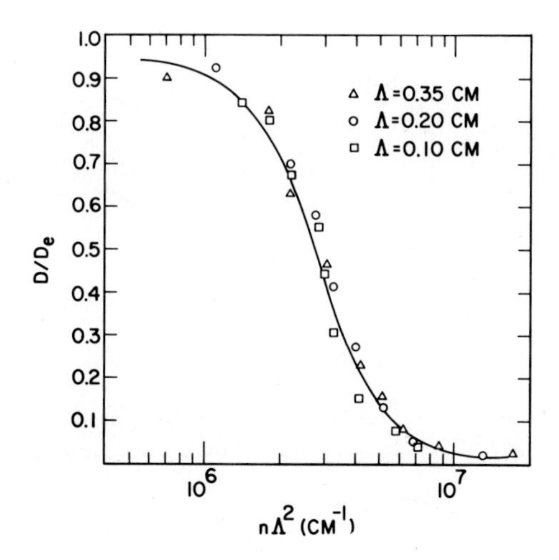

Fig. 15. Ratio of transition to free diffusion coefficients in neon.

polar diffusion coefficient has been estimated to be about one fortieth that of free diffusion in the analysis of Whitmer and MacDonald (1961). Using this value we obtain the curves shown. Direct measurements of this effect in air are very difficult because of the complex nature of the phenomenon, but there has been a detailed study of the effect of various degrees of pre-ionization breakdown in neon at 3.06 GHz (Bandel and MacDonald, 1969). This study, along with a study of the transition from free to ambipolar diffusion (MacDonald and Bandel, 1969) in neon, gives results consistent with the considerations used in the analysis above. Figure 15 shows the experimental data on the transition in neon and it should be noted that this transition is much sharper than theoretical analyses suggest.

III. MAINTAINING AND STEADY STATE DISCHARGES

The electric field required to break down a gas is much higher than that required to maintain the discharge after it has started. This is consistent with the fact that ambipolar diffusion coefficients are much smaller than free diffusion coefficients since the electron concentrations after breakdown are high enough so the electron loss is primarily by ambipolar diffusion: thus smaller fields are required to produce the ionization to replace the loss. The amount of the reduction in the field varies considerably depending on the gas and other experimental parameters. An example of the type of variation to be expected is shown in Fig. 16, which is based on the theoretical analysis

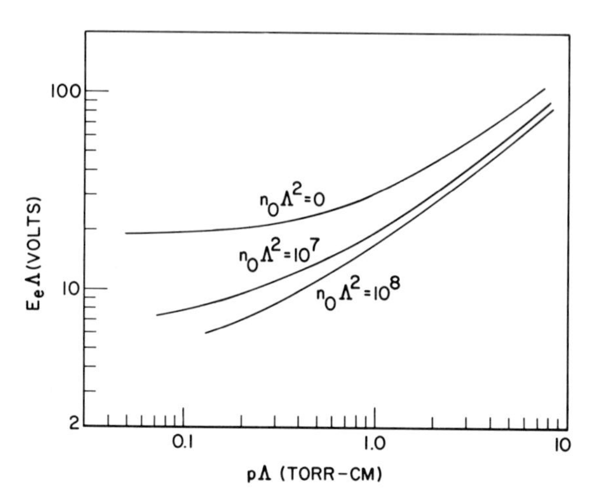

Fig. 16. Steady-state values of field as a function of $p\Lambda$ and electron concentration.

of Rose and Brown (1955). The gas in this case is hydrogen so that the effective field concept applies.

Both experiment and theory for steady state discharges are very difficult, particularly for the higher electron concentrations. Allis *et al.* (1951) studied the electron density distribution in a microwave steady-state discharge in helium. They were able to obtain electron density distributions across a microwave cavity for a variety of experimental conditions. They also calculated the electric field and the ionization rate across the cavity. The basic experimental data and the method of making these calculations for a wide range of variation of conditions in helium are given. Krasik *et al.* (1949) made careful and extensive measurements of the maintaining fields for argon at 2.95 GHz for a variety of experimental conditions and also obtained voltage-current characteristics for a considerable range of pressure. Rose and Brown (1955) measured the field required to maintain a hydrogen plasma as a function of both pressure and electron concentration. They also developed a theory based on a solution of the Boltzmann transport equation to relate all the variables. This theory was based on the assumption that electron and ion concentrations were not so high as to require consideration of recombination, or of electron–electron interactions. Their calculations yielded a value for the effective diffusion coefficient as well as distribution functions which made possible calculations of the electric field in the cavity as a function of pressure and of electron concentration. The curves of Fig. 16 are based on this calculation. Rose and Brown also did careful experimental measurements of the field over a range of variables within which the conditions of theory were met. The experimental data are not shown on the figure but agreed well with theory (within about 10%) over most of the range of pressure variation.

There has been quite a number of recent investigations of maintaining field discharges, but for the most part, although they contain information about specific cases, they do not alter our understanding of the basic processes.

IV. MICROWAVE GAS-DISCHARGE APPLICATIONS

A. Switches

There are many applications in which it is desired to switch microwave power in a controlled manner. Such applications include the switching of a transmitter to either of two antennas, the switching of a number of transmitters to a single antenna, such as in standby operation or to increase the system bandwidth, and duplexing in radar sets. The switching element generally has two states, a low impedance and a high impedance state. The switches can take many different forms, such as mechanical switches, solid state semiconductor and ferrite switches and gas discharge switches. The latter may be either of the dc discharge type or be self-activated by the incident high frequency electromagnetic waves. The different types of switches have relative advantages and disadvantages, depending upon the required isolation, insertion loss, firing and recovery times, power capability, and lifetime. In this section we will be concerned primarily with those devices in which a gas discharge is produced by the incident microwave power.

1. TR AND ATR TUBES

Most radars use a common antenna for both transmission and reception. The circuits which allow rapid low-loss switching between the transmitter and the antenna during transmission while protecting the sensitive input elements of the receiver from damage, and which allow fast, low-loss switching between the antenna and the receiver during reception, are called duplexers (Smullin and Montgomery, 1948; Kraszewski, 1967). Many such circuits use gas discharge devices called transmit–receive (TR) tubes and anti-transmit–receive (ATR) tubes. A typical TR tube providing a 10% bandwidth is shown in Fig. 17. The structure consists of two resonant filter sections each consisting of truncated cones as the capacitive element and an inductive iris. The relatively high electric field in the gap between the cones causes a rapid ionization of the gas. Resonant windows are located at each end, the spacing between successive resonant elements being $\lambda_g/4$.

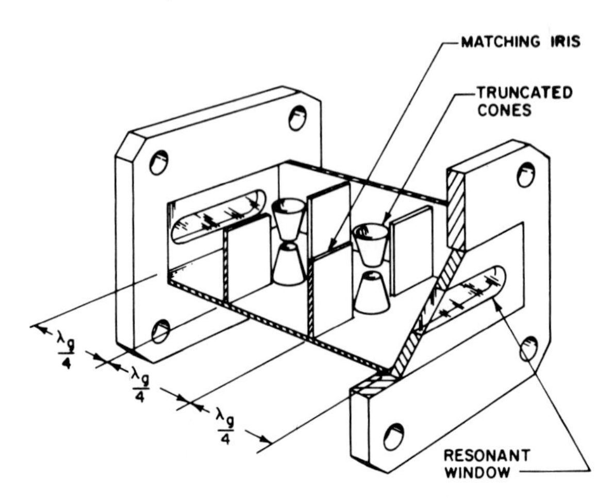

Fig. 17. Typical broadband TR tube.

The cone assembly nearest the output window usually has an ignitor or "keep-alive" electrode, coaxial with one of the cones. This provides a low-level dc glow discharge to ensure that a microwave discharge will be initiated within a few nanoseconds after the arrival of a high-power microwave pulse. The localized microwave discharge quickly expands, and in a time of the order of 10 nsec, an intense arc is created behind the input window, providing an effective short circuit. Another type of gas discharge device used in duplexing is the pre-TR tube. It is generally used in conjunction with a low-level TR tube called a crystal-protector (see below). Pre-TR tubes consist of a $\lambda_g/4$ section of waveguide sealed at each end with microwave windows. TR and pre-TR tubes produce an arc discharge in shunt with the appropriate waveguide line. One can also produce an arc in series with the waveguide line by means of an ATR tube. It consists of a $\lambda_g/4$ section of guide with a window on the input end and a metallic short at the other end. In a duplexer using ATR tubes, the window is located on the broad wall of the rectangular waveguide.

When very high power, long life operation is required, it is usually desirable to prevent the gas filling and gas discharge products from being in contact with metallic walls. Small, compact quartz tubes designed to fit snugly into appropriate waveguide mounts are often used to provide the desired short circuits (Ward *et al.*, 1965).

Most TR tubes are filled with a gas mixture consisting of various combinations of argon, water vapor, hydrogen, ammonia, and nitrogen. Krypton, chlorine and oxygen are also occasionally used. The total gas pressure is of the order of 10 Torr. Different gas fills will optimize one or another of the duplexer parameters such as the arc loss (microwave power absorbed by the gas when tube is fired), the recovery time, the leakage and the tube life. Two types of leakage are important, the spike leakage energy occurring at the start of the pulse and the flat leakage power which is present for the remainder of the pulse. The particular fill used depends upon the requirements of the radar system and is based on a compromise among the various parameters.

During duplexing, the TR tube must switch and protect the receiver simultaneously. It is advantageous to use separate tubes operating at widely different power levels for each of these functions. This allows optimum design for each function. The relatively low power TR tube is called a crystal or receiver protector. The high power switching function in gas discharge duplexers is usually performed by dual tubes with a common gas fill which are placed in a balanced waveguide configuration. Typical balanced duplexers are shown in Fig. 18. In Fig. 18a, the incident high power pulse fires the dual pre-TR and the power leaking through it fires the low-threshold power receiver protector. The effective short circuits, in conjunction with the left-hand coupler, switch the pulse to the antenna, while preventing excessive power from reaching the receiver. At the end of the transmitted pulse, the discharges decay and the tubes act like low-loss sections of wave-

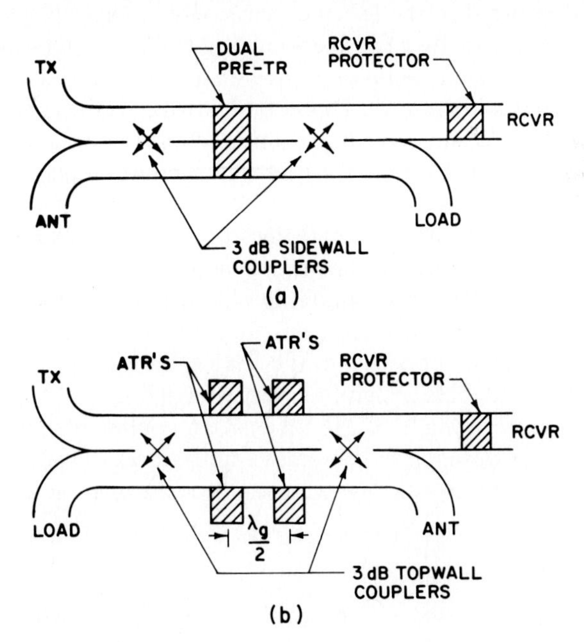

Fig. 18. Typical balanced duplexer configurations. (a) Balanced pre-TR duplexer. (b) Balanced ATR duplexer.

guide so that the radar return passes from the antenna through both couplers to the receiver. In Fig. 18b, during transmission, all the tubes fire. The effective short circuits at the ATRs provide a low impedance waveguide path and, in conjunction with both couplers, switch the pulse to the antenna. The protector short circuit limits the power to the receiver. During reception, the ATRs present a high impedance. The return signal passes from the antenna through the right-hand coupler, is reflected back through this coupler and passes through the receiver protector to the receiver.

Although many of the older radars still use gas discharge tubes for the high power switching function, the majority of modern radar systems use a ferrite circulator due to its long life and transmitter protection capability. The use of the balanced ATR duplexer in very high power radars is an exception to this trend because of its superior power handling capacity.

Whether the high power switching is done by gas discharge or ferrite devices, receiver protection is still required because of the vulnerability of present day receivers to unwanted signals from the radar transmitter or a nearby transmitter.

The basic limitation of the receiver protector TR is the dc "keep-alive" discharge, which tends to be unstable and to cause gas cleanup due to cathode sputtering (Walsh *et al.*, 1956). Two major improvements in receiver protectors have occurred in the last few years. One is the replacement of the active ignitor electrode by a passive ignitor. The ignitor takes the form of a metallic tritide radioactive film deposited on a rod or sheet close to the TR breakdown gap (Goldie, 1972). The second advance is the combining of the TR tube with a semiconducting diode limiter to form a device called a plasma limiter or a TR limiter (Brown, 1974). The TR limiter is superior to the active TR receiver protector in its leakage and recovery time characteristics and has a ten times greater useful life. It does not generate noise and provides receiver protection during shutdown, whereas the active TR tube usually requires a mechanical switch or shutter to isolate the receiver when the "keep-alive" is shut off. The TR limiter has found wide acceptance in modern radar systems. Typical properties of this device in the frequency bands from approximately 1–18 GHz are given by Brown (1974).

Studies have been made of TR tubes in an applied magnetic field near electron cyclotron resonance (Abramova and Starek, 1971). Such devices have the potential for drastically reducing the spike and flat leakage.

2. Multipactor

The multipactor or secondary electron resonance discharge (Hatch and Williams, 1958; Brown, 1959) is a form of high frequency and microwave discharge which occurs at very low pressures. It depends upon a sufficiently large secondary electron emission from the walls or electrodes of the evacuated volume.

Although this type of discharge has been used in devices for duplexing and the controlled switching of very high microwave power (Forrer and Milazzo, 1962; Ferguson and Dokken, 1974) and as an electron source (Gallagher, 1969), it is of practical importance primarily because one wishes to avoid its deleterious effects when designing electronic components and systems. For example, during multipacting, large amounts of energy can be dissipated in a small volume, and the resulting temperature rise can damage or destroy electrodes and rf windows. Multipacting can cause excessive loading in high power tubes such as klystrons (Priest and Talcott, 1961) and magnetrons (Vaughan, 1968), and in linear accelerators (Hubbard, 1968). Multipactor breakdown can occur on satellite antennas (Taylor *et al.*, 1972) and rocket-borne antennas (August *et al.*, 1967). Even relatively weak multipactor discharges can cause undesirable electronic noise.

Priest (1963) and Schrader (1968) have discussed many different types of multipactors. These can be of the one or two surface types, can occur with or without applied dc magnetic fields, and can have various orientations of the rf electric field, the static magnetic field, and the surfaces. The most common type of multipactor is the two-surface, no magnetic field type. Multipacting can occur when the electron transit time between the secondary electron emitting surfaces is close to an odd integral multiple of half-cycles of the applied electric field. Hatch and Williams (1958) have determined the conditions under which multipacting occurs between two parallel plate electrodes. They find that the frequency f, electrode separation d, and the applied voltage must fall within ranges or zones appropriate to the required electron dynamics. Figure 19, taken from their paper, shows the theoretical multipacting zones for the 1/2- through 9/2-cycle modes, the modes which have been observed experimentally. It can be seen that multipacting is possible under the condition

$$70 \lesssim fd \lesssim 10{,}000 \quad \text{MHz-cm.}$$

This condition should be avoided in the design of electronic components. If this is not feasible, one can resort to a number of techniques which have been devised for suppressing multipacting. These fall into two general categories: (1) alteration of the electron dynamics, and (2) reduction in the secondary emission ratio. Those which cause changes in the electron dynamics include biasing of the electrode with a dc potential (Zager and Tishin, 1964), superimposing a low-level rf field at a different frequency from the main field (Andreev and Zardin, 1971), applying a static magnetic

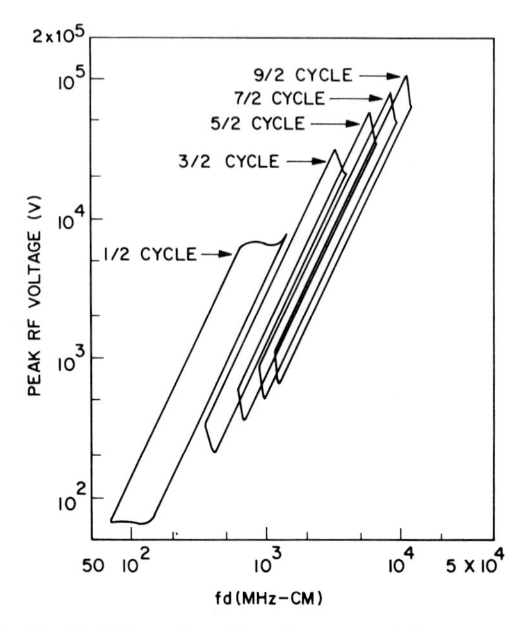

Fig. 19. Multiple mode multipacting theory; $\frac{1}{2}$–$\frac{9}{2}$-cycle modes.

field,* applying the rf power rapidly enough so that the voltage rises above the multipacting zone before the current has time to build up (Hubbard, 1968), and modifying the geometry if not incompatible with electrical and mechanical requirements. Those which involve a reduction in the secondary emission ratio include properly choosing and conditioning the surfaces, and using special coatings (Lesensky *et al.*, 1973).

3. Electron Cyclotron Resonance

In Section II microwave breakdown with applied magnetic field was discussed. It was pointed out that there is a large decrease in the breakdown field at electron cyclotron resonance ($B_c = (m/e)\omega$) compared to that in the absence of a magnetic field, when the pressure is sufficiently low, i.e., in the range below 1 Torr. This effect has been used to develop high-power magnetic-field-controlled switches (Tetenbaum and Hill, 1959).

The switch tube consists of a section of waveguide sealed at each end with microwave windows. One can also use encapsulated windows or a quartz tube fitting snugly into a waveguide section. The tube is filled with a gas at a sufficiently low pressure that with no magnetic field or only a small magnetic

* Under the proper magnetic field conditions, one can get enhanced multipacting (Brown, 1959; Priest, 1963).

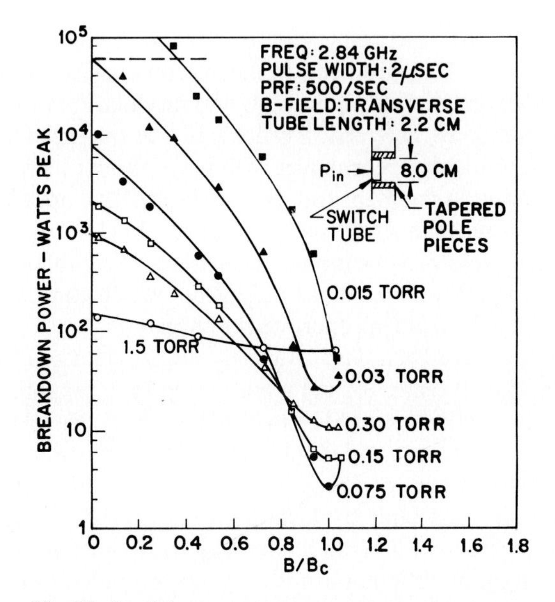

Fig. 20. Breakdown power versus magnetic field in argon.

field, the incident power cannot cause breakdown. However, when a magnetic field near cyclotron resonance is switched on, an intense discharge occurs within the tube. Figure 20 shows the breakdown power in an argon filled quartz tube in an S-band waveguide as a function of magnetic field (Tetenbaum and Hill, 1959). The magnetic field is parallel to the broad face of the waveguide. Similar results are obtained with the magnetic field oriented along the direction of propagation. It is seen that for pressures of the order of 10^{-2} to 10^{-1} Torr, the input power required for breakdown at cyclotron resonance is almost four orders of magnitude less than at zero magnetic field. The dashed line meets the 0.03 Torr breakdown curve at a point corresponding to 60 kW and $B = 0$. This means that the tube can be used for the controlled switching of a source power of 60 kW peak, if one uses a pressure below 0.03 Torr.

B. Frequency Converters

There is a large and continuing literature on the generation of harmonics and combination frequencies in gas discharges (Sodha and Kaw, 1969). Frequency conversion depends upon the nonlinear characteristics of the ionized gas. Sources of nonlinearities can be seen in the Boltzmann equation, which contains terms involving spatial gradients of the distribution function, $\mathbf{v} \times \mathbf{B}$ terms, and terms due to energy dependent collision frequencies. One can also have gradients in the microwave electric field intensity and in the static magnetic field, if present, and modulation of the ionization frequency. The frequency conversion process can be enhanced if the incident power is coupled to internal resonances within the plasma such as the Tonks-Dattner resonances (Asmussen and Beyer, 1968), the upper hybrid and electron cyclotron resonances (Tetenbaum *et al.*, 1964; Cano *et al.*, 1973), and the electron cyclotron harmonic resonances (Tetenbaum, 1967). We will be concerned primarily with gas discharges which can be self-initiated and maintained by the incident microwave power.

Gas discharges have been used as the nonlinear elements in microwave harmonic generators because of their high cw power capabilities and reasonable conversion efficiencies. Much of this work has been motivated by the desirability of obtaining relatively simple and inexpensive millimeter and sub-millimeter wave power sources. The first microwave plasma harmonic generator was built by Uenohara *et al.* (1957). The microwave discharge was created in the gap of a cylindrical post in S-band waveguide. Since then many different kinds of harmonic generators using cw or pulsed incident power have been built and investigated. For gas discharges with no applied magnetic fields, of the many gas fills tested, air appears to give the best results. The optimum pressure is generally near the value for which $\nu_c = \omega$. For gas discharge harmonic generators having an applied static magnetic field, best results are generally obtained in argon and at much lower pressures. Experimental results through 1965 have been tabulated by McIntosh (1968). The devices operated at fundamental frequencies of approximately 3, 10, and 35 GHz. Since 1965, a number of other harmonic generators at fundamental frequencies near 35 and 55 GHz have been studied (Yen and Lauks, 1966; Tamaru *et al.*, 1968, 1969). A second harmonic conversion efficiency of -15 dB was obtained by Yen and Lauks (1966). Using 14.7 W of cw power at 34 GHz, they obtained 440 mW at 68 GHz. Tamaru *et al.* (1968, 1969) used 5–10 W of cw power at a fundamental frequency of 54 GHz to obtain second, third, and fourth harmonic powers at conversion efficiencies of -28, -38, and -64 dB, respectively. Although these results are impressive, at the present time it is not clear whether they can be incorporated into practical and stable harmonic generators having reasonable lifetimes.

The devices discussed above were designed to optimize the harmonic output. In contrast, some consideration has been given to reducing or eliminating spurious harmonic generation in TR tubes (Maddix, 1969). Such generation can cause interference in nearby radar sets operating at the appropriate frequencies.

Gas discharges have been used as the nonlinear element in microwave mixers which generate the sum and difference frequencies of two incident signals (Tetenbaum *et al.*, 1964). Baird and Coleman (1961) built an X-band mixer and generated a 20 GHz sum frequency signal from incident 9 and 11 GHz signals. The amplitude of the sum signal was 28 dB down from the smaller input signal. Up to the present time very little additional work has been done on the design and construction of practical gas discharge mixers.

C. Sources for Physical and Chemical Studies

It is convenient to divide microwave discharge sources into those occurring in an applied magnetic field and those with no applied magnetic field. The former generally occur at pressures in the 10^{-6}–10^{-2} Torr range and are important in controlled thermonuclear fusion studies. They will be discussed below. This section will be primarily concerned with discharges with no applied magnetic field. Such discharges generally occur at pressures in the 10^{-1}–10^{3} Torr range.

Direct current and low frequency gas discharges have been used for many decades as sources for physicochemical studies and for chemical processing. Microwave discharges began to be used after the end of World War II, and during the past two decades there has been an expansion in their use for certain applications. This is due to their desirable properties relative to other kinds of discharges, and because the increased uses of microwave ovens and diathermy machines have made stable, high power, cw sources readily available. Microwave discharges generally provide a higher degree of ionization and dissociation than other types of discharges, can be used over a wide range of pressures, and generally provide a very high ratio of electron temperature to ion and gas temperatures. In addition, the absence of internal electrodes removes a possible source of contamination.

The discharges are almost always energized by cw magnetrons operating at 2.45 GHz with power outputs of up to a few kilowatts. In a few cases modulated power is used. Alger *et al.* (1972) found that a 20 kHz modulation of the 2.45 GHz source improved the stability and life of an electrodeless discharge lamp (EDL) and Callear *et al.* (1970) used S-band single pulses of 25 to 500 kW peak for flash-spectroscopy studies. The magnetron power is fed to a microwave structure containing the cylindrical discharge tube. The most common structures are coaxial cavities (Fehsenfeld *et al.*, 1965). Tapered rectangular cavities (Busch and Vickers, 1973), antennas (Mansfield *et al.*, 1968), cylindrical cavities (Asmussen *et al.*, 1974), and slow-wave structures (Bosisio *et al.*, 1972) are also used to couple the microwave energy to the discharge tube. The discharge tubes are usually made of quartz and are usually part of a gas flow system. Sealed off EDLs are generally used for atomic and molecular spectrometry and atomic fluorescent spectrometry studies.

Microwave discharges can efficiently generate electrons as well as positive and negative ions, free atoms and radicals, and excited atoms, molecules, and ions. They can also generate both continuous and line radiation. Although particle and radiation processes occur together, it is convenient to consider microwave discharge sources separately in terms of their primary applications.

1. Radiation Sources

Microwave discharges are being used as spectral line radiation and continuous radiation sources in many physical and chemical studies. These include studies of photoionization, atomic and molecular emission and absorption, and atomic fluorescence. Freeman and Wentworth (1971) used a microwave discharge as an atomic line source for photoionization detection in gas chromatography. Xenon and krypton vacuum ultraviolet (vuv) microwave discharge sources (Wilkinson and Tanaka, 1955; Wilkinson, 1955) provide strong continuous emission in the 0.125–0.225 μm wavelength region for absorption studies. Intense vuv atomic line sources have been obtained in flowing systems (Kikuchi, 1971). Helium and neon electron cyclotron resonance discharges have been used as sources for photoemission studies at wavelengths as short as 30 nm (Vorburger *et al.*, 1976). Campbell *et al.* (1971) used a sealed off tube as a continuous emission source in the vuv, the near uv, and the visible regions of the spectrum.

In recent years there has been much work done on EDLs as primary light sources for atomic absorption spectrometry and particularly for atomic fluorescence spectrometry. In comparison to other nonmicrowave discharge sources, they have the advantages of greater atomic line intensity with less background noise, narrow spectral lines with freedom from self-reversal, and greater stability and life. A description of the applications of these tubes has been given by Dagnall and West (1968), and Haarsma *et al.* (1974) have given a critical review of their preparation and operation. The sealed off quartz tubes have typical diameters of 0.2–1 cm and lengths of 2–10 cm. The tubes contain a metal or metal salt weighing of the order of 1 mg and are filled with argon or other rare gases at a pressure of the order of 1 Torr. Multielement EDLs have also been developed (Norris and West 1973).

EDLs almost always use a working frequency of 2.45 GHz. However, a high intensity radiation source has been built which uses a 35 GHz klystron as the excitation source (Kikuchi, 1972).

2. Particle Sources

A microwave discharge can produce electrons and a wide variety of neutral, excited and ionized atoms and molecules. Such particle sources have found use in gaseous electronics and plasma chemistry studies. A number of studies have utilized the afterglow of the discharge. The gas to be investigated is usually contained in a cylindrical or rectangular S-band cavity and is energized for a period of the order of 0.1–10 msec. One of the earliest studies dealt with ambipolar diffusion (Biondi and Brown, 1949). The many electron–positive ion recombination studies have been reviewed by Bardsley and Biondi (1970). Continuous-wave discharges have been used to produce atomic and molecular beams (Brink *et al.*, 1968).

Plasma chemistry is concerned with the identification of chemical species and the study of chemical reactions in a plasma as well as with studies of chemical reactions outside the active discharge region. McTaggart (1967) has summarized this field through 1964 with emphasis on microwave discharges. A survey of work on the chemical effects of microwave discharges covering the period from 1965 to 1973 has been given by Wightman (1974). These studies generally use cylindrical tubes in microwave cavities energized by 2.45 GHz cw power.

Microwave discharges have also been used to provide the active medium for laser pumping. Ahmed and Kocher (1964) used a 2.45 GHz cw magnetron to pump a He–Ne laser. The laser tube was placed in a waveguide T-junction shorted at both ends to form a resonant cavity. When a transverse static magnetic field at electron cyclotron resonance was applied at right angles to the microwave electric field, an additional 3 dB in output power was obtained. Electron cyclotron resonance discharges have also been used as the active medium in an argon ion laser (Goldsborough, 1966). A 2.45 GHz, 1 kW cw magnetron fed a meander line structure which produced an electric field parallel to the axis of the laser cavity and an electromagnet provided the perpendicular magnetic field. Experiments have shown that there is no significant improvement in laser output by employing electron cyclotron resonance. A pulsed argon ion laser has been investigated by Paik and Creedon (1968). The laser tube was incorporated into a dual-TR structure and excited by 1.26 GHz, 400 kW peak pulses at a duty factor of 0.001. Tuma (1970) has discussed a microwave discharge for xenon ion laser applications which is uniform and has a low fluctuation level. The gas was contained in a quartz tube used in conjunction with a cylindrical cavity excited by a 3.03 GHz, 100 W cw magnetron.

Microwave discharges have been used in chemical lasers, i.e., those where the population inversion is achieved by the direct production of excited species in chemical reactions. For example, in the cw CO chemical laser (Suart *et al.*, 1972) the chemical reactions are initiated by mixing CS_2 in a fast flow system with streams of oxygen atoms which have been produced in a microwave discharge.

Up to the present time, the use of microwave discharges for laser applications has been quite limited in comparison with other laser excitation techniques. With the possible exception of chemical lasers, their future use does not appear promising.

D. Controlled Thermonuclear Fusion

Microwave discharges have found application in the field of controlled thermonuclear fusion. Conventional fusion machines use magnetic fields to confine the charged particles and the plasmas of interest are usually located in magnetic fields appropriate to toroidal or magnetic mirror configurations.

Microwave plasmas are generally employed in two ways: (1) as an initial plasma source, to provide the plasma particles which are then available for the additional heating and confinement required to achieve practical fusion operation; and (2) as a medium for the study of wave propagation, conversion, and absorption, plasma instabilities, diffusion and turbulence.

The plasmas are usually initiated under electron cyclotron resonance conditions, although they are often maintained under other conditions (Budnikov *et al.*, 1970; Bernabei *et al.*, 1973). At the very low pressure used, one can obtain a plasma which is almost collisionless and fully ionized and has a low fluctuation level. The microwave power, which may be either pulsed or cw, is coupled to the flowing gas in a number of different ways. The use of sections of waveguide and of microwave cavity structures and of antennas has already been mentioned in a previous section. A common coupling scheme makes use of an interdigital or helical "Lisitano Coil" (Lisitano, 1966; Lisitano *et al.*, 1970), and can provide very efficient energy transfer.

There have been numerous studies of the physical properties of these discharges and of their interactions with incident electromagnetic waves. Measurements have been made of the electron and ion concentrations and energies, and their spatial and temporal variations. Studies of the collisionless absorption properties of these plasmas are of great importance because of the need to heat the plasma to very high temperatures. Efficient heating can be obtained by coupling the incident waves to a natural resonace of the plasma (Stix, 1962). The electromagnetic waves are converted to plasma waves which can then be absorbed in the plasma. Resonances which have been used for heating plasmas are the electron cyclotron resonance (Musil and Zacek, 1971) and the upper hybrid and electron cyclotron harmonic resonances (Grek and Porkolab, 1973). The corresponding ion resonances have been used to heat the positive ions in the plasma directly.

REFERENCES

Abramova, T. S., and Starik, A. M. (1971). *Sov. Phys.–Tech. Phys.* **15,** 1323–1326.
Ahmed, S. A., and Kocher, R. (1964). *Proc. IEEE* **52,** 1737–1738.
Alger, D., Dagnall, R. M., Sylvester, M. D., and West, T. S. (1972). *Anal. Chem.* **44,** 2255–2256.
Allis, W. P., and Rose, D. J. (1954). *Phys. Rev.* **93,** 84–93.
Allis, W. P., Brown, S. C., and Everhart, E. (1951). *Phys. Rev.* **84,** 519–522.
Anashkin, G. A. (1968). *Sov. Phys.–Tech. Phys.* **12,** 1076–1080.
Anashkin, G. A. (1970). *Sov. Phys.–Tech. Phys.* **15,** 972–976.
Andreev, V. G., and Zardin, D. G. (1971). *Instrum. Exp. Tech.* **14,** 845–846.
Asmussen, J., Jr., and Beyer, J. E. (1968). *J. Appl. Phys.* **39,** 2963–2964.
Asmussen, J., Jr., Mallavarpu, R., Hamann, J. R., and Park, H. C. (1974). *Proc. IEEE* **62,** 109–117.
August, G., Chown, J. B., and Nanevicz, J. E. (1967). *Digest Int. Antennas Propagat. Symp., Ann Arbor*, p. 147.
Baird, J. R., and Coleman, P. D. (1961). *Proc. IRE* **49,** 1890–1900.
Bandel, H. W., and MacDonald, A. D. (1969). *J. Appl. Phys.* **40,** 4390–4394.
Bandel, H. W., and MacDonald, A. D. (1970). *J. Appl. Phys.* **41,** 2903–2905.
Bardsley, J. N., and Biondi, M. A. (1970). *Adv. At. Mol. Phys.* **6,** 1–54.
Bernabei, S., DeDionigi, R., and Fontanesi, M. (1973). *Appl. Phys. Lett.* **22,** 85–86.
Biondi, M. A. (1963). *Phys. Rev.* **129,** 1181–1188.
Biondi, M. A., and Brown, S. C. (1949). *Phys. Rev.* **75,** 1700–1705.
Bleakney, W. (1930). *Phys. Rev.* **36,** 1303–1308.
Bosisio, R. G., Weissfloch, C. F., and Wertheimer, M. R. (1972). *J. Microwave Power* **7,** 325–346.
Bradbury, N. E. (1933). *Phys. Rev.* **44,** 883–890.
Bradford, H. M., Fraser, D. M., Langstroth, G. F. O., and MacDonald, A. D. (1959). *Can. J. Phys.* **37,** 1166–1170.
Brink, G. O., Fluegge, R. A., and Hull, R. J. (1968). *Rev. Sci. Instrum.* **39,** 1171–1172.
Brown, S. C. (1959). "Basic Data of Plasma Physics." MIT Press, Cambridge, Massachusetts.
Brown, N. J. (1974). *Microwave J.* **17,** 61–64.
Buchel'nikova, I. S. (1959). *Sov. Phys. JETP* **8,** 783–791.
Budnikov, V. N., Golant, V. E., and Obukhov, A. A. (1970). *Sov. Phys.–Tech. Phys.* **15,** 97–100.
Burch, D. S., and Geballe, R. (1957). *Phys. Rev.* **106,** 183–187.
Busch, K. W., and Vickers, T. J. (1973). *Spectrochim. Acta* **28B,** 85–104.
Callear, A. B., Guttridge, J., and Hedges, R. E. M. (1970). *Trans. Faraday Soc.* **66,** 1289–1296.
Campbell, J. P., Spisz, E. W., and Bowman, R. L. (1971). *Appl. Opt.* **10,** 2555–2557.
Cano, R., Fidone, I., Schwartz, M. J., and Zanfagna, B. (1973). *Plasma Phys.* **15,** 81–84.
Craggs, J. D., Thorburn, R., and Tozer, B. A. (1957). *Proc. Roy. Soc. London* **A240,** 473–483.
Crompton, R. W., Huxley, L. G. H., and Sutton, D. J. (1953). *Proc. Roy. Soc. London* **A218,** 507–519.
Dagnall, R. M., and West, T. S. (1968). *Appl. Opt.* **7,** 1287–1294.
Dawson, E. F., and Lederman, S. (1973). *J. Appl. Phys.* **44,** 3066–3073.
Fehsenfeld, F. C., Evenson, K. M., and Broida, H. P. (1965). *Rev. Sci. Instrum.* **36,** 294–298.

Ferguson, P., and Dokken, R. D. (1974). *Microwaves* **13** (7), 52–53.
Forrer, M. P., and Milazzo, C. (1962). *Proc. IRE* **50,** 442–450.
Fox, R. E. (1961). *J. Chem. Phys.* **35,** 1379–1382.
Freeman, R. R., and Wentworth, W. E. (1971). *Anal. Chem.* **43,** 1987–1991.
Gallagher, W. J. (1969). *Proc. IEEE* **57,** 94–95.
Gill, D. H., and Dougal, A. A. (1965). *Phys. Rev. Lett.* **15,** 845–847.
Golant, V. E. (1959). *Izv. Acad. Nauk SSR Ser. Fiz.* **23,** 952.
Goldie, H. (1972). *IEEE Trans. Electron Devices* **ED-19,** 917–928.
Goldsborough, J. P. (1966). *Appl. Phys. Lett.* **8,** 218–219.
Gould, L. J., and Roberts, L. W. (1956). *J. Appl. Phys.* **27,** 1162–1170.
Grek, B., and Porkolab, M. (1973). *Phys. Rev. Lett.* **30,** 836–839.
Haarsma, J. P. S., DeJong, G. J., and Agterdenbos, J. (1974). *Spectrochim. Acta* **29B,** 1–18.
Harrison, M. A., and Geballe, R. (1953). *Phys. Rev.* **91,** 1–7.
Hatch, A. J., and Williams, H. B. (1958). *Phys. Rev.* **112,** 681–685.
Herlin, M. A., and Brown, S. C. (1948). *Phys. Rev.* **74,** 1650–1656.
Hubbard, E. L. (1968). *Adv. Electron. Electron Phys.* **25,** 62.
Huxley, L. G. H., and Crompton, R. W. (1974). "Diffusion and Drift of Electrons in Gases." Wiley (Interscience), New York.
Kikuchi, T. T. (1971). *Appl. Opt.* **10,** 1288–1295.
Kikuchi, T. T. (1972). *Appl. Opt.* **11,** 687–688.
Krasik, S., Alpert, D., and McCoubrey, A. D. (1949). *Phys. Rev.* **76,** 722–730.
Kraszewski, A. (1967). "Microwave Gas Discharge Devices." Illiffe, London.
Lax, B., Allis, W. P., and Brown, S. C. (1950). *J. Appl. Phys.* **21,** 1297–1304.
Lesensky, L., Tiernan, R. J., and Tisdale, L. H. (1973). *Conf. Electron Device Tech., New York*, pp. 84–86.
Lisitano, G. (1966). *Proc. Conf. Phenomena Ionized Gases, 7th, Beograd* **1,** 464–467.
Lisitano, G., Fontanesi, M., and Sindoni, E. (1970). *Appl. Phys. Lett.* **16,** 122–124.
MacDonald, A. D. (1966). "Microwave Breakdown in Gases." Wiley, New York.
MacDonald, A. D., and Bandel, H. W. (1969). *Int. Conf. Phenomena Ionized Gases, Bucharest, 9th,* p. 427.
MacDonald, A. D., and Betts, D. D. (1952). *Can. J. Phys.* **30,** 565–576.
MacDonald, A. D., and Brown, S. C. (1949a). *Phys. Rev.* **75,** 411–418.
MacDonald, A. D., and Brown, S. C. (1949b). *Phys. Rev.* **76,** 1634–1639.
MacDonald, A. D., and Brown, S. C. (1950). *Can. J. Res.* **A28,** 168–174.
MacDonald, A. D., and Matthews, J. H. (1955). *Phys. Rev.* **98,** 1070–1073.
MacDonald, A. D., and Matthews, J. H. (1956). *Can. J. Phys.* **34,** 395–397.
MacDonald, A. D., Gaskell, D. U., and Gitterman, H. N. (1963). *Phys. Rev.* **130,** 1841–1850.
Maddix, H. S. (1969). *IEEE Trans. Electron Devices* **ED-16,** 278–283.
Maier-Leibnitz, H. (1935). *Z. Phys.* **95,** 499–523.
Mansfield, J. M., Bratzel, M. P., Norgordon, H. O., Knapp, D. O., Zacha, K. E., and Winefordner, J. D. (1968). *Spectrochim. Acta* **23B,** 389–402.
Massey, H. S. W., and Burhop, E. H. S. (1952, 1969). "Electronic and Ionic Impact Phenomena," Oxford Univ. Press, London and New York.
McDaniel, E. W. (1964). "Collision Phenomena in Ionized Gases." Wiley, New York.
McIntosh, R. E. (1968). *Proc. IEEE* **56,** 1210–1212.
McTaggart, F. K. (1967). "Plasma Chemistry in Electrical Discharges." Elsevier, Amsterdam.
Mentzoni, M. (1973). *J. Phys. Appl. Phys.* **6,** 490–497.
Mentzoni, M. (1974). *J. Phys. Appl. Phys.* **7,** 374–377.
Meyerand, R. G., and Haught, A. F. (1963). *Phys. Rev. Lett.* **11,** 401.
Musil, J., and Zacek, F. (1971). *Plasma Phys.* **13,** 471–476.

Norris, J. D., and West, T. S. (1973). *Anal. Chem.* **45,** 226–230.

Paik, S. F., and Creedon, J. E. (1968). *Proc. IEEE* **56,** 2086–2087.

Pim, J. A. (1949). *Proc. Inst. Elec. Eng.* **96,** 117–129.

Priest, D. H. (1963). *Microwave J.* **6,** 55–60.

Priest, D. H., and Talcott, R. C. (1961). *IRE Trans. Electron Devices* **ED-8,** 243–251.

Raizer, Yu. P. (1966). *Sov. Phys.-Usp.* **8,** 650–673.

Ramien, H. (1931). *Z. Phys.* **70,** 353–374.

Reder, F. H., and Brown, S. C. (1954). *Phys. Rev.* **95,** 885–889.

Rogers, W. A., and Biondi, M. A. (1964). *Phys. Rev.* **134,** A1215–A1225.

Rose, D. J., and Brown, S. C. (1955). *Phys. Rev.* **98,** 310–316.

Schrader, W. J. (1968). *Physica* **40,** 223–228.

Schulz, G. J. (1962). *Phys. Rev.* **128,** 178–186.

Smullin, L. D., and Montgomery, C. G. (1948). "Microwave Duplexers." McGraw-Hill, New York.

Sodha, M. S., and Kaw, P. K. (1969). *Adv. Electron. Electron Phys.* **27,** 187–293.

Stix, T. H. (1962). "The Theory of Plasma Waves." McGraw-Hill, New York.

Suart, R. D., Dawson, P. H., and Kimbell, G. H. (1972). *J. Appl. Phys.* **43,** 1022–1032.

Tamaru, T., Chiyoda, K., and Maejima, Y. (1968). *Jp. J. Appl. Phys.* **7,** 767–770.

Tamaru, T., Chiyoda, K., and Maejima, Y. (1969). *Proc. IEEE* **57,** 851–852.

Tate, J. T., and Smith, P. T. (1932). *Phys. Rev.* **39,** 270–277.

Taylor, W. C., Scharfman, W. E., and Morita, T. (1971). *Adv. Microwaves* **7,** 59–130.

Taylor, W. C., August, G., and Chown, J. B. (1972). *Digest Int. Antennas Propagat. Symp., Williamsburg,* pp. 44–46.

Tetenbaum, S. J. (1967). *Phys. Fluids* **10,** 1855–1856.

Tetenbaum, S. J., and Hill, R. M. (1959). *IRE Trans. Microwave Theory Tech.* **MTT-7,** 73–82.

Tetenbaum, S. J., Whitmer, R. F., and Barrett, E. B. (1964). *Phys. Rev.* **135,** A374–A381.

Tetenbaum, S. J., MacDonald, A. D., and Bandel, H. W. (1971). *J. Appl. Phys.* **42,** 5871–5872.

Tetenbaum, S. J., MacDonald, A. D., and Bandel, H. W. (1973). *IEEE Trans. Plasma Sci.* **PS-1,** 55–57.

Tuma, D. T. (1970). *Rev. Sci. Instrum.* **41,** 1519–1520.

Uenohara, M., Uenohara, M., Masutani, T., and Inada, K. (1957). *Proc. IRE* **45,** 1419–1420.

Vorburger, T. V., Waclawski, B. J., and Sandstrom, D. R. (1976). *Rev. Sci. Instrum.* **47,** 501–504.

Vaughan, J. R. M. (1968). *IEEE Trans. Electron Devices* **ED-15,** 883–889.

Walsh, D., Bright, A. B., and Bridges, T. J. (1956). *Brit. J. Appl. Phys.* **7,** 31–35.

Ward, C. S., Jellison, F. A., Brown, N. J., and Gould, L. (1965). *IEEE Trans. Microwave Theory Tech.* **MTT-13,** 801–805.

Whitmer, R. F., and MacDonald, A. D. (1961). *In* "Electromagnetic Effects of Re-Entry" (W. Rotman and G. Meltz, eds.), pp. 149–154. Oxford Univ. Press (Pergamon), London and New York.

Wightman, J. P. (1974). *Proc. IEEE* **62,** 4–11.

Wilkinson, P. G. (1955). *J. Opt. Soc. Am.* **45,** 1044–1046.

Wilkinson, P. G., and Tanaka, Y. (1955). *J. Opt. Soc. Am.* **45,** 344–349.

Yen, J. L., and Lauks, V. (1966). *Electron. Lett.* **2,** 20–21.

Zager, B. A., and Tishin, V. G. (1964). *Sov. Phys. Tech. Phys.* **9,** 234–241.

Chapter 4

Corona Discharges

M. GOLDMAN and A. GOLDMAN
LABORATOIRE DE PHYSIQUE DES DÉCHARGES
DU CENTRE NATIONAL DE LA RECHERCHE SCIENTIFIQUE
ECOLE SUPÉRIEURE D'ELECTRICITÉ
GIF-SUR-YVETTE, FRANCE

I. INTRODUCTION

By the term "corona discharge," one generally refers to the ensemble of phenomena which occur in a gaseous medium in the vicinity of conductors of small radius of curvature, subjected to intense, but not disruptive, electric fields. These can occur, for example, along electrical power transmission

ISBN 0-12-349701-9

lines, along a wire surrounded by a coaxial cylinder, or in the vicinity of irregularities in the form of sharp points, on the surface of a conductor at high voltage. The corona is a self-sustaining discharge, that is, it requires no external source of ionization for the discharge to be initiated or maintained.

In the book by Loeb (1965) dealing with the corona discharge, one will find a review of the progressive evolution of our understanding of the subject which has taken place since the earliest studies were carried out at the end of the last century. The persistent improvement of experimental techniques in electronics and electro-optics, the increasing acquisition of basic data in the physics of particle collisions and of ionized mediums, and the generalized employment of computers have permitted, during the past two decades, a significant step to be taken between the macroscopic properties of the corona on the one hand and its fundamental mechanisms on the other. This step has been taken by the elaboration of quantitative models starting from a detailed study of microscopic collision properties.

In the process, one might say that the corona discharge has lost its individuality as a result of these studies. In fact, in spite of its very specific aspects, it appears that, in terms of the physical mechanisms which it brings into play, the corona discharge belongs to the general class of glow discharges (see Chapter 2), as has been established by Loeb's school for the negative corona, and more recently by Marode (1972) for the positive discharge. Besides, the domains of pressure of the two types of discharge converge at several Torr and, by extrapolation, it has already been possible to apply to the corona discharge certain results relative to the glow discharge, for example, those referring to cathode phenomena.

The number of works still being carried out on the corona discharge bears witness to the complexity of the fundamental studies on the subject, which is related to the diversity of the relevant parameters, but displays equally well the interest which it bears because of its applications. Although responsible for power losses and radio frequency noise in the domain of electrical transport as well as for insulation faults in apparatus with gaseous dielectrics, corona discharge is utilized for many beneficial ends in numerous applications, such as dust precipitation and electrostatic painting, the commercial generation of ozone, the surface treatment of cellulosic and polymeric materials, and telecopying.

The extent of the subject explains the gaps in this chapter, where the authors must leave these applications as well as certain aspects of the discharge mechanism itself practically untreated. This applies especially to chemical reactions in the gaseous medium, to the influence of the nature of the gas and its pressure and humidity (studies which have been insufficiently developed), to the role of the electrical wind, and to the influence of overvoltages.

II. ONSET OF THE CORONA DISCHARGE

A. Phenomena below Corona Threshold

The threshold voltage V_s of the corona discharge is characterized by an abrupt increase in the current between the electrodes (from about 10^{-14} to 10^{-6} A) (Bandel, 1951), and the appearance of a feeble glow in the vicinity of the electrode with the smaller radius of curvature (we shall call this the "active" or "stressed" electrode). Below V_s, the current consists of charges resulting from natural ionization in the gas, notably by cosmic rays and radiations from radioactive substances. These produce on the average of 7 to 20 electron-positive ion pairs per cubic centimeter per second in air at ground level (Schonland, 1953; Kuffel, 1959). In air at atmospheric pressure and for low electric fields, the electrons thus produced are rapidly converted into negative ions, mainly of oxygen (Lécuiller *et al.*, 1972), either of the type O_3^- (O_3^-, CO_3^-) or the type O_2^- (O_2^-, O_4^-, CO_4^-, $N_2O_2^-$), clustered with neutral molecules, notably H_2O. The positive ions seen are predominantly of the type $H^+(H_2O)_n$ ($n = 0, 1, 2, \ldots$) (Shahin, 1969).

The positive and negative ions diffuse through the gas under the influence of the applied field, and some of them recombine with each other. In a volume containing no externally applied electric field, the concentrations n_+, n_- of the positive and negative ions are essentially equal: $n_+ \approx n_- \equiv n$. The ion densities approach a steady state value characterized by (Thomson and Rutherford, 1896):

$$dn/dt = \gamma_i - \alpha_R n^2 = 0, \tag{1}$$

where γ_i is the rate of production, and α_R the rate coefficient for the recombination of positive and negative ions; this leads to a steady-state concentration $n_s = (\gamma_i/\alpha_R)^{1/2}$ which depends strongly on the composition and the state (temperature, humidity, wind, etc.) of the atmosphere. In practice, one can divide the ions grossly into two categories: "small" ions, of density n, characterized by high mobility ($K \cong 10^{-4}\ m^2\ s^{-1}\ V^{-1}$), and "large" ions, of density N, with mobilities on the order of 10^3 times smaller (Langevin, 1905). Because of the variation of α_R with ion identity, the large ions seem to be formed at the expense of the small ions, such that the product nN remains approximately constant (Bricard and Pradel, 1966); at ground level, for example, one may find at the same place, in two successive measurements, $n = 400\ cm^{-3}$, $N = 2 \times 10^3\ cm^{-3}$ and $n = 40\ cm^{-3}$, $N = 3 \times 10^4\ cm^{-3}$. It is common to find such large densities of these heavy ions in the atmosphere.

It is interesting to emphasize that the production of one electron-positive ion pair results from the average expenditure of about 35 eV by the ionizing source, although the actual ionization energies for nitrogen and oxygen are only 15.6 and 12.2 eV, respectively. More than half the energy supplied by the ionizing source must be dissipated in other forms, notably kinetic energy, molecular dissociation, and excitation, which can all lead to chemical reactions in the volume of the gas and at surfaces (e.g., corrosion).

B. Threshold Criteria

1. Threshold Field at the Stressed Electrode

The corona discharge proper begins with an avalanche. This avalanche must be initiated by an electron. If this electron cannot arise from the cathode, and if only negative ions and no electrons are initially present, the electron must first be detached from the ion before avalanche formation can occur. The above two statements will be seen to account for the variation of the gross behavior of the corona threshold with electrode polarity.

The Townsend (1914a) criterion for a self-sustaining discharge is written as

$$\gamma\left[\exp\int_l (\alpha - \eta)\,dl - 1\right] = 1, \tag{2}$$

where α and η are the primary ionization and attachment coefficients, and γ the secondary ionization coefficient, as defined in Chapter 2, for the glow discharge. These coefficients depend on the ratio of the effective local field E at each point x along the trajectory l of the avalanche to the gas pressure p (E is here a superposition of the applied electric field and the space-charge field produced by the development of the avalanche itself). In modern usage, the ratio E/N is frequently used in place of E/p, where N is the number density of air molecules; since much of the literature refers to E/p, however, we continue that practice here.

In an inhomogeneous field, avalanches develop near those regions where the field is greatest, i.e., where the radius of curvature of the electrodes is smallest. To lead to corona discharges, the avalanches must satisfy two criteria which will be described below. Apart from particular cases (e.g., Beattie and Cross, 1974), this occurs only if the ratio E_r/E_R of the fields at the stressed electrode E_r and the unstressed electrode E_R is sufficiently large; what constitutes "sufficiently large" depends on the electrode geometry. For example (Cobine, 1958), a corona does not form on parallel wires with radius r and gap distance d in air, if the ratio d/r is less than 5.85, a spark occurring first.

The threshold phenomenology varies with the polarity of the active electrode.

(a) *Maximum Field Occurring in the Vicinity of the Negative Electrode (Cathode)* Avalanches develop in a direction away from the cathode, out to that distance at which $\alpha(E/p) = \eta(E/p)$, which limits the active multiplication zone. For atmospheric air, this corresponds to a critical field E_c on the order of 26 kV cm^{-1} (Driver, 1969). Equation (2) can then be written

$$\gamma\left[\exp\int_{(E_r)_s}^{E_c}(\alpha-\eta)\,dx - 1\right] = 1, \tag{3}$$

where $(E_r)_s$ represents the field at the stressed electrode at corona onset. For this case, the coefficient γ reflects, for the most part, secondary processes on the cathode (Hosokawa *et al.*, 1970), and luminous phenomena associated with multiplication remain localized near the cathode.

(b) *Maximum Field Occurring in the Vicinity of the Positive Electrode (Anode)* In this case, avalanches develop toward the anode, beginning at a location between the electrodes which depends on the mode of production of the "seed" electrons. If free electrons are present, the minimum field required for avalanches in air is 26 kV cm^{-1}, as above. If there are no free electrons, the avalanches in air are created from electrons detached from negative ions, primarily O_2^-. One should note that this process, which requires higher fields for maximum efficiency, only affects the discharge at corona threshold in terms of a reduced cross section which is reflected in the time lag.

Since the cathode is far away, the essential role in the γ coefficient is now played by secondary processes in the gas, although some observations of secondary cathode processes above threshold have been observed, for example, when the discharge is stabilized into a glow discharge (Le Ny *et al.*, 1974). Visually, the discharge now manifests itself by luminous filaments which develop in the gap outward from the active electrode. These are the streamers, of which we shall speak in greater detail later; for the moment, we are content to indicate that they occur when the space-charge field created by the cloud of positive ions remaining behind the head of the avalanche is of the same order of magnitude as the applied field (Meek criterion, 1940; Loeb and Meek, 1940). This occurs when the number of electrons created by the avalanche reaches a critical value given in terms of Eq. (2) by

$$N_e = \int_{E_c}^{(E_r)_s}(\alpha-\eta)\,dx \approx 1 - 5\times 10^8 \quad \text{electrons} \tag{4}$$

(Raether criterion, 1940).

2. Calculation of Threshold Field

Peek's law (1929) was developed to calculate the threshold field at the surface of the inner conductor for the case of coaxial cylindrical geometry. This field is given by

$$(E_r)_s = E_0 m\delta[1 + k/(\delta r)^{1/2}], \tag{5}$$

where $(E_r)_s$ is the threshold field (kV cm^{-1}), $E_0 = 31$ kV cm^{-1} (value close to the sparking field strength for air under STP and uniform field, in gaps on the order of 1 cm), m is a coefficient describing the conductor surface ($0.6 < m < 1$), $\delta = 0.392p/(273 + T)$ ($\delta = 1$, for $p = 760$ Torr, $T = 25°$C), $k = 0.308$ cm$^{1/2}$, and r is the radius of the conductor (cm). In fact, this empirical law has been shown to be valid for air at pressures from several Torr to several atmospheres, for conductor radii between 0.1 mm and several centimeters, and for applied voltages from dc to several kilohertz. In its general form, it is also valid for other gases (Hackam *et al.*, 1976).

If we know the functions $E(x)$, $\{\alpha - \eta\}(E/p)$, and eventually $\gamma(E/p)$, the critical field at the surface of the active electrode should be calculated by either (3) or (4), according to whether the stressed electrode is negatively or positively polarized. This was done semianalytically by Gary *et al.* (1972) for a negatively polarized stressed electrode, by using for $\alpha(E/p)$, in place of the expression of Townsend

$$\frac{\alpha}{p} = A \exp\left(-\frac{B}{E/p}\right), \tag{6}$$

the following relation which better agrees with experiment in the interesting range of electric fields encountered in electrotechnical engineering ($E/p < 150$ V cm^{-1} Torr^{-1}):

$$\frac{\alpha}{p} = \frac{C}{p_0}\left[\left(\frac{E/p}{E_0/p_0}\right)^2 - 1\right], \tag{7}$$

where, for air under STP, $E_0 = 31$ kV cm^{-1} as in Peek's formula, and $C/p_0 = 0.14$ electron-positive ion pair cm^{-1} Torr^{-1}. In this way, they obtained the Peek expression analytically for electrodes presenting a single curvature (coaxial cylinders or parallel wires); for electrodes presenting a double curvature (paraboloidal point-to-plane, concentric spheres, etc.), they obtained a formula which differs from (5) only by a factor of 1/2 applied to the radius of curvature:

$$(E_r)_s = E_0\left[1 + \frac{k}{(r/2)^{1/2}}\right], \tag{8}$$

where E_0 and k are as given for the Peek formula, m and δ being here assumed equal to unity. The experiments which they carried out show that these relations apply equally to stressed anodes and stressed cathodes, the critical fields being nearly identical for the two polarities, despite the difference between the corresponding criteria.

3. Threshold Voltage

Microscopically, the discharge threshold condition is defined by $(E_r)_s$; however, one can easily determine the corresponding threshold voltage V_s, if one can express the field E_x at a point x in the gap as a function of the applied voltage V. For example, consider the following geometries:

(a) *Coaxial cylinders, radii r and R, at a distance x from the axis*

$$E_x = V[x \ln(R/r)]^{-1}, \tag{9a}$$

from which $V_s = (E_r)_s r \ln(R/r)$.

(b) *Concentric spheres, radii r and R, at a distance x from the center*

$$E_x = VrR[x^2(R - r)]^{-1}, \tag{9b}$$

from which $V_s = (E_r)_s[r(R - r)/R]$. This formula leads, for the sphere-to-plane gap ($r \ll R$): $V_s = (E_r)_s r$.

(c) *Paraboloidal point-to-plane gap, point radius r and gap distance d, along the gap axis at a distance x from the tip of the point*

$$E_x = \frac{2V}{(r + 2x) \ln\{(r + 2d)/r\}} \tag{9c}$$

from which $V_s = (E_r)_s(r/2) \ln\{(r + 2d)/r\}$.

The effect of pressure on V_s has been studied by Llewellyn-Jones and Williams (1953) for a coaxial wire-cylinder geometry. They showed that the similarity principle (see Chapter 2) is obeyed for clean electrodes, and that the variations of V_s are described by Paschen-type curves (Llewellyn-Jones, 1957). We call attention to the fact that V_s can vary significantly with the state of the electrode surfaces.

C. Time Lags

The time interval between application of the voltage to the electrodes and the appearance of a corona (which can be seen by the abrupt increase in collected current) is called the time lag. This time lag can be measured by applying a voltage step across the gap at time $t = 0$ which is equal to or greater than the threshold voltage. Foreknowledge of this time lag may be of practical interest for designing breakdown protection in critical apparatus; understanding of the time lags is of fundamental interest in unraveling the most probable mechanisms of corona formation.

Until now, only measurements in positive discharges (i.e., in which the anode is the stressed electrode) have given useful results; in negative discharges, the measured time lags display great scatter. Loeb attributes this fact to difficulties inherent in the conditioning to which the cathode is subjected when it is the active electrode. "The breakdown, being in this case a cathode-controlled phenomenon, is very sensitive to the surface properties of the metal cathode, i.e., to the state of the surface, which changes in time" (Loeb, 1965).

On the other hand, for positive polarities, the cathode does not play any role in the formation of the filamentary streamer at threshold. This is precisely one of the results to which Menes and Fisher (1954) arrived by measuring time lags shorter than the transit time of electrons from the cathode across the total gap, so cathode electrons could not be responsible. In this case, streamer generation occurs in two phases:

Phase 1 A seed electron, coming from natural processes (cosmic rays, natural radioactivity) or artificial ones (artificial radioactivity, uv light, additional corona), appears in a suitable place in the gap where it can multiply; to this first phase corresponds a delay time t_s referred to as the "statistical" time lag.

Phase 2 The avalanche born from this electron attains the critical density for streamer formation; to this second phase corresponds a time t_f called the "formative" time lag.

We show first that t_f plays a negligible role in the measured establishment time. In effect, the order of magnitude of t_f, assuming it to be the time for the seed electron to travel the distance l covered by the avalanche, is given by

$$t_f = \int_l \frac{dl}{v_e}. \tag{10}$$

For small distances (≈ 1 cm), one obtains t_f smaller than 10^{-9} *s*, compared with measured time lags in the 10^{-8}–10^{-6} *s* range.

Let us consider now the statistical time lag. If the seed electron is too close to the electrode, the avalanche strikes the electrode before the streamer has fully formed; if it is too far, in a region where attachment dominates ionization ($E < 26$ kV cm^{-1}), the streamer cannot form. This defines an "initiation volume" (Fig. 1a), which at threshold should theoretically be reduced to a point on the axis of the electrode system, and which should grow with increasing applied voltage, both in length along the axis and in lateral extent. In fact, if one calculates this volume for the observed threshold voltages, one obtains initiation volumes of nonvanishing size: for example, 0.36 cm^3 for a hyperboloidal point electrode with an 8° cone angle at a distance of 25 mm from a plane (Saint-Arnaud, 1969).

Because of their short lifetimes due to attachment (Waters and Jones, 1964a), electrons produced by natural events can be available only if created in the initiation volume, but the probability for a seed electron to be created directly by this process during the corresponding mean time lag (0.7 μs for the case referred to above) is definitely negligible. Even in the case of a long gap, e.g., $d = 10$ m, $r = 1$ cm, for which the initiation volume at threshold is almost 0.5 m^3, this probability is only 5% (Berger and Goldman, 1974).

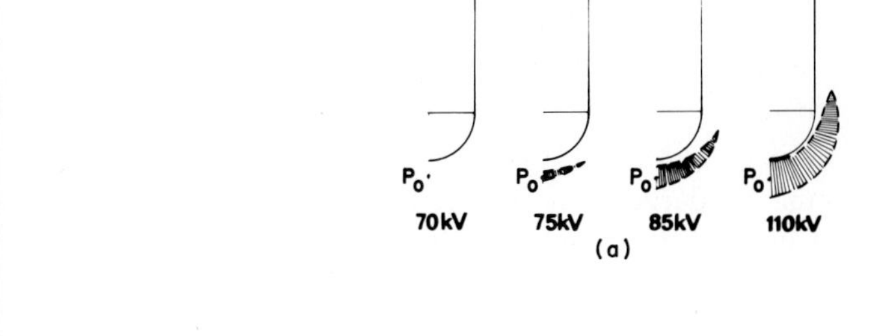

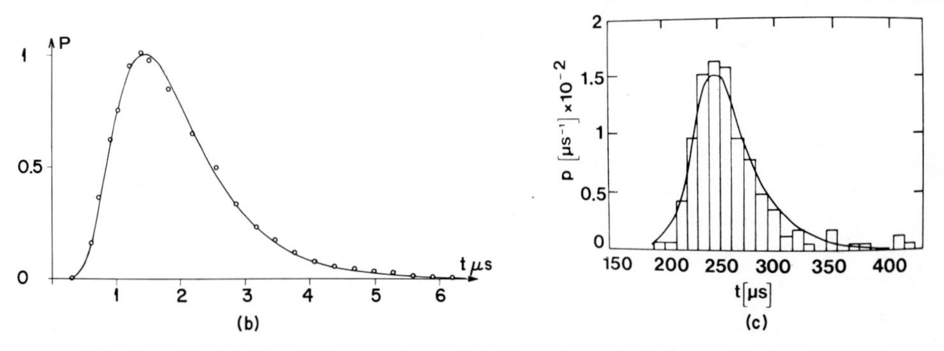

Fig. 1. Initiation of positive corona. (a) Variation of the "initiation volume" as function of applied voltage; 2 cm diameter rod (Baldo, 1974). (b) and (c) Distribution of time lags for the initiation of positive corona: (b) Point-to-plane air gap in the centimeter range (d = 15 mm, point radius r = 40 μm, 5 kV applied pulse with 10 ns rise time and 100 μs duration). Experimental points; distribution computed from Eq. (14) (Berger *et al.*, 1972; Berger, 1973). (c) Long air gap (d = 2 m, hemispheric tip, 30 cm radius, 350/10,000 μs impulse voltage, pulse height 1000 kV). Experimental histogram; distribution computed from Eq. (15) (Renardières Group, 1974).

To explain the appearance of the seed electron, it is necessary to invoke detachment from negative ions (Saint-Arnaud, 1969), notably from O_2^- ions, which occur with maximum efficiency at 68.5 kV cm^{-1} (Loeb, 1935). However, the negative ions present in the atmosphere being predominantly hydrated ions, and the probability of direct detachment from one of these ions being small, a retarding effect could intervene due to declustering processes (Renardières Group, 1974).

The problem can be treated theoretically by considering the probability of streamer formation $P(t)$ as the product of three probabilities:

$$P(t) = p_1(t) \cdot p_2(t) \cdot p_3(t), \tag{11}$$

where p_1 is the probability that a negative ion comes into the useful volume, p_2 the probability of electron detachment from this ion, and p_3 the probability that the electron thus created gives birth to an avalanche of sufficient size to initiate a streamer. The relative importance of these three factors is strongly dependent on the parameters defining the nature of the discharge. Thus, for small active volumes, a model which only considers p_1, and for large active volumes, a model which only considers p_2, both give satisfactory agreement with experiment.

(1) First, it was found by Berger *et al.* (1972) that the best fit to their experimental measurements using a point-to-plane gap in the centimeter range ($d = 15$ mm) is given by a lognormal distribution (Aitchison and Brown, 1969) of the form

$$dN(t) = [(t - a)\sigma(2\pi)^{1/2}]^{-1} \exp\{(-1/2\sigma^2)[\ln(t - a) - \mu]^2\}\, dt \tag{12}$$

where a is a formative time lag. A transformation $Y = \ln(t - a)$ gives a normal or gaussian distribution with parameters μ, σ^2.

The problem has been treated theoretically by Berger (1973), who related the probability density p_1 for obtaining the first discharge to the negative ion current $i_-(t)$ flowing toward the highly stressed anode by

$$p_1(t) = [i_-(t)/e] \exp\left\{-\int_0^t [i_-(t')/e]\, dt'\right\}. \tag{13}$$

Applying this general equation, which is independent of geometry, to the case of a point-to-plane gap, he obtained, with the help of some simplifying assumptions, a relation of the form

$$p_1(t) = bt^2 \exp[-bt^3/3], \tag{14}$$

where b is a function of the concentration and the drift velocity of negative ions, which we assume remain constant, at least to first approximation.

Figure 1b shows the agreement between experimental and calculated results. By relating $p_1(t)$ with $i_-(t)$, one can explain:

(a) The decrease and the stabilization of the time lags which have been observed when using a radioactive source (Aked *et al.*, 1972; Saint-Arnaud and Jordan, 1974) or when irradiating the cathode with ultraviolet photons or applying overvoltage (Menes and Fisher, 1954). These techniques enhance the dark current, thus increasing p_1.

(b) The increase in the time lag with increasing humidity (Blair and Farish, 1965; Hahn *et al.*, 1974). If one ascribes the effect to the lower mobility of the hydrated negative ions, this would decrease p_1. An alternative explanation could consist of an increase in the mean life of the ions due to hydration (Renardieres Group, 1974) which would decrease p_2, as was seen before.

(2) The second case concerns discharges at large distances (between 5 and 10 m). These are generally distinguished from small-gap discharges by a smaller curvature of the electrodes, hence smaller field gradients and larger initiation volumes, and by the long rise time of the voltage pulses used to produce them (rates of voltage increase frequently encountered are in the range 1 to 100 kV μs^{-1}). The criterion adopted for this case by the Renardières group (1972, 1974) is expressed in terms of the mean lifetime τ_- of negative ions contained in the initiation volume $v_i(t)$, by the relation

$$p_2(t) = \left[\int_{v_i} n_-/\tau_- \, dv_i\right]\left[1 - \int_{t_i}^{t} \bar{P}(t)\, dt\right], \tag{15}$$

where n_- is the negative ion number density, t_i the time at which the voltage reaches threshold, and $\bar{P}(t)$ is the probability that the corona does not occur before time t.

Satisfactory agreement has been obtained with relation (15) between theory and experimental results over a large domain of configurations (e.g., cone-plane and hemisphere-plane) and pulse rise times dV/dt (2 to 70 kV μs^{-1}). Figure 1c shows an example for a hemisphere-plane system with 2 m gap.

III. CONTINUOUS GLOW DISCHARGE

A. I(V) Characteristics

1. General Remarks

From the beginning of this century, Townsend (1914,a,b) carried out studies of the space-charge limited current in cylindrical geometry (a wire surrounded by a concentric cylinder), which is well suited for theoretical treatment. He *only* considered the conduction in the region exterior to the multiplication volume, characterized by the critical distance x_c from the cylinder axis. The current there consists of charges of only one polarity, swept toward the outer electrode by the applied field: positive ions if the wire is positive, negative charges in the other case (either essentially negative ions or only electrons, whether or not the gas is electronegative).

Townsend took the charge density ρ to satisfy the Poisson equation (written below in cylindrical coordinates, with x the axial distance):

$$\frac{1}{x}\frac{d}{dx}(xE) = \frac{\rho}{\epsilon_0} \tag{16}$$

and the motion of the charges to be related to their density and to the field E by Ohm's law:

$$j = K \cdot \rho(x) \cdot E(x), \tag{17}$$

where the mobility K was assumed to be independent of E in the region of interest. By making some approximations valid for small currents, he obtained the well-known equation giving the current per unit length of the conductor as a function of the voltage applied between the two cylinders, of radii r and $R(r < R)$, and the threshold voltage V_s:

$$I = 2KV(V - V_s)/R^2 \ln(R/r) \qquad (V \geq V_s). \tag{18}$$

A relation of the same form

$$I = CKV(V - V_s) \tag{19}$$

can be applied to other geometries as well, such as the point-to-plane geometry, by proper choice of the value and dimensions of C. For example, in their systematic study of negative needle-to-plane coronas made in ambient air using very small tip radii ($0.003 < r < 0.045$ mm), and gaps from 4 to 16 mm, Lama and Gallo (1974) have obtained the relation:

$$I = (52/d^2)V(V - V_s)$$

with I in microamperes, V in kilovolts, d the gap length in millimeters, and V_s ($\sim$2.3 kV) independent of d over the domain investigated.

Generally speaking, for a given geometry and sufficiently small currents, one has nearly the same $I(V)$ curves in all cases for which the ion mobilities are of the same order. Notably, this will be the case for both positive and negative discharges in electronegative gases such as air, where the average positive and negative ion mobilities are about the same. In nonelectronegative gases (or mixtures) such as nitrogen or argon, the ion-dominated positive characteristic differs noticeably from the negative characteristic, where the charge is carried by electrons (Weissler, 1943).

2. Electronegative Gases

In gases or mixtures with at least one electronegative component, three types of charge carriers are involved in the conduction process: electrons, negative ions, and positive ions. Many workers have tried to understand the existence of continuous glow discharges, with steady currents instead of pulses in these gases.

One important example of work in this area deals with the highly stressed positive electrode, for which Hermstein (1960a,b) has shown that transient filamentary discharges of the "streamer" type can be interrupted, and converted to a positive glow discharge, by injecting negative ions from the cathode.

Hermstein's explanation of this effect, which had been generally accepted until a few years ago, was to assume that the injected negative charges establish a negative space charge sheath in the vicinity of the point anode. This space charge would increase the field near the anode, and decrease it in the exterior region to a value too low to support streamer formation, thus extinguishing the pulses observed at the cathode. This explanation does not agree with the experimental results of Buchet and Goldman (1969a), presented in Fig. 2, which show the relative importance of glow and streamer discharges in binary gas mixtures in which the fractional concentration of the electronegative component can be varied. It was found that:

(a) In N_2, there is only a glow discharge; this is also true for the rare gases (Das, 1961).

(b) For sufficient concentrations of the electronegative component, only impulse coronas are observed. For $N_2:O_2$ mixtures, this requires a concentration of O_2 somewhat greater than its atmospheric value ($\gtrsim 25\%$); in mixtures of N_2 with CCl_2F_2 (freon), the required concentration of freon is much smaller than this, because of its greater electronegativity (more rapid electron attachment to form negative ions).

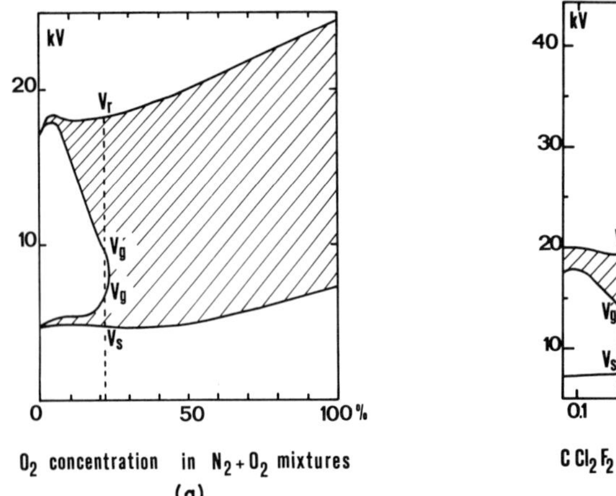

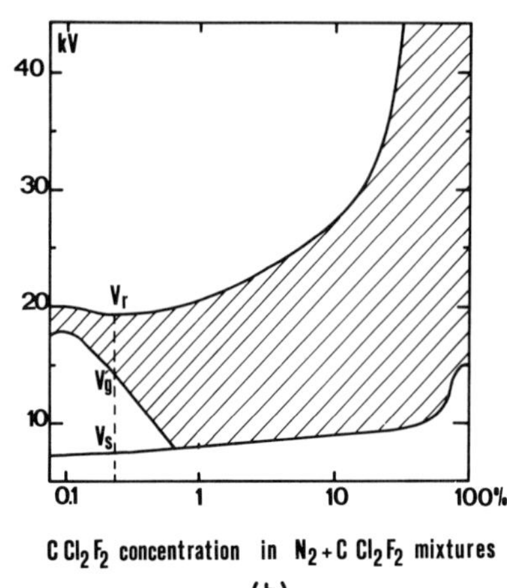

Fig. 2. Graph illustrating the domains of existence of positive impulse corona (shaded zones) and glow discharge (nonshaded zones) regimes in a point-to-plane gap ($d = 20$ mm, point radius $r = 70\ \mu$m) as a function of the concentration of the electronegative component in two different gaseous mixtures. (a) Nitrogen–oxygen. (b) Nitrogen–freon (Buchet and Goldman, 1969a). V_s = corona threshold voltage, $V_g - V_g'$ = glow discharge zone, V_r = spark breakdown voltage.

One can show by a very simple argument that it is not possible to maintain a negative space charge in a gap where a continuous positive corona is set up (Buchet and Goldman, 1969b). We note that the total current must be constant through all equipotential surfaces:

$$I = i_+ + i_-, \tag{20}$$

where i_+, i_- are the partial currents of positive and negative charge carriers. In particular, at the limit of the multiplication zone:

$$I = (i_+)_{x_c} + (i_-)_{x_c} \tag{21}$$

and on the point anode, where the electrons are neutralized:

$$I = (i_-)_{x_c} \exp \int_{x_c}^{r} \alpha\, dx, \tag{22}$$

where the notations are as previously defined. These yield

$$(i_+)_{x_c} = (i_-)_{x_c}\left[\left(\exp \int_{x_c}^{r} \alpha\, dx\right) - 1\right]. \tag{23}$$

Thus at the level x_c one has $i_+ \gg i_-$. Since both kinds of charge carriers have about the same mobility, then the positive charge greatly dominates the negative charge at x_c; this must also be the case inside the multiplication region. Exterior to the multiplication region, the ratio i_-/i_+ must be very low, on the order of γ, where γ includes all secondary ionization effects in the discharge, both in the gas and at the cathode.

This view is supported by Buchet and Goldman, who solved the equations for $V(I)$ for a discharge under these conditions. They studied the conduction in a spherical geometry, by dividing the entire space into two regions:

(a) A zone of ionic conduction [the only zone considered by Townsend in Eqs. (16) and (17)], in which $\rho = n_+ e$.

(b) A multiplication zone of radius x_c around the anode sphere (of radius r) which satisfies the self-sustaining condition:

$$\gamma \left[\exp \int_{x_c}^{r} \alpha(x)\, dx - 1 \right] = 1. \tag{24}$$

Here the current results from a movement of both ions and electrons produced by the avalanche.

From the continuity equation I = constant, from the Poisson equation (16), and from Ohm's law (17), one finds that the electric field in the gap is given by

$$E = (E_0/u^2)[1 + (E_I/E_0)^2 u^3]^{1/2}, \tag{25}$$

where

$$E_0{}^2 = E_R{}^2 - E_I{}^2; \qquad E_I{}^2 = I(6\pi K \varepsilon_0 R)^{-1}.$$

Here $u = x/R$ is the spatial variable and R the radius of the spherical cathode. E_R is the field at this outer electrode.

With the help of the above equations, and with expression (6) for $\alpha(E)$, they showed that, for every value of current I, one can evaluate the applied voltage V, in the form

$$V = RE_0 \int_1^{r/R} -\frac{du}{u^2} \left[1 + \left(\frac{E_I}{E_0} \right)^2 u^3 \right]^{1/2}, \tag{26}$$

where the integration is now carried out over the complete interelectrode space. This result shows that a stable continuous regime can be established above threshold, in spite of positive space–charge accumulated in the multiplication zone. This reinforces the work of Colli *et al.* (1954) on amplification in Geiger-Muller counters and that of Sarma and Janischewskyj (1969) on losses in power lines.

Thus, it becomes clear that Hermstein's experiments with the disappearance of pre-onset streamers (see Section IV.B.1) by the injection of negative ions from the cathode do not prove that a negative space charge sheath is formed in the gap, but simply signify that the negative ions strongly detached in the multiplication zone give birth to supplementary electrons, which increase the positive space charge by "feeding" the avalanches.

But, if the injection of negative ions has the effect of maintaining the continuous regime (that is, of reinforcing the Townsend regime in the vicinity of the point), they produce a stabilizing effect on the discharge; this is because, as we shall see, pre-onset streamers can only be born after the space charge is partially swept out. This result underlines how important it is (for obtaining a self-sustaining discharge) to provide secondary electrons. These could be produced near the point, according to the initial theory of Loeb, or at the cathode, according to the work of the Swansea school (Llewellyn-Jones, 1967). It explains the spark breakdown voltage increase observed by Hermstein in his experiments. This result has been generalized by Davidson *et al.* (1976) who have shown that the presence (or absence) of a glow regime increases (or decreases) the spark breakdown voltage at high pressures. It also explains the establishment of the continuous regime observed during the positive phase for conductors of large diameter (e.g. 4 cm) subjected to alternating voltage. Residual negative charges from the negative half-cycle could play the stabilizing role described above (Waters *et al.*, 1972). It is interesting to point out that this inhibiting effect has been utilized to reduce radio-frequency interference from the positive corona discharge by Popkov (1962) and Waters *et al.* (1972), who used a screening wire and a radioactive source as a source of supplementary electrons, respectively.

B. Current Density Distribution

In cylindrical geometry, where the electric field only varies with radial distance from the axis, the current density is uniform over the surface of the outer cylinder. This property is used to advantage in a number of applications, for example, in determining the mobilities of ions in their parent gases by measuring the current collected by the outer electrode (Townsend, 1914b; Couralet, 1973; Albrecht and Wagner, 1973). In the point-to-plane corona discharge the current density distribution over a plane normal to the system axis depends on the electric field distribution over it, the field at each point on the plane varying with the distance d of the plane from the point electrode, and with the distance y of the field point from the system axis. As early as the end of the last century, Warburg (1899, 1927) established experimentally that the current density to the plane for air gaps in the centimeter range follows a law of the form

$$j(y) = j_0 \cos^m \theta, \tag{27}$$

where $\theta = \arctan(y/d)$, j_0 is the current density on the axis, and $m \approx 5$.

The factor $\cos^5 \theta$ is purely geometrical, so that the total current is proportional to j_0:

$$j_0 = (m - 2)I/2\pi d^2. \tag{28}$$

The current appears to be contained within a cone of half-angle of opening θ_M ($\sim 60°$), independent of I (and of the curvature of the point, at least over the domain investigated).

To provide a theoretical basis for the experimental $\cos^5 \theta$ law of Warburg, one can refer to various theoretical studies of the field in the presence of space charge in a stationary regime. The solutions proposed in the cases of symmetry lower than that treated by Townsend present some noticeable differences, due as much to the choice of initial conditions (Félici, 1963a,b, 1964) and to the methods of solution as to the physical hypotheses introduced (Deutsch, 1933; Popkov, 1949). Atten (1974) calculated the distribution of field, current, and volume density of ionic charge at the surface of the ion-collecting electrode for various geometries, including the point-plane configuration. The work of Canadas *et al.* (1973) and Canadas (1976), following that of Dupuy (1958) and Dupuy and Berseille (1969), explain the existence of a unique current distribution by showing that the network of field lines evolves rapidly toward an asymptotic configuration whatever the geometry of the electrode system, when one increases the voltage applied until one approaches the arc transition.

Experiments by Dupuy and Goldman (1974) to study the range of validity of the $\cos^5 \theta$ law in the impulse regimes of the point-to-plane corona were carried out in air at pressures between 300 and 1000 Torr. They showed that, in the domain of gap distances studied ($0.5 < d < 3$ cm) the law remains valid for all regimes, continuous or impulsive, of the negative corona up to the arc transition. In the positive corona, the law is obeyed down to weak currents, with a profile which narrows about the axis when breakdown streamers occur. However, since we know that the streamer only contributes to the mean current I by localized impulses at the center of the plane (Marode, 1970; Peyrous, 1974), and that it only makes a partial contribution to the current (Buchet *et al.*, 1966; see Fig. 3a), the continuous component of I could still satisfy the Warburg law. This hypothesis remains to be tested.

C. Neutral Species Associated with the Ions

It is interesting to note that, even in the discharge regimes where the plane only contributes to the conduction as a simple collector of charges, it can be the seat of important chemical and energetic effects caused by the neutral species which accompany the ions. A study of the effective energy balance by Goldman (1974), in which the plane was used as a calorimetric detector to study the incident species, has shown, for example, that in a negative corona discharge, where incident particles arrive with negligible kinetic energy, the plane can absorb more than half of the energy injected into the discharge, the ions only making a negligible contribution to this transfer. Under these conditions, the significant transfer of energy observed must be ascribed to deactivation processes (vibrational and electronic metastables deexcitation, heterogeneous recombination) involving the neutral gas particles which reach the surface with potential energy.

This agrees with the results of Lecuiller *et al.* (1972) who have shown that the chemical activity of neutrals on the plane can be of an order of magnitude larger than that of the charged species (Fig. 19c). We note that if ozone plays an important role in this activity, it is apparent that other species could be still more active. Among others, the singlet oxygen molecule in the state ($^1\Delta_g$) should be mentioned (Rippe *et al.*, 1974).

IV. TRANSIENT CORONA

A. General Aspects of the Impulsive Corona

It is possible to treat the corona discharge as a superposition of two types of discharge: the Townsend discharge, characterized by electrical and visual stability, and the streamer discharge, which is inherently a transient phenomenon. The general properties of the steady behavior have been treated in the preceding section; in the treatment we now present, we will consider essentially the impulsive aspects of the principal regimes which occur as one increases the voltage across the gap.

Given the number of significant parameters which can favor one or another type of process, involving the electrodes, the gaseous medium, and the external circuit, one can see that the discharge may assume a large variety of aspects. However, one can treat the subject most simply by considering only the discharge in the point-plane geometry, using points of small radius of curvature, for small gap distances, established in air at atmospheric pressure and under continuous voltage. In its transient behavior, this discharge appears in the form of individual discharges which recur with very good reproducibility in space and time, at high repetition rates (up to about 10^4 and 10^6 Hz respectively for positive and negative discharges; see Figs. 3b and 20b). This permits one to isolate the basic phenomena, on the one hand, and on the other hand, one can then utilize pulse-counting and accumulation techniques in the laboratory, to overcome the experimental difficulties arising from the rapid evolution of the phenomena and the extremely weak level of the accompanying luminous emission. We note that:

(a) The underlying mechanisms of the pulsed-voltage discharge, the study of which is more difficult, especially for the positive discharge, because of the large number of streamers which branch out as they propagate toward the cathode, are generally similar to those which apply here.
(b) The results can be extended to electronegative gases of many varieties and pressures.
(c) One can study many of the fundamental mechanisms of large-gap phenomena more easily in the simpler geometry of the small-gap point-to-plane discharge.
(d) Corona discharges induced by steady or low-frequency voltages have many practical applications.

In general, Loeb and his school have explained the impulsive aspects of the corona discharge in terms of ion accumulation effects in the vicinity of the stressed electrode. The space charge which forms during an impulse reduces the local field to a value insufficient to permit further multiplication; for a new pulse to appear, the space charge must be swept out toward the opposite electrode by the electric field. To sweep the space charge out over the necessary distance (called the "clearing length") requires a certain time (called the "clearing time"). This explains the repetition period of the corona: It is tied to a relaxation time for the movement of ions, and not to the time constant for the external circuit. For a negative point, the negative ions (which result from the attachment of electrons which are multiplied as they leave the point) play the major role. For the positive point, the positive ions accumulated in the tail of the avalanches constitute the space charge controlling the evolution of the discharge. For either polarity, one can modify the discharge development by the application of gas flows. The various observed effects (Sujak, 1956; Nygaard, 1966; Fieux and Boutteau, 1970; Farish and Davidson, 1974; Davidson and Farish, 1975) can be explained by the action of the gas flows in sweeping the space charge out of the active zone.

As we show below, there exist a number of empirical formulas to characterize the different discharge regimes which for a given point seem to occur at essentially the same current values for all gap distances. Meanwhile, more and more quantitative models are being developed, in accord with experimental results, to describe the phenomena.

B. Positive Corona

1. The Different Regimes

The different corona regimes which one observes between the electrodes of the point-plane system, as the voltage is increased from the threshold voltage V_s to the spark breakdown voltage V_r, are illustrated in Fig. 3. In terms of their functional characteristics, three voltage regimes can be distinguished:

(a) the domain V_s–V_g of autostabilization (burst pulses, pre-onset streamers);

(b) the domain V_g–V_g' of the pulseless glow corona;

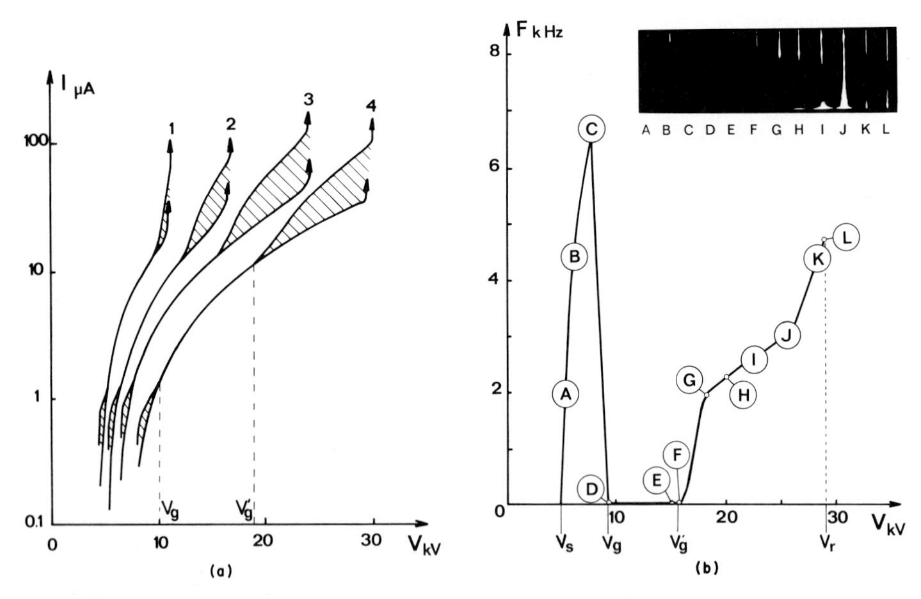

Fig. 3. Functional characteristics of the positive point corona in ambient air under continuous voltage: (a) Current-potential curves for different gap lengths, from 10 to 30 mm; the shaded regions characterize the impulsive contribution to the current, and the arrows indicate the spark transition (Buchet *et al.*, 1966). (b) Pulse frequency as a function of the applied potential for a given gap (d = 31 mm; r = 170 μm) illustrating the different discharge regimes. In the inset at the top: the luminous aspect of the discharge at each of the reference marks on the frequency curve (Hartmann, 1964).

(c) the domain $V_g'-V_r$ of the stable impulsive discharge (pre-breakdown and breakdown streamers) with pulses regularly spaced in time.

The relative importance of these different domains depends in particular on geometrical factors, especially the radius of curvature of the point.

(a) *Pre-onset Streamers and Burst Pulses in the Autostabilization Domain* ($V_s < V < V_g$) Like all other discharges, the positive corona is unstable at its threshold (Kip, 1938). This behavior is principally due to two reasons (Llewellyn-Jones, 1967):

(i) The avalanches created in the auto-stabilizing discharge are statistical phenomena (Legler, 1956; Davies and Jones, 1963; Maurel *et al.*, 1968), and certain of them do not attain the critical value necessary for streamer formation.

(ii) Since the coefficient α varies exponentially with the field (Eq. (6)), it is sensitive to small fluctuations in voltage.*

The phenomena may be interpreted qualitatively in the following manner: When the field is sufficiently high, a streamer (often called a "pre-onset streamer") forms, and develops away from the point in the form of a filament along the gap axis (Fig. 3b, frames A and B), until the potential energy stored in the space charge has been dissipated in the gap by electron collisions. Meanwhile, the space charge developed near the anode diminishes the field in this region. This field only permits the formation of a very unstable discharge of reduced intensity, which is composed of small-amplitude pulses (called "burst pulses"), accompanied by a permanent glow which envelopes the tip of the point and which will persist alone in the region V_g-V_g' of the glow discharge (Fig. 3b, frames D and E). The time interval T_n^{n+1} between two burst pulses is related to the amplitudes A_n and A_{n+1} of these pulses by an empirical formula (Fieux and Boutteau, 1970):

$$T_n^{n+1} = kA_n + T_0 \ln[A_0/(A_0 - A_{n+1})], \tag{29}$$

where k, T_0, and A_0 are constants.

* The exterior circuit could intervene at this level. If one operates with a high-value resistor R (e.g., 30 MΩ) to avoid breakdown, one can show that the instantaneous current is generally larger than the current from the source. The energy must then be furnished from the stray capacitance ($Q = CV = \int i\,dt$), which recharges between pulses with the time constant RC. In general, it is sufficient to furnish the necessary energy, but very good stabilization can be obtained by adding an external capacitor in parallel.

These considerations explain why the circuit does not usually affect the voltage signal for very large gaps, which have a large stray capacitance ($>10^{-9}$ F) (Newi, 1974). For R very great, however, an effect has been observed even in this case (Allibone and Meek, 1938).

Suppose now that T_n^{n+1} becomes large enough that the discharge ceases momentarily, and the field can again become large enough to initiate another pre-onset streamer. This behavior is illustrated in Fig. 4; the pre-onset streamers (large current pulses) appear after a period of more or less complete extinction of the burst pulses (small current pulses). (The oscillogram in the figure which displays the autostabilization regime of the negative discharge shows the similarity between the two discharge polarities.) The interdependence between pre-onset streamers and burst pulses becomes less pronounced as the radius of curvature of the point is increased; then the discharge extends laterally more and more, until for $r \gtrsim 2$ mm, the corona threshold V_s is often characterized by the direct appearance of only burst pulses, without pre-onset streamers.

Buchet (1976) explained the simultaneous existence of these two phenomena by considering the negative image charge of the positive space charge developed by the first avalanches in the vicinity of the point M of the anode (assumed to be a sphere centered at O and of radius r small with respect to the gap distance). He assumed that these two types of charges constitute an electric dipole at the point M of moment $\mathbf{p}$, lying along the direction OM (Fig. 5). For this condition, the resulting field outside the sphere is the same as that which would be created by a dipole $\mathbf{p}$ associated with a point charge $q = 4\pi\varepsilon_0 r^2 E_a$, placed at O, where E_a is the local component of the applied field. A map of the field (Durand, 1953) is shown in Fig. 5. It can be seen that the space is divided into two separate regions by a critical surface Σ:

Inside this surface, the field lines all converge to the point M on the axis of the initial avalanches. This gives rise to a "funneling" effect which will be the origin of the pre-onset streamers.

Outside Σ, on the contrary, the field lines terminate on different points of the electrode. Photoelectrons created inside this zone initiate the burst pulses.

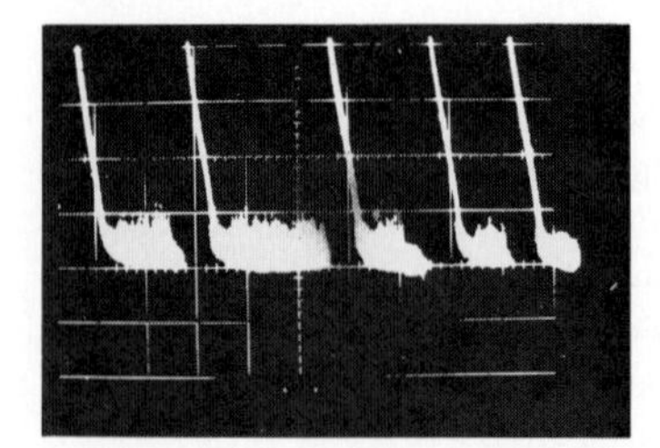

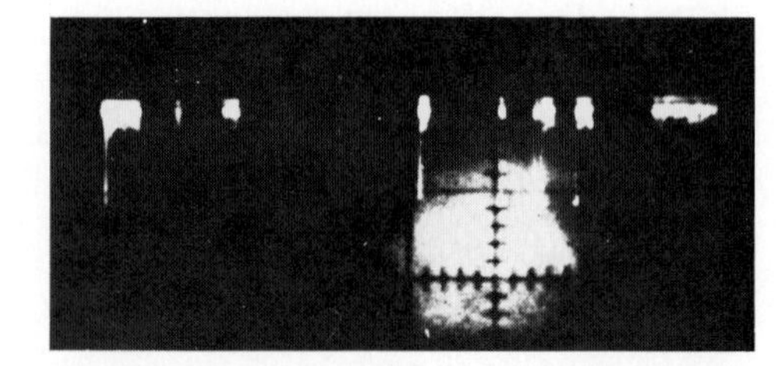

Fig. 4. Corona pulses in the autostabilization regime under continuous voltage. Oscillogram at left (Fieux and Boutteau, 1970): positive polarity, with $d = 50$ mm and $r = 3$ mm; sweep speed 1 ms/div. Oscillogram at right (Buchet *et al.*, 1966): negative polarity, with $d = 15$ mm and $r = 0.2$ mm, sweep speed 2 ms/div. In both cases, each pulse packet is initiated by a large pulse.

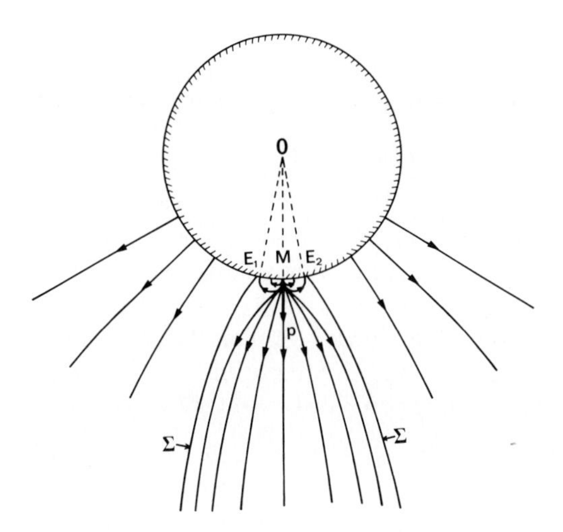

Fig. 5. Field map for the dipole model proposed by Buchet (1976) to characterize the axial propagation of the pre-onset streamers and the lateral spread of the burst pulses in the autostabilization region of the positive point corona.

(b) *The Positive-Glow Corona Discharge* ($V_g < V < V_g'$) We have commented at length on the continuous aspect of this discharge in Section III.A.2. Now we shall determine, first, its mode of inception, and second, the small amplitude fluctuations which coexist with it.

The positive-glow corona discharge is characterized by:

(i) a voltage region $V_g < V < V_g'$ bracketed by mean current values which are nearly independent of the gap distance for a given point ($I_g \sim 1\ \mu$A and $I_g' \sim 15\ \mu$A for the case of Fig. 3a), and also independent of pressure, at least between 40 Torr and 1 atm;

(ii) an essentially constant field at the point (Waters *et al.*, 1972), too weak for streamers to develop;

(iii) a permanent glow localized at the point (Figure 3b, frames D and E).

In Section IV.B.1(a), it was mentioned that after the onset of the discharge the formation of the pre-onset streamers requires at least a partial extinction of the pseudopermanent current of the burst pulses. This leads us naturally to consider the positive glow corona discharge as an extension of the preceding regime into a regime which is stable. In fact, a fine analysis of the current reveals the presence of ripples, which Colli *et al.* (1954) explain as a response of the external circuit, which has a definite resonant frequency, to the statistical fluctuations of the photoelectric current; they consist of high frequency pulses of amplitude much lower than the continuous current in the glow. Amplitude and frequency increase with the applied voltage, but depend predominantly on the radius r of the point. These variations can be described by the following empirical formulas, for r between 0.2 and 2.75 mm and gap distances between 2.5 and 20 cm (Fieux and Boutteau, 1970):

$$(F - 0.5)r = 1.25 \tag{30}$$

$$A_M = 24\{1 - \exp(-1.2r)\} \tag{31}$$

where r is expressed in millimeters, F in megahertz, and A_M, the maximum amplitude attained at the end of this voltage regime ($V = V_g'$), in microamperes.

We recall that the reduction of the field at the point inhibits streamer formation. A voltage $V > V_g'$ is required in order for streamers to develop in spite of the space charge produced by the glow corona.

(c) *Prebreakdown and Breakdown Streamers in the Stable Pulse Domain* ($V_g' < V < V_r$) In the vicinity of the threshold of this second pulse domain, the current pulses have essentially the same characteristics as those of the pre-onset streamers; visually, the luminous filaments are superimposed on the permanent glow at the point (Fig. 3b, frame F). As the voltage is increased, the impulsive component of the current assumes increasing importance (Fig. 3a). As is the case for the glow-discharge regime, the voltage range V_g'–V_r is bracketed by current values which are nearly constant for a given point (about 15 and 100 μA for the case in Fig. 3a).

The variation of $V_{r'}$ with the several parameters characterizing the gaseous medium has been the object of numerous studies. We cite here some of the authors who have studied in particular the effects of

(i) Pressure: At pressures below 1 atm, Akazaki and Tsuneyasu (1974) have shown that the variation of V_r with gap distances from 1 to 10 cm in air follows laws of the same form as the Paschen law. For pressures above 1 atm, see, for instance, Boulloud (1955) and Popkov *et al.* (1973);

(ii) Temperature: Thomas and Wong (1958);

(iii) Humidity: Waidmann (1963), Govinda Raju and Hackam (1973), Gosho (1974), Hahn *et al.* (1974).

2. Temporal Development of Ionization

The development of the luminous filamentary phenomena (preonset streamers for $V_s < V < V_g$, prebreakdown and breakdown streamers for $V > V_g'$) associated with the current pulses has been well studied, either for pulsed or continuous voltage, by various authors. To mention only the earliest work, we cite the photomultiplier studies of Hudson and Loeb (1961) for small gaps and Saxe and Meek (1948, 1955; Loeb, 1954) for large gaps, and the streak camera measurements of Hartmann *et al.* (1963) for small gaps and Stekolnikov (1960a,b) for large gaps.

Figure 6 shows the growth in total light and in current of three of the individual discharges (Fig. 3b). Frame a shows the development of a streamer ("pre-breakdown streamer") at a voltage insufficient to permit it to reach the plane. In frame b, the streamer (which we designate in the following as the "primary streamer," this term including both the primary streamer tip and its trajectory) has traversed the entire gap. It is followed by a luminous channel called a "secondary streamer" which forms at the point. The impact of the primary streamer on the plane is accompanied by an abrupt increase in the current signal. In frame c, the channel forms a junction between the two electrodes, and the transient arc follows. A minimum in the emitted light within the body of the secondary streamer is observed, related to a corresponding minimum in the current; both disappear as the voltage is increased.

(a) *Emission Spectrum of the Primary Streamer* Streamers have been studied by emission spectroscopy, a very powerful tool for discharge studies. In the study of the emission spectrum of the positive corona discharge in air, the first feature identified was the second positive system of nitrogen (from the neutral N_2 molecule) by Kip (1939) and English (1948), as well as the first negative system (from N_2^+) by Byer and Waidmann (see Loeb, 1965, p. 235).

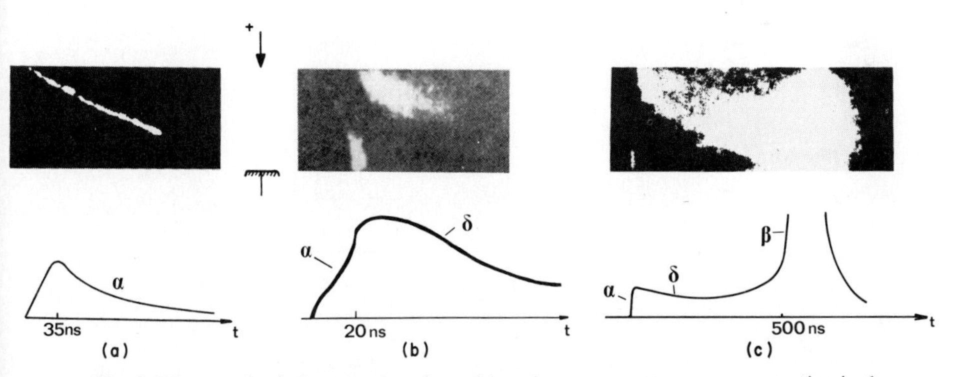

Fig. 6. Time resolved photographs taken with an image converter camera operating in the streak mode, and oscillograms of the associated current, illustrating the evolution of the individual discharges of the positive-point corona under continuous voltage as the voltage is increased. (a) Propagation of the primary streamer, denoted by α (Hartmann, 1964). (b) Propagation of the primary streamer, followed by development of the secondary streamer, denoted by δ (Marode, 1975a). (c) Complete development from primary streamer to transient arc, labeled β (Marode, 1975a).

Later, by utilizing working conditions in which the streamers regularly repeat themselves with high frequency (Fig. 3b), permitting the collection of data with high sensitivity and time resolution by individual photon counting, Hartmann (1970, 1977) obtained detailed spectra of the primary streamers, and identified 154 molecular bands which show in particular the existence of fifteen electronic states of N_2 (Fig. 7). By analyzing the relative intensity of radiation within both the 2nd positive and 1st negative systems in nitrogen, he measured the rotational temperature characterizing the dis-

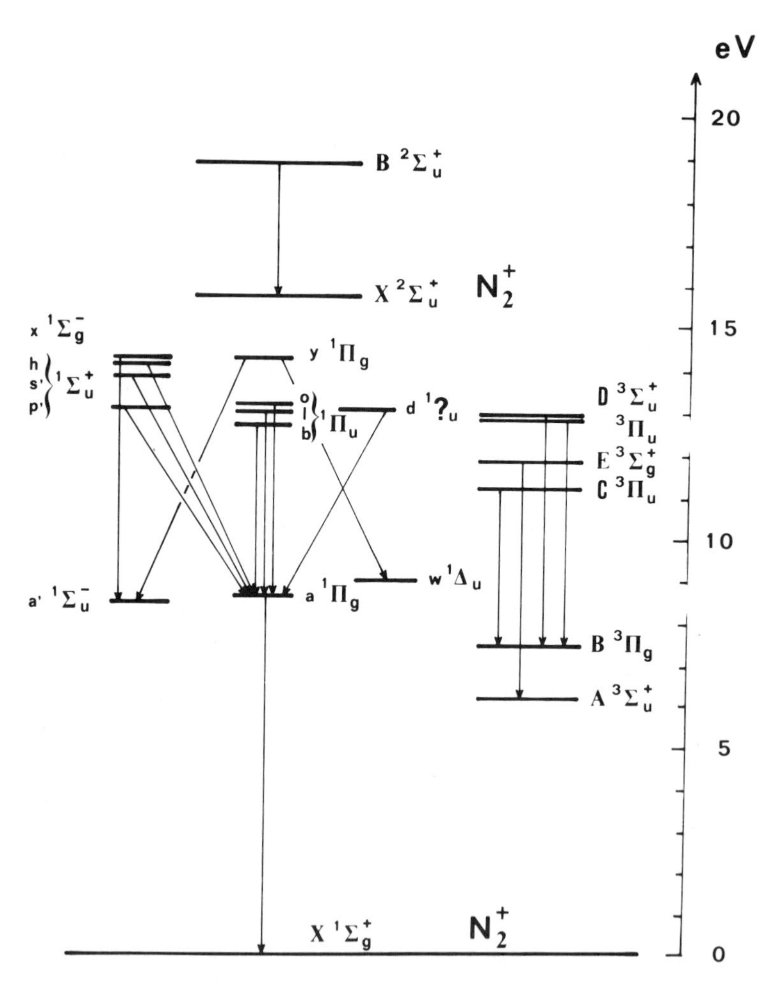

Fig. 7. Diagram indicating the radiative states of the nitrogen molecule populated by the passage of the primary streamers in air at small gaps (Hartmann, 1977).

tribution of components within each band (330°K for N_2 and 420°K for N_2^+). The analysis of the population distribution which he obtained for the various vibrational levels of the states C $^3\Pi_u$ and B $^2\Sigma_u^+$ in the primary streamer (Hartmann, 1972) leads to three important conclusions (Hartmann, 1977):

(a) The C $^3\Pi_u$ and B $^2\Sigma_u^+$ states are populated directly from the ground state X $^1\Sigma_g^+$. This explains the absence of emission from the oxygen molecule since, according to the Franck–Condon principle, radiating states of this molecule cannot be populated "vertically" from the X $^1\Sigma_g^+$ state.

(b) The distribution can only be explained by supposing that the vibrational levels of the ground state are distributed according to a "temperature" different from the ambient, which increases with the current (e.g., $T_v = 1200$°K for $I = 70\ \mu$A). The "heating" is in reality only apparent; it can be attributed to significant production of the v = 1, 2, 3, . . . metastable vibrational levels of the ground state, not by thermal but by electronic collisions (Hartmann and Gallimberti, 1975).

(c) The mean electron energy $\bar{\varepsilon}$ in the streamer, calculated on the basis of either a Maxwellian or Druyvesteyn distribution (Hartmann, 1974b) is found to be mildly sensitive to the velocity of propagation of the primary streamer v_S. For example, $\bar{\varepsilon}$ increases from 12 to 16 eV when v_S varies from 2×10^6 to 5×10^7 cm s^{-1} (Hartmann, 1976). These elevated electron temperatures should correspond to a reduced field E/p on the order of 200 to 350 V cm^{-1} Torr^{-1}, thus confirming the model of the primary streamer (see Section IV.B.2(c) below).

(b) *Physical Mechanisms in the Development of the Discharge* Streak photographs (Fig. 6) show that there is no luminescent emission between the tip of the primary streamer (PS) and the point anode. However, Marode (1975a,b) has shown that a conducting filament connects the tip of the PS to the point, forming a bridge between the two electrodes after the PS has reached the cathode. Experimental investigation of the nature of this filament has shown that the discharge possesses all the properties of a glow discharge.

The light emitted by the secondary streamer (SS) after arrival at the cathode of the PS, has been analyzed spectroscopically (Marode and Hartmann, 1969). The emission is essentially from excited states of the neutral nitrogen molecule. An electron energy of about 1.4 eV is found from the measured production ratios of excited states of N_2 and N_2^+, after correction for pressure quenching. Such energy implies the existence of a space charge field in which the attachment coefficient η is greater than the ionization coefficient α. The SS thus does not represent an increase in conductivity, as was previously thought.

In fact, the SS proves to be only the luminous trace of a conducting filament produced by the PS during its propagation. This has been shown from an analysis of the current induced on the electrodes by the discharge, which permitted the conduction and displacement currents to be separated. Figure 8 shows that the conduction currents collected at the point (i_p) and the plane (i_{pl}) must be interpreted in terms of three phases of evolution of the discharge.

(a) Propagation phase of PS ($0 < t < t_1$), in which the filamentary conductor is created; it carries a positive charge, as attested to by the collected current i_p.

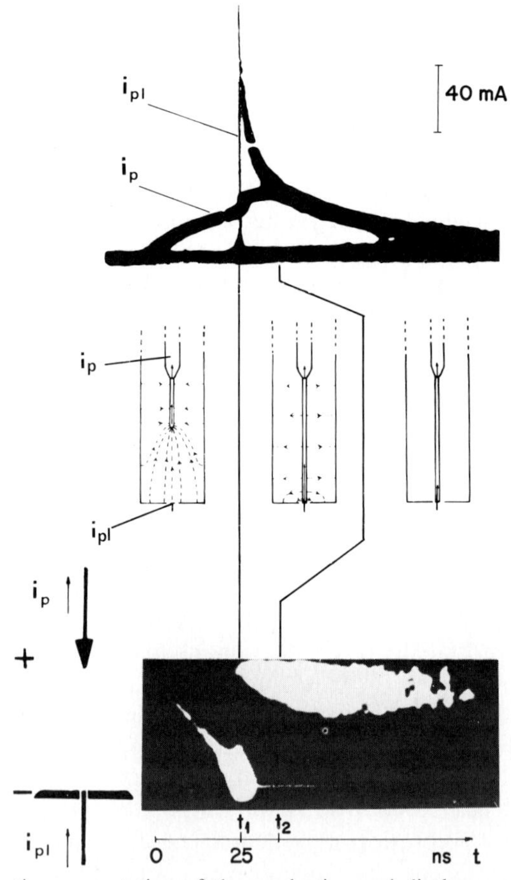

Fig. 8. Schematic representation of the conduction and displacement currents during successive phases of the positive point corona correlated with, above, an oscillogram showing the conduction currents i_p and i_{pl} flowing through the point and through a central section of the plane, respectively, and, below, a streak photograph (Marode, 1975a).

(b) Compensation phase ($t_1 < t < t_2$), which begins upon arrival of the PS at the plane, by the emission of a larger current from the plane than that collected at the point. The difference between these two currents, which escapes by displacement current, corresponds to a storage of negative charge which "compensates" for the positive charge deposited in the preceding phase.

(c) Resistive phase ($t > t_2$), in which the potential at various points along the filament is stabilized, and the current is constant all along the length of the filament.

Regarding its nature, the current flowing in the conducting filament is essentially electronic. Marode has demonstrated, by varying the oxygen partial pressure, that the decrease in current results from the loss of electrons by attachment. But since the secondary emission coefficient γ at the cathode is clearly lower than unity, the current cannot be purely electronic near the cathode. This proves the existence of a cathode region similar to that which in the low pressure glow discharge gives rise to ionization phenomena.

The overall picture is then the following. At each instant, the discharge consists of an active region, producing electrons, and a passive region collecting particles, a situation similar to the glow discharge. During the first phase of development, the active zone (PS), requires a voltage drop ΔV_f of the order of several kilovolts to propagate, whereas after the PS arrives at the cathode, the active zone only requires the normal cathode fall ΔV_c; the situation becomes as shown in Fig. 9, which compares the corona and the

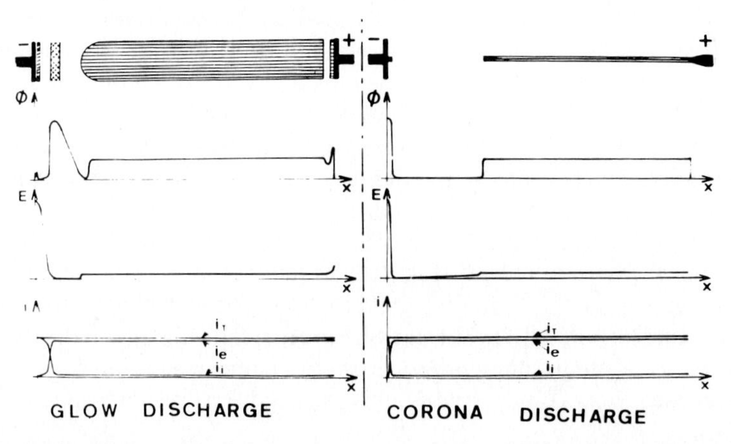

Fig. 9. Comparison of the features of the positive corona discharge after the formation of a conducting filament between point and plane (*i.e.*, for $t > t_1$; see Fig. 8) with those of the glow discharge. Distribution along the gap of light intensity ϕ, field intensity E, and the various currents represented by total current i_T, electronic current i_e, and ionic current i_i (Marode, 1975a).

glow discharges. There are two consequences of the arrival of the PS at the cathode. First, the difference $\Delta V_f - \Delta V_c$ must be distributed along the length of the filament; this distribution occurs with the aid of a potential wave, the "return wave" or "return stroke," particularly easily seen at lower pressures (Oshige, 1967; Ikuta *et al.*, 1970; Ikuto and Kondo, 1976). Second, the filament is now connected to the two electrodes, and the equilibrium potential distribution along its length adjusts itself to the electron density distribution created by the PS. Since the current $I = n_e(x) \cdot e \cdot K \cdot E(x)$ is constant along the length of the filament in the resistive phase, it is clear that where the electron density n_e is low (that is, where the primary streamer had the smallest luminosity), the field E is large. Since the photon excitation coefficient $\delta(E)$ (Legler, 1963) increases with E, the luminosity, which is proportional to $I \cdot \delta(E)$, appears in the SS where the electron density is low: the spatial distribution of the luminous intensity of the PS and the SS are exactly complementary. In terms of the glow discharge, the two luminous zones (PS and SS) can now be interpreted in the following way: The PS is an active zone of glow discharge, moving rapidly toward the cathode, whereas the SS corresponds to the luminous emission from a positive column. This positive column is cold, that is, the neutral and ionic temperatures are slightly elevated, on the order of 10^3 °K, while that of the electrons is several times 10^4 °K. In the course of time, the discharge can either disappear by the collection of the charges, within about 10^{-6} s, or, if the applied voltage is sufficient, transform itself into a hot positive column with a hot cathode spot, characteristic of a highly conductive arc discharge.

Experimental confirmation of the analogy between corona and glow discharges has been obtained by observing cathode sputtering beyond the propagation phase of the PS. The phenomenon was first studied by Johnson *et al.* (1972), then by Belbel (1976) over an extended range of gap lengths (1–20 mm and 0.5–2 m), both groups using a spectroscopic technique in which individual photons were counted. They measured the intensity $\phi(t)$ of a selected line from the emission spectrum of the cathode metal [for example, with aluminum cathodes, the 396.15 nm line, which is well separated from the lines in the gas (Fig. 10a)]. One sees in Fig. 10b, showing the time evolution of the luminous emission $\phi(t)$ correlated with that of the cathode current $I(t)$, that $\phi(t)$ begins with the arrival of the PS on the cathode, taken as $t = 0$, and becomes proportional to I after a maximum which occurs at that time $t = t_N$. The luminous emission can be expressed by a relation of the form

$$\phi(t) \propto \delta_r(t) \cdot I(t) \cdot \int_0^t I(t')\, dt'. \tag{32}$$

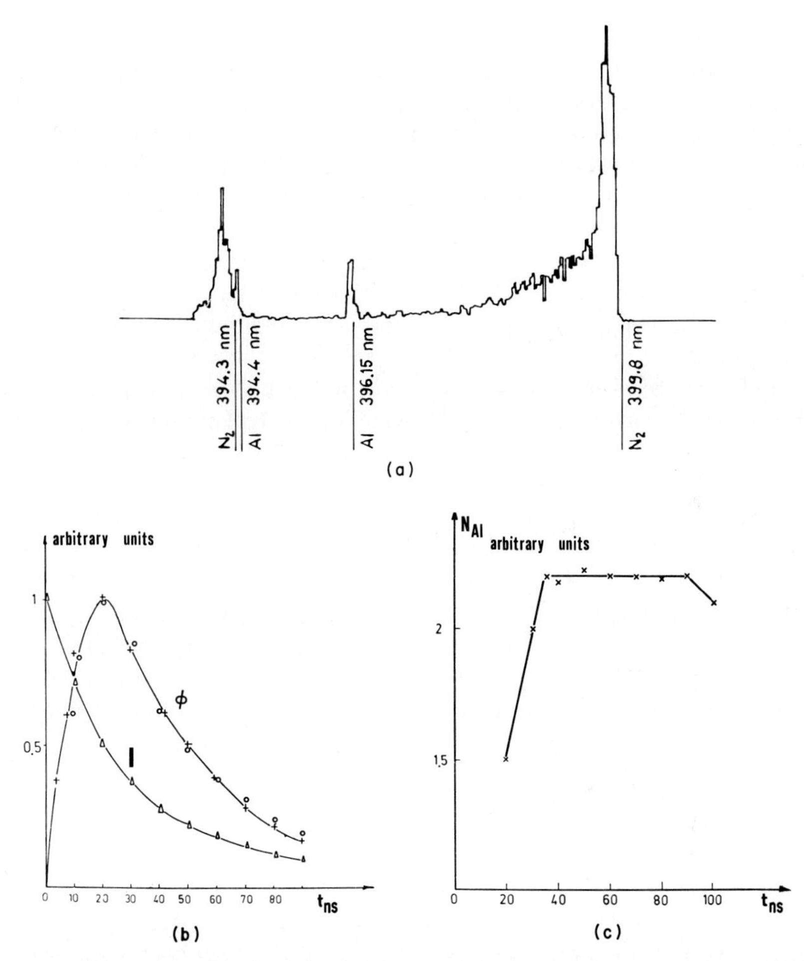

Fig. 10. Cathode sputtering by the primary streamer of the positive-point corona in air on an aluminum plane, with steady applied potential (Johnson *et al.*, 1972; Belbel, 1976), d = 10 mm. (a) Spectral emission in the zone of impact of the primary streamer on the plane. (b) Time variation of the emission $\phi(t)$ of the 396.15 nm line of the aluminum emission spectrum, correlated with the time variation of the current $I(t)$ ($t = 0$ corresponds to t_1 of Fig. 8, that is, to the arrival of the primary streamer on the plane). (c) Temporal variation of the ratio $\phi(t)/I(t)$, showing the variation, in the course of an individual discharge, of the number N of metal atoms ejected between t_1 and t.

Here the integral represents the amount of metal sputtered as being proportional to the number of ions received by the cathode; being less mobile, the metal particles are assumed to amass in the vicinity of the cathode, starting from the origin of the emission. The factor $I(t)$ indicates that the sputtered material is excited by the electrons emitted by the cathode, proportional to their flux, as is verified in Fig. 10b for $t > t_N$. The factor $\delta_r(t)$ accounts for lateral diffusion of the metal particles; it reduces the emission by the ratio $\{r_S/[r_S + (4Dt)^{1/2}]\}^2$, which can be neglected in a first approximation (r_S = radius of the PS; D = diffusion coefficient). Under these conditions, the curve $\phi(t)/I(t)$ of Figure 10c represents the variation of the number of atoms sputtered starting from time $t = 0$. It shows that at $t = t_N$ the discharge passes from an abnormal regime to a normal one; the transition is characterized by a significant reduction in the amount of sputtering (approximately from 10^{-9} atoms/incident ion to several times 10^{-11}), corresponding to a diminution of the mean ion energy from about 8 to 5 eV. Practically, sputtering is observed in the time interval $0 < t < t_N$, which could correspond to the compensation phase observed by Marode ($t_1 < t < t_2$ in Fig. 8).

We note that calculation of the local cathode heating at the point of impact of the PS, made under the assumption that all the energy transported by the ions is transformed into heat, shows that this factor can make a significant contribution to the sputtering only for large interelectrode distances. This agrees with the fact that for this case, as opposed to that for short distances, the traversal of the gap by the streamer can generally be considered a precursor to the spark breakdown.

(c) *Simulation of the Development of the Discharge* In the light of the results presented above, the streamer filament appears as a low-conductivity, almost neutral channel (zone of weak field, where the mean electron energy is low, 1.4 eV in the case considered), with a highly concentrated positive charge at its tip (zone of high field, $200 < E/p < 350$ V cm^{-1} Torr^{-1}, where the mean electron energy is on the order of 12 to 16 eV) which propagates from the anode toward the cathode with a velocity ($v_S \sim 10^7$ cm s^{-1}) much higher than velocities compatible with bridging the interelectrode gap by electrons under the influence of the applied field.

According to the streamer theory mentioned above (Loeb, 1954), which is equally valid for space-charge controlled discharges in uniform field geometry (Raether, 1964), the streamer propagates in regions of low applied field because of the effect of its own field. This hypothesis agrees with various experimental observations: the persistence of the streamer after the end of the voltage pulse (Dawson, 1965) which, however, marks a change in its luminescence and its propagation velocity (Fig. 11a), and the independence of the

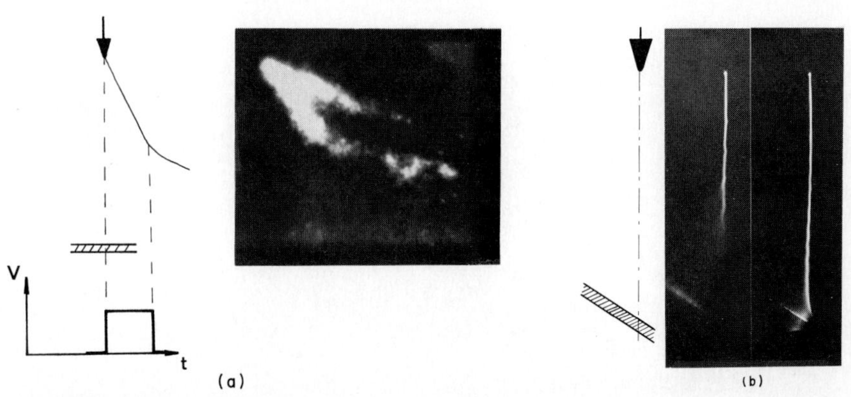

Fig. 11. Frames illustrating the concept of the self-propagating primary streamer. (a) Streak photograph of primary streamers of a positive point corona discharge induced by pulsed applied voltage; their reproduction on the trace, correlated with the applied voltage waveform, shows, for the most vigorous of them, that the streamer continues to propagate toward the plane even after the voltage is no longer applied (Marode, not published). (b) Static photographs of a corona discharge induced by a steady potential applied to the point-to-plane gap; the inclination of the plane with respect to the axis of the point affects the streamer trajectory only near the end of its course, and that the later the higher the applied voltage, that is, the more vigorous the streamer (Hartmann, not published).

trajectory with respect to field lines (Fig. 11b). The propagation is supposed to occur through a series of steps, one of which is shown schematically in Fig. 12. In Fig. 12a, one sees the streamer at the instant t_n of its propagation. The streamer tip, at a distance $x = x_n$ from the anode, can then be represented as a spherical zone of radius r_S, containing N_S positive ions and N_* excited molecules which are the result of previous steps. The decay of the excited states has produced, by photoionization, a distribution of secondary electrons around the charged sphere. Some of them, which are in the "active region" around it (frame d) again defined by $\alpha > \eta$ (i.e., $E \gtrsim 26$ kV cm^{-1} in atmospheric air), develop avalanches, under the combined action of the external applied field and of the space charge field created by the positive charge N_S localized at the streamer tip. Photoionization of a gas by its own radiation has been studied particularly by Przybylski (1958, 1962) and Sroka (1968, 1969). In air, molecular oxygen, for which the ionization potential is only 12.2 eV, can be photoionized by photons from the deexcitation of nitrogen, some of which have higher energy than this. However, a process of collisional transfer of the type

$$M^* + M \rightarrow M^+ + M + e$$

could contribute to the production of secondary electrons, rather than act as the principal mechanism of streamer propagation as has been proposed by

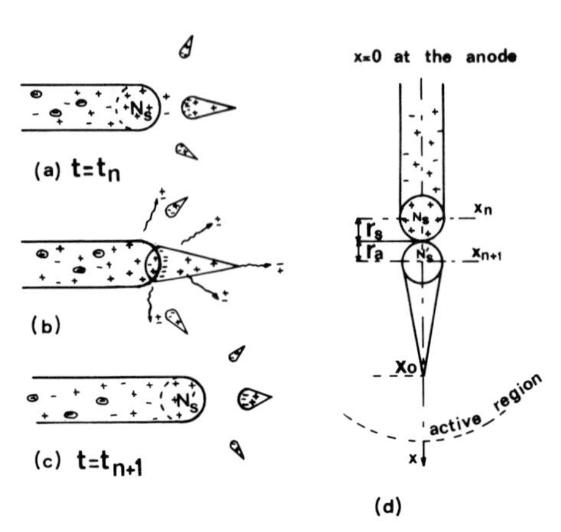

Fig. 12. Schematic representation of the propagation mechanism of the primary streamer. Frames (a) to (c): one step of the process. Frame (d): simplified model with a single equivalent avalanche, used by Gallimberti (1972) and others to simulate the process.

Losansky and Firsov (1969). Rodin and Starostin (1973). In (b) the heads of new avalanches have rejoined the streamer tip. First, their electrons neutralize the positive space charge in front of them and build up a new streamer tip with N_S' positive ions and N_*' excited molecules. Second, these excited molecules produce, by deexcitation of the gas ahead of the new streamer tip, new photoelectrons which renew the process. In (c), the streamer is shown at the time t_{n+1} of its propagation; photoelectrons produced at (b), having found at the new streamer tip $x = x_{n+1}$ a field sufficiently large for them to multiply, have formed new avalanches. The situation (c) thus reproduces that at (a), except that close to the filament it has advanced by one step of length $x_{n+1} - x_n \simeq 2\,r_S$ in the direction of the cathode.

Streamer propagation has been the object of diverse approaches seeking to establish quantitative models which agree with the experimental results. Dawson and Winn (1965) made one of the first attempts, by assuming that the streamer tip is autonomous, bound to the anode by a filament of negligible conductivity, and propagating because of the potential energy stored in its positive space charge. They made calculations from this model which indicate that the criterion for streamer propagation in zero external field is that the number of ions in the tip be $N_S \simeq 10^8$, and that its radius $r_S \sim 30\ \mu$m, in agreement with previously predicted values. They calculated streamer lengths for the assumption that the streamer ceases to propagate as a result of the losses of streamer tip potential energy in the formation of the secondary avalanches.

Also considering energy balance as a criterion for streamer propagation Gallimberti (1972) improved Dawson and Winn's model, by assuming that the energy is not supplied only by the potential energy of the streamer tip but also by the external circuit. The energy balance equation can then be written:

$$W_l = \Delta W_p + W_E, \tag{33}$$

where W_l is the energy lost by the electrons during the formation of a new series of avalanches, through ionization, excitation, attachment, etc.; ΔW_p is the change in the potential energy of the streamer tip between two successive steps (supplied through the space-charge forces) and W_E is the energy supplied by the external circuit to the moving charges (through the forces of the applied field). If one takes into account that the energy exchange stops when the field is reduced by the space-charge distortion produced by the electron-ion separation in the avalanches themselves, the energy balance becomes a synthetic condition through which charge multiplication can be estimated without a complete solution of the continuity and Poisson equations. The model was first constructed by considering, for each step of the propagation, a single "equivalent avalanche" producing the same space charge as the series of avalanches (Fig. 12d). Figure 13 shows that the results obtained with this model are in good agreement with experiment for the variation with time of both the streamer length (frame a) and the current pulses associated with the streamers (frame b). Furthermore, the model predicts that in a uniform field, for a critical value, the streamer propagation can become energetically stable, that is, it can persist indefinitely. The existence of such a stable field was established experimentally by Phelps (1971), who found a value of the critical field ~ 7 kV cm^{-1}, similar to that predicted by the model. Finally, from a Monte Carlo calculation of the probability of two equivalent avalanches developing simultaneously, Badaloni and Gallimberti (1973) obtained a representative model for the three-dimensional branched structure of corona streamers occurring in a medium unmodified by previous discharges, which is the case particularly for pulsed voltages. Figure 13c presents a two-dimensional computer picture of the spatial development of a branched streamer obtained with this model, compared with a photograph of actual streamer development. It is interesting to note that in this case the current increases by several factors of order unity, while the streamer lengths only undergo a small reduction. This shows that streamer propagation is determined mainly by the distribution of the applied field, and not by the redistribution of the charge in the tip among the various branches.

Wright (1964) has tried to obtain a solution for the complete discharge, PS and SS, by taking account of the filament traced out by the PS. With the

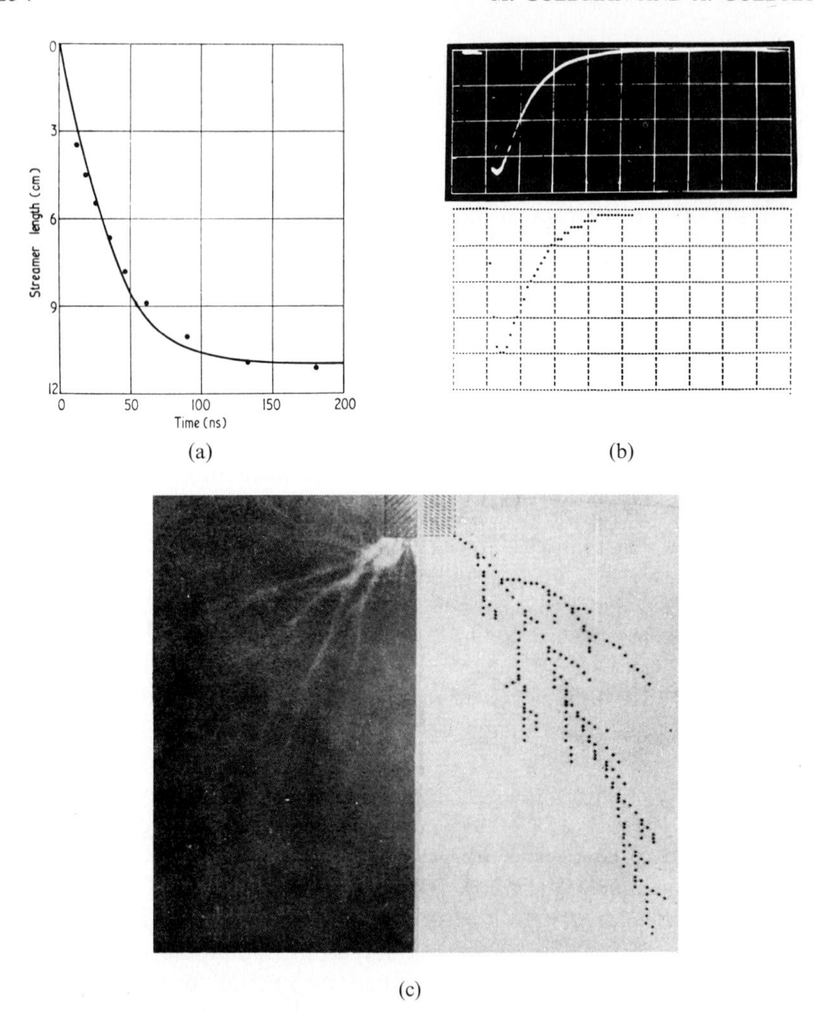

Fig. 13. Computed results for the streamer propagation characteristics in a 150 cm gap by Gallimberti (1972) and Badaloni and Gallimberti (1973) using their models of one and two equivalent avalanches, respectively. (a) Temporal development of the streamer: computed curve, experimental points. (b) Current oscillograms: computed curve, experimental oscillogram above (sweep speed-150 ns/div). (c) Two-dimensional spatial development of streamers leaving a 2 cm diameter rod: right-computer simulation, left-experimental picture.

aid of the Poisson and the continuity equations, he has succeeded in formulating the mechanisms governing the discharge as a stationary problem, independent of time, by utilizing a reference frame attached to the propagation of the PS, and by adopting for the filament the greatly simplifying assumptions of a constant electron density along its entire length and a negligible attachment rate.

Marode (1975a, b) has repeated the treatment of the filament in a manner much closer to reality, by utilizing his experimental results to simplify the continuity equation which can be written completely as:

$$\frac{\partial \rho_{\mathrm{e}}}{\partial t} + \nabla j + \frac{D_{\mathrm{e}}}{\mathrm{e}} \nabla^2 \rho_{\mathrm{e}} = (\alpha - \eta) j + \frac{\rho_-}{\tau_-} + \alpha_{\mathrm{Re}} \frac{\rho_+}{\mathrm{e}} \frac{\rho_{\mathrm{e}}}{\mathrm{e}} + \sum S_i, \tag{34}$$

where j is the electron current density; ρ_{e}, ρ_-, ρ_+ are the charge densities of electrons, negative ions, and positive ions; D_{e} is the electron diffusion coefficient; α and η are the ionization and attachment coefficients; τ_- is the lifetime of the negative ions; α_{Re} is the electron-positive ion recombination coefficient; S_i represents other possible source terms, such as photoionization, electron production by collisions with excited particles, etc. In fact, as we have already seen, experimental evidence has been obtained by Marode that the leading terms are the current divergence term ($\nabla j \neq 0$) and the electron attachment term ηj. If, further, one notices that the filamentary aspect makes possible a one-dimensional treatment, this leads to

$$s \frac{\partial \rho_e}{\partial t} + \frac{\partial i}{\partial x} = -\eta i, \tag{35}$$

where s and i represent the cross sectional area of the filament and the circulating current. Thus reduced, the continuity equation holds only on the filament. To solve it, we must specify boundary conditions, those dealing with the head of the filament being determined by the properties of the PS. The electron density imposed by the passage of the PS is deduced from the emitted light by an approximate treatment. The evolution of the discharge current predicted from the model reproduces the measured current quite well (Fig. 14, frames d and e). Moreover, the interpretation of the formation of the SS by a reorganization of the potential along the length of the filament is confirmed by the spatio-temporal distribution of emitted light, determined from the model (Fig. 14, frames f and g). However, this model could be further refined by making use of the formulation of Wright which, while it is not very productive at the level of comparison with experiment, does present the advantage of describing the PS not as a step-by-step process but as a continuous mechanism which could make an easy connection with the filament.

3. Transition to the Transient Arc ($V = V_{\mathrm{r}}$)

Whether it be in an inhomogeneous positive point-to-plane, or a homogeneous plane-to-plane structure, under the conditions of the field giving rise to the development of streamers, the formation of an arc shows the same phases, as schematized in Fig. 15 (Marode, 1973).

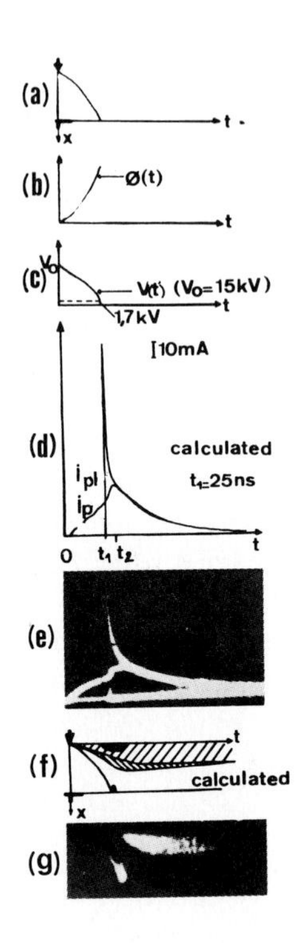

Fig. 14. Computed (Marode, 1975b) spatio-temporal evolution of filamentary streamer track in a 15 mm gap. Frames (a) to (c): experimental results constituting the boundary conditions for the computer calculations; they are related to the primary streamer tip, its position $x(t)$ in the gap, the light $\phi(t)$ it emits, and the potential distribution defined by it in the gap. Frames (d) and (e): computed and experimental currents; i_{pl} and i_p designate the currents flowing in the streamer track near the plane and the point, respectively. Frames (f) and (g): computed and experimental streak images.

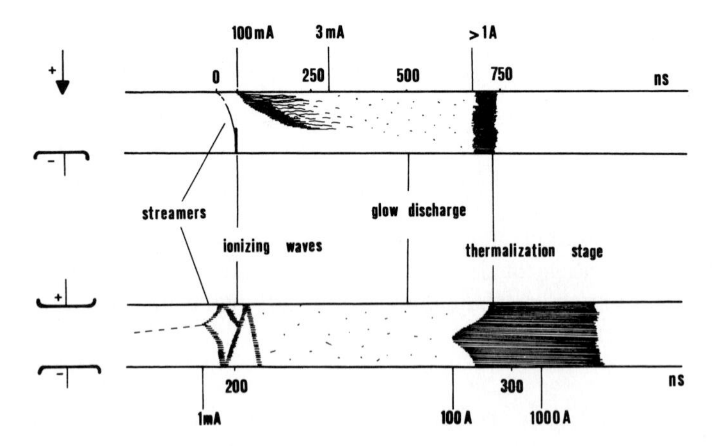

Fig. 15. Schematic representation of arc formation by the streamer mechanism in positive point-to-plane and plane–plane geometries (Marode, 1973).

First, space charges develop in the gap, accompanied by more or less luminous fronts. When they reach an electrode, a reorganization of potential occurs along the discharge channel; this is accomplished by the very rapid propagation ($\sim 10^9$ cm s^{-1}) of a potential wave which, according to the value of the field associated with its front, excites or ionizes the medium in its passage. At the cathode, the arrival of the space charge constitutes a cathode zone which injects electrons into the discharge; in this case, the potential wave could be induced by those electrons which compensate the positive charge developed earlier. Whether they be multiple (Kritzinger, 1963) or single, these waves lead to a stabilization of the potential and the currents along the length of the discharge channel, giving it a glow-discharge structure. We have seen that the positive column of the positive corona discharge at this stage is cold, i.e., the temperature T_i of the ions and T_n of the neutrals are both on the order of 10^3 °K, while that of the electrons reaches several times 10^4 °K.

But we know that the transient arc toward which this discharge evolves when the applied voltage is sufficient is characterized by a hot column ($T_e = T_i = T_n > 5000$°K) and a hot cathode spot leading to a vaporization process which injects metallic ions into the cathode zone. Therefore, both the positive column and the cathode region must be transformed during the transition.

The mechanisms of heating and thermalization which take place in the gas and on the cathode are not yet well understood. However, for the formation of an arc between parallel planes through a streamer process, Tholl (1970) has found that molecular dissociation plays an important part in the heating of the discharge column and that, when the electron density is of the order of 10^{17} cm^{-3}, the number of coulombic collisions becomes so large that thermalization occurs rapidly (Spitzer, 1956) despite the small energy loss sustained by the electron per collision.

Bastien *et al.* (1976) have measured the Stark broadening of the H_α and H_β lines emitted from traces of hydrogen introduced into a corona discharge in oxygen at a pressure in the vicinity of 300 Torr. From these measurements, and from a theory of the Stark effect taking account of a space charge field superposed on the macroscopic applied field, an estimation of the radius of the discharge (~ 20 μm) and of the density of charged particles (several times 10^{15} electrons cm^{-3}) has been made (Bastien and Marode, 1977). This density is too small for coulombic collisions to play a large role. However, we have seen that the development of the secondary streamer leading to the transient arc is, at least for $V \sim V_g'$, accompanied by a current decrease (Fig. 6c) due to electron attachment, whereas the field is too weak for strong ionization to take place. The transition to the arc could then imply a process of detachment, still unknown, within the secondary streamer (Marode, 1972).

Let us now consider the relation found for small gaps by Marode (1968) for steady voltages, and by Marode and Sulkowski (1969) for pulse voltages, which relates the time Δt for formation of the transient arc which elapses between the corona pulse (α) and the transient arc pulse (β) to the amplitude i_m of the current pulse of the corona discharge; the latter shows a critical threshold i_c below which no transient arc occurs:

$$(i_m - i_c)/i_c = A/\Delta t^{\nu}. \tag{36}$$

Here $\nu = 1.62$, $A = 8 \times 10^3$, and Δt is expressed in nanoseconds, for the case represented in Fig. 16, corresponding to a steady applied voltage, for which i_c is equal to 80 mA; for a pulse-induced discharge $\nu = 1$, $A = 1.4 \times 10^3$ and $i_c = 1$ A for the same conditions. In fact, all the various detachment processes are compatible with this relation.

We remark that, for large gaps, a critical threshold Q_c of charge which must be injected into the gap for arc formation has been observed by the Renardières Group (1972); for a 10 m gap subjected to 130 μs rise time pulses, they found $Q_c \simeq 225$ μC.

For small gaps, once attachment is overcome, the current increase should be controlled by the cathode region; in relation with the heating of this zone, a cathode spot should be formed which emits an increasing current of electrons. In parallel-plane geometry in air, Driver (1969) has shown that a

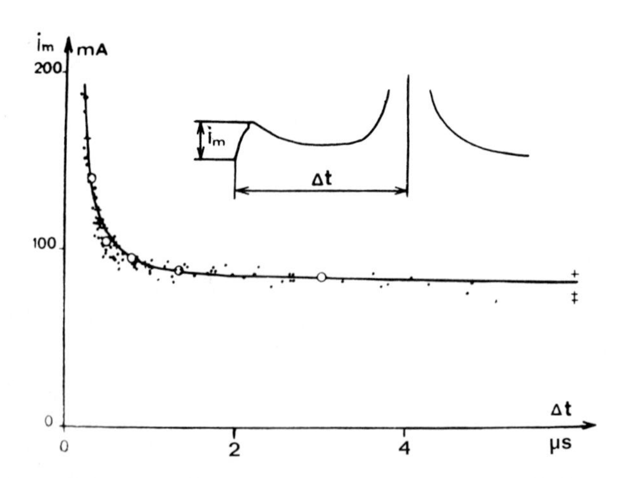

Fig. 16. Curve representing Eq. (36) between the maximum amplitude i_m of the corona pulse and the formation time Δt of the transient arc in a positive point corona discharge with $d = 10$ mm (Marode, 1975a). (○): mean values of 20 experimental measurements. (+): values for which no transient arc occurred.

positive space charge progresses toward the cathode due to the separation of positive and negative charges. Such space charge will be established as well when the arc is produced at $V = V_g'$, i.e., with a marked decrease of current between the primary streamer and the arc; the space charge field will then intensify the ion bombardment of the cathode, favoring the formation of a cathode spot.

The problem of the arc transition has also been approached by energy balance considerations. Aleksandrov (1966) has evaluated the order of magnitude of the temperature rise ΔT in the long discharge channel (see Section VI.A), by assuming that the injected energy is totally transferred to gas molecules as kinetic energy, i.e.:

$$N_e e E_s dl = \tfrac{1}{2} Z \cdot N k \Delta T \cdot \pi r_s^2 dl, \tag{37}$$

where N_e is the total number of electrons in the channel, r_s the radius of the channel, E_s the field in the region dl, N the number of molecules of gas per unit volume, and k the Boltzmann constant. The coefficient Z gives the number of degrees of freedom of the molecules, and has the value 5 for diatomic molecules. For an estimate of ΔT, he assumed $E_s = 20$ kV cm^{-1} and $r_s = 0.5$ mm; he obtained $\Delta T = 5500$°K from Eq. (37).

Independently, Suzuki (1971) explained the transition to the arc for short gaps (~ 1 cm) by a calculation analogous to that of Aleksandrov and by taking account of return waves to explain the increase in current.

At last, research workers in the authors' laboratory are currently exploring, by means of Schlieren photography and by theory, the role which might be played in the transition to the arc by the decrease in gas density within the discharge channel resulting from its heating, under the assumption that the pressure there remains constant. The resulting increase in E/N could lead to a decisive runaway phenomenon.

4. Influence of Previous Discharges on the Development of a Discharge

Since each individual discharge leaves in its wake some ions and neutral excited species, the gaseous medium takes a certain time after each discharge to recover its initial properties.

The most directly discernable effect of a discharge on the development of the following discharge deals with the lateral extension of the streamers (Fig. 17a). When they are produced by isolated voltage pulses, many streamers develop from the point and branch out as they propagate toward the cathode. Interesting information regarding their spatial development under these conditions has been obtained by the technique of Lichtenberg figures, a technique which consists of obtaining a plane representation of the spread-

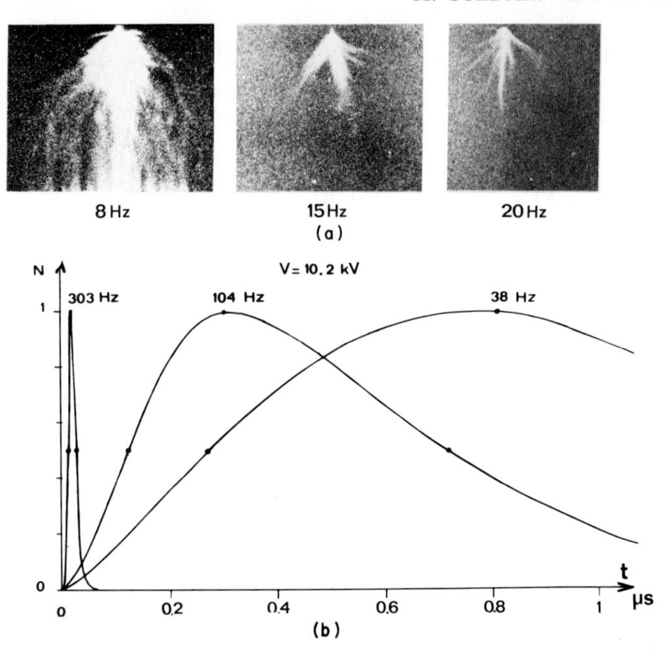

Fig. 17. Influence of previous discharges on the formation and development of a positive-point corona discharge, studied with variable repetition rate voltage pulses applied to a 15 mm gap (Berger *et al.*, 1972; Berger, 1974). (a) Photographs showing the influence of prior discharges on the width of the discharge according to the repetition rate. (b) Time-lag distributions for different pulse frequencies illustrating the respective domains of influence of ions (narrow distribution at 303 Hz) and excited neutral species (broad curves at 104 and 38 Hz).

ing of the streamers by intercepting them on a photographic plate in the discharge gap (Merrill and von Hippel, 1939; Nasser, 1968, 1971). By contrast, we have seen that under continued voltage the streamers succeed one another along a privileged channel consisting of the gap axis. The passage from one mode of production to the other has been studied by using steeply rising voltage pulses, but with a sufficiently long flat top to obtain several successive discharges (Marode and Sulkowski, 1968). The first discharge then develops with maximum current and a branched structure, in a medium which has not been perturbed by any preceding discharge. Since each of the following discharges is influenced by a preceding discharge, they are born in a reduced field, and develop as though initiated by a steady potential, i.e., in unique streamers directed along the axis and draining less current (e.g., $i_1 = 1$ A; $i_2 = i_3 = \cdots = 0.1$ A) like burst pulses induced by a pre-onset streamer. However, the evolutionary scheme of the discharges remains almost unchanged between the first discharge and the following ones, and for both steady and pulsed applied voltages the transient arc occurs along a unique breakdown channel.

(a) *Influence of Residual Ions* When discharges carry significant current, the influence of residual ions may be marked by a weakening of the dielectric rigidity of the gaseous medium, lowering the spark breakdown voltage. However, large-gap experiments, performed by the Renardières Group (1974) with 30 s intervals between two consecutive discharges under normal atmospheric conditions with humidity lower than 7 g m^{-3}, have not shown perceptible effects on the breakdown voltage, despite the accumulated charges still remaining in the gas which have not had time to diffuse and recombine, as different experiments have demonstrated (Waters *et al.*, 1968).

Yet, the residual ions have a marked effect on the discharge formation time. In their small-gap (d = 15 mm) experiments with variable repetition-frequency voltage pulses, Berger *et al.* (1972) have found that the effect of ions can last for times of the order of 5 ms. As one can see from Fig. 17b, when the time intervals between pulses are shorter (time lag distribution spectrum at 303 Hz), the time lags are noticeably smaller and less dispersed than for longer time intervals (spectra at 38 and 104 Hz). Moreover, in place of the lognormal distribution of time lags observed for pulses which can be considered isolated (Section II,C), the spectrum presents an exponential decrease characteristic of an accumulation of negative charges in the vicinity of the point; these charges, which increase the probability of production of seed electrons able to initiate new streamers, explains the very short formation times which one obtains under these conditions.

(b) *Influence of Residual Neutral Excited Species* We have already seen that, from spectroscopic measurements of the vibrational temperature of nitrogen molecules in the X $^1\Sigma_g{}^+$ ground state, Hartmann (1974b) has established, for the passage of successive discharges, that a localized heating of the gas takes place. This heating which varies as a function of the repetition frequency, is explained by the significant number of metastable species produced in the discharge, especially in the first few vibrational levels of the ground state X $^1\Sigma_g{}^+$ ($v > 0$) of nitrogen; from Hartmann's results, these represent for example about 10% of the N_2 molecules when the gas temperature is of the order of 1200°K, obtained by him for a discharge current of 70 μA (see Section IV.B.2.(a)).

Due to their long lifetimes, the metastable species can affect the development of succeeding discharges: first, because they require less energy to be ionized than the other particles, and second, because they can serve as an energy reservoir for the electrons in superelastic collisions (Hartmann and Gallimberti, 1975).

In practice, it has been shown that the presence of metastables leads to an enhancement of the streamer propagation velocity (Acker and Penney, 1968), with a consequent reduction in the rise time of the current pulse. Moreover, one can invoke their presence to account for the spatial configuration of a single filament almost axially located, which represents the influence of earlier discharges on the discharge. In effect, one passes progressively from the widespread structure of the streamers to an axial structure (Fig. 17a) when using voltage pulses of increasing frequency and insuring the sweeping out of ions between successive discharges. According to the experiments carried out by Berger (1974), it can be supposed that metastables produced by one discharge will have an influence on the following discharges for a duration on the order of 0.1 s in air with a 1 cm gap. This influence is reduced when one increases the gap length.

The influence of metastables also manifests itself in the time lags. In the absence of residual ions from the preceding discharge, their statistical distribution remains lognormal, but as one increases the repetition rate of the discharge, the distribution is translated to shorter times and becomes more and more compressed, as shown in the time lag distribution spectra at 38 and 104 Hz of Fig. 17b.

C. Negative Corona

Although the negative corona discharge has been the object of many studies, the physical mechanisms which underlie it are as a whole more poorly understood than those in the positive discharge. One can attribute this in part to the greater practical interest in solving problems associated with the positive discharge, including problems both of electrical insulation and of electromagnetic and acoustic parasitics in the electrical transport. Nonetheless, the negative discharge forms the basis for our understanding of lightning, and is a source of numerous interesting applications (plasma chemistry, telecopying *etc.*).

In the negative corona (as in the positive case) when one increases the voltage applied to the electrodes of a point-to-plane gap from the threshold voltage V_s to the spark breakdown voltage V_r, different domains of the discharge are observed. The relative importance of these domains depends in particular on geometrical factors, especially the radius of curvature of the point. One can distinguish in this way:

(1) a domain V_s–V_s' of autostabilization
(2) a domain V_s'–V_g of discharge with regular pulses
(3) a domain V_g–V_r of continuous-current discharge.

1. The Autostabilizing Discharge ($V_s < V < V_s'$)

This unstable regime is reminiscent of the corresponding positive regime (Fig. 4). As one increases the voltage, the current pulses, at first isolated and repeating randomly about some mean frequency, tend to group themselves into individual packs, rather bushy and separated by variable time intervals. Each pack always begins with a large pulse, followed by others, smaller and more or less regular (Fig. 18a). With further increase in voltage the packets enlarge; the small pulses become more regular and more numerous until the packets merge at $V = V_s'$, and a regime of only small pulses is then observed.

As in the positive autostabilization regime, the simultaneous existence of two types of pulses is explained by the motion of space charge. The field at the point is governed by a double movement of space charge. First, the field in the ionization region is enhanced by positive ions which have been formed along the path of the avalanches, moving relatively rapidly toward the cathode because of the high field in which they find themselves, but not yet having reached it. On the other hand, the electrons moving away from the cathode in the lower field region attach to neutral atoms or molecules, generating a negative-ion space charge which moves relatively slowly toward the anode; this weakens the field in the direct vicinity of the positive space charge zone and limits the development of the pulse. This is illustrated by the curves of Fig. 22, calculated by Aleksandrov (1963), which give an approximate idea of the situation; for simplicity, he took the space charge field to be that due to the charge distribution of the first avalanche initiating the pulses, and neglected subsequent distortion of this distribution by the applied field. Spectra of the light emitted by the Trichel pulse discharge show metal lines from the point, due to its sputtering by the incident positive ions. Since it is known that the rate of sputtering depends strongly on the impact energy which the field imparts to the ions, then the variations observed by Buchet *et al.* (1962a) in the mean sputtering rate reflect variations in the field in the immediate vicinity of the point. Taking account of the interruptions of the discharge, which permits the accumulated negative ions to be completely swept out between the packs of pulses, the field at the point in the autostabilizing regime can reach values more than an order magnitude larger

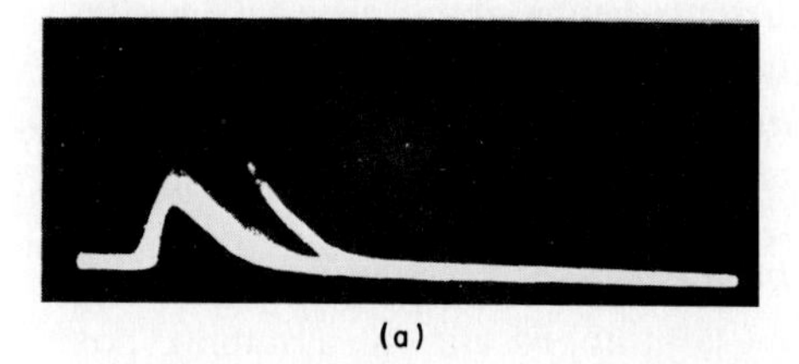

(a)

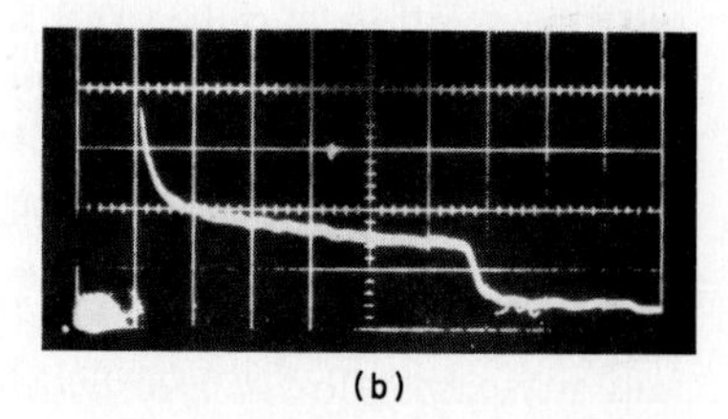

(b)

Fig. 18. Current pulses from negative-point corona in air under continuous voltage. (a) Large and small pulses in the autostabilization domain observed on the oscilloscope with internally triggered sweep (Buchet *et al.*, 1966). Compare with the oscillogram corresponding in Fig. 4, recorded with a single slow sweep. (b) Trichel pulses showing a step in the decreasing edge (Fieux and Boutteau, 1970).

than the applied field (Buchet *et al.*, 1962b). Beyond this regime, as a result of the accumulation of negative ions from one pulse to the next, the sputtering rate decreases as the large current pulses disappear (Fig. 19a,b), since both the sputtering rate and the pulse amplitude are governed by the field at the point (Buchet and Goldman, 1970). Owing to the relatively low values obtained in nitrogen by Weissler and Schindler (1952) for the impact energy of the ions ($\sim$15 eV) using a calorimetric method, compared to those (till 40 eV) obtained by Buchel *et al.* in air, one may consider that the first ones correspond to a steady discharge, beyond the autostabilizing regime which should also exist in the case considered.

The transition V_s' marking the end of the domain of autostabilization is also accompanied by a change in the chemical nature of the discharge products (Lécuiller *et al.*, 1972). In oxygen in particular, one finds mainly ions related (see Section II.A) to O_3^- in the autostabilizing regime, and to O_2^- above that regime; also, the number of neutral oxidizing particles undergoes a diminution at the transition voltage (Fig. 19c).

2. The Regular Pulse Discharge ($V_s' < V < V_g$)

(a) *Impulsive Aspects of the Discharge* The stable discharge is established at the transition voltage V_s' with the disappearance of the large pulses of the preceding regime, but without either the shape or the amplitude of the small pulses being affected. This regular regime carries the name of Trichel (1938), who first studied it. The pulses which characterize it have generally the same shape as the small pulses of Fig. 18a. Sometimes they display a step in their trailing edge (Fig. 18b); this is notably the case for points of weak curvature.

For a given point, only the repetition frequency F of the pulses varies sensibly with voltage and consequently with current. It is generally stated that F is proportional to the mean current I:

$$F = kI \tag{38}$$

but this relation is only valid for currents below about 30 μA (Fig. 20a); $F(I)$ increases more rapidly than I above this current value (Peyrous, 1974). On the other hand, Fieux and Boutteau (1970) find a linear variation over the entire domain, provided they use for the current I only the continuous component I_c on which the pulses are superposed (Fig. 20b). These results could be compatible if the ratio I_C/I keeps a constant value at low currents and increases slowly afterwards, the continuous current assuming dominance as one approaches the glow discharge. Figure 20b shows also that the gap length exerts no significant influence on the frequency-current curves; this can be explained by the fact that the quenching of the pulses is due to phenomena confined to a region close to the point.

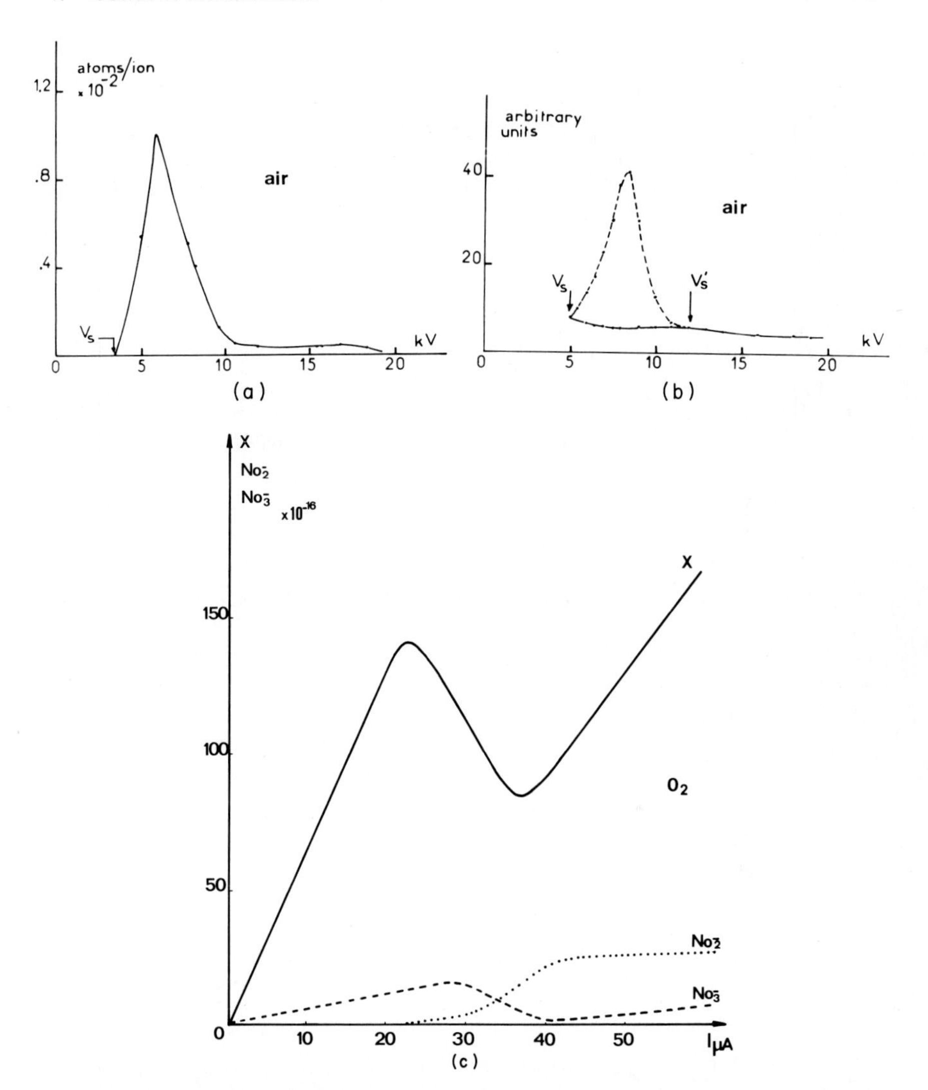

Fig. 19. Properties of the negative-point corona under continuous applied voltage; $d = 20$ mm. (a) Variation in sputtering rate with applied voltage in air. (b) Distribution of heights of large and small current pulses for the same experiment, in the autostabilization regime $V_s - V_s'$ and for Trichel pulses beyond V_s' (Buchet and Goldman, 1970). (c) Chemically active species formed by the negative corona in oxygen as a function of average discharge current (Lécuiller *et al.*, 1972). The curves show the number of particles formed in 15 min of discharge.

$N_{O_3^-}$: O_3^- ions and their derivatives of comparable chemical activity (CO_3^-, NO^-, . . .)
$N_{O_2^-}$: O_2^- ions and their derivatives (O_4^-, CO_4^-, $N_2O_2^-$, . . .)
X: total of neutral oxidizing species.

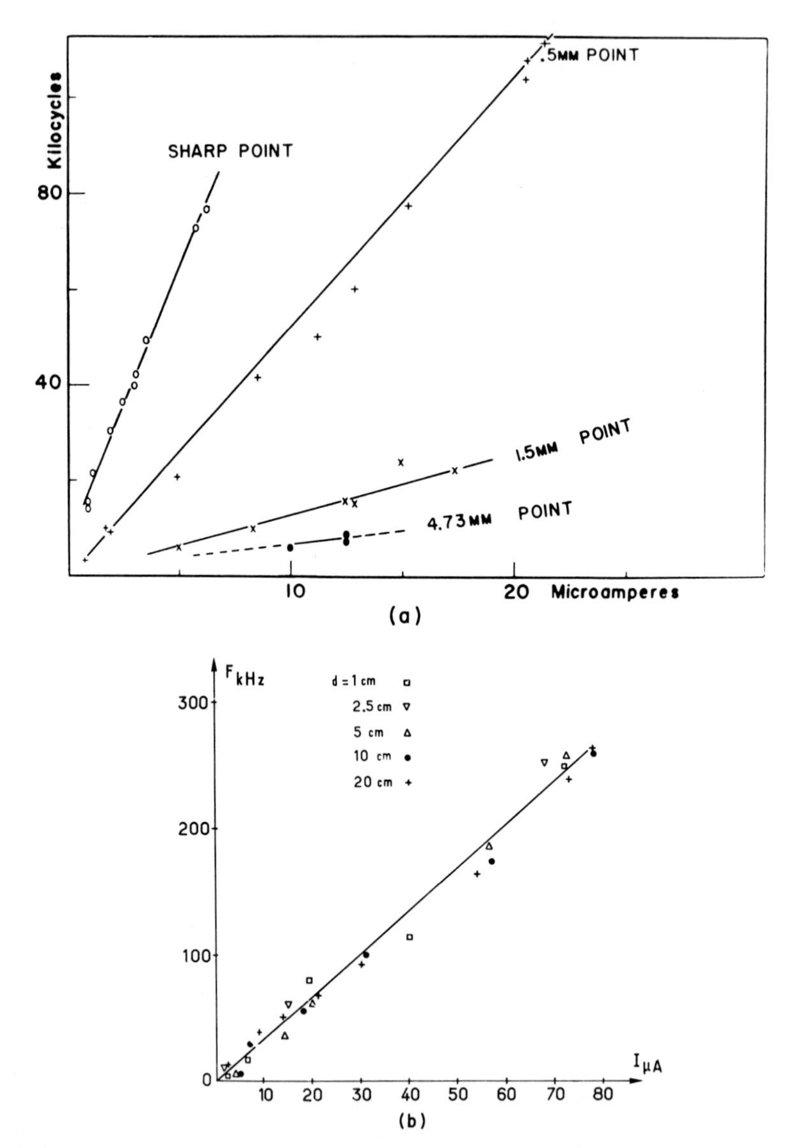

Fig. 20. Trichel pulse frequency versus current. (a) Trichel's curves for different points as function of mean current (Loeb and Kip, 1939). (b) Curves for different gap lengths as function of direct current (Fieux and Boutteau, 1970).

The parameter with the greatest influence on the pulses is the radius r of the point. On the one hand, it controls the pulse amplitude and consequently the charge per pulse (which in fact depend relatively weakly on current), and, on the other hand, the pulse repetition frequency. Fieux and Boutteau (1970) characterize the relationship between the frequency F and the point radius r in the domain $r > 0.125$ mm for gap lengths $d > 10$ mm by the empirical relation:

$$dF/dI = 2.27 \quad r^{-1} \tag{39}$$

where F is expressed in kilohertz, I in microamperes, r in millimeters. The decrease of F with increasing r is accompanied by an increase in the pulse amplitude i_m; for example, for V near V_g, one finds $i_m \sim 0.4$ mA for $r = 0.2$ mm, and $i_m \sim 6$ mA for $r = 3$ mm.

An important characteristic of Trichel pulses which can be observed experimentally is their very short rise time (on the order of 10^{-9} to 10^{-8} s) which is nearly independent of the radius of the point (English, 1950; Amin, 1954; Denholm, 1960), depending essentially on the gas, on its pressure, and on the applied voltage (Zentner, 1970a,b). This observation has served as a critical test for the principal theories developed to explain the formation of the pulses. The first such model was proposed by Loeb (1952), and was based on a mechanism which Aleksandrov (1963) refers to as one of "successive development" of electron avalanches, relative to his own model of "parallel development." In fact, to describe the rapid growth of the current pulses, Loeb proposed a mechanism in which the number of avalanches increases in a geometrical progression: each pulse avalanche produces q photoelectrons, which initiate q new avalanches, the duration of each of the consecutive avalanche cycles being equal to the time required for the development of one avalanche. If the pulse rise time which one calculates thus by taking $q = 3$ ($q = n_a \gamma$, where n_a, the number of ions produced in one avalanche, is about 6×10^4 and γ, the secondary multiplication coefficient, is 5×10^{-5}) is compatible with the values measured for sharp cathodes, for example $r = 0.2$ mm, the results obtained in the same manner for points of smaller curvature turn out to be much too large.

The parallel development mechanism of Aleksandrov gives better insight into the observed phenomena, both for sharp and blunt points. According to this model, secondary avalanches form and develop in parallel with primary ones as soon as the number of electrons in the developing avalanche n exceeds n_0, the number of electrons in the initial avalanche, i.e., from the moment that the condition for a self-sustaining discharge, $n/n_0 = 1$, is satisfied. The field increase in the ionization zone, due to the rapid accumulation of space charge along the path of the developing avalanche, leads to an increase in n. Aleksandrov divided the avalanche path into individual segments, in each of which he supposed n/n_0 to be constant, with values of that ratio becoming higher and higher as the discharge develops. This led him to distinguish two stages in the development. During the first stage, corresponding to the smallest values of n/n_0 (<2) (see Fig. 21), the number of avalanches developing in parallel increases relatively slowly to a value of about 1000. Also during this stage, for which the duration is relatively long ($\sim 5 \times 10^{-7}$ s), the pulse current is extremely small, about two orders of magnitude less than the pulse amplitude, and so was not detected till now in the oscilloscope observation of the pulses. However, Bugge and Sigmond (1969) have observed that the Trichel pulses are preceded by a slowly rising cathode glow emission of duration at least ten times the Trichel pulse 10–90% rise time, and Hosokawa *et al.* (1969) have shown that this glow is formed by successive multiavalanches. According to the Aleksandrov calculations, the pulse waveforms which were observed on the oscilloscope corresponded to the second stage. Since Loeb set $n/n_0 = 3$, he was considering essentially this second stage of the pulse but, since he did not take into account the large number of avalanches developing in parallel, he obtained pulse rise times for large points significantly greater than the observed values.

As his calculations would lead to an unlimited increase in the pulse current, Aleksandrov explained the quenching of the pulse by invoking on the one hand the effects of distortion of the space-charge distribution in the ionization zone by the applied field, and on the other hand the reduction of the enhanced field region by the negative ions, as seen by Morton (1946). This is illustrated by Fig. 22, already presented in Section IV.C.1.

The effects of field distortion were confirmed by the experiments of Torsethaugen and Sigmond (1973), which showed that the superposition of a very small voltage pulse (1% of the applied voltage) was enough to destroy the discharge equilibrium around the self-sustaining condition. In

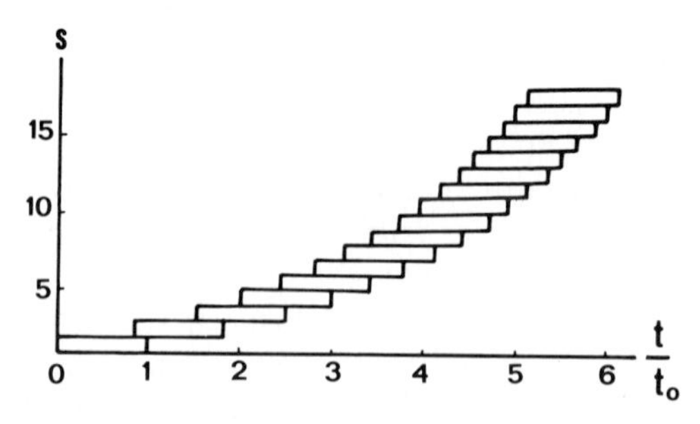

Fig. 21. Schematic representation of the overlapping of avalanches in the Aleksandrov (1963) model of Trichel pulse formation. The diagram is made for a small value of n/n_0 corresponding to the first state of the development. $n/n_0 = 1.2$, t_0 = duration of development of one avalanche, s = number of avalanches starting from the beginning of the cycle.

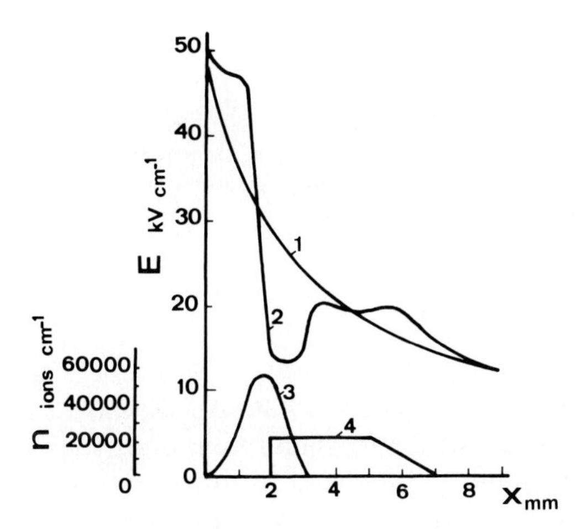

Fig. 22. Distribution of field intensity and ion concentration in the negative corona ionization zone, close to a wire conductor of 6 mm diameter. Calculated curves for x = 0 at the conductor surface (Aleksandrov, 1963). Curves 1 and 2, respectively, show the distribution of field intensity at the time of formation of the pulse and after the formation of 0.25×10^9 ions of each polarity, the negative ions being concentrated near the ionization zone boundary. Curves 3 and 4 illustrate the distribution of the positive and negative ions, respectively, after the passage of an avalanche.

fact, while it aids the establishment of the Townsend discharge which follows the plateau of the Trichel pulse (Fig. 18b) if its application coincides with this plateau, on the contrary it destroys the Townsend discharge, to the profit of a new Trichel pulse, if it is applied afterwards. The spatio-temporal evolution of the field distortion bears a certain resemblance to the mechanism of advance of the primary streamer in the positive-point case, where the propagation occurs by the projection into the gap of a zone of field distortion, leaving behind it a field too feeble for the electrons to be able to make ionizing collisions (Marode and Hartmann, 1969). These hypotheses find experimental support in the observations of Ushita *et al.* (1967, 1968) and Ikuta and Kondo (1976) during the early stage of the Trichel pulse, thanks to a very detailed spatio-temporal analysis of the luminous phenomena during that period. In particular, they have shown that a glow (which one can associate with a primary streamer owing to the simultaneous presence in the emission of the second positive and first negative systems of the nitrogen molecule) is born at a certain distance from the cathode point and propagates toward it, while a positive column (emission spectrum containing only the second positive system of N_2) develops in the opposite direction (Fig. 23). The rapid growth of the pulse current, which covers about 50 ns, overlaps the propagation of the "streamer" toward the cathode. This hypothesis

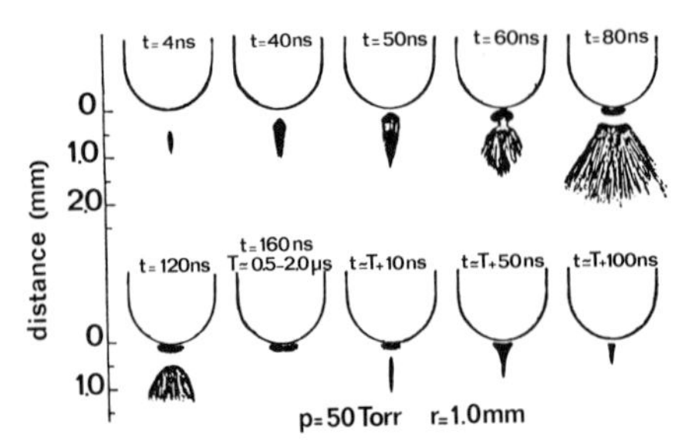

Fig. 23. Spatio-temporal evolution of negative corona discharge under continuous voltage during a Trichel pulse (Ushita *et al.*, 1967).

of the streamer has been confirmed by the experiments of Hosokawa *et al.* (1970) which showed that the transition from avalanches to Trichel pulse occurs when the number of electrons created in the avalanches gives rise approximately to the critical field for initiation of a streamer.

(b) *Continuous Aspect of the Discharge* As has already been mentioned, the Trichel pulses are accompanied by a permanent current I_c. One can see in Fig. 24.b that, at the transition V_g between the Trichel and the glow discharge regions, I_c decreases with the gap distance d, very slowly at first, then rapidly to zero at a value $d = d_{min}$ (<2.5 mm in the case considered); for gap distances smaller than d_{min} one obtains direct spark breakdown without preliminary corona discharge (Miyoshi *et al.*, 1963).

A self-sustained negative corona discharge is thus always accompanied by a permanent current in the gap, which can be considered to be the result of two phenomena; a flux of ions and a permanent self-sustained discharge current leaving the point.

(i) The negative ion space charge, which is responsible for the intermittent character of the discharge, builds up relatively close to the point. For a paraboloidal point, if one neglects space charge effects and assumes a constant ionic mobility K in the gap, then, the transit time t_d, that is, the time required for the negative space charge to cross the gap, can be roughly evaluated from:

$$t_d = \int_0^d \frac{dx}{KE} = \int_0^d \frac{r + 2x}{KE_r r}\, dx \approx \frac{d^2}{KE_r r} \tag{40}$$

with E_r, the field at the point, deduced from relation (9c). At the discharge threshold V_s, one finds thus that the transit time t_d is of the same order as the interval $T = 1/F$ between two successive pulses. This is to be correlated with a clearing length, which one can assimilate into d_{min}, i.e., the distance which the negative ions must traverse to restore the threshold field at the cathode point. As the voltage is increased from V_s to V_g, the ratio t_d/T increases. There then forms in the gas, beyond the multiplication zone, a reserve of negative ions, nourished by successive discharges, which drift slowly toward the plane (Loeb, 1965; Buchet *et al.*, 1966; Lama and Gallo, 1974).

(ii) The work of Zentner (1970a) indicates that ionization in the cathode region never ceases at all. This result was confirmed by the photomultiplier measurements of Sigmond and Torsethaugen (1973), which established that a Townsend discharge is sustained at the point by a cathode electron current.

Thus, as was suggested at the beginning of Section IV for the corona discharge in general, the negative discharge is composed of a quasisteady Townsend discharge, characterized by a permanent weak current, and of an impulsive discharge which appears as a transient phenomenon with a glow discharge structure (see Loeb, 1965) as for the positive corona streamers, but in a manner still more evident since the Trichel pulses involve secondary γ effects at the cathode and develop according to the Townsend mechanism (Hosokawa *et al.*, 1970).

One is thus led to a harmonious generalization permitting one to join the ensemble of corona discharges to the general class of glow discharges.

3. The Pulseless Glow Discharge ($V_g < V < V_r$)

The transition to the pulseless discharge at $V = V_g$ occurs when the frequency F, increasing with V, reaches a limiting frequency which is only a function of r. This in turn defines a critical current within the limits of validity of Eq. (39). Fieux and Boutteau (1970) express this limiting frequency by

$$F_{max} = (170/r) + 15 \quad \text{kHz}, \tag{41}$$

where r is in millimeters; the validity of this expression has been verified for $0.4 < r < 5$ mm and $d > 10$ mm. When F_{max} is reached, the dead time between two pulses is reduced to the order of the pulse width or even smaller.

Unlike the establishment of the positive glow discharge, the transition to the negative pulseless glow discharge is marked by the pulse current being incorporated into the continuous current, rather than by its disappearance. In fact, there is an abrupt increase in the continuous current (by about a factor of two in the case of Fig. 24b) which is seen as a much reduced variation in the total current, either positive, negative, or zero (Fig. 24a), of which the amplitude and sign depend on geometric factors (Miyoshi, 1958).

The visual appearance of the discharge on both sides of the transition is shown in Fig. 25. In the Trichel regime, it is characterized by a glow in the vicinity of the point which looks like a shaving brush (frame a); in the pulseless glow discharge, a narrow central spike extending far out into the gap predominates (frame b).

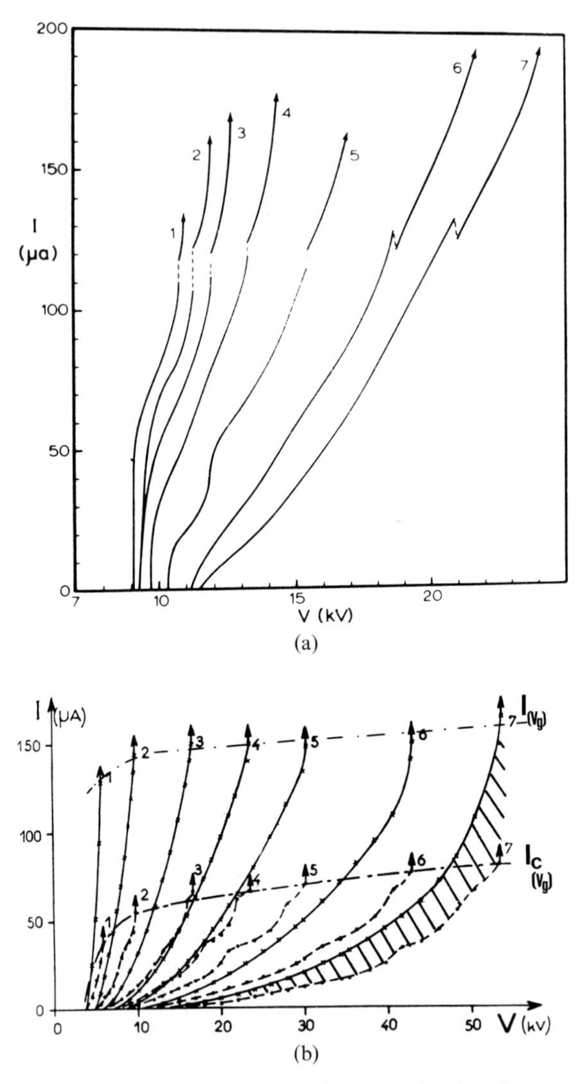

Fig. 24. Current-voltage curves for negative point corona in air under continuous voltage, for different gap lengths. (a) 6 to 15 mm. (b) 2.5 to 40 mm. (a) Miyoshi's curves showing the changes of mean current at transition voltage V_g from Trichel regime to glow (Loeb, 1965, p. 391). (b) Correlated values of continuous current I_c (---) and of mean total current I (——) as a function of voltage up to the transition voltage V_g (Buchet *et al.*, 1966). The hatched zone characterizes the impulsive part of the current for the largest gap length.

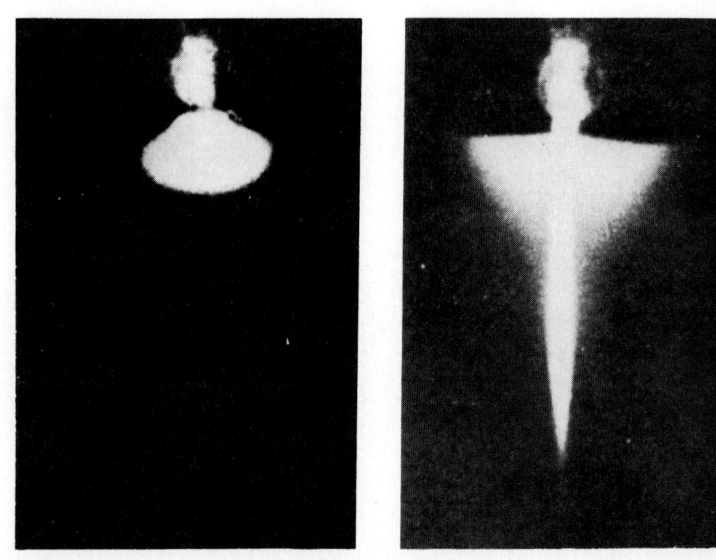

Fig. 25. Luminous appearance of negative-point corona in air under continuous voltage (Fieux and Boutteau, 1970). $r = 2$ mm, $d = 10$ cm. Left—in the Trichel regime. Right—in the glow regime. From the photographic enlargement one can deduce a real spike length ~13 mm.

4. Transition to the Transient Arc ($V = V_r$)

When a steady discharge is established in air, the transition to the transient arc is marked by a critical average current on the order of 200 to 250 μA, which is more or less independent of geometry and also of the pressure, at least below atmospheric pressure (Fig. 26a). The determining physical parameter for the transition seems to be a critical field at the plane sufficient to initiate a "positive" streamer (Goldman *et al.*, 1965); in fact, when the current reaches its critical value, one observes a luminous spot appearing on the plane (Fig. 26b) which, at pressures near atmospheric, develops into a streamer.

Now the resultant of the applied electrostatic field $\mathbf{E}_e$ and the local space-charge field $\mathbf{E}_s$ is insufficient to explain this phenomenon. To calculate the true field $\mathbf{E}$, one should take account of a local amplification factor β due to surface irregularities on the plane which, even for a well-polished surface, form rapidly:

$$\mathbf{E} = \beta\mathbf{E}_e + \mathbf{E}_s. \tag{42}$$

The value of β depends on the form of the irregularities, but usually lies between 1 and 10. The discharge behavior could also be influenced to some extent by the surface oxidation due to previous discharges.

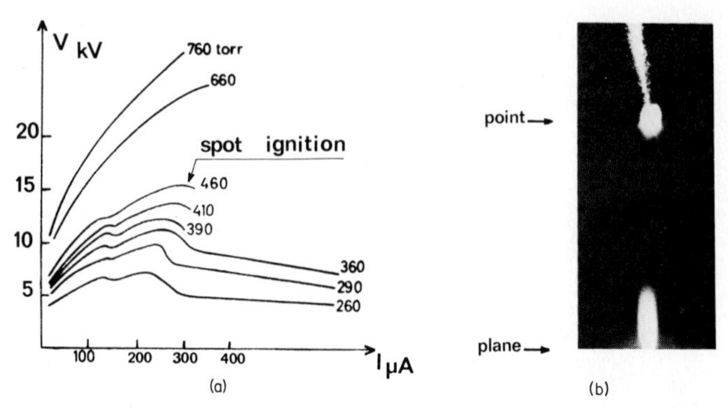

Fig. 26. Transition from negative-point corona to arc under continuous voltage in air at different pressures (Goldman *et al.*, 1965). (a) Potential curves as function of mean current, showing the ignition of anode spots for $p < 500$ Torr. At atmospheric pressure, the arc transition occurs at essentially the same values of mean current. (b) Photograph of anode spot at $p = 500$ Torr.

If the spot ignition leads to a voltage drop due to the external circuit, the discharge cuts itself off, and Trichel pulses reappear while the voltage is re-establishing itself (Fig. 27). The phenomena then repeat according to a relaxation mechanism quite comparable to the classical relaxation observed with discharge tubes.

Using pulse voltage, Akazaki and Tsuneyasu (1975) showed that in air, for pulses having a 1 μs rise time and 150 μs duration, and for the ranges of pressures (20 to 760 Torr) and gap lengths (1 to 10 cm), the 50% spark breakdown voltage (that is, the voltage corresponding to a 50% breakdown probability) follows a law of the same form as the Paschen law, as for positive discharges (see Section IV.B.1.(c)):

$$V_{50} = 0.5 + 0.025 \quad kpd \tag{43}$$

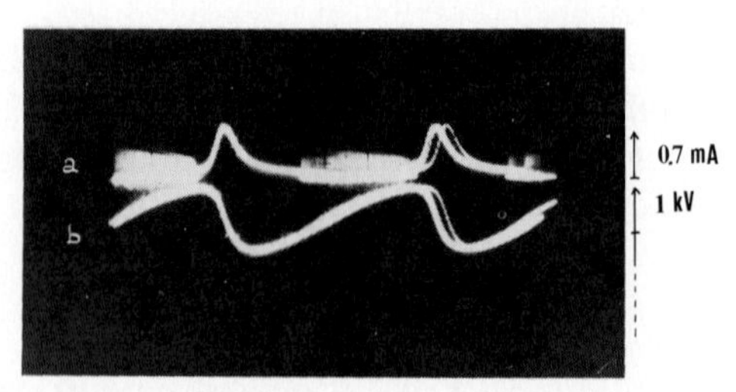

Fig. 27. Current (trace a) and voltage (trace b) oscillograms of the negative discharge after formation of anode spot seen in Fig. 26b (sweep time = 2 ms). After the voltage falls, series of Trichel pulses take the place of the low-frequency pulses associated with the anode spot.

with p in Torrs, d in centimeters, V_{50} in kilovolts, and k a quantity weakly dependent on d ($k \sim 1$ for $d = 1$ cm; $k \sim 1.3$ for $d = 10$ cm).

Image-converter camera studies of the spatio-temporal development of the discharge to the transient arc have shown that:

(i) negative primary streamers develop away from the point, in a form which is sometimes diffuse, sometimes filamentary (Akazaki and Tsuneyasu, 1975)

(ii) among these primary streamers, one can distinguish two groups differing by their velocities (Suzuki, 1973): slow streamers with speeds of the order of 5 to 7×10^7 cm s^{-1}, and fast ones with speeds in the 2 to 12×10^8 cm s^{-1} range.

(iii) in the development of the discharge, one also sees phenomena such as secondary streamers, return waves, *etc.* (Suzuki, 1973; Akazaki and Tsuneyasu, 1975).

But the physical nature of those phenomena, as well as the processes of heating and thermalization governing the transition to the arc, remain still to be studied.

V. ALTERNATING VOLTAGE CORONA DISCHARGE

Corona discharge under applied alternating voltage presents itself as a composite regime consisting of alternating periods in which the corona is successively positive and negative; in each of these regimes the discharge development follows the applied voltage, which is by definition variable.

Since the electric wind induced by the discharge is always directed from the stressed to the unstressed electrode, independently of the polarity of the field, aerodynamic equilibrium is established as for continuous applied voltage. But the ions here play a different role from one half-cycle to the next. Also, one must expect that the behavior of the discharge under alternating voltage should differ from the continuous discharge if the working conditions are such that the ions produced during one half-cycle leave their imprint on the next half-cycle. This occurs if the interelectrode distance and the applied frequency are both sufficiently large that at the end of one half-cycle there are still ions left in the gap whose motion will be reversed when the applied voltage changes polarity periodically.

Based on what has already been said in Section III.A.2 about the establishment of the positive glow-discharge corona, we already know that in the positive half-cycle, residual negative ions from the previous half-cycle inhibit the impulsive component of the current in favor of the continuous component. In particular, above a certain critical interelectrode distance for a given applied frequency, the burst pulses and preonset streamers are seen to disappear. On the contrary, there is a lack of information on the possible influence of residual ions from a positive half-cycle on the next negative period. The analysis of the perturbing effect played by the negative ions produced during a negative half-cycle on the following positive half-cycle has been utilized profitably by Hutzler (1971) to measure the mean mobility K of the negative ions; this mobility plays a determining role in the movement of the ions, since they are displaced with a velocity $\mathbf{v}$ proportional to the field $\mathbf{E}$ and their mobility, and in this way, it intervenes in the generation conditions of the perturbing effect:

$$\mathbf{v} = K\mathbf{E}. \tag{44}$$

Without attempting to present an exhaustive bibliography on the subject, we cite:

(a) in the domain of industrial frequencies, the works of Cobine (1958), Gary (1974) and Gary and Moreau (1976), in which are treated, in particular, the problems of radio-frequency noise produced by the alternating-voltage corona discharge; also among others, an interesting calculation of power line losses taking account of the harmonic motion of the ions imposed by the applied alternating voltage, presented in the above-cited review article of Gary (1974).

(b) in the medium and high frequency regimes, the specialized book of Aronov *et al.* (1969), as well as the work of Korge and Kudu (1973), Laan and Kudu (1975), Larionov *et al.* (1975).

VI. CORONA DISCHARGES IN LONG AIR GAPS

Studies of corona discharges in long air gaps are primarily motivated by their application to the study of lightning, and to the production and transport of electrical energy. Just as there exist competing techniques for the production of electrical energy, because of the continually increasing consumption, the voltage of power distribution networks is increasing, and very high voltage aerial lines have been still projected. Because the corona discharges which form in the vicinity of the conductors lead to power loss and radio frequency interference, they are of great concern to the energy producers.

A. Spatio-Temporal Development of the Discharge for Positive Polarity

We consider first the general scheme of the spatio-temporal evolution of the discharge (Fig. 28) (cf. Baldo, 1974; Berger *et al.*, 1974); for this, we employ laboratory conditions which well simulate the overvoltage conditions in those practical situations which cause the greatest problem for maintaining voltage on networks: a point-to-plane gap from 1 to 25 m, subjected to positive voltage pulses with rise times on the order of several hundreds of microseconds.

When the field at the point reaches the critical value with the statistical time lag (t_0, t_i) with respect to the instant at which the voltage pulse is applied (Section II.C), at first a "first corona" develops; this looks like a

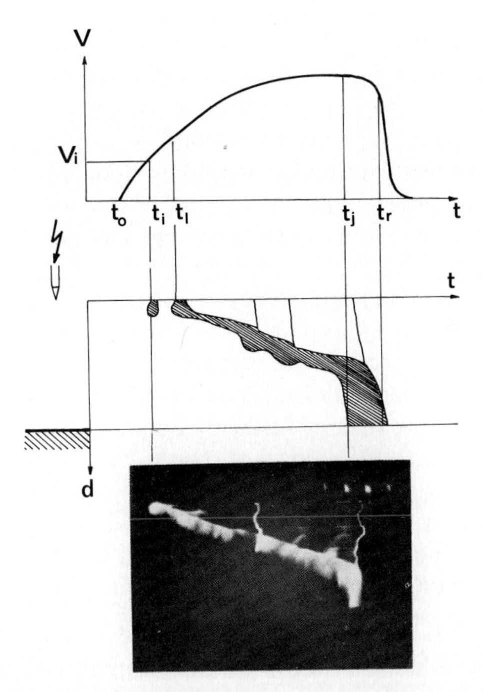

Fig. 28. Spatio-temporal development of a positive discharge leading to spark breakdown in a long air gap (Leroy, 1974). Above, applied voltage signal; below, time-resolved picture of the development of the discharge, taken with an image converter camera operated in the streak mode and shuttered just before spark breakdown to avoid film exposure. Center: schematic reproduction of the typical features appearing in the streak picture.

glow, more or less homogeneous because of the numerous streamers of which it is composed. The space charge developed by this corona cuts off the discharge, and it dissipates during the time (t_i, t_1), the "dark period," before the field at the point again reaches high enough values to cause ionization. This could take place by the drift of the space charge toward the plane, in the same manner in which it occurs in short gaps for the successive discharges induced by steady applied voltages. But there are two additional factors which can act here: First, the applied voltage itself can increase, since first corona often occurs during the rise time of the applied pulse. Second, a thermal ionization effect can take place in the "stem" of the discharge, the part of the streamer near the electrode which often appears more brilliant.

At time t_1, streamers again develop, but behind them there forms a new phenomenon, specific to large distances: the "leader channel," an almost dark channel, often traversed by abrupt reilluminations of short duration, which could be due to the propagation of potential waves as observed for return strokes with small gaps. This channel elongates into the gap with nearly constant velocity when measured along its tortuous path (1.5–2 cm s^{-1}) and almost constant current (several tenths of an ampere); its progression appears more or less continuous depending upon whether the rate of rise of voltage is more or less high.

Depending on the value of the applied voltage, the leader can either stop after having crossed only a portion of the gap or reach the ground electrode, causing spark breakdown. In the latter case, the leader tip has only traversed 60 to 70% of the interelectrode gap when the corona emanating from it, the "leader corona," reaches the ground electrode; this occurs at time t_j. But at this stage, its development accelerates by an irreversible process referred to as the "final jump:" the velocity of the tip and the current both increase very rapidly, causing spark breakdown at time t_r, only a few microseconds after t_j.

However, in certain conditions the discharge develops more rapidly, and the arc occurs without the leader being able to develop in the manner just described. This type of direct spark breakdown is observed:

(a) with steeply rising pulses such that the conditions for the final jump are filled in a time less than the time t_i for formation of the first corona;

(b) with electrode configurations which have a smaller field gradient; namely, when the radius of the active electrode is of the order of one-tenth of the gap length or larger. In this case, which favors a large flow of charge, the arc transition is made directly by the first corona streamers which cross the entire gap.

The very nature of the mechanisms controlling the development of the leader is still not well known. However, a fundamental difference between the leader channel and the leader corona has very recently been put into relief by spectroscopy (Renardières Group, 1974; Hartmann, 1974a). In fact, as Fig. 29 shows, the leader channel emits primarily in the red and infrared, while the leader corona emits in the ultraviolet. More precisely:

(i) The spectrum of the leader channel is essentially limited to 7.1 eV, the upper state energy of the 1st positive system in N_2. (In the second positive system, the upper state of which lies at higher energies, some radiation is seen, but very little; this is in contrast to the spectrum from the secondary streamer at small distances where the electron energy is greater, although still small.) The spectrum also reveals the presence of excited CO_2 and of molecules resulting from dissociation and recombination, such as NO, CO, and OH, which are not observed in small gaps.

(ii) In the spectrum of the leader corona, one finds the important emission bands observed at small intervals, i.e., the bands of N_2 and $N_2{}^+$, along with some $O_2{}^+$ bands not seen in spectra obtained for short gaps. These bands imply that in the active region in front of the leader tip there are electrons of elevated mean energy, corresponding to local fields (50 to 180 kV cm^{-1}) compatible with the experimentally observed propagation characteristics. The leader channel, by contrast can be considered to be a region with a field nearly constant and low (e.g., 0.5 to 2 kV cm^{-1} for long leader lengths).

According to the similarities which one observes in the behavior of the discharges for large and small gaps, one can expect electron attachment to attain a significant rate in the leader corona, in front of whose head the

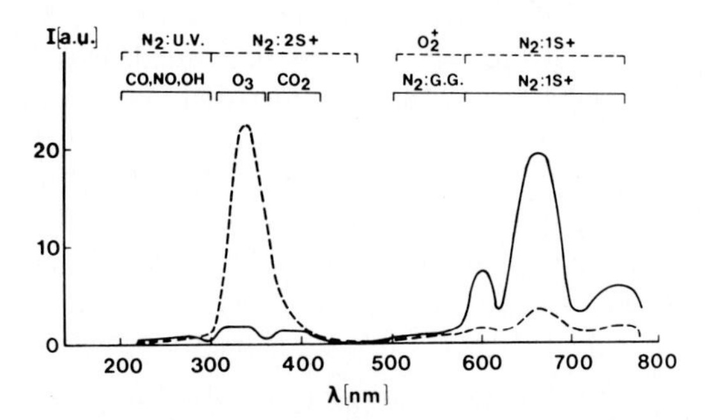

Fig. 29. Gross spectral variation of light emitted respectively by the leader channel (——) and the leader corona (----) during the development of the discharge in long air gaps in positive polarity (Renardières Group, 1974).

ionization occurs. One finds again the conditions already encountered for small gaps (Section IV. B.3): The increase of current in the latter stages of the discharge implies that there is electron detachment, suggesting a local temperature in excess of about 1200°K. Since the leader is hot, the observed phenomena could be explained if we suppose that these conditions are met in the leader tip, from which one can consider the velocity of advance to be like the phase velocity corresponding to the propagation of such a zone (Goldman and Berger, 1975).

To rationalize the high field values needed at the leader tip to cause corona with the low mean electric stress at spark breakdown which is, in some cases, below 2 kV cm^{-1}, the leader channel must be considered as a high conductivity path. Now the reduced field E/p in this leader channel is too weak to give rise to collisional ionization; we can however invoke a thermal ionization mechanism (Waters and Jones, 1964a,b), justified *a priori* by the spectroscopic results which indicate an elevated temperature. Thus the first phenomena of heating and thermalization should be established (Section IV.B.3).

To explain the increase of ionization, Waters and Jones used the Saha equation, which gives the rate of ionization as a function of the temperature. In doing this, they consider the gaseous volume occupied by the discharge to be constant along the duration of its development.

One can also treat the problem in a different fashion, by taking into account that the development of the discharge is accompanied by a shock wave (Few *et al.*, 1967; Kekez and Savic, 1974), which implies that the discharge channel will expand. As for the short gaps (Section IV.B.3), we suppose that the expansion occurs with the speed of sound (300 m s^{-1}) or quicker and without change of pressure (which must of course yet be verified). A rough calculation, based on the expansion, at sonic speeds, in 1 μs of a channel of 30 μm initial diameter, shows that the density of molecules in the channel should decrease by a factor of the order of 400, while the diameter goes from 30 to about 600 μm. The resulting increase in E/p would make possible significant ionization by collision, which could lead to a runaway phenomenon. But, in spite of the recent studies on the ultimate stages of the discharge, the physical mechanisms which they call into play are still not well understood.

B. Parameters Used in Determining Insulation Distances for Maintaining Voltage on Networks

On the practical level, one hopes to gain from the study of the physics of discharges at large gaps some information useful in reducing the work of the testing laboratory, while at the same time improving the validity of its results. Thus, it would not be necessary to follow through laboratory studies to spark breakdown if one could find a simple breakdown criterion based on leader propagation. Unfortunately, this objective has not yet been realized at the present time, and predictions of insulation requirements for very high voltage transport networks must be made from measurements and empirical relations made for breakdown under "worst case condition," which, let us remember, is for pulsed positive voltage. (See Leroy, 1974.)

Among the disruptive parameters is the time interval (t_0, t_r) (Fig. 28). Since it represents the delay time between the application of high voltage and the breakdown it has evident practical value for the design of protective devices, which must break down before the equipment which they are protecting. But the most important disruptive parameter which is practically considered is the 50% spark breakdown voltage V_{50}, i.e., the voltage which corresponds to a breakdown probability of 50%.

Curves showing the variation of V_{50} as a function of the time to crest t_{cr} of the voltage pulse are given in Fig. 30a. These curves show the existence of a critical value of t_{cr} (of the same order of magnitude as that of the practical overvoltage pulses observed on networks) for which the spark breakdown voltage shows a minimum $(V_{50})_{min}$. In light of the results presented in the preceding paragraph, this minimum represents the "best conditions" for the progression of the leader; in this spirit, we can make the following interpretation. On the left part of the curve, the rapid rate of rise of the pulse voltage could enhance the local field increase in the active region of the discharge due to the drift of space charge; on the right part of the curve, the slowly increasing voltage must be compensated by a higher value of the applied voltage to satisfy the field conditions for leader development. The minimum in the spark breakdown voltage is given by the following empirical relation (Leroy *et al.*, 1973):

$$(V_{50})_{min} = \frac{3400}{1 + 8/d} \tag{45}$$

with V in kilovolts and d in meters. This relation, valid for point-to-plane geometry under pulse voltage conditions, can be extended to other geometrical configurations and other wave shapes, notably alternating voltages, by introducing empirical correction factors which have been determined in the laboratory. Furthermore, a saturation effect appears (Fig. 30b) which gives a glimpse of a voltage limitation to transmission lines. One is tempted to explain this effect in terms of leader progression mechanisms, as resulting from an increase in conductivity in its channel, due to an increase of the charge injected by the leader corona when its length increases.

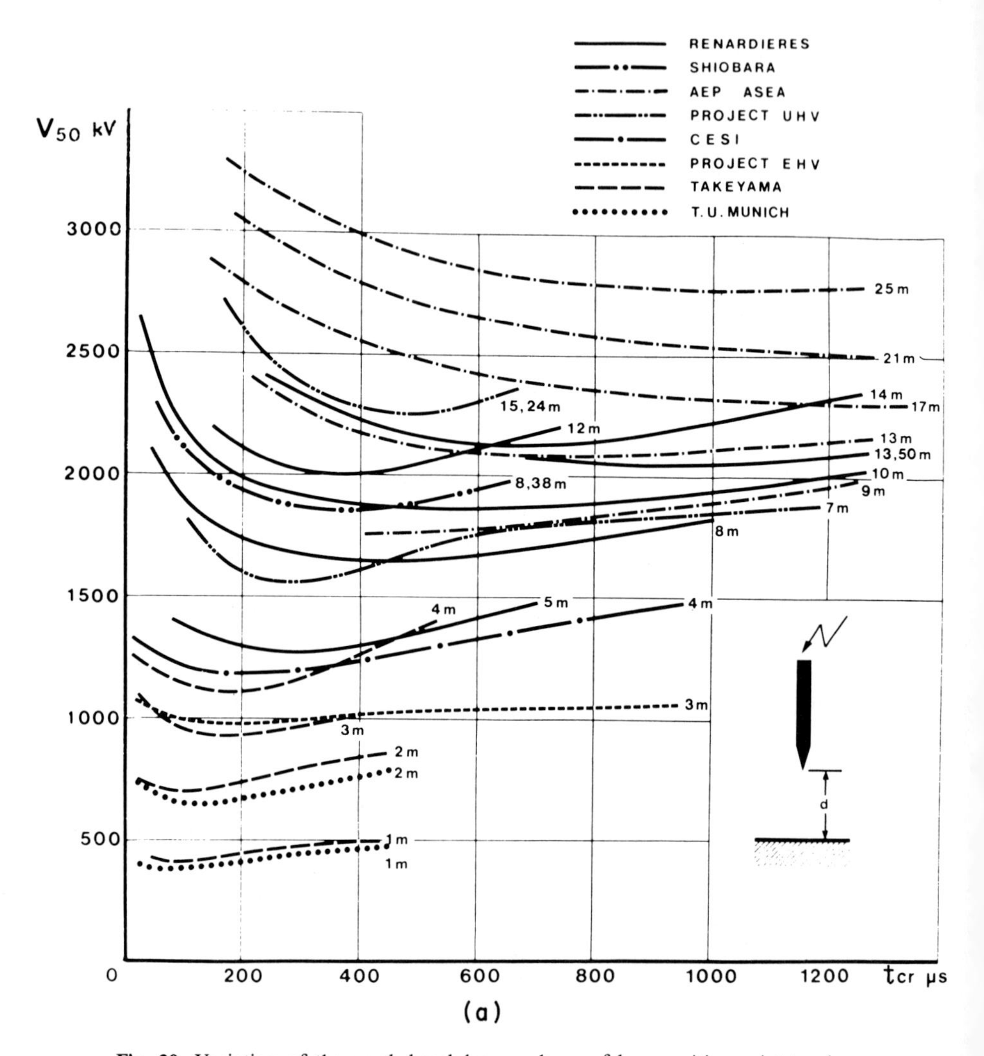

Fig. 30. Variation of the spark breakdown voltage of long positive point-to-plane gaps in air (Leroy, 1974). (a) Variation of the 50% spark breakdown voltage V_{50} versus time to crest t_{cr} of the applied voltage impulses. The results which come from different laboratories show that there is a minimum value of V_{50} for critical values of t_{cr}. (b) Variation of the minimum value $(V_{50})_{min}$ versus gap length; experimental points and curve estimated from Eq. (45).

To compensate for the insufficiency of physical data accumulated so far, several models have been constructed for practical purposes, taking into account the observed electrical phenomena. A review of these models has been made by Klewe *et al.* (1974).

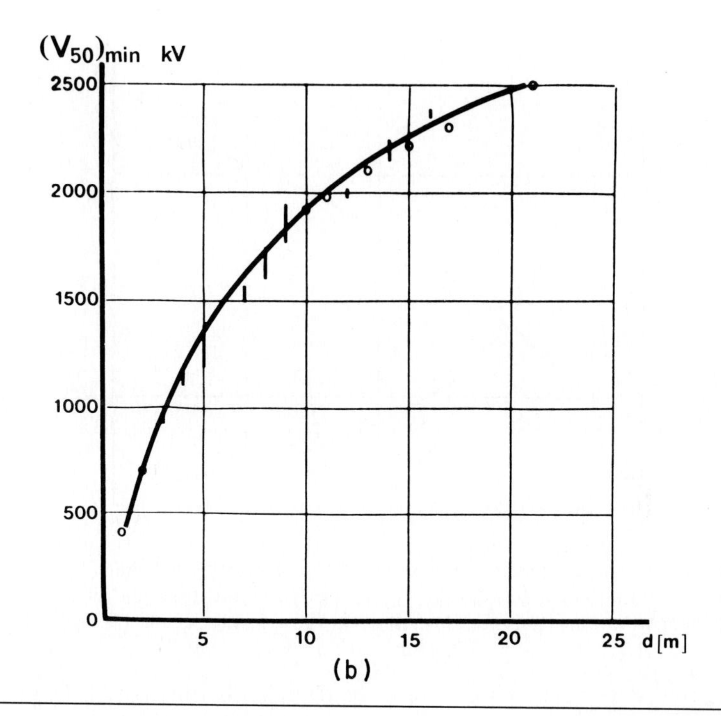

C. Negative Coronas in Long Gaps

Negative large-gap discharges have so far been studied less than the positive discharges (see Gruber *et al.*, 1975), primarily because of the ability of configurations having the same geometry to hold negative voltage better than positive, as illustrated in Fig. 31. Although the ratio V_{50-}/V_{50+} of the 50% spark breakdown voltages corresponding to the two polarities diminishes as the gap length increases, one sees in the figure that it is still in excess of 2 for distances of the order of 7 m in the conditions considered (i.e.—a cone-shaped rod with a 300/10,000 μs impulse, that is, 300 μs to crest and 10,000 μs duration). As in the positive case, the 50% spark breakdown voltage saturates at large gap lengths (Fig. 31), and also varies with the time to crest t_{cr} of the applied voltage pulse, showing a minimum $(V_{50})_{min}$ for a certain critical value of t_{cr}.

From the point of view of the luminous aspect, a "first corona" forms early in the vicinity of the active electrode, then the discharge progresses into the gap again in a more or less continuous fashion, depending on whether the rising edge of the applied voltage pulse is more or less steep. For the continuous progression, Gruber *et al.* (1975) have observed the formation of

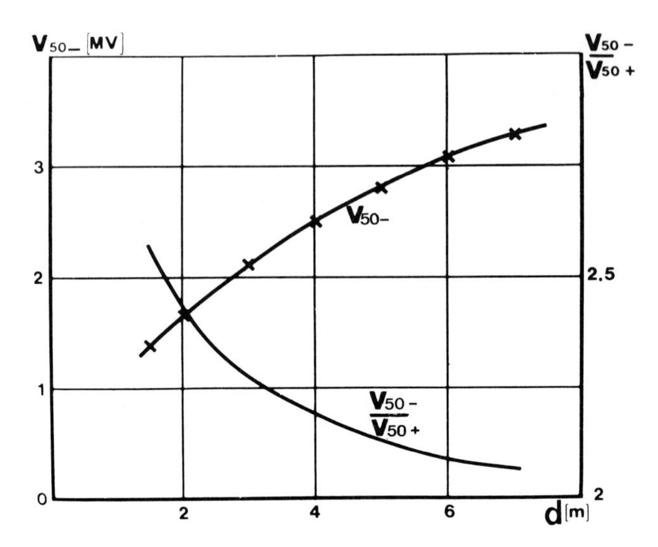

Fig. 31. Variation with gap length of the 50% spark breakdown voltage for long air gaps of negative polarity, and of the ratio V_{50-}/V_{50+}, which illustrates the greater ability of negative-polarity systems to withstand breakdown than positive-polarity systems (Gruber *et al.*, 1975).

mid-gap leaders developing in the gap from both ends (Fig. 32) as do the mid-gap streamers occurring in uniform-field geometry for high-overvoltage pulses. A sudden lengthening of the "negative" leader, developing outward from the cathode, occurs when its tip joins the upper tip of the mid-gap leader, accompanied by a strong negative corona at the leader head.

The process may stop at any stage in the gap, but if the negative streamers reach the plane, spark breakdown will take place through a final jump; as soon as they do, one can observe the growth of ascending "positive" leaders, causing breakdown when they join the negative ones.

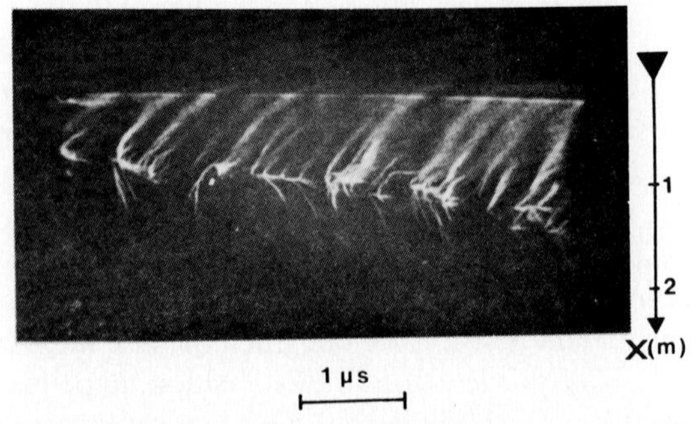

Fig. 32. Time-resolved picture of discharge development in a long air gap of negative polarity taken with an image converter camera operating in the streak mode (Gruber *et al.*, 1975). Conical tip, $d = 2$ m. 300/10,000 μs voltage impulse.

ACKNOWLEDGMENTS

The authors cannot pass silently over the significant contribution brought to this chapter by M. N. Hirsh who, after having encouraged them to undertake the work, participated in its realization by performing a translation of the text which was both detailed and intelligent. Imperfections, however, surely remain in this text; they are of course chargeable only to the authors who assure the translator of their gratitude.

The authors also wish to take this opportunity to gratefully acknowledge Prof. L. B. Loeb's continued interest in their research, and to thank him for the many useful discussions they have had with him over a period of about fifteen years.

REFERENCES

Acker, F. E., and Penney, G. W. (1968). *J. Appl. Phys.* **39,** 2363.

Aitchison, J., and Brown, J. A. C. (1969). "The Lognormal Distribution." Cambridge Univ. Press, London and New York.

Akazaki, M., and Tsuneyasu, I. (1974). *Proc. Int. Conf. Gas Discharges, 3rd, London,* I.E.E. Conf. Publ. No. 118, pp. 220–223.

Akazaki, M., and Tsuneyasu, I. (1975). *Proc. Int. Conf. Phenomena Ionized Gases, 12th, Eindhoven, Netherlands* **1,** 152.

Aked, A., Dale, S. J., and McAllister, I. W. (1972). *Proc. Int. Conf. Gas Discharges, 2nd, London* I.E.E. Conf. Publ. No. 90, pp. 24–26.

Albrecht, H., and Wagner, E. (1973). *Proc. Int. Conf. Phenomena Ionized Gases, 11th, Prague* p. 192.

Aleksandrov, G. N. (1963). *Sov. Phys.–Tech. Phys.* **8,** 161–166.

Aleksandrov, G. N. (1966). *Sov. Phys.–Tech. Phys.* **10,** 948–951.

Allibone, T. E., and Meek, J. M. (1938). *Proc. Roy. Soc. London* **A166,** 97.

Amin, M. R. (1954). *J. Appl. Phys.* **25,** 358.

Aronov, M. A. *et al.* (1969). "Electrical Discharge in Air at Audio Frequency Voltage." Moscow.

Atten, P. (1974). *Rev. Gén. Elec.* **83,** 143–153.

Badaloni, S., and Gallimberti, I. (1973). *Proc. Int. Conf. Phenomena Ionized Gases, 11th, Prague* p. 196.

Baldo, G. (1974). *I.E.E.E. Summer Meeting, Anaheim, California* Publ. No. 74, CH 0910-0-PWR, pp. 13–22.

Bandel, H. W. (1951). *Phys. Rev.* **84,** 92–99.

Bastien, F., and Marode, E. (1977). *J. Quant. Spectrosc. Radiat. Trans.* **70,** 453–469.

Bastien, F., Fertil, B., and Marode, E. (1976). *J. Phys. D Appl. Phys.* **9,** L. 155–158.

Beattie, J. E., and Cross, J. D. (1974). *Proc. Int. Conf. Gas Discharges, 3rd, London,* I.E.E. Conf. Publ. No. 118, pp. 279–283.

Belbel, E. (1976). Docteur-Ingénieur Thesis, Univ. Paris-Sud, France.

Berger, G. (1973). Docteur-Ingénieur Thesis, Univ. Paris-Sud, France.

Berger, G. (1974). *Proc. Int. Conf. Gas Discharges, 3rd, London* I.E.E. Conf. Publ. No. 118, pp. 294–297.

Berger, G., and Goldman, M. (1974). *I.E.E.E. Summer Meeting, Anaheim, California* Publ. No. 74, CH 0910-0-PWR, pp. 84–86.

Berger, G., Johnson, P. C., and Goldman, M. (1972). *Proc. Int. Conf. Gas Discharges, 2nd, London* I.E.E. Conf. Publ. No. 90, pp. 236–238.

Berger, G. *et al.* (1974). *Rev. Gén. Elec.* **83,** 761–789.

Blair, D. T. A., and Farish, O. (1965). *Proc. Int. Conf. Phenomena Ionized Gases, 7th, Beograd,* pp. 597–601.

Boulloud, A. (1955). *Rev. Gén. Elec.* **64,** 283–299.

Bricard, J., and Pradel, J. (1966). *In* "Aerosol Science," pp. 87–109. Academic Press, New York.

Buchet, G. (1976). not published.

Buchet, G., and Goldman, M. (1969a). *Proc. Int. Conf. Phenomena Ionized Gases, 9th, Bucharest* p. 291.

Buchet, G., and Goldman, M. (1969b). *Proc. Int. Conf. Phenomena Ionized Gases, 9th, Bucharest* p. 292.

Buchet, G., and Goldman, A. (1970). *Proc. Int. Conf. Gas Discharges, 1st, London* I.E.E. Conf. Publ. No. 70, pp. 459–462.

Buchet, G., Goldman, M., and Fakiris-Zeitoun, A. (1962a). *C.R. Acad. Sci. Paris* **255,** 79–81.

Buchet, G., Goldman, M., Goldman, A., and Reinhardt, J. (1962b). *C.R. Acad. Sci. Paris* **255,** 480–482.

Buchet, G., Goldman, M., and Goldman, A. (1966). *C.R. Acad. Sci. Paris* **263B,** 356–359.

Bügge, C., and Sigmond, R. S. (1969). *Proc. Int. Conf. Phenomena Ionized Gases, 9th, Bucharest* p. 289.

Canadas, P. (1976). Docteur-Ingénieur Thesis, Univ. de Toulouse, France.

Canadas, G., Canadas, P., Dupuy, J., Genêt, J., and Marsan, J. (1973). *Proc. Int. Conf. Phenomena Ionized Gases, 11th, Prague* p. 202.

Cobine, J. D. (1958). "Gaseous Conductors." Dover, New York.

Colli, L., Facchini, U., Gatti, E., and Persano, A. (1954). *J. Appl. Phys.* **25,** 429–435.

Couralet, J. L. (1973). CNAM Thesis, Univ. de Pau, France.

Das, M. K. (1961). *Z. Angew. Phys.* **13,** 410–415.

Davidson, R. C., and Farish, O. (1975). *Proc. Int. Conf. Phenomena Ionized Gases, 12th, Eindhoven, Netherlands* **1,** 156.

Davidson, R. C., Lynch, B. R., and Farish, O. (1976). *Proc. Int. Conf. Gas Discharges, 4th., Swansea, Great Britain* I.E.E. Conf. Publ. No. 143, pp. 242–245.

Davies, D. K., and Jones, E. (1963). *Proc. Phys. Soc.* **82,** 537–542.

Dawson, G. A. (1965). *Z. Phys.* **183,** 172–183.

Dawson, G. A., and Winn, W. P. (1965). *Z. Phys.* **183,** 159–171.

Denholm, A. S. (1960). *IEEE Trans. Power Apparatus Syst.* **50,** 698.

Deutsch, W. (1933). *Ann. Phys.* **16,** 588–612.

Driver, C. (1969). *Z. Phys.* **27,** 326.

Dupuy, J. (1958). *Rev. Gén. Elec.* **67,** 85–104.

Dupuy, J., and Berseille, J. (1969). *C. R. Acad. Sci. Paris* **268,** 1689.

Dupuy, J., and Goldman, A. (1974). *Proc.* "*Journées sur les interactions décharge-surface.*" *Orsay, France* pp. 41–46 (not published).

Durand, E. (1953). "Electrostatique et Magnétostatique." Masson, Paris.

English, W. N. (1948). *Phys. Rev.* **74,** 170.

English, W. N. (1950). *Phys. Rev.* **77,** 850.

Farish, O., and Davidson, R. C. (1974). *Proc. Int. Conf. Gas Discharges, 3rd, London* I.E.E. Conf. Publ. No. 118, pp. 361–365.

Félici, N. (1963a). *Direct Current* **8,** 252.

Félici, N. (1963b). *Direct Current* **8,** 278.

Félici, N. (1964). *C. R. Acad. Sci. Paris* **259,** 2967.

Few, A. A., Dessler, A. J., Latham, D. J., and Brook, M. (1967). *J. Geophys. Res.* **72,** 6149.

Fieux, R., and Boutteau, M. (1970). *Bull. Direction Etudes Rech. d'E.D.F.*, France **2-B,** 55–88.

Gallimberti, I. (1972). *J. Phys. D Appl. Phys.* **5,** 2179–2189.

Gary, C. (1974). *In* "Techniques de l'Ingénieur," Vol. D, Sect. 640. 3; 4; 5 (French engineering treatise edited by M. Postel, Paris).

Gary, C., and Moreau, M. (1976). "L'effet de couronne en tension alternative." Eyrolles, Paris.

Gary, C., Hutzler, B., and Schmitt, J. P. (1972). *IEEE Summer Meeting*.
Goldman, A. (1974). *Proc. Int. Conf. Gas Discharges, 3rd, London* I.E.E. Conf. Publ. No. 118, pp. 275–278.
Goldman, M., and Berger, G. (1975). *C. R. Acad. Sci. Paris.* **280-B,** 167–169.
Goldman, A., Goldman, M., Rautureau, M., and Tchoubar, C. (1965). *J. Phys.* **26,** 486–489.
Gosho, Y. (1974). *Proc. Int. Conf. Gas Discharges, 3rd, London* I.E.E. Conf. Publ. No. 118, pp. 284–287.
Govinda Raju, G. R., and Hackam, R. (1973). *Proc. Int. Conf. Phenomena Ionized Gases, 11th, Prague*, p. 186.
Gruber, G., Hutzler, B., Jouaire, J., and Riu, J. P. (1975). *Proc. Int. High Voltage Symp., Zürich*, pp. 519–523.
Hackam, R., Raja Rao, C., and Govinda Raju, G. R. (1976). *Proc. Int. Conf. Gas Discharges, 4th, Swansea, Great Britain*, pp. 209–211.
Hahn, G., Zacke, P., Fischer, A., and Boecker, H. (1974). *IEEE Summer Meeting, Anaheim, California* Publ. No. 74, CH 0910-0-PWR, pp. 74–80.
Hartmann, G. (1964). C.N.A.M. Thesis, Paris, France.
Hartmann, G. (1970). *C. R. Acad. Sci. Paris.* **B270,** 309–312.
Hartmann, G. (1972). *C. R. Acad. Sci. Paris.* **B275,** 311–314.
Hartmann, G. (1974a). *IEEE Summer Meeting, Anaheim, California* Publ. No. 74, CH 0910-0-PWR, pp. 62–64.
Hartmann, G. (1974b). *Proc. Int. Conf. Gas Discharges, 3rd, London* I.E.E. Conf. Publ. No. 118, pp. 634–638.
Hartmann, G. (1976). Not published.
Hartmann, G. (1977). Thesis, Univ. Paris-Sud, France.
Hartmann, G., and Gallimberti, I. (1975). *J. Phys. D Appl. Phys.* **8,** 670–680.
Hartmann, G., Goldman, A., Buchet, G., and Zeitoun, A. (1963). *Proc. Int. Conf. Phenomena Ionized Gases, 6th, Paris* **2,** 301–303.
Hermstein, W. (1960a). *Arch. Elektrotech.* **45,** 209.
Hermstein, W. (1960b). *Arch. Elektrotech.* **45,** 279.
Hosokawa, T., Kondo, Y., and Miyoshi, Y. (1969). *Elec. Eng. Jpn.* **89,** 120–127.
Hosokawa, T., Kondo, Y., and Miyoshi, Y. (1970). *Elec. Eng. Jpn.* **90,** 1123–1131.
Hudson, G. G., and Loeb, L. B. (1961). *Phys. Rev.* **123,** 29.
Hutzler, B. (1971). *C. R. Acad. Sci. Paris* **272B,** 1123–1126.
Ikuta, N., and Kondo, K. (1976). *Proc. Int. Conf. Gas Discharges, 4th, Swansea, Great Britain* I.E.E. Conf. Publ. No. 143, pp. 227–230.
Ikuta, N., Ushita, T., and Ishiguro, Y. (1970). *Elec. Eng. Jpn.* **90,** 52–60.
Johnson, P. C., Berger, G., and Goldman, M. (1972). *Proc. Int. Conf. Gas Discharges, 2nd, London* I.E.E. Conf. Publ. No. 90, pp. 239–241.
Kekez, M. M., and Savic, P. (1974). *J. Phys. D Appl. Phys.* **7,** 620.
Kip, A. F. (1938). *Phys. Rev.* **54,** 139.
Kip, A. F. (1939). *Phys. Rev.* **55,** 549.
Klewe, R. C., Jones, B., and Waters, R. T. (1974). *IEEE Summer Meeting, Anaheim, California* Publ. No. 74, CH 0910-0-PWR, pp. 29–40.
Korge, H., and Kudu, K. (1973). *Proc. Int. Conf. Phenomena Ionized Gases, 11th, Prague*, p. 200.
Kritzinger, J. J. (1963). *Proc. Int. Conf. Phenomena Ionized Gases, 6th, Paris* pp. 295–299.
Kuffel, E. (1959). *Proc. IEE* **106C,** 133.
Laan, M., and Kudu, K. (1975). *Proc. Int. Conf. Phenomena Ionized Gases, 12th, Eindhoven, Netherlands* **1,** 172.

Lama, W. L., and Gallo, C. F. (1974). *J. Appl. Phys.* **45,** 103–113.
Langevin, P. (1905). *Ann. Chim. Phys.* **5,** 245.
Larionov, V. P., Koletchitsky, E. S., and Sergeev, J. G. (1975). *Proc. Int. Conf. Phenomena Ionized Gases, 12th, Eindhoven, Netherlands* **1,** 173.
Lécuiller, M., Julien, R., and Pucheault, J. (1972). *J. Chim. Phys.* **9,** 1353–1359.
Legler, W. (1956). *Ann. Physik* **18,** 5.
Legler, W. (1963). *Z. Phys.* **173,** 169.
Le Ny, A. M., Le Ny, R., and Boulloud, A. (1974). *Proc. Int. Conf. Gas Discharges, 3rd, London* I.E.E. Conf. Publ. No. 118, pp. 306–310.
Leroy, G. (1974). *Bull. Direction Etudes Rech. d'E.D.F.* (*France*) **4-B,** 5–28.
Leroy, G., Gallet, G., Lacey, R., and Kromer, I. L. (1973). *IEEE Summer Meeting* Trans. paper T 73-408.2.
Llewellyn-Jones, F. (1957). "Ionization and Breakdown in Gases." Methuen, London.
Llewellyn-Jones, F. (1967). "Ionization Avalanches and Breakdown." Methuen, London.
Llewellyn-Jones, F., and Williams, G. C. (1953). *Proc. Phys. Soc.* **B66,** 345.
Loeb, L. B. (1935). *Phys. Rev.* **48,** 684.
Loeb, L. B. (1952). *Phys. Rev.* **86,** 256.
Loeb, L. B. (1954). *Phys. Rev.* **94,** 227–232.
Loeb, L. B. (1965). "Electrical Coronas." Univ. of California Press, Berkeley, California.
Loeb, L. B., and Kip, A. F. (1939). *J. Appl. Phys.* **10,** 142.
Loeb, L. B., and Meek, J. M. (1940). *J. Appl. Phys.* **11,** 438–459.
Losansky, E. D., and Firsov, O. B. (1969). *Zh. ETF.* **56,** 670.
Marode, E. (1968). *J. Phys.* **29-C3,** 103–105.
Marode, E. (1970). *Proc. Int. Conf. Gas Discharges, 1st, London* I.E.E. Conf. Publ. No. 70, pp. 525–529.
Marode, E. (1972). Thesis, Univ. Paris-Sud, France.
Marode, E. (1973). *Conf. "Séminaire sur les arcs électriques," Milly-la-Forêt,* France (not published).
Marode, E. (1975a). *J. Appl. Phys.* **46,** 2005–2015.
Marode, E. (1975b). *J. Appl. Phys.* **46,** 2016–2020.
Marode, E., and Hartmann, G. (1969). *C. R. Acad. Sci. Paris.* **B269,** 748–751.
Marode, E., and Sulkowski, J. (1968). *C. R. Acad. Sci. Paris.* **B267,** 1199–1202.
Marode, E., and Sulkowski, J. (1969). *C. R. Acad. Sci. Paris.* **B268,** 1711–1714.
Maurel, J., Ségur, P., and Blanc, D. (1968). *C. R. Acad. Sci. Paris,* **266B,** 1390–1393.
Meek, J. M. (1940). *Phys. Rev.* **57,** 722–728.
Menes, M., and Fisher, L. H. (1954). *Phys. Rev.* **94,** 1–6.
Merrill, F. H., and von Hippel (1939). *J. Appl. Phys.* **10,** 873–887.
Miyoshi, Y. (1958). *Elec. Eng. Jpn.* **78,** 1413.
Miyoshi, Y., Kosokawa, T., and Hayashi, M. (1963). *Proc. Int. Conf. Phenomena Ionized Gases, 6th, Paris* **2,** 327.
Morton, P. L. (1946). *Phys. Rev.* **70,** 358.
Nasser, E. (1968). *IEEE Spectrum* **5,** 127–134.
Nasser, E. (1971). "Fundamentals of gaseous ionization and plasmas electronics." Wiley (Interscience), New York.
Newi, G. (1974). *IEEE Summer Meeting, Anaheim, California* Publ. No. 74, CH 0910-0-PWR, pp. 81–83.
Nygaard, K. J. (1966). *J. Appl. Phys.* **37,** 2850–2852.
Oshige, T. (1967). *J. Appl. Phys.* **38,** 2528.
Peek, F. W. (1929). "Dielectric Phenomena in High-Voltage Engineering." McGraw-Hill, New York.

Peyrous, R. (1974). CNAM Thesis, Univ. de Pau, France.
Phelps, C. T. (1971). *Geophys. Res.* **76,** 5799–5806.
Popkov, V. I. (1949). *Elektrichestvo* **1,** 33.
Popkov, V. I. (1962). *Proc. Int. Conf. Gas Discharges and Electricity Supply Industry, Leatherhead, Great Britain,* pp. 225–237.
Popkov, V. I., Lyapin, A. G., and Shevtsov, E. N. (1973). *Proc. Int. Conf. Phenomena Ionized Gases, 11th, Prague,* p. 189.
Przybylski, A. (1958). *Z. Phys.* **151,** 264.
Przybylski, A. (1962). *Z. Phys.* **168,** 504–515.
Raether, H. (1940). *Naturwissenschaften* **28,** 749–752.
Raether, H. (1964). "Electron Avalanches and Breakdown in Gases." Butterworths, London.
Renardières Group (1972). *Electra* **23,** 53–157.
Renardières Group (1974). *Electra* **35,** 49–156.
Rippe, B., Lécuiller, M., and Koulkès-Pujo, A. M. (1974). *J. Chim. Phys.* **9,** 1185–1190.
Rodin, A. V., and Starostin, A. N. (1973). *Proc. Int. Conf. Phenomena Ionized Gases, 11th, Prague* p. 191.
Saint-Arnaud, R. (1969). *Proc. Int. Conf. Phenomena Ionized Gases, 9th, Bucharest* p. 290.
Saint-Arnaud, R., and Jordan, I. B. (1974). *Proc. Int. Conf. Gas Discharges, 3rd, London* IEE Conf. Publ. No. 118, pp. 320–323.
Sarma, M. P., and Janischewskyj, W. (1969). *Proc. IEE* **116,** 161–166.
Saxe, R. F., and Meek, J. M. (1948). Allied Brit. Ind. and Res. Assoc. Rep. L/T, p. 183.
Saxe, R. F., and Meek, J. M. (1955). IEE Monograph No. 124 M.
Schonland, B. F. J. (1953). "Atmospheric Electricity." Methuen, London.
Shahin, M. M. (1969). *Adv. Chem. Ser.* **80,** 48.
Sigmond, R. S., and Torsethaugen, K. (1973). *Proc. Int. Conf. Phenomena Ionized Gases, 11th, Prague* p. 194.
Spitzer, L. (1956). "Physics of Fully Ionized Gases." Wiley (Interscience), New-York.
Sroka, W. (1968). *Z. Phys.* **23a,** 2004.
Sroka, W. (1969). *Z. Phys.* **24a,** 398.
Stekolnikov, I. S. (1960a). "The Nature of the Long Spark." Acad. Sci. USSR.
Stekolnikov, I. S. (1960b). *II All-Union Conf. of Gaseous Electron.,* USSR.
Sujak, B. (1956). *Nature (London)* **178,** 485–486.
Suzuki, T. (1971). *J. Appl. Phys.* **42,** 3766–3777.
Suzuki, T. (1973). *J. Appl. Phys.* **44,** 4534–4544.
Tholl, H. (1970). *Z. Naturforsch.* **25A,** 420.
Thomas, J. B., and Wong, E. (1958). *J. Appl. Phys.* **29,** 1226–1230.
Thomson, J. J., and Rutherford, E. (1896). *Phil. Mag.* **42,** 392.
Torsethaugen, K., and Sigmond, R. S. (1973). *Proc. Int. Conf. Phenomena Ionized Gases, 11th, Prague,* p. 195.
Townsend, J. S. (1914a). "Electricity in Gases." Oxford Univ. Press, London and New York.
Townsend, J. S. (1914b). *Phil. Mag.* **28,** 83.
Trichel, G. W. (1938). *Phys. Rev.* **54,** 1078–1084).
Ushita, T., Ikuta, N., and Yatsuzuka, M. (1967). *Bull. Faculty Eng. Tokushima Univ., Jpn.* **4,** 89–99.
Ushita, T., Ikuta, N., and Yatsuzuka, M. (1968). *Elec. Eng. Jpn.* **88,** 45–53.
Waidmann, G. (1963). *Dielectrics* **1,** 81–90.
Warburg, E. (1899). *Wied. Ann.* **67,** 69.
Warburg, E. (1927). *In* "Handbuch der Physik," Vol. 14, pp. 154–155. Springer-Verlag, Berlin and New York.
Waters, R. T., and Jones, R. E. (1964a). *Phil. Trans. Roy. Soc.* **256-A,** 185.

Waters, R. T., and Jones, R. E. (1964b). *Phil. Trans. Roy. Soc.* **256-A,** 213.
Waters, R. T., Rickard, T. E., and Stark, W. B. (1968). *Proc. Roy. Soc.* **A304,** 187–200.
Waters, R. T., Rickard, T. E., and Stark, W. B. (1972). *Proc. Int. Conf. Gas Discharges, 2nd, London* IEE Conf. Publ. No. 90, pp. 188–190.
Weissler, G. L. (1943). *Phys. Rev.* **63,** 96–107.
Weissler, G. L., and Schindler, M. (1952). *J. Appl. Phys.* **23,** 844.
Wright, J. K. (1964). *Proc. Roy. Soc.* **A280,** 23.
Zentner, R. (1970a). *Z. Angew. Phys.* **29,** 294–301.
Zentner, R. (1970b). *Elektrotech. Z.* **A91,** 303–305.

Chapter 5

Electric Arcs and Arc Gas Heaters

E. PFENDER
HEAT TRANSFER DIVISION
DEPARTMENT OF MECHANICAL ENGINEERING
UNIVERSITY OF MINNESOTA
MINNEAPOLIS, MINNESOTA

I. INTRODUCTION

This chapter is written with those arc applications in mind which use the arc for heating of gases or vapors, known as arc gas heaters. This includes the so-called arc plasma torches wich have found widespread applications over the past 20 years. There is a wide spectrum of other arc applications, as for example in the area of electric circuit breakers and lighting, which will not be covered in this survey. Since arc gas heaters predominantly make use of thermal arcs, the emphasis will be on arcs in which the thermodynamic state of the plasma approaches *local thermodynamic equilibrium* (LTE).

Although electric arcs have been known for more than 150 years, many arc phenomena remain unexplained or are still poorly understood. In spite

ISBN 0-12-349701-9

of the wealth of information on electric arcs in the literature, many observations and their interpretations are still contradictory. The retrograde motion of the cathode attachment in a transverse magnetic field is a typical example. It seems that the arc is capable of producing a vast variety of different phenomena induced by minor changes of the arc parameters which are frequently difficult to control. Small impurities of the electrode surfaces or of the working fluid, or minor changes of the mechanical properties of the electrodes or of their geometry may give rise to substantial changes in the behavior and appearance of the arc. This situation imposes an extremely difficult task on the theoretical description of arcs which is still far from complete.

Since there are many unsolved problems, it is not surprising that there is still a strong and continuing interest in arc physics and technology. Present research trends are primarily directed towards a better understanding of the interaction of arcs with flow and/or magnetic fields, and the interaction with walls or electrodes. High temperature transport properties and deviations from LTE are other areas which are of great concern.

As with many other fields of science, arc physics experienced severe fluctuations in emphasis and direction over the past years. Most of the newer, fundamental work in arc physics has its origin in Europe, initiated after World War II. With the advent of space flight, arc physics and technology experienced a dramatic upswing in research activities, particularly in the U.S. and in Russia. There seem to be indications that the decline in emphasis which this field suffered in the late sixties and early seventies, at least in the U.S., may be reversed again, precipitated by the energy crisis. Both the availability of nuclear fuels and the abundant supplies of coal suggest that an *electric energy economy* may be at least part of the solution to the long term energy problem. A shift to electricity as the major energy base will have a profound impact on high temperature material processing for which arc gas heaters may find increasing applications.

As previously mentioned, the literature on electric arcs is vast and, due to space limitations, it is impossible to include in this survey a comprehensive literature review. Reference will only be made to newer (up to 1975) and selected publications which are of immediate interest in the context of this chapter. Fortunately, the developments up to 1955 are summarized in an excellent, comprehensive review "Electric Arcs and Thermal Plasmas" (see Finkelnburg and Maecker 1956). Also, a number of textbooks and monographs should be mentioned which provide more details on the subject of electric arcs and arc components (Sommerville, 1959; Ecker, 1961; Rieder, 1967; Hoyaux, 1968; Kuhn, 1956; von Engel, 1965). It is unfortunate that the rapidly increasing Russian literature cannot be adequately covered in this survey because of translation problems.

Throughout this survey the International System of Units (S.I.) will be used.

II. PHYSICS OF ELECTRIC ARCS

The basic features of electric arcs manifest themselves in dc as well as in ac arcs. Since the change of polarity in ac arcs results frequently in complex electrode phenomena which obscure those effects which are significant for a basic understanding of arc behavior, only dc arcs will be discussed in this chapter.

All gases at room temperature are excellent electrical insulators. In order to make them electrically conducting, a sufficient number of charge carriers have to be generated. Although there are already a certain number of charge carriers present at room temperature (in atmospheric air approximately 10^6 electrons/m^3) this number is by far too small to produce a measurable electrical conductivity. This small number of charge carriers is, however, responsible for an electrical breakdown which occurs if a sufficiently high electric field is applied to a pair of electrodes. Breakdown of the originally nonconducting gas establishes a conducting path between the electrodes. Passing of electrical currents through the electrode gap leads to an array of phenomena known as gaseous discharges.

In such a gaseous discharge a more or less electrically conducting plasma is generated which consists of a mixture of ions, electrons, and neutral particles. The composition and distribution of the plasma between the electrodes is a function of the existing discharge mode and of other discharge parameters. Figure 1 shows schematically the various types which may occur in a discharge vessel filled with a gas at a pressure of the order of 0.1 kPa* and having an electrode gap of a few centimeters. Many types of discharges require a stabilizing resistor in series as indicated in the sketch inserted in Fig. 1.

After breakdown is achieved, thè current flowing between the electrodes may be gradually increased over many orders of magnitude, starting from typical breakdown currents in the order of 10^{-10} to 10^{-8} A. In this process the discharge passes through a number of well-known modes as indicated in Fig. 1. These discharge modes with the exception of the arc mode are not the subject of this chapter; they are discussed in other chapters of this book. After the discharge characteristic reaches a voltage maximum, a steep

* 1 pascal (Pa) = 1 newton/meter^2.

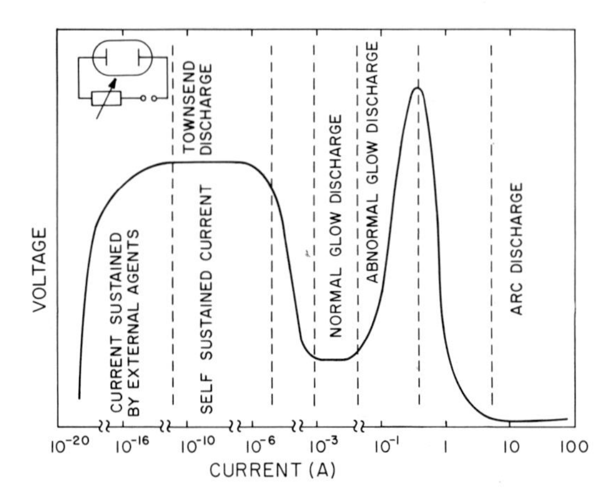

Fig. 1. Classification of dc discharge modes.

descent follows into the arc region. The voltage required to sustain an arc is, in general, substantially lower than for other types of discharges.

Establishing an arc is, of course, not limited to the described procedure; there are many other and technically superior ways to initiate an arc (see Section II.B).

A. Definition of an Arc

Although there is no clear-cut definition of an arc, there are a number of features which distinguish an arc from other discharge modes. For the sake of simplicity the following discussion will be restricted to steady-state arcs. A comparison with other discharge forms suggests singling out three typical features which are characteristic for arcs and which are not found in other discharge modes.

1. Relative High Current Density

The current density in an arc column (see Section II.C) may reach values of 100 A/cm^2 and higher whereas the current density in the preceding glow discharge (Fig. 1) is on the order of 1–10 mA/cm^2. The situation is even more pronounced at the electrodes. Arcs may attach to the electrodes, and in particular to the cathode, in the form of tiny spots in which current densities can be as high as 10^6 A/cm^2. The associated heat flux densities are in the order of 10^6 to 10^7 W/cm^2, which requires special precautions to ensure integrity of the electrodes.

Another frequently mentioned criterion for the existence of an arc is that the total arc current must be ≥ 1 A. This is certainly a necessary, but by no means sufficient criterion, because the total current in a glow discharge may well exceed 1 A if a sufficiently large cathode surface is provided.

2. Low Cathode Fall

The potential distribution in an electric arc shows a peculiar behavior as indicated in Fig. 2. The potential changes rapidly in front of the electrodes forming the so-called cathode and anode fall. In this context the cathode fall is of particular interest; it assumes values of around 10 V in contrast to cathode falls in glow discharges which usually exceed 100 V. This relatively low cathode fall is a consequence of the more efficient electron emission mechanisms at the cathode compared with those prevailing in glow discharges. For a discussion of the emission mechanisms and their influence on the cathode fall the reader is referred to Section II.C. It should be pointed out that the thicknesses d_c' and d_a' of the electrode regions are exaggerated in Fig. 2, particularly if high pressure arcs are considered. Also, V_c' does not in general, represent the true cathode fall. The actual cathode fall V_c may be several volts less than V_c' (see Section II.C).

Although the overall arc voltage in a given discharge vessel is lower than that of a glow discharge in the same vessel and at the same pressure, the overall voltage drop over the discharge does not provide a useful criterion for distinguishing an arc from other types of discharges. Depending on the arc length and the energy balance of the arc column, the overall voltage drop of an arc may be very high. In vortex-stabilized, high pressure arc gas heaters, for example, arc voltages may reach 15 kV (see Section III).

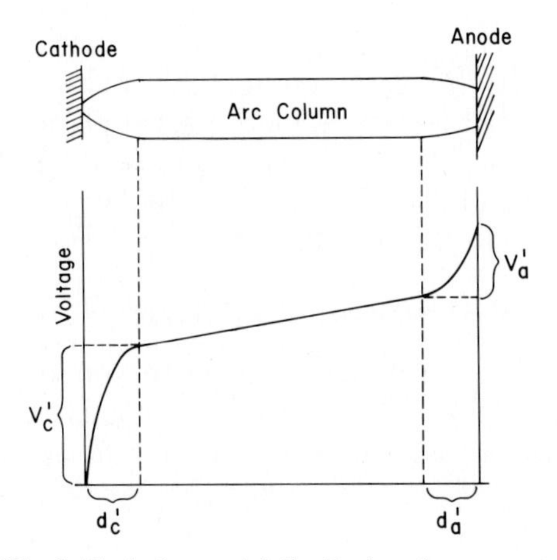

Fig. 2. Typical potential distribution along an arc.

3. High Luminosity of the Column

This criterion represents a useful distinction between an arc column and the column of other discharge modes, provided the pressure is sufficiently high ($p \geq 1$ kPa). The extremely high luminosity of the column of high pressure ($p \geq 100$ kPa) thermal arcs, finds many applications in the illumination field.

B. Initiation of Arcs

Electric arcs may be initiated in three basically different ways. The choice of the method for striking an arc depends mainly on the arc arrangement and, in particular, on the electrode configuration.

1. By Electrode Contact

If one or both electrodes are movable, electrode contact may be established after an electric potential is applied to the electrodes. The short circuit current flowing over the contact bridge between the electrodes heats the contact point to temperature levels sufficient for thermionic emission from the cathode. At the same time electrode material is evaporated and ionized at the contact point providing the required charge carriers for developing an arc as soon as the electrodes are separated (drawing of an arc). There is, however, a lower current limit for drawing an arc which is mainly a function of the electrode materials. If the discharge circuit allows at least this minimum current to be drawn, an arc may be established in the described way.

This method of drawing an arc may still be applied even when the electrodes cannot make contact with each other. A metal or graphite rod may be used as an auxiliary electrode to establish contact between cathode and anode. As the rod touches the cathode, a small plasma volume develops, initiating an arc between the rod and the cathode. As the rod is withdrawn, the arc is transferred to the anode. In the described process the polarity of rod and fixed electrode may be exchanged.

2. By Preionization of the Discharge Gap

Preionization of the electrode gap reduces to convenient values the voltage required for striking an arc. There are a number of different ways to make the electrode gap electrically conducting. In the most common approach a high frequency spark generates the necessary charge carriers between the electrodes.

Another frequently used method for initiating an arc employs a thin wire stretched across the electrode gap, making contact with both electrodes. The applied voltage leads to a wire explosion supplying the necessary charge carriers for establishing an arc.

In general, any auxiliary plasma source (plasma jet, dc spark, etc.) may serve as a means for preionizing the arc gap. In addition, ionizing radiation (α, β, γ, and x rays, uv and laser radiation) may also provide the desired ionization level. Finally, chemical processes may be useful for initiating an arc. An arc may be started, for example, by holding a burning match below a pair of carbon electrodes.

3. High-voltage Breakdown

This method is based on the usual dielectric, high-voltage breakdown of a discharge gap under steady state conditions. The breakdown voltage is an increasing function of the gas pressure and the size of the electrode gap. In atmospheric air approximately 30 kV are necessary to break down a gap of 1 cm. For most high pressure arc applications this method cannot be used because of the extremely high voltages involved.

C. Arc Components

The potential distribution shown in Fig. 2 is typical for long arcs (arc length $\gg$ arc diameter) and suggests dividing the arc into three regions: the cathode region, the arc column, and the anode region.

1. The Arc Column

(a) *Definition of the Arc Column* Since cathode and anode regions may be considered as thin "boundary layers" overlying the electrodes, the column with its comparatively small potential gradient represents the main body of an arc. In contrast to the regions immediately in front of the electrodes in which net space charges exist, the arc column represents a true plasma in which quasineutrality prevails. The pressure in the arc column is uniform and equal to the pressure in the surrounding fluid with the exception of arcs operated at extremely high current levels. In such arcs the interaction of the arc current with the self-magnetic field produces a pressure gradient in the radial direction (pinch effect) so that the pressure becomes elevated in the axis of the arc column.

For a given arc current the conditions in the column (temperature distribution and associated distribution of thermodynamic and transport properties) adjust themselves in such a way that the field strength required for driving this current becomes a minimum (Peters, 1956).

The relatively small field strength prevailing in the arc column may also be interpreted as a consequence of the favorable energy balance which, in turn, is to a large degree determined by the charge carrier balance.

Let us consider a small control volume in a rotationally symmetric, fully developed arc column (Fig. 3). The applied electric field imposes a drift velocity on electrons and ions in the column which gives rise to a certain current flow. The contribution of the ions to this current flow is almost negligible due to the imbalance of ion and electron mobilities. Steady state requires that the same number of electrons entering the control volume through surface A per unit time must again leave through surface B. The same argument holds for the ions traveling in the opposite direction.

There is, however, a continuous loss of charge carriers by ambipolar diffusion across surface C, accompanied by recombination outside of the control volume. Neutral particles diffuse in the opposite direction maintaining the mass balance within the control volume. Neutral particles must be ionized in the control volume at the same rate as charge carriers are lost in order to maintain steady state conditions. Enhanced cooling of the arc fringes by convection, for example, increases not only the temperature gradient but also the charged particles density gradient, resulting in a corresponding increase of diffusion losses of charged particles. These losses must be compensated by a correspondingly higher rate of ionization in the control volume, i.e., by a higher field strength. In summary, the arc column responds to increased cooling of its fringes by an increase of the field strength and, therefore, of the energy dissipation IE per unit length of the arc column. This higher energy dissipation leads to higher temperatures in the core of the arc column. The net effect of cooling of the arc fringes is, therefore, an increase of the core temperature, provided that the current is kept constant.

Although the previous arguments indicate the proper trend of the described effects, the picture is not quite complete because the charge carrier

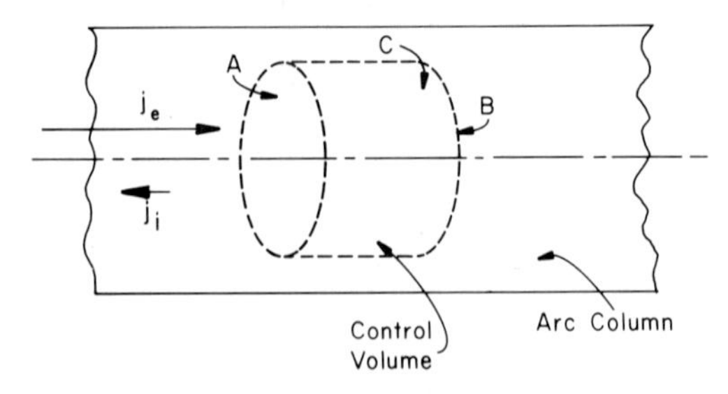

Fig. 3. Control volume within arc column.

losses account only for part of the total energy losses in the arc column. The actual field strength or power dissipation in the arc column is determined by the entire energy balance, including losses by heat conduction/convection and radiation.

The dominating process responsible for ionization in the arc column is due to electron impact. The field strength in the arc column in the case of high pressure arcs ($p > 10$ kPa) is by far insufficient for an electron to accumulate enough kinetic energy over a mean free path to make an ionizing collision, i.e.,

$$e\lambda_e E \ll \chi_i, \tag{1}$$

where E represents the field strength, λ_e the mean free path length, e the elementary charge, and χ_i the ionization energy. In this inequality it is assumed that the electron travels in the field direction, accumulating the maximum possible energy from the electric field. Charge carrier production in this situation must be accomplished by thermal ionization rather than field ionization. Electrons in the tail of the Maxwellian distribution possess sufficient energy for making ionizing collisions.

Thermal processes in the arc column and the associated thermodynamic state of the plasma will be discussed in the following paragraph.

(b) *Thermodynamic State of the Arc Plasma—the Concept of LTE* It is useful to consider first a plasma which is in a state of perfect *thermodynamic equilibrium* (TE) although this state cannot be realized under laboratory conditions. For the sake of simplicity it will be assumed that the plasma is generated from a monatomic gas or a mixture of monatomic gases.

Thermodynamic equilibrium prevails in a uniform, homogeneous plasma volume if kinetic and chemical equilibria as well as every conceivable plasma property are unambiguous functions of the temperature. The temperature, in turn, has to be the same for all plasma constituents and their possible reactions. More specifically, the following conditions must be met:

(α) The velocity distribution functions for particles of every species r which exist in the plasma, including the electrons, follow a Maxwell–Boltzmann distribution

$$f(v_r) = \frac{4v_r^2}{\sqrt{\pi(2kT/m_r)^3}} \exp\left(-\frac{m_r v_r^2}{2kT}\right) \tag{2}$$

v_r is the velocity of particles of species r, m_r their mass, and T their temperature, which is the same for every species r, and which is, in particular, identical to the plasma temperature.

(β) The population density of the excited states of every species r follows a Boltzmann distribution

$$n_{r,s} = n_r(g_{r,s}/Z_r)\exp(-\chi_{r,s}/kT), \tag{3}$$

where n_r is the total number density of ions of species r, Z_r is their partition function, $\chi_{r,s}$ is the energy of the sth quantum state and $g_{r,s}$ is the statistical weight of this state. The excitation temperature T which appears explicitly in the exponential term and implicitly in the partition function Z_r is identical to the plasma temperature.

(γ) The particle densities (neutrals, electrons, ions) are described by the Saha equation which may be considered as a mass action law

$$n_{r+1}n_e/n_r = (2Z_{r+1}/Z_r)[(2\pi m_e kT)^{3/2}/h^3]\exp(-\chi_{r+1}/kT), \tag{4}$$

where χ_{r+1} represents the energy which is required to produce an $(r + 1)$-times ionized atom from an r-times ionized atom (ionization energy). The ionization temperature T in this equation is identical to the plasma temperature. Lowering of the ionization potential has been disregarded in Eq. (4).

(δ) The electromagnetic radiation field is that of blackbody radiation of the intensity B_ν as described by the Planck function

$$B_\nu = \frac{2h\nu^3}{c^2}\frac{1}{e^{h\nu/kT} - 1}. \tag{5}$$

The symbol ν stands for the frequency, h represents Planck's constant, and c the light velocity. The temperature of this blackbody radiation is again identical to the plasma temperature.

In order to generate a plasma which follows this ideal model as described by Eqs. (2)–(5), the plasma would have to dwell in a hypothetical cavity whose walls are kept at the plasma temperature, or the plasma volume would have to be so large that the central part of this volume, in which TE prevails, would not sense the plasma boundaries. In this way the plasma would be penetrated by blackbody radiation of its own temperature. An actual plasma will, of course, deviate from these ideal conditions. The observed plasma radiation, for example, will be much less than the blackbody radiation because most plasmas are optically thin over a wide wavelength range. Therefore, the radiation temperature of a gaseous radiator deviates appreciably from the kinetic temperature of the plasma constituents or the already mentioned excitation and ionization temperatures. In addition to radiation losses, plasmas suffer irreversible energy losses by conduction, convection, and diffusion which also disturb the thermodynamic equilibrium. Thus, laboratory arc plasmas as well as some of the natural plasmas cannot be in a perfect TE state. In the following sections, deviations from TE and the associated concept of LTE will be discussed.

The following considerations will be restricted to optically thin plasmas, a situation which is frequently approached by laboratory arc plasmas. In contrast to a complete TE situation, LTE in optically thin plasmas does not require a radiation field which corresponds to the blackbody radiation intensity of the respective LTE temperature. It does require, however, that collision processes and not radiative processes govern transitions and reactions in the plasma and that there is a microreversibility among the collision processes. In other words, a detailed equilibrium of each collision process with its reverse process is necessary. Steady-state solutions of the respective collision rate equations will then yield the same energy distribution pertaining to a system in complete thermal equilibrium with the exception of the rarefied radiation field. LTE further requires that local gradients of the plasma properties (temperature, density, heat, conductivity, etc.) are sufficiently small so that a given particle which diffuses from one location to another in the plasma finds sufficient time to equilibrate, i.e., the diffusion time should be of the same order of magnitude or larger than the equilibration time. From the equilibration time and the particle velocities an equilibration length may be derived which is smaller in regions of small plasma property gradients (for example in the center of an electric arc). Therefore, with regard to spatial variations, LTE is more probable in such regions. Heavy particle diffusion and resonance radiation from the center of a nonuniform plasma source help to reduce the effective equilibrium distance in the outskirts of the source.

In the following, a systematic discussion of the important assumptions for LTE based on arc plasmas will be undertaken.

(α) *Kinetic equilibrium* It may be safely assumed that each species (electron gas, ion gas, neutral gas) in a dense, high temperature arc plasma column will assume a Maxwellian distribution (excluding regions close to walls and electrodes). However, the temperatures defined by these Maxwellian distributions may be different from species to species. Such a situation which leads to a two temperature description will be discussed in the following.

The electric energy fed into an arc is dissipated in the following way: the electrons, according to their high mobility, pick up energy from the electric field which they partially transfer by collisions to the heavy plasma constituents. Because of this continuous energy flux from the electrons to the heavy particles, there must be a "temperature gradient" between these two species, so that $T_e > T_g$. (T_e is the electron temperature and T_g the temperature of the heavy species, assuming that ion and neutral gas temperatures are the same.)

In the two-fluid model of a plasma defined in this manner, two distinct temperatures T_e and T_g may exist. The degree to which T_e and T_g deviate from each other will depend on the thermal coupling between the two species. The difference between these two temperatures can be derived from an energy balance leading to

$$\frac{T_e - T_g}{T_e} = \frac{\pi m_g}{24 m_e} \frac{(\lambda_e e E)^2}{(k T_e)^2}, \tag{6}$$

where m_g is the mass of the heavy plasma constituents, λ_e the mean free path length of the electrons, and E the electric field strength. Since the mass ratio $\pi m_g/24 m_e$ is already 240 for hydrogen, the amount of (directed) energy $(\lambda_e e E)$ which the electrons pick up along one mean free path length has to be very small compared with the average thermal energy (kT_e) of the electrons. Low field strengths, high pressures $(\lambda_e \sim 1/p)$ and high temperature levels are favorable for a kinetic equilibrium among the plasma constituents. At low pressures, for example, appreciable deviations from kinetic equilibrium may occur. Figure 4 shows in a semischematic diagram how electron and gas temperatures separate in an electric arc with decreasing pressure. For an atmospheric argon high intensity arc with $E = 13$ V/cm, $\lambda_e = 3 \times 10^{-4}$ cm, $m_A/m_e = 7 \times 10^4$, and $T_e = 30 \times 10^3$ K, the deviation between T_e and T_g is on the order of 1% (Finkelnburg and Maecker, 1956).

(β) *Excitation equilibrium* In order to determine the excitation equilibrium, every conceivable process which may lead to excitation or deexcitation has to be taken into account. For simplicity only the most prominent

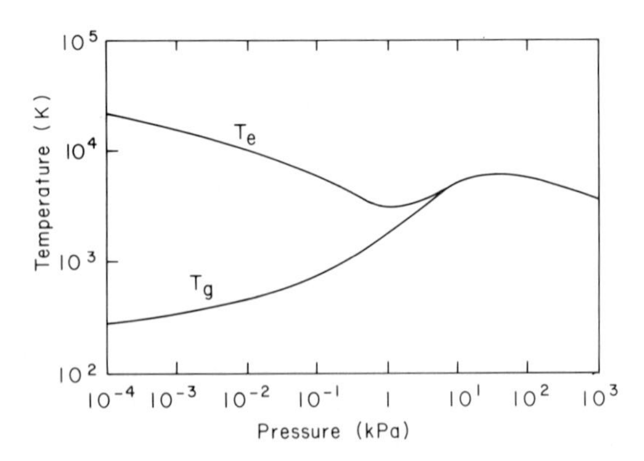

Fig. 4. Electron and heavy particle temperatures as a function of pressure.

mechanisms, which are collisional and radiative excitation and deexcitation, will be considered.

Excitation: (1) electron collisions; (2) photoabsorption.
Deexcitation: (1) collisions of the second kind; (2) photoemission.

In the case of TE, microreversibilities would have to exist for all processes, i.e., in the above scheme, excitation by electron collisions would have to be balanced by the reverse process, namely, collisions of the second kind. Also, excitation by the photoabsorption process would have to be balanced by photoemission processes which include spontaneous and induced emission. Furthermore, the population of excited states would have to follow a Boltzmann distribution [see Eq. (3)].

The microreversibility for the radiative processes holds only if the radiation field in the plasma reaches the intensity B_ν of blackbody radiation. However, actual plasmas are frequently optically thin over most of the spectral range, so that the situation for excitation equilibrium seems to be hopeless. Fortunately, if collisional processes dominate, photoabsorption and emission processes do not have to balance; only the sum of the left- and right-hand sides of the scheme above have to be equal. Since the contribution of the photoprocesses to the number of excited atoms is almost negligible when collisional processes dominate, the excitation process is still close to LTE.

(γ) *Ionization equilibrium* For the ionization equilibrium again only the most prominent mechanisms which lead to ionization and recombination will be considered.

Ionization: (1) electron collision; (2) photoabsorption.
Recombination: (1) three body recombination; (2) photorecombination.

In a perfect thermodynamic equilibrium state with cavity radiation, a microreversibility among the collisional and radiative processes would exist and the particle densities would be described by the Saha equation. Without cavity radiation, the number of photoionizations is almost negligible requiring, instead of the microreversibility, a total balance of all processes involved. Photorecombinations, especially at lower electron densities, are not negligible. The frequency of the three remaining elementary processes is a function only of the electron density leading, for a certain electron density, to the same order of magnitude of frequency of these elementary processes. The result is an appreciable deviation between actual and predicted values [from Eq. (4)] of the electron densities. Only for sufficiently large electron densities does the Saha equation predict correct values. For smaller electron densities the corona formula (Elwert, 1952b) has to be used, which considers ionization by electron impact and photorecombination only. The particle concentrations in low intensity arcs at atmospheric pressure, for example, must be calculated with this formula. Significant deviations of the electron density predicted by the Saha equation from the true electron density may also occur in the fringes of high intensity arcs and plasma jets.

In summary, it has been found that LTE exists in a steady state, optically thin plasma when the following conditions are simultaneously fulfilled:

(a) The different species which form the plasma have a Maxwellian distribution.
(b) Electric field effects are small enough, and the pressure and the temperature are sufficiently high, so that $T_e = T_g$.
(c) Collisions are the dominating mechanism for excitation (Boltzmann distribution) and ionization (Saha equation).
(d) Spatial variations of the plasma properties are sufficiently small.

Besides the conditions for the two extreme cases, namely LTE (based on Saha ionization equilibrium) and corona equilibrium, conditions in the regions between these two limiting cases are also of interest. In this range three body recombinations as well as radiative recombination and de-excitation are significant. A large number of investigations have been reported over the past 15 years on the subject of radiative-collisional processes, LTE and deviations from LTE. The results of these studies up to 1966 are summarized in two books on plasma diagnostics (Huddlestone and Leonard, 1965; Lochte-Holtgreven, 1968). In a more recent survey Drawin (1970a) presents a comprehensive review of the validity conditions for LTE including a discussion of *complete local thermodynamic equilibrium* (CLTE) and *partial local thermodynamic equilibrium* (PLTE). This distinction is associated with the population of excited levels which may deviate from ideal Boltzmann distributions. In the case of optically thin plasmas, the lower-lying excited energy levels tend to be underpopulated with respect to the ground state. This situation is referred to as PLTE provided that all the other conditions for LTE are met. The electron densities n_e required for CLTE in an optically thin plasma are substantially higher than those needed for the less stringent requirement of PLTE.

Griem (1964) established the following criterion for the existence of CLTE in an optically thin homogeneous plasma

$$n_e \geq 9 \times 10^{23}(\chi_{21}/\chi_H)^3(kT/\chi_H) \quad (\mathrm{m}^{-3}), \tag{7}$$

where χ_{21} represents the energy gap between the ground state and the first excited level, $\chi_H = 13.58$ eV is the ionization energy of the hydrogen atom and T the plasma temperature. This criterion shows the sensitivity of the required electron density for CLTE on the energy of the most critical, first excited state.

It is obvious that deviations from CLTE or even from PLTE will occur in regions of low electron densities as, for example, in plasma regions adjacent to walls or in arc fringes and in all types of low density plasmas of laboratory dimensions.

For many years, the existence of LTE in atmospheric pressure, high current arcs has not been questioned. Only recently, deviations from CLTE have been found in such arcs. Evans *et al.* (1970) as well as Bober and Tankin (1970) have shown that a pressure of approximately 300 kPa is necessary to reach a state of CLTE in the central portion of a free-burning argon arc at currents of 300 to 400 A. These conditions correspond to an electron density of approximately 10^{24} m^{-3} in the center. Deviations from CLTE still persist in the outer regions of such arcs where the electron density drops substantially below 10^{24} m^{-3}.

Numerous analytical as well as experimental studies over the past years (Kruger, 1970; Taussig, 1970; Uhlenbusch *et al.*, 1970a,b; Drawin, 1970b, 1973; Drawin and Emard, 1973; Drawin *et al.*, 1973; Uhlenbusch and Fisher, 1971; Giannaris and Incropera, 1971; Clark and Incropera, 1972; Incropera and Murrer, 1972; Wiese *et al.*, 1972; Giannaris and Incropera, 1973; Garz, 1973; Nizovsky *et al.*, 1973; Eddy *et al.*, 1973) demonstrate that LTE in high intensity arcs is rather the exception than the rule. Several of the previously mentioned studies show that besides the underpopulation of lower-lying energy levels deviations from LTE may be frequently attributed to strong gradients in the arc plasma and the associated diffusion effects.

(c) *Properties of Arc Plasmas* For the behavior of thermal arcs a number of thermophysical properties are of importance which are usually subdivided into thermodynamic and transport properties. In this survey, these properties will be briefly discussed considering atmospheric pressure arc plasmas. For many basic investigations arc plasmas have been generated from pure monatomic gases, which facilitates comparisons with theoretical predictions. In actual arc applications molecular gases or gas mixtures are frequently used. This introduces additional complexities which cannot be covered in this survey. For the following discussion of plasma properties, argon will be chosen as a typical example of a monatomic gas; for demonstrating the influence of chemical reactions on certain plasma properties, nitrogen, which has been extensively studied as an arc fluid, will be considered.

(α) *Plasma composition and thermodynamic properties* For the calculation of the chemical composition of a plasma in thermodynamic equilibrium (or LTE) as a function of the temperature, a system of simultaneous nonlinear equations has to be solved. This system of equations remains essentially the same for any given gas or gas mixture from which a plasma may be generated. The appropriate equations are obtained from the conservation-of-mass law

$$n^{(a)} = \sum_r n_r^{(a)}, \tag{8}$$

where $n^{(a)}$ represents the total number density of particles of a given species (a), $n_r^{(a)}$ is the number density of atoms $(r = 0)$ or ions $(r > 1)$ of the same species (a), and r indicates the ionization state, and the condition for quasineutrality

$$n_e = \sum_{r,a} r n_r^{(a)} \tag{9}$$

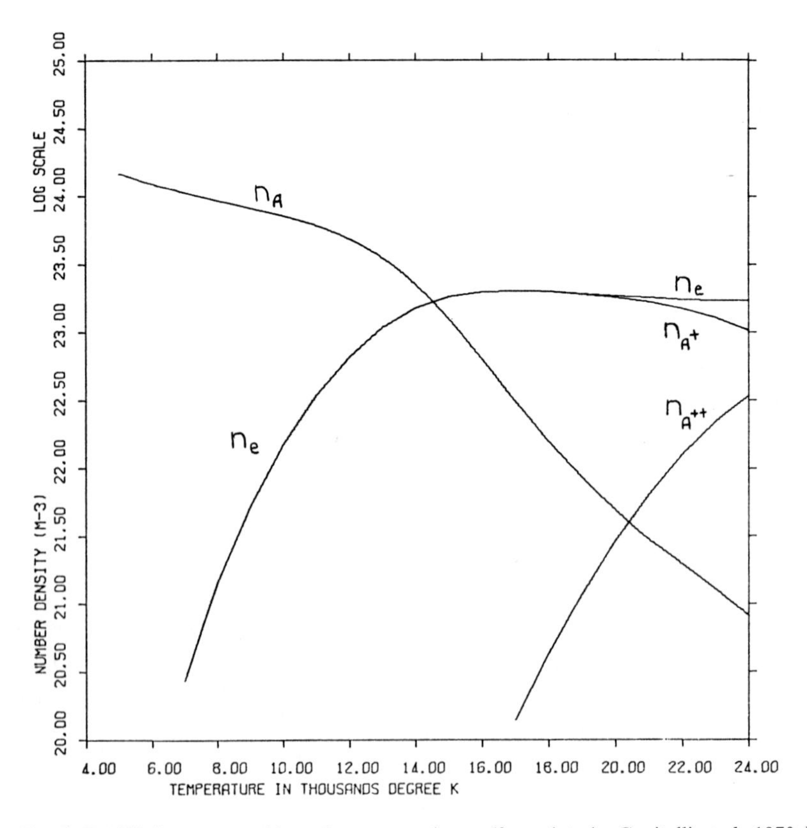

Fig. 5. Equilibrium composition of an argon plasma (from data by Capitelli *et al.*, 1970a).]

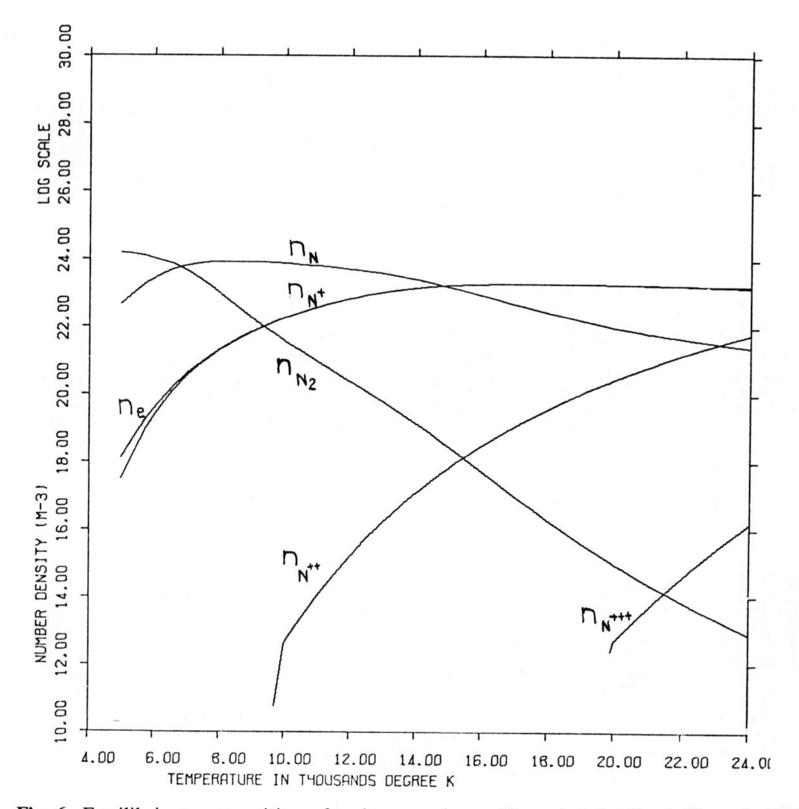

Fig. 6. Equilibrium composition of a nitrogen plasma (from data by Capitelli *et al.*, 1970a).]

and the mass-action law. In the case of monatomic gases, ionization represents the sole mechanism for chemical reactions.

The particle densities of the various ionization stages are described by the Saha equation

$$\frac{n_{r+1} n_e}{n_r} = \frac{2Z_{r+1}}{Z_r} \frac{(2\pi m_e kT)^{3/2}}{h^3} \exp\left(-\frac{\chi_{r+1}}{kT}\right). \tag{10}$$

If the plasma contains r different ionization stages of a certain species (a) and if there are N different species in the plasma, then the number of unknown particle densities is $rN + 1$ where the electron density is the additional unknown. Since there are $(r - 1)$ Saha equations for each species and one conservation-of-mass relation, the total number of equations for all N species is rN. The last required relation is provided by the condition for quasi-neutrality. From this complete set of equations the equilibrium composition of any given plasma can be calculated regardless of whether or not the plasma is generated from a pure or a mixture of monatomic gases.

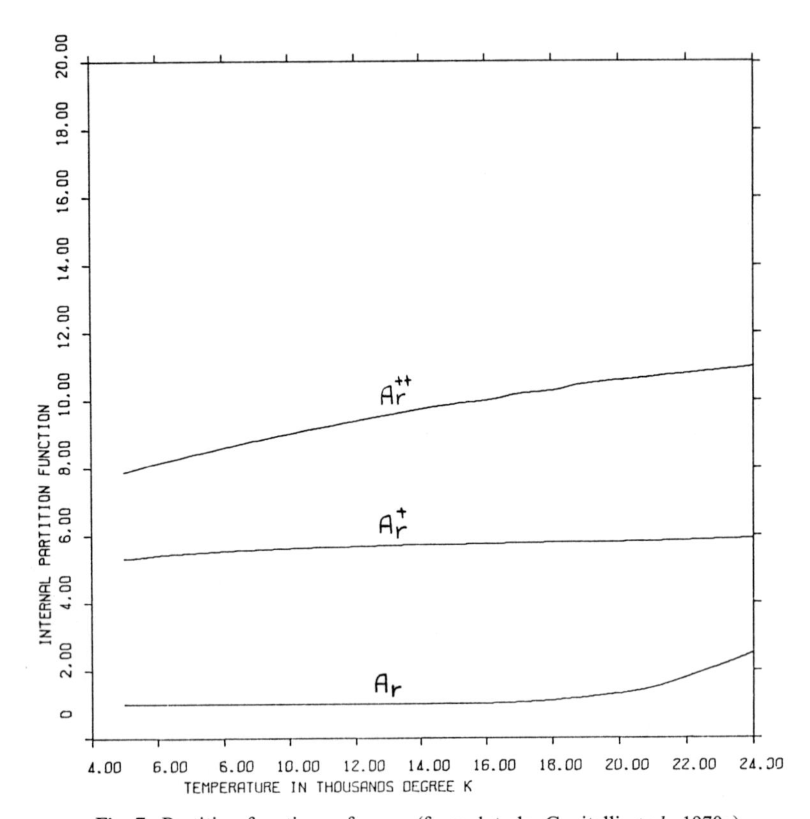

Fig. 7. Partition functions of argon (from data by Capitelli *et al.*, 1970a).

In the case of diatomic or polyatomic gases additional equations similar to the Saha equation are needed to describe the dissociation process. Figures 5 and 6 (Capitelli *et al.*, 1970a) show the equilibrium composition of argon and nitrogen plasmas at atmospheric pressure, respectively. The contribution of molecular ions for temperatures below 7000°K is responsible for the deviation between n_{N^+} and n_e. As pointed out in the previous paragraph, actual plasmas deviate more or less from the ideal state of TE which, of course, will also cause deviations from the equilibrium plasma composition.

With the calculated plasma composition, thermodynamic properties of a plasma can be calculated (mass density, internal energy, enthalpy, entropy, and specific heat) provided that the partition functions for the various species are known. Examples of such partition functions are shown in Figs. 7–9 and the corresponding thermodynamic properties of atmospheric pressure plasmas are illustrated in Figs. 10 and 11 for argon and nitrogen respectively (Capitelli *et al.*, 1970a). Calculated plasma compositions, densities, enthal-

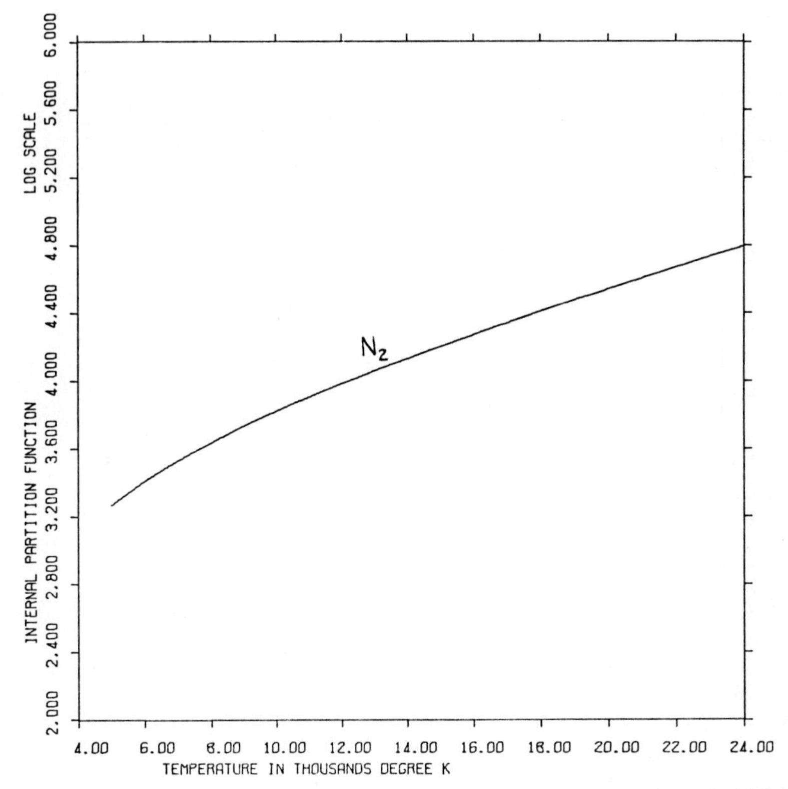

Fig. 8. Partition function of the nitrogen molecule (from data by Capitelli *et al.*, 1970a).

pies, and specific heats have been reported for pure gases as well as for gas mixtures (Burhorn and Wienecke, 1960a,b,c; Drellishak *et al.*, 1965; Knoche, 1968; Capitelli *et al.*, 1970a,b, 1972; Nelsen, 1972).

(β) *Transport properties* Transport phenomena in plasmas encompass the flow situation of every plasma constituent, namely electrons, ions, and neutrals, including radiation fluxes under the influence of driving "forces" as, for example, electric fields, temperature, pressure, density and velocity gradients. In order to describe the transfer of electrical charge, mass, momentum, and energy within a plasma and from a plasma to its surroundings, characteristic transport properties have been defined as, for example, electrical conductivity, heat conductivity, viscosity, diffusivity, and optical emission coefficient. Various methods for the calculation of these transport properties have been proposed in conjunction with presentation of results of such calculations for selected plasmas (Finkelnburg and Maecker, 1956; Amdur and Mason, 1958; Cambel, 1963; Meador and Staton, 1965; Devoto,

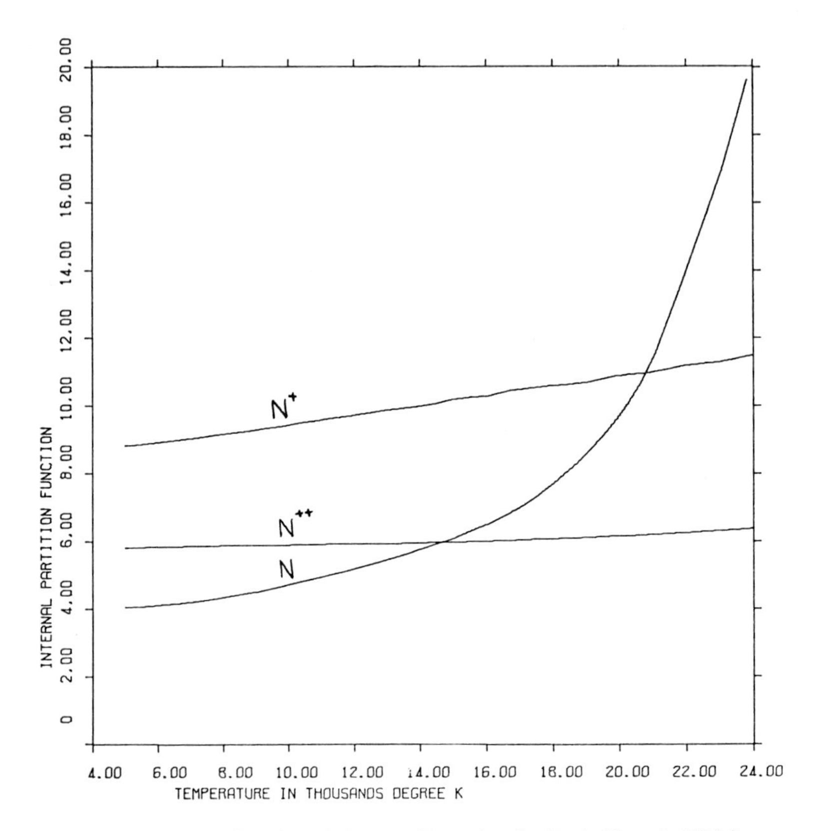

Fig. 9. Partition function of nitrogen (from data by Capitelli *et al.*, 1970a).

1966, 1967, 1973; Mason *et al.*, 1967; Nelson and Goudard, 1968; Capitelli and Ficocelli, 1970; Uhlenbusch, 1972a; Capitelli and Devoto, 1973).

For measurements of transport properties, electric arcs and in particular wall-stabilized arcs are well suited. The general procedure for deriving transport properties from arc measurements has been described by Emmons (1967), Bauder and Maecker (1971), Uhlenbusch (1972b), and by Devoto and Mukherjee (1973).

Over the past years there has been an increasing awareness of the significance of radiative transport in arc plasmas. For determining radiative transport coefficients, the various mechanisms responsible for the emission and absorption of radiation in such plasmas must be taken into account. Inspection of the spectrum of a typical high intensity arc generated in a monatomic gas reveals continuous as well as line radiation. Electronic transitions from higher to lower energy states of excited atoms or ions give rise to the emission of spectral lines. Since the electron involved in the

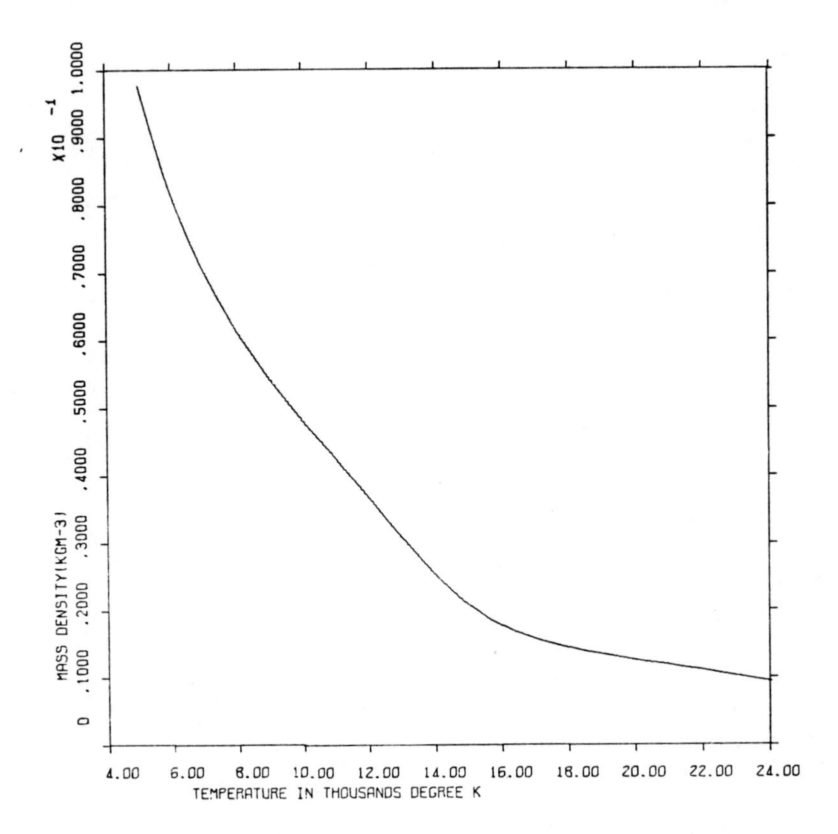

Fig. 10a. Mass density of an equilibrium argon plasma (from data by Capitelli *et al.*, 1970a).

radiation process remains in a bound state, this radiation is also referred to as bound–bound radiation. The total energy transport by line radiation from a plasma is frequently only a small fraction of the total radiated energy, depending on the number and wavelength of the emitted lines which in turn depend on the nature of the arc fluid including the number of possible species for a given temperature. For a given gas an arc may be a "strong" or "weak" line radiator depending on the plasma density and composition which are functions of pressure and temperature.

Continuous radiation in an arc under the previously specified conditions originates from recombination of ions with electrons (free–bound radiation) and from bremsstrahlung (free–free-radiation). In the process of radiative recombination a free electron is captured by a positive ion into a certain bound energy state and the excess energy is converted into radiation. Recombination may occur into all possible energy levels of an ion so that the number of continuous spectra for a particular species will coincide with the

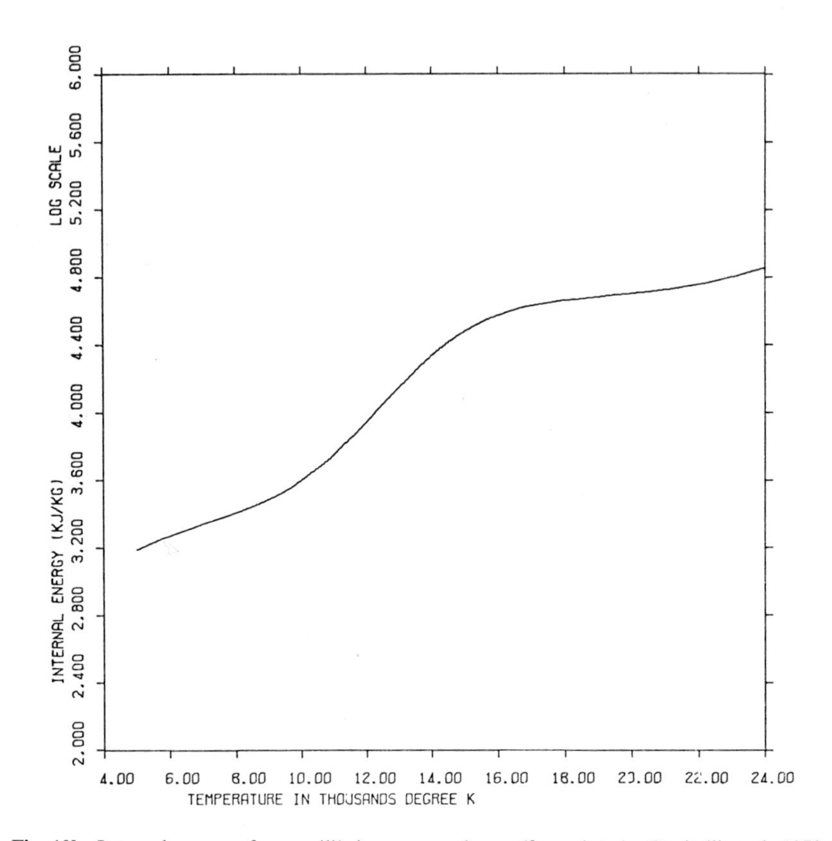

Fig. 10b. Internal energy of an equilibrium argon plasma (from data by Capitelli *et al.*, 1970a).

number of electronic energy states of this ion. The entire free-bound continuum consists, therefore, of a superposition of all continuous spectra emitted by the different species which exist in the plasma.

Bremsstrahlung has its origin in the interaction of free electrons with other charged particles, i.e., a free electron may lose kinetic energy in the Coulomb field of an ion and this energy is readily converted into radiation. Since the initial as well as the final state of the electrons are free states in which the electrons may assume arbitrary energies within the Maxwellian distribution, the emitted radiation is of the continuum type.

The total radiation continuum consisting of free–free and free–bound radiation frequently dominates the radiative balance in high pressure arcs ($p \geq 100$ kPa).

If molecular species are present in the arc, there will be, in addition, radiation bands in the spectrum due to the excitation of vibrational and rotational energy levels of molecules.

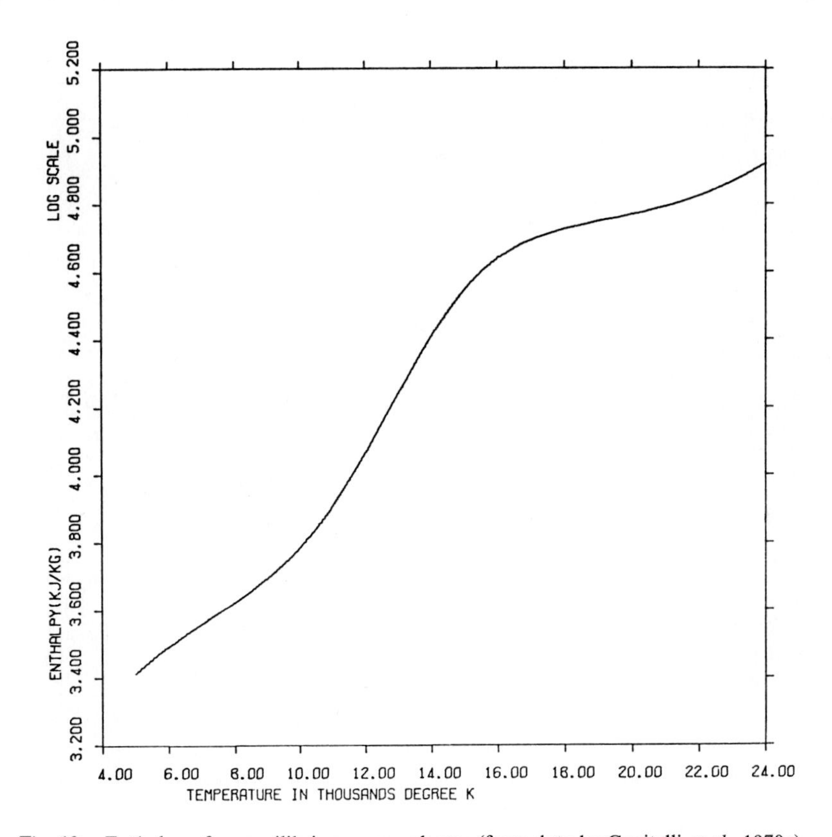

Fig. 10c. Enthalpy of an equilibrium argon plasma (from data by Capitelli *et al.*, 1970a).

The total radiation which has its origin in various emission mechanisms as previously described, leaves the arc without appreciable attenuation as long as the plasma may be considered as optically thin. This assumption may fail for line and band radiation, as well as for continuum radiation. Very strong absorption occurs, for example, for resonance lines. In general, absorption effects become more pronounced as the pressure increases. Plasmas at very high pressures become optically thick and may approach the radiation intensity of a blackbody radiator if the temperature is sufficiently high. An argon arc, for example, will behave as a blackbody radiator in a certain wavelength range for pressures $p \geq 10^4$ kPa and $T > 2 \times 10^{4}$°K (Finkelnburg and Peters, 1957).

Radiation emitted by arcs has been extensively used for diagnostic purposes. Over the past years plasma spectroscopy has become a highly sophisticated and powerful diagnostic tool which plays an extremely important role in arc physics (Finkelnburg and Maecker, 1956; Griem, 1964; Huddle-

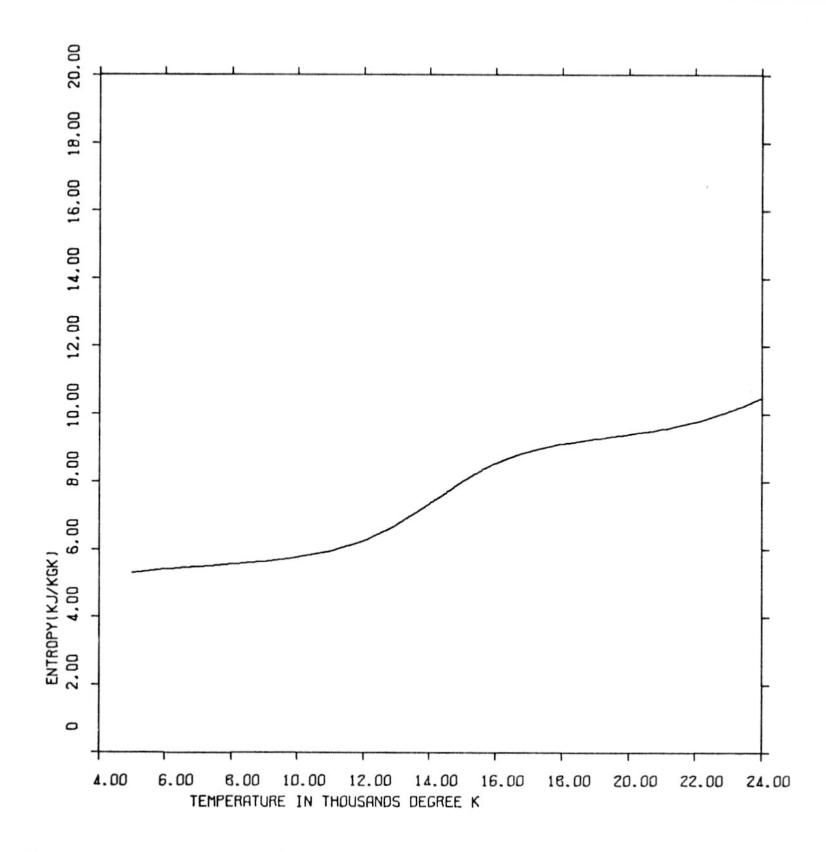

Fig. 10d. Entropy of an equilibrium argon plasma (from data by Capitelli *et al.*, 1970a).

stone and Leonard, 1965; Lochte-Holtgreven, 1968; Marr, 1968; Griem, 1974).

Experimentally determined transport properties of arc plasmas generated from pure gases as well as from gas mixtures have been reported, including comparisons with pertinent theories (Burhorn, 1959; Maecker, 1959b; Olsen, 1959; Kühn and Motschmann, 1964; Knopp and Cambel, 1966; Uhlenbusch and Detloff, 1966; Motschmann, 1967a,b; Behringer *et al.*, 1968; Fischer *et al.*, 1969; Plantikow, 1969, 1970; Hermann and Schade, 1970; Morris *et al.*, 1970; Plantikow and Steinberger, 1970; Kopainsky, 1971a; Jordan and Swift, 1973). Examples of such transport properties for argon at 100 kPa are shown in Figs. 12–14 (Cambel, 1963).

For many years arc plasmas have been treated as optically thin even at higher pressures. Only in recent years the significance of reabsorption of radiation in the arc has been recognized as an important mechanism which

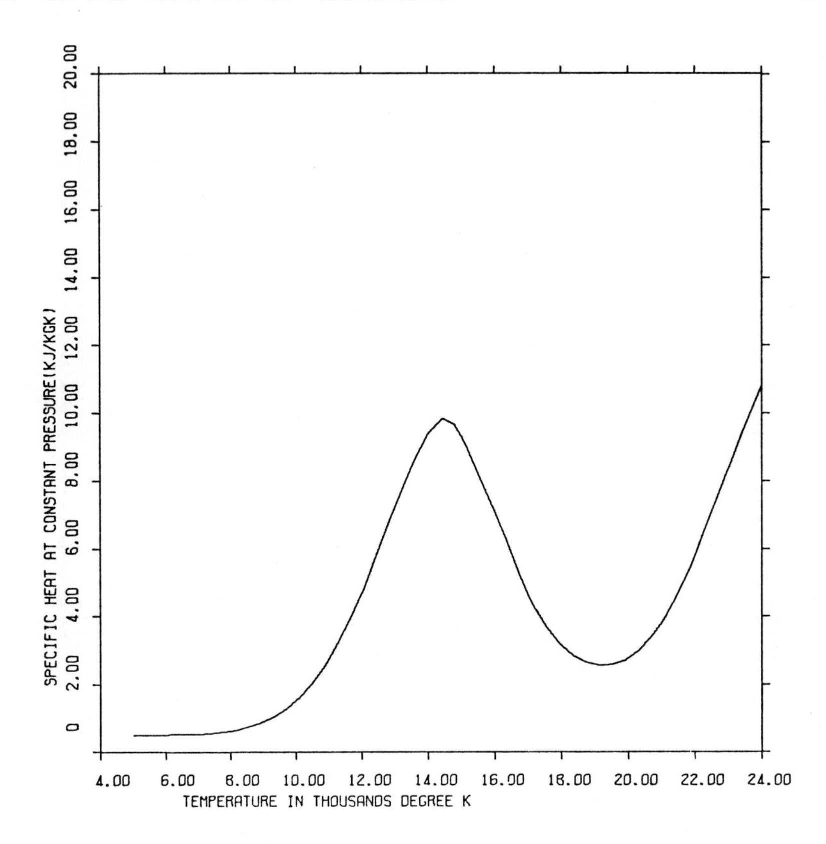

Fig. 10e. Specific heat of an equilibrium argon plasma (from data by Capitelli *et al.*, 1970a).]

may have a strong influence on transport properties, particularly on the thermal conductivity in high pressure arcs. Figure 15 shows a schematic diagram of the radiative balance in high pressure argon arcs (Kopainsky, 1971a). At low pressures and/or temperatures, the contribution of radiation to the energy balance is negligible (region I). Since only Ohmic heating and heat conduction are involved, the resulting temperature profiles are relatively narrow. As the temperature increases, radiation can no longer be neglected, in particular at higher pressures.

In general, radiation for wavelengths $\lambda > 2000$ Å may be considered as optically thin whereas radiation for $\lambda < 2000$ Å (uv) will be partially or totally absorbed in the arc depending on the mean free path length of the photons (λ_{ph}). For uv radiation λ_{ph} is inversely proportional to the absorption coefficient which, in turn, is proportional to the number density of the atoms. Therefore, λ_{ph} decreases with increasing pressure for a given tempera-

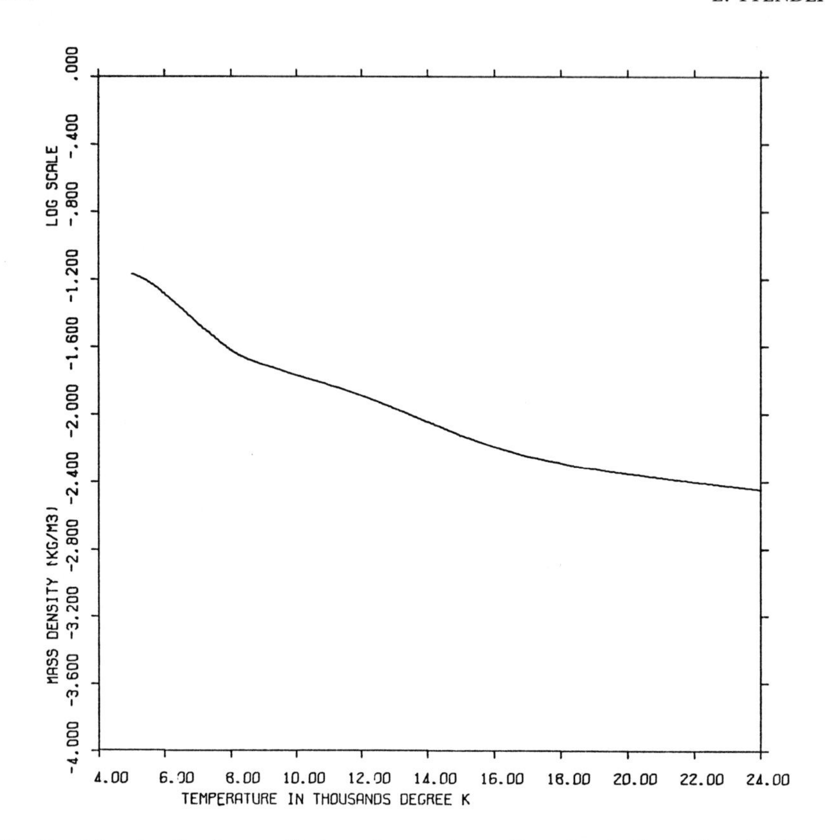

Fig. 11a. Mass density of an equilibrium nitrogen plasma (from data by Capitelli *et al.*, 1970a).

ture. At pressures $p > 300$ kPa and temperatures $T < 15{,}000\,^\circ$K, $\lambda_{ph} \ll R$ (arc radius), i.e., uv radiation is immediately reabsorbed without contributing to the energy transport in the arc (region II). In this situation the energy balance of the arc is determined by Ohmic heating, heat conduction and optically thin radiation.

At higher temperatures ($T > 15{,}000\,^\circ$K) in the same pressure region ($p \geq 300$ kPa) $\lambda_{ph} < R$ and the transport of uv radiation may be described by ordinary diffusion of radiation (region III). In this region, the energy balance is governed by Ohmic heating, optically thin radiation and a modified heat conduction term which includes radiative transport (emission and reabsorption of uv radiation).

As the temperature further increases (or at relatively low pressures) $\lambda_{ph} \simeq R$ and radiative transport can no longer be described by local properties in the arc. An integral expression is needed which depends not only on local temperature and pressure in the arc but also on the field strength. The

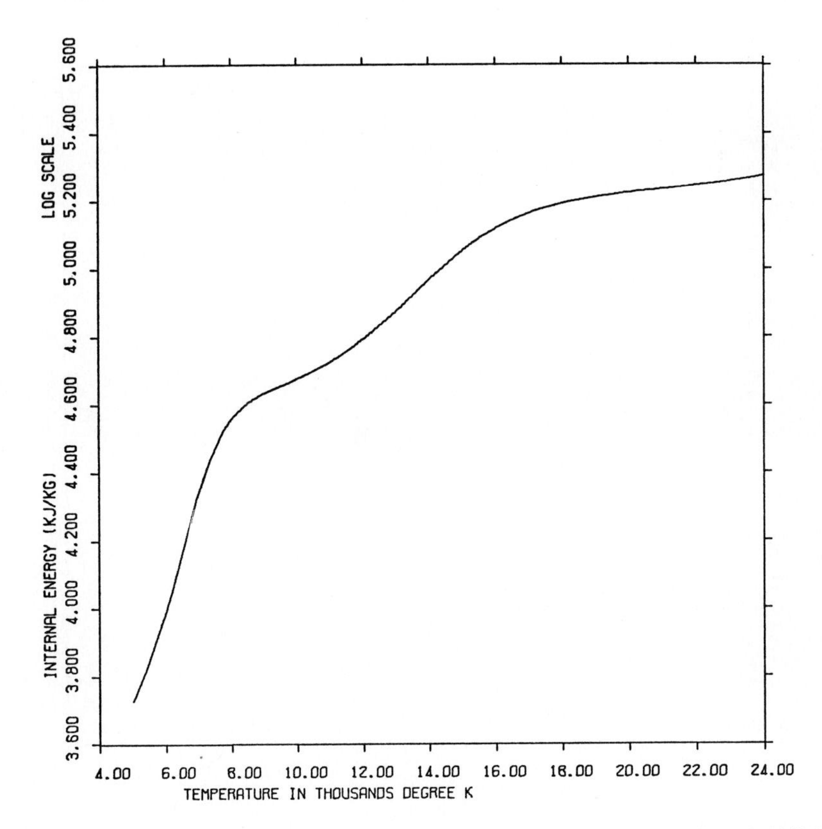

Fig. 11b. Internal energy of an equilibrium nitrogen plasma (from data by Capitelli *et al.*, 1970a).

resulting radiation term is denoted as "far-reaching diffusion of radiation" (region IV) in contrast to the ordinary diffusion of radiation. Optically thin as well as the previously described far-reaching radiation exert a strong influence on the temperature profiles of arcs. Radiative energy transport toward the arc fringes enlarges the arc diameter and in the case of confined (wall-stabilized) arcs, this energy transport leads to almost rectangular shapes of the temperature profiles.

Figures 16 to 18 show electrical (σ) and thermal (κ) conductivities and transparent emission of atmospheric pressure nitrogen arcs, respectively (Ernst *et al.*, 1973a).

In the case of molecular gases the heat conductivity is a complicated function of the plasma composition as shown in Fig. 19 (Burhorn, 1959). Besides possible contributions to the heat conductivity by molecules (κ_m), atoms (κ_{at}), electrons (κ_e), and ions (κ_i), there is a substantial contribution due to chemical reactions. In the temperature range between 5000 and

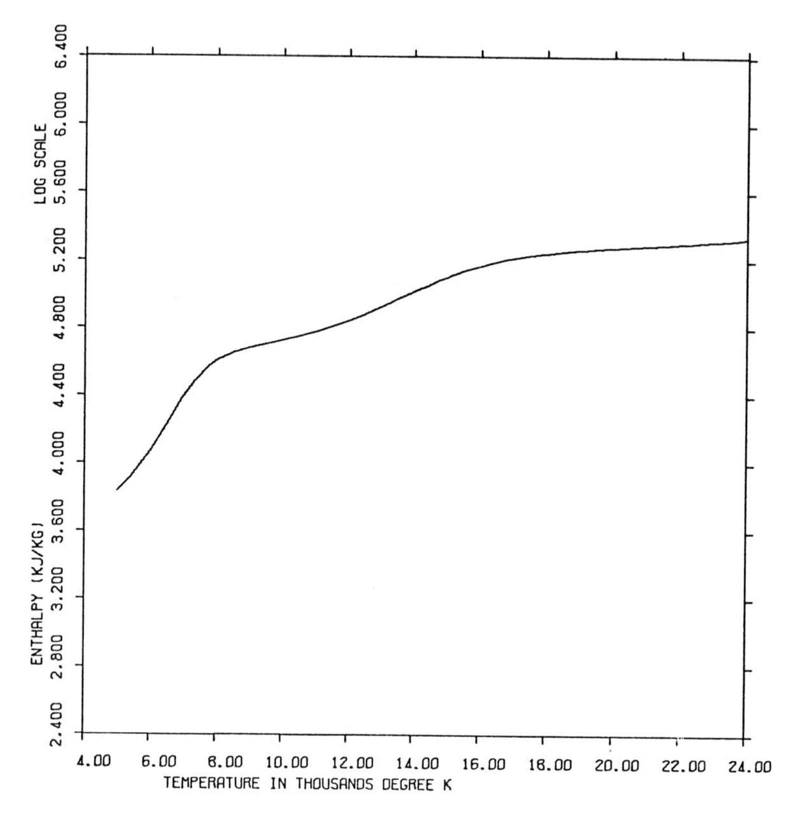

Fig. 11c. Enthalpy of an equilibrium nitrogen plasma (from data by Capitelli *et al.*, 1970a).

8000 °K the heat conductivity is essentially determined by the dissociation process and the associated transport of dissociation energy to the arc fringes (κ_D). Due to the existing density gradients of atoms and molecules in a nitrogen arc, atoms diffuse continuously from hotter arc regions toward the cooler arc fringes where they recombine. Molecules diffuse at the same rate in the opposite direction maintaining the mass balance. Since the nitrogen atoms carry the energy of dissociation along in the form of potential energy, there is a continuous flow of dissociation energy from hotter to colder arc regions. A similar situation exists for the ionization process in the temperature range around 15,000 °K leading to a net flux of ionization energy from the arc core toward the arc fringes (κ_I).

It is interesting to notice that, for a given pressure, energy transport by radiation increases sharply with temperature. At axis temperatures of 26,000 °K, for example, approximately 95% of the energy input to the arc core of an atmospheric pressure nitrogen arc is dissipated by radiation (Her-

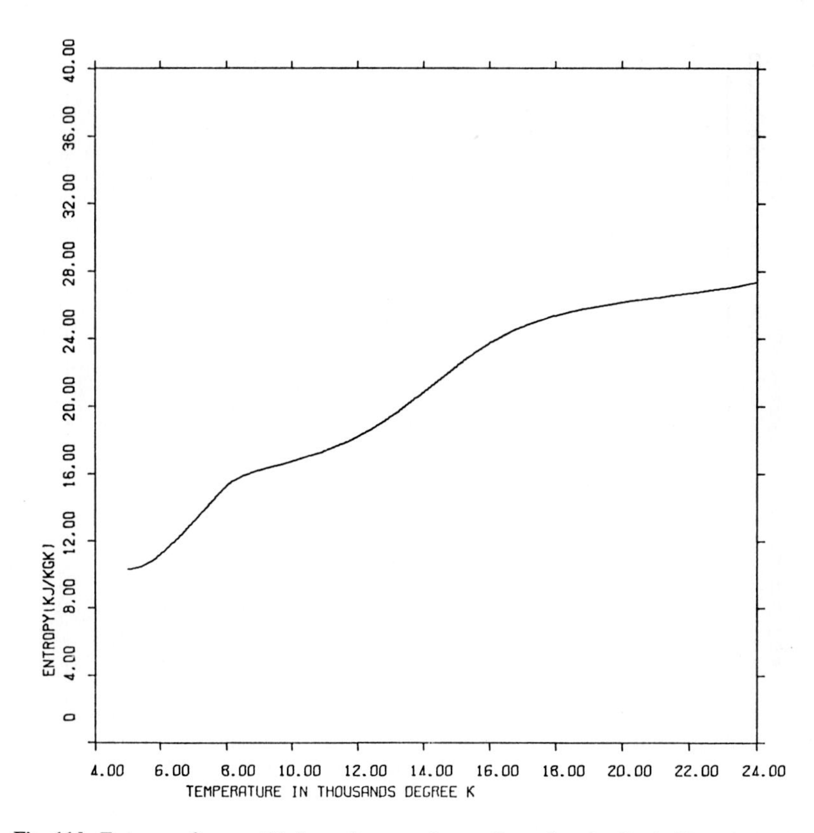

Fig. 11d. Entropy of an equilibrium nitrogen plasma (from data by Capitelli *et al.*, 1970a).

mann and Schade, 1970). Ernst *et al.* (1973a) present a detailed discussion of the energy transport in nitrogen arcs, including emission and absorption of radiation. For temperatures above 13,000 °K the emission and reabsorption of radiation play a governing role in the energy transport within atmospheric pressure nitrogen arcs. A similar situation is to be expected for other working gases or gas mixtures especially at higher pressure levels (Bauder and Bartelheimer, 1973).

(d) *Modeling of the Arc Column* In principle, the behavior of any arc column may be determined by solving the conservation equations with appropriate boundary conditions, provided that the thermodynamic state of the plasma and the transport coefficients are known. Even if the assumption of LTE throughout the arc column can be justified, specification of realistic boundary conditions imposes a serious problem, in addition to the mathematical difficulties of solving a system of coupled nonlinear differential

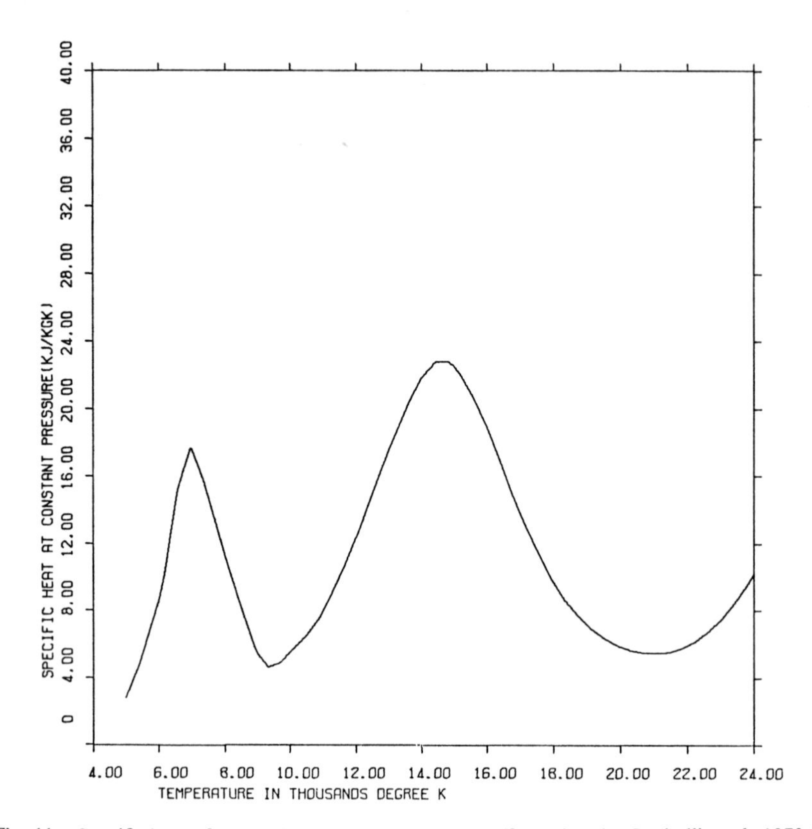

Fig. 11e. Specific heat of an equilibrium nitrogen plasma (from data by Capitelli *et al.*, 1970a).

equations. It is customary in such situations to introduce simplifications which facilitate solutions of the governing equations. Although such solutions cannot describe the actual behavior of the arc column, they frequently reveal important physical trends. Comparisons of analytical solutions with pertinent experiments may then serve as a basis for an improvement of the initial model. By "iterating" between analytical solutions and experimental results, more realistic models may be established, leading finally to the desired agreement between analysis and experiment.

The first attempts to solve the conservation equations for an arc column were reported by Elenbaas (1935) and Heller (1935). Elenbaas and Heller considered an arc column in an asymptotic equilibrium flow regime which leads to a decoupling of the energy equation from the momentum equation. As far as the energy equation is concerned this situation is identical with the case of no flow. Neglecting radiation from the arc entirely, the energy balance may be written as

$$\operatorname{div} \mathbf{W} - \sigma E^2 = 0, \tag{11}$$

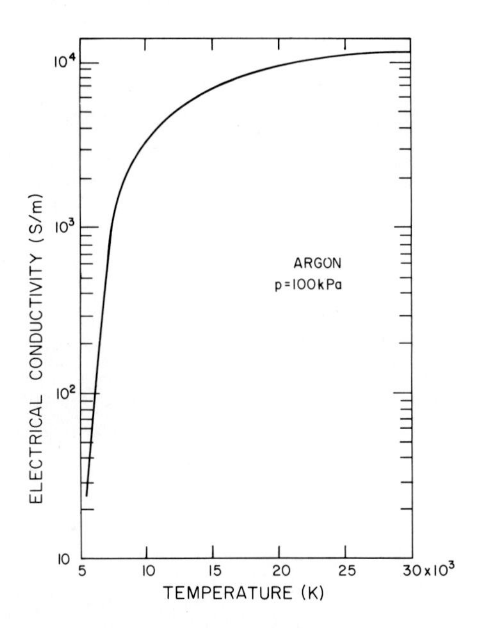

Fig. 12. Electrical conductivity of an argon plasma (Cambel, 1963).

where

$$\mathbf{W} = -\kappa \operatorname{grad} T; \tag{12}$$

W is the heat flow vector, κ the thermal conductivity, σ the electrical conductivity, and E the electrical field strength. According to this equation the heat source term σE^2 is balanced by heat conduction, i.e., heat transfer by thermal diffusion effects is also neglected. For a rotationally symmetric arc column Eq. (11) transforms (in cylindrical coordinates r, ϕ, z) into

$$\frac{1}{r}\frac{d}{dr}\left(r\kappa\frac{dT}{dr}\right) + \sigma E_z{}^2 = 0 \tag{13}$$

which is known as the Elenbaas–Heller equation. E_z represents the field strength in the axial direction. By introducing the heat flux potential

$$S = \int_{T_0}^{T} \kappa \, dT \tag{14}$$

Eq. (13) reduces to

$$\frac{1}{r}\frac{d}{dr}\left(r\frac{dS}{dr}\right) + \sigma E_z{}^2 = 0, \tag{15}$$

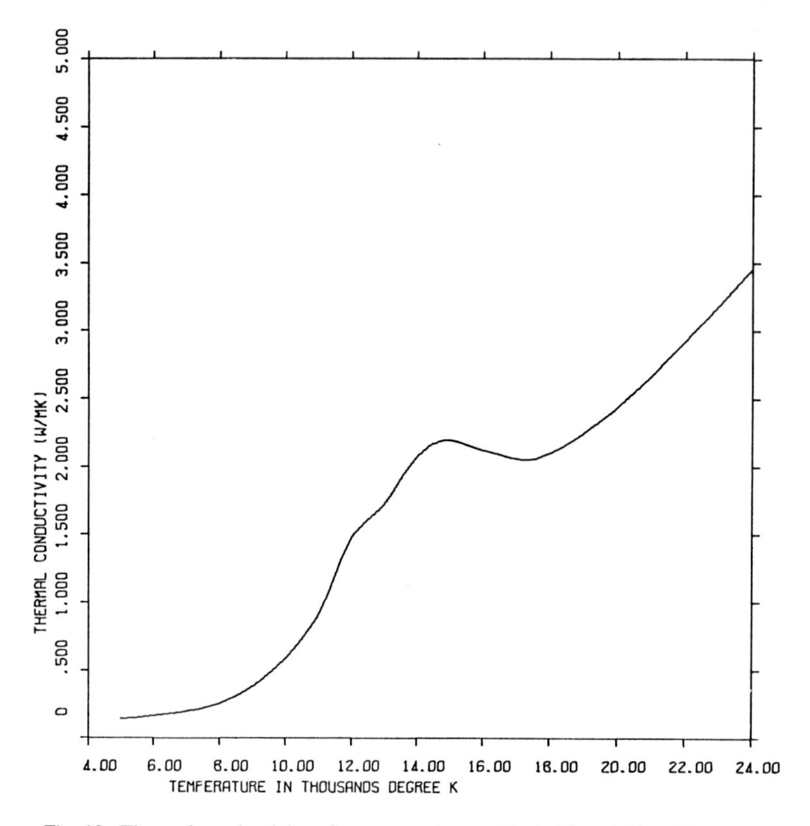

Fig. 13. Thermal conductivity of an argon plasma (Capitelli and Ficocelli, 1970).

where S may be considered as a function of σ. Conservation of current in the arc column may be expressed by Ohm's law

$$I = 2\pi E_z \int_0^R \sigma r \, dr, \tag{16}$$

where R represents the arc periphery.

In spite of the severe simplifications of the Elenbaas–Heller model, solutions of Eqs. (15) and (16) are still complex because of the strong nonlinearities of the transport coefficients κ and σ. In order to facilitate closed form solutions, various approximations for $S(\sigma)$ have been proposed, ranging from linear to high order polynomial approximations, and corresponding solutions of Eqs. (15) and (16) have been reported (Maecker, 1959a; Goldenberg, 1959) which indicate some of the basic trends in arc columns. Figure 20 shows, as an example of such trends, the maximum temperatures

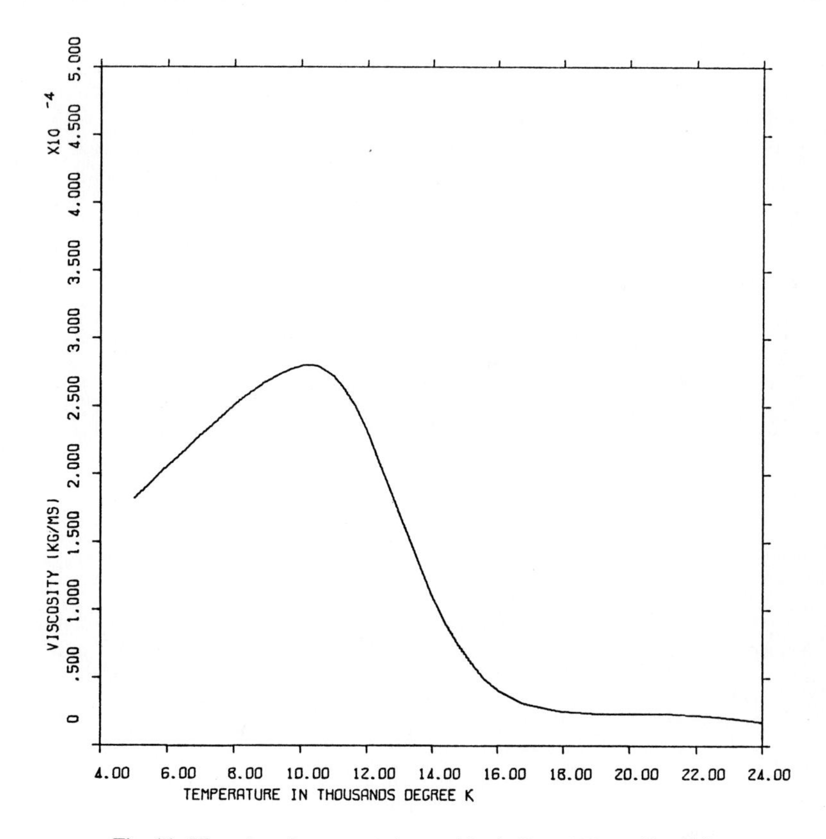

Fig. 14. Viscosity of an argon plasma (Capitelli and Ficocelli, 1970).

which may be reached in an arc as a function of the power input per unit length (Maecker, 1959a).

An extension of the Elenbaas–Heller model to include radiation losses results in the following energy balance equation

$$\frac{1}{r}\frac{d}{dr}\left(r\frac{dS}{dr}\right) + \sigma E_z^{\,2} - P_r = 0, \tag{17}$$

where P_r represents radiative energy losses per unit volume and unit time. It is assumed in this model that the arc column is optically thin, i.e., there is no appreciable reabsorption of radiation within the arc column. Attempts were also made to approximate the radiation source term in Eq. (17) for facilitating closed form solutions of this equation combined with Eq. (16) (Maecker, 1960a; Marlotte *et al.*, 1964).

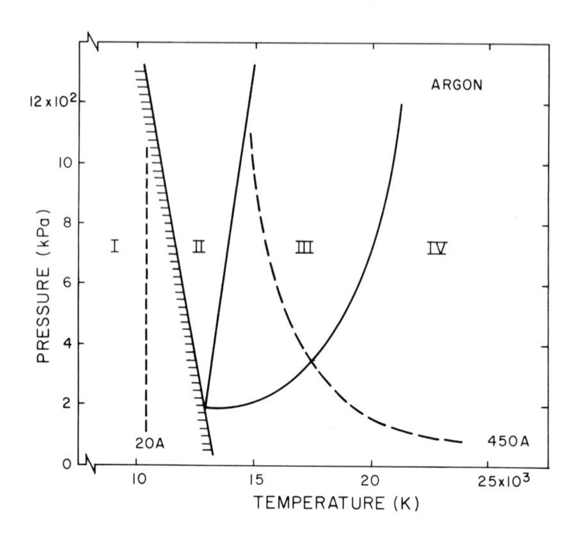

Fig. 15. Schematic diagram of radiative properties of high pressure argon arcs (Kopainsky, 1971a).

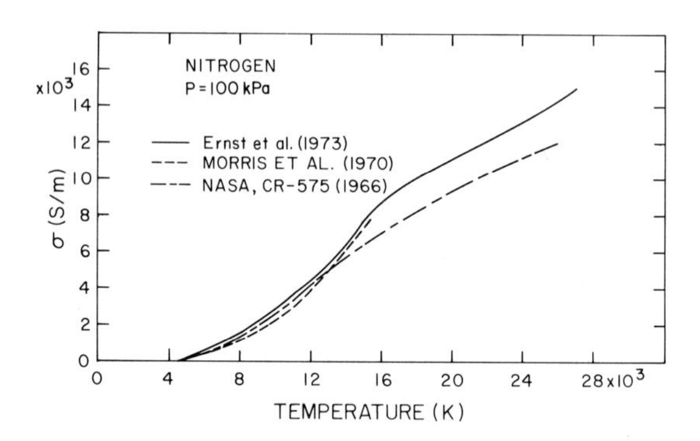

Fig. 16. Electrical conductivity of a nitrogen plasma (Ernst *et al.*, 1973a).

For an accurate assessment of the behavior and properties of an arc column exact values of the transport coefficients must be introduced which necessitates numerical solutions of Eqs. (16) and (17).

Although arcs with little or no superimposed gas flow are frequently used in the laboratory, arcs exposed to substantial flows are of great practical interest as for example in the development of arc gas heaters. The wall-stabilized cascaded arc with superimposed laminar flow (to be discussed in more detail in Section II.E.3) received particular attention because it offers the opportunity to apply scaling laws. Figure 21 shows a schematic arrangement of a wall-stabilized arc.

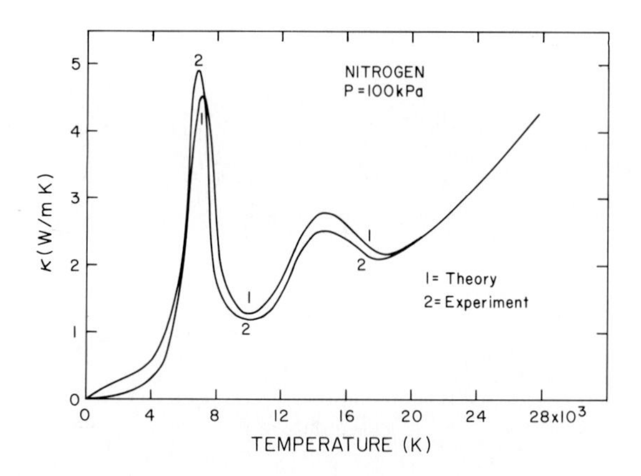

Fig. 17. Thermal conductivity of a nitrogen plasma (Ernst *et al.*, 1973a).

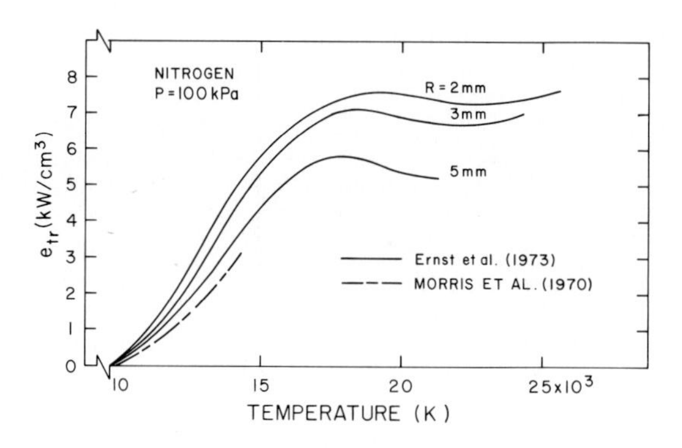

Fig. 18. Transparent emission from nitrogen arcs.

A simple, single-fluid description applies for modeling of the arc if the arc plasma may be assumed to be in LTE. For this situation the conservation equations expressed in cylindrical coordinates may be written as

Mass:
$$\frac{\partial}{\partial z}(\rho u) + \frac{1}{r}\frac{\partial}{\partial r}(r\rho v) = 0 \tag{18}$$

Momentum:
$$\rho\left(u\frac{\partial u}{\partial z} + v\frac{\partial u}{\partial r}\right) = -\frac{\partial p}{\partial z} + \frac{1}{r}\frac{\partial}{\partial r}\left(r\mu\frac{\partial u}{\partial r}\right) \tag{19}$$

Energy:
$$\rho\left(u\frac{\partial h}{\partial z} + v\frac{\partial h}{\partial r}\right) = \frac{1}{r}\frac{\partial}{\partial r}\left(r\frac{\kappa}{c_p}\frac{\partial h}{\partial r}\right) + \sigma E_z^2 - P_r \tag{20}$$

Current:
$$I = 2\pi E_z \int_0^R \sigma r\, dr. \tag{21}$$

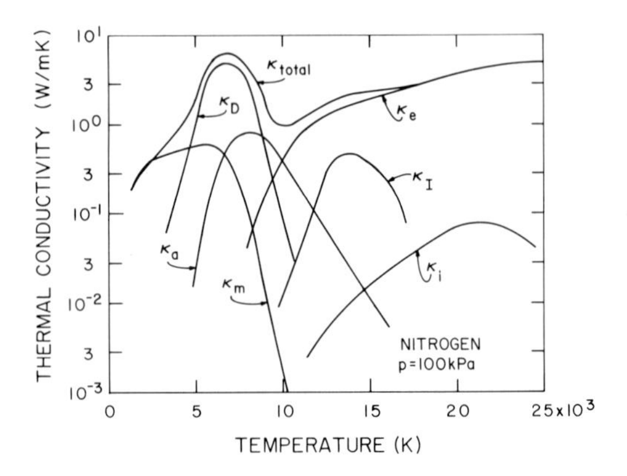

Fig. 19. Individual contributions to the heat conductivity of a nitrogen plasma (Burhorn, 1959). Contribution due to molecules κ_m, atoms κ_{at}, electrons κ_e, ions κ_i, dissociation κ_D, ionization κ_I.

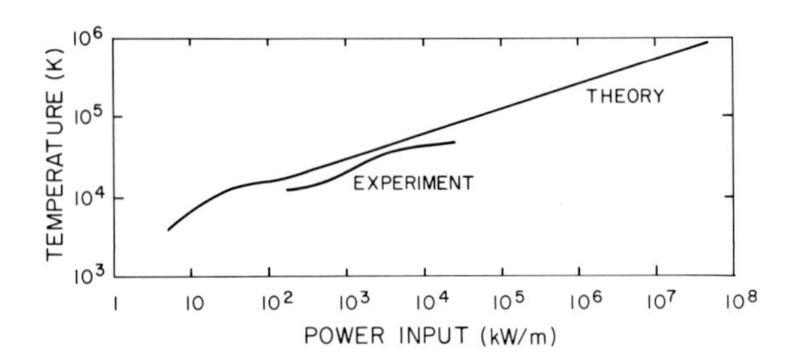

Fig. 20. Highest possible axis temperature in a hydrogen arc (Maecker, 1959a).

The mass density of the plasma is expressed by ρ; u and v are the velocity components in axial (z) and radial (r) direction, respectively; p is the pressure, E_z the axial field strength, h, μ, κ, c_p, σ, and P_r are the plasma enthalpy, viscosity, thermal conductivity, specific heat at constant pressure, electrical conductivity, and radiative energy emitted per unit volume and unit time, respectively. Equation (21) implies that the radial component of the current is negligible. The plasma is treated as a perfect gas so that

$$h - h_0 = \int_{T_0}^{T} c_p \, dT \tag{22}$$

and

$$p = \sum_r n_r kT \tag{23}$$

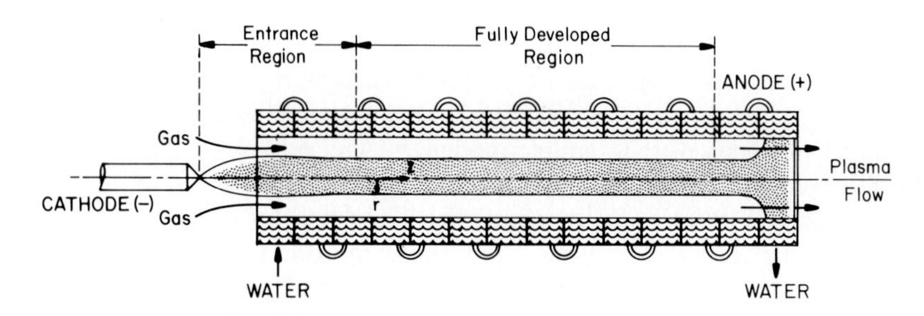

Fig. 21. Schematic of a wall-stabilized arc in axial flow.

where k represents the Boltzmann constant and n_r the particle densities (electrons, ions, neutrals).

In addition to the previously stated assumptions, viscous dissipation in the plasma is neglected as well as self-magnetic field effects. The flow is assumed to be steady and axially symmetric without swirl components, and reabsorption of radiation within the arc is neglected. For establishing momentum [Eq. (19)] and energy [Eq. (20)] equations the usual hydrodynamic and thermal boundary layer approximations have been introduced.

For solving the conservation equations the temperature dependence of thermodynamic and transport properties must be known. Examples of such properties have been shown in the previous paragraph.

Because of severe gradients of the temperature and the associated particle densities in an arc, diffusion effects play an important role. As discussed previously, energy transport due to chemical reactions in the plasma (for example, dissociation and ionization) may be the dominating contribution to the total energy transfer in certain temperature intervals. These contributions, however, are included in the thermal conductivity which, in general, becomes a strongly nonlinear function of the temperature.

It should be pointed out that the conservation equations (18–21) apply to the entrance region of the arc. In the fully developed region where $\partial h/\partial z = \partial u/\partial z = 0$ and also $v = 0$, a corresponding modification of Eqs. (18–20) adapts these equations for the fully developed (asymptotic) region of the arc. The energy equation, for example, reduces to Eq. (17) and becomes decoupled from the momentum equation.

Analytical solutions of the conservation equations for the entrance region have been reported by Stine and Watson (1962). In their original arc model, radiation has been entirely neglected among other simplifications. Although the results of this analysis are only qualitative without accurate predictions of local property variations, they provide valuable guidelines for general trends (Watson, 1965). During the Sixties many attempts were made to remove some of the simplifications in the Stine–Watson model in order to improve the still lacking agreement between experiment and theory. Unfortunately, these attempts had only limited success because the previously mentioned strong variations of thermodynamic and transport properties do not lend themselves to simple modeling. Accurate predictions can only be expected from numerical solutions of the unaltered system of conservation equations. Such solutions have been reported by Watson and Pegot (1967) for the entrance as well as for the asymptotic region and by Bower and Incropera (1969) for the asymptotic region of arcs in laminar flow. Figure 22 shows as a typical example, selected results of the calculations by

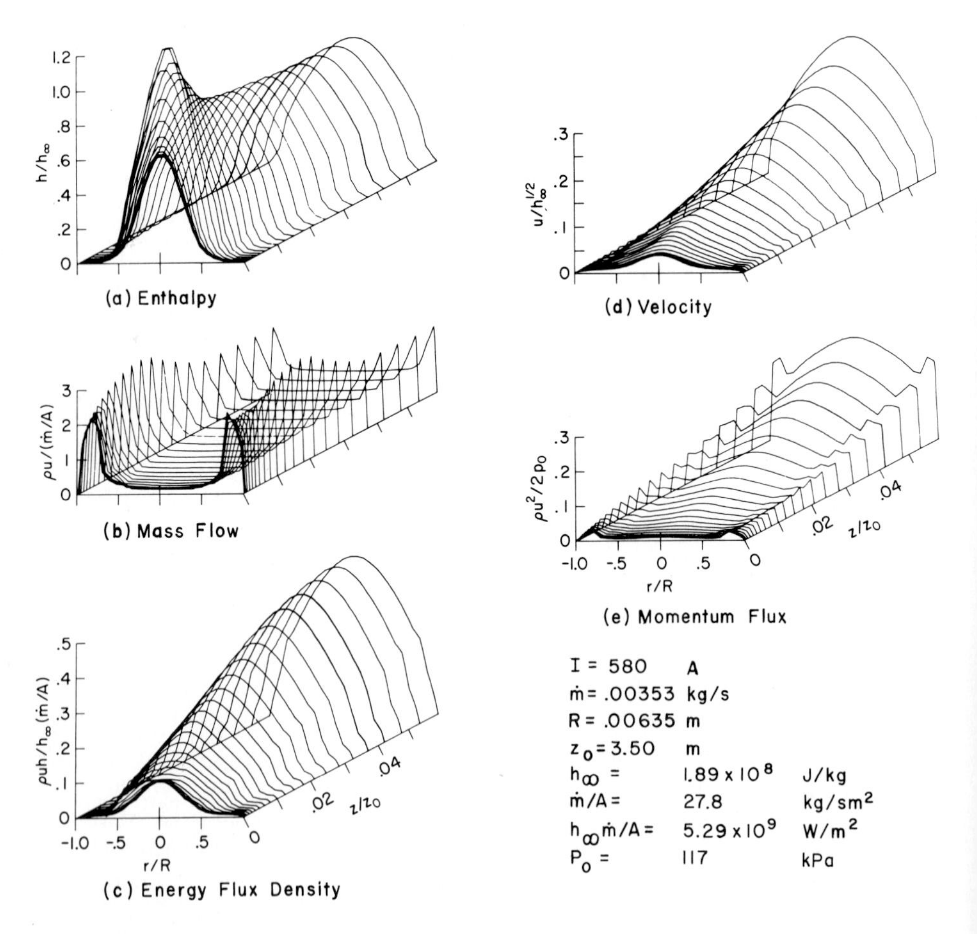

Fig. 22. Calculated performance of a wall-stabilized nitrogen arc in axial flow (Watson and Pegot, 1967).

Watson and Pegot (1967) for a nitrogen arc. Part (a) of this diagram indicates that the plasma enthalpy increases rapidly in the entrance region, reaches a peak and then levels off towards the fully developed (asymptotic) region of the arc. In spite of the high axial velocities, the mass flow within the constrictor [Part (b)] is essentially confined to a relatively cold layer close to the wall, especially in the vicinity of the entrance. This effect is due to the low mass density in the arc core which is a consequence of the high enthalpies (temperatures) in the arc axis. With increasing distance from the entrance, more and more of the cold gas permeates into the arc.

The crucial test for the validity of these predictions is, of course, a comparison with pertinent experiments. Taking into account the uncertainties involved in the determination of the transport properties, the agreement between theory and experiment is reasonable as long as the assumption of LTE holds.

As previously mentioned, the existence of LTE in an arc is more of an exception rather than the rule. For those types of arcs which are of primary interest in this survey, deviations from kinetic equilibrium ($T_e > T_g$) and/or deviations from chemical equilibrium (charged particle concentrations) are of importance. Possible deviations from excitation equilibrium involving lower lying energy levels (PLTE) are less significant, although they may play an important role for plasma diagnostics.

If, for example, $T_e > T_g$, modeling of the arc requires adoption of a two-fluid system which introduces additional equations to the set of conservation equations. Procedures for modeling of wall-stabilized arcs which are either in LTE or which deviate from kinetic and chemical equailibrium have been recently discussed in a survey by Incropera (1973). Figure 23 shows a comparison of measured and calculated characteristics of a fully developed argon arc. The measurements deviate substantially from the calculated characteristic based on LTE but they are in reasonable agreement with nonequilibrium calculations.

2. The Electrode Regions

In the following discussion of the electrode regions those types of arcs will be emphasized which find widespread applications in arc gas heaters and plasma torches. The first part of this paragraph will be devoted to a number of features of cathode and anode regions, followed by a more detailed, separate discussion of the cathode and anode regions.

The electrode region (cathode or anode region) of an arc is defined as that part of the discharge path which contains the electrode surface, the region of the net space charge (sheath or fall zone) immediately in front of the electrode, and the transition zone towards the arc column. The sheath

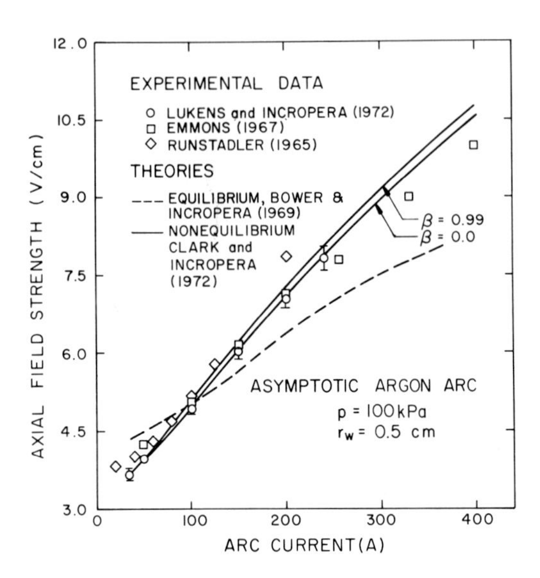

Fig. 23. Comparison of measured and calculated characteristics of a wall-stabilized, fully-developed argon arc (Incropera, 1973).

may be considered as an electrical boundary layer accommodating the transition between metallic and gaseous plasma conduction. The transition or contraction zones which are characteristic for high intensity, high pressure arcs may be interpreted as thermal boundary layers in which strong axial gradients of the plasma properties (temperature, potential, particle densities, current density, radiation) exist, in contrast to the arc column in which axial gradients are small or even negligible. The influence exerted by the electrodes on the plasma is no longer felt in the arc column. This fact is illustrated in Fig. 24, which shows schematically the potential distribution of an arc. The thickness of the electrode regions is exaggerated in this sketch. The thickness of the cathode fall (d_c) as well as of the anode fall (d_a) is in the order of one mean free path length of the electrons. For atmospheric pressure high intensity arcs this thickness is in the range from 10^{-4} to 10^{-3} cm whereas the thickness of the transition regions II and IV is in the order of 10^{-1} cm under the same conditions. These thicknesses are, of course, functions of the pressure and temperature in the electrode regions. The potential drop over the transition regions may well be of the same order of magnitude as the electrode falls themselves. In the anode region $V_a' - V_a$ may even exceed the anode fall.

The electrode regions are characterized by much higher field strengths, temperature gradients, and current densities than experienced in arc columns.

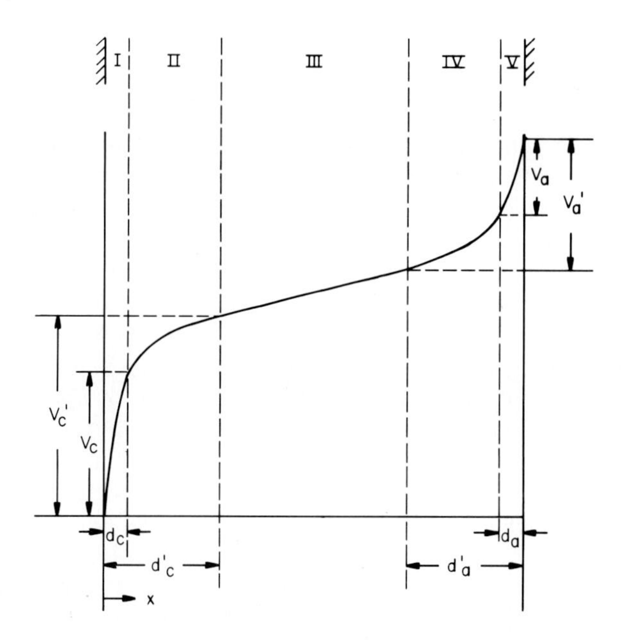

Fig. 24. The electrode regions of an arc.

In addition, fluid dynamic effects in the electrode regions give rise to the so-called plasma jet phenomena.

In spite of intensive research over the past decades the electrode regions are still poorly understood. This lack of understanding is a consequence of the complexities prevailing in these regions caused by the interaction of electrical, magnetic, thermal, and fluid dynamic effects, in addition to surface effects which are difficult to assess and frequently impossible to control. Considering these facts, it is not surprising that there is still no comprehensive theory capable of describing all the observed phenomena in these regions.

Gathering of experimental data from the electrode regions faces similar problems. The extreme values of temperature, current density, field strength, etc. prevailing in these regions and the steep gradients of these parameters makes diagnostics a formidable task. Furthermore, these extreme conditions in the electrode regions may cause deviations from LTE which creates another problem for the interpretation of experimental data. It is even doubtful whether or not the velocity distribution of the various plasma components (electrons, ions, neutrals) is still Maxwellian close to the electrodes.

In the case of electrode spots (extremely constricted current attachments), the highly disturbed surface state contributes further to this complexity. One may consider the location of an electrode spot as a highly disturbed metal lattice or as an extremely dense metal plasma. The electrode surface at the location of such a spot does not exist any more; it becomes a transition zone between solid state and a probably nonideal gaseous plasma. These facts explain why there are only very few and rather scattered experimental data available in the literature about the electrode regions. Many findings are more or less concerned with phenomenological observations only. Nevertheless, the literature dealing with the electrode regions of electric arcs is very extensive; in particular the cathode region has attracted great interest because of the strong impact of effects in the cathode region on the entire discharge. Pertinent literature references on the electrode regions up to 1955 are given by Finkelnburg and Maecker (1956). A more comprehensive review listing 645 references has been published by Ecker (1961) summarizing relevant work on this subject up to 1959. Ecker stresses in his treatment the basic physics involved in the electrode regions with emphasis on low intensity arcs, whereas Guile (1971) in a later review ("Arc-Electrode Phenomena," which includes 238 references) emphasizes those aspects which are of interest for engineers involved in arc applications, particularly of high intensity arcs. In a recent review by Kimblin (1974), experimental aspects of the anode region are discussed for vacuum and atmospheric pressure arcs with emphasis on the effect of plasma flow from the cathode.

(a) *The Cathode Region* Phenomenologically, the current attachment at the cathode of arcs may be divided into two broad categories:

(1) "Diffuse attachment" without evidence of single or multiple cathode spots.
(2) Attachment in the form of one or several distinct spots.

The term "diffuse attachment" characterizing the first mode requires further explanation because this mode is frequently referred to in the literature as "spot attachment." The arc may indeed reveal in this mode an appreciable constriction in front of the cathode so that the actual current transition zone appears as a "spot" with current densities in the range from 10^3 to 10^4 A/cm^2, at least one order of magnitude higher than in the arc column. This mode, however, may be clearly distinguished from the second mode for which the constriction is much more severe, with current densities in the range from 10^6 to 10^8 A/cm^2. In addition to the entirely different cathode electron emission mechanisms for these two modes, the cathode attachment in the first mode is stationary or slowly moving in contrast to the second mode which frequently shows one or several spots moving with high velocities and randomly over the cathode surface.

In the first mode thermionic emission of electrons is the governing mechanism for liberation of electrons from the cathode. This mode is, however, restricted to cathode materials with very high boiling points (carbon and refractory metals). In general, thermionic electron emission from a cathode surface is a function of the surface temperature and of a characteristic property of the cathode material, known as the work function ϕ_c.

Richardson derived an expression from thermodynamic considerations for the electron emission from metallic surfaces, assuming that the electrons in the metal obey the Boltzmann statistics. Based on this assumption he found the following expression for the electron saturation current density

$$j_s = A'T^{1/2} \exp(-e\phi_c/kT), \tag{24}$$

where A' is a constant and $e = 1.6 \times 10^{-19}$ is the elementary electron charge. By using the proper Fermi statistics for the electrons in the metal v. Laue and Dushman derived the following equation

$$j_s = AT^2 \exp(-e\phi_c/kT) \tag{25}$$

that is, besides T^2 instead of $T^{1/2}$ and the constant, identical with Eq. (24). Equation (25) is known today as the Richardson–Dushman equation for thermionic emission. The constant A has a value of $\simeq 60$ A/cm^2 K^2.

As previously mentioned, only certain materials with sufficiently high boiling points are suitable for thermionically emitting cathodes. Figure 25 shows the materials which are proven thermionic emitters in an argon atmosphere, as well as those metals for which no appreciable thermionic emission has been observed. Also included in this diagram are three metals for which information is still lacking (Guile, 1971).

Since thermionic emission from an arc cathode requires very high surface temperatures, the energy transfer mechanism responsible for maintaining this high surface temperature must be identified. For a basic understanding of this mechanism a simplified energy balance for the cathode will be considered.

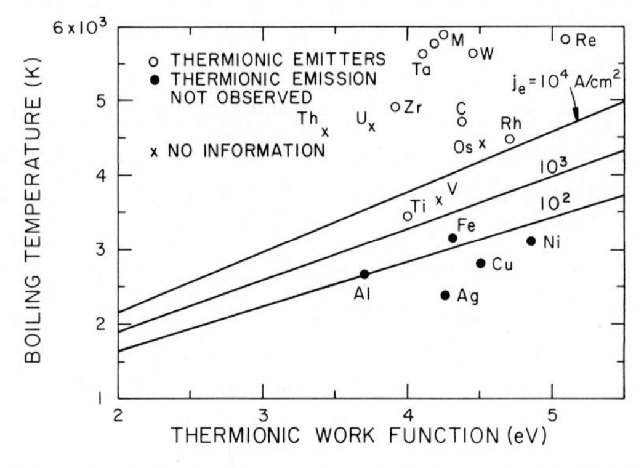

Fig. 25. Thermionically emitting and nonemitting materials (Guile, 1971).

Electrons emitted thermionically from the cathode are accelerated by the field in the cathode fall zone away from the cathode, and ions entering the anode fall zone or generated in this zone are accelerated towards the cathode. The energy released per unit area and unit time by the ions impinging on the cathode surface is given by

$$j_i[V_c + V_I - \phi_c + \tfrac{5}{2}(kT_i/e)], \tag{26}$$

where j_i is the ion current density, V_c the potential drop in the cathode fall zone, V_I the ionization potential of the working fluid, ϕ_c the cathode work function, and T_i the temperature of the ions entering the cathode fall zone. For simplicity, an accommodation coefficient of unity is assumed for the ions impinging on the cathode surface. The energy (per unit area and unit time) required for liberating electrons comprising an electron current of density j_e is

$$-j_e\phi_c. \tag{27}$$

Neglecting secondary effects including heat losses from the cathode by conduction/convection and radiation (Shih and Pfender, 1970), the energy balance yields the ratio of ion current density to the total current density j at the cathode surface

$$\frac{j_i}{j} = \frac{\phi_c}{V_c + V_I + \frac{5}{2}(kT_i/e)}. \tag{28}$$

According to Eq. (28) the ion current density at the cathode is in the range from 15 to 50% of the total current density. The cathode fall for an arc with thermionically emitting cathode is typically around 10 V for high intensity arcs, revealing some minor variation with current (Finkelnburg and Maecker, 1956). For low intensity arcs the cathode fall may be substantially higher, approaching values close to the ionization potential of the working fluid.

In a refined model, Lee and Greenwood (1963) and Lee *et al.* (1964) subdivided the cathode region of a thermionically emitting cathode into a number of zones which allow the calculation of the fraction of the ion current to the total current over the entire thickness of the cathode region. They found for a 200 A carbon arc that this fraction varies from 15% at the cathode surface to zero at the column end of the cathode region.

An increase of the fraction of the ion current density toward the cathode is consistent with the formation of a net positive space charge which determines the cathode sheath or cathode fall zone. According to the Poisson equation, this space charge gives rise to an electric field which enhances electron emission from the thermionically emitting cathode, provided that this field is sufficiently strong. This process may be described by an equation similar to Eq. (25) with a correction term (Schottky, 1914) in the exponent which accounts for the field effect

$$j_s = AT^2 \exp\left\{\frac{-[e\phi_c - (e^3E/4\pi\epsilon_0)^{1/2}]}{kT}\right\}, \tag{29}$$

where the constant A is the same as in Eq. (25), E is the field strength in V/m, and $\epsilon_0 = 8.86 \times 10^{-12}$ A sec/V m is the permittivity constant. As a modification to Eq. (25) this equation is only valid for high cathode temperatures and moderate field strengths.

For completeness it should be mentioned that the true spot attachment (second category) may occur as a stationary or rapidly moving current attachment which can be clearly observed as one or several bright spots on the cathode surface. Stationary spots have been observed in Hg and noble gas arcs with refractory metal cathodes at moderate current levels ($\sim$10 A). By increasing the current this mode frequently shows a sudden transition to the first mode which is an indication that the current transition at the cathode is governed by a different mechanism which has been extensively discussed in Ecker's (1961) review. Since this mode is of minor importance for the application of arcs in arc gas heaters and plasma torches, this mode will not be further discussed in this survey.

In contrast, arcs attaching to the cathode in the form of single or multiple, rapidly moving, extremely small spots are of great practical importance for applications in arc gas heaters and plasma torches. Since the current density of thermionic emission depends critically on the cathode surface temperature, electron emission can no longer be ascribed to thermionic emission because the size of the attachment spot and the thermionically feasible current densities are not consistent with the observed currents. In this case a different mechanism for electron liberation from the cathode, known as field emission must be postulated. Cathodes for which the electron emission is ascribed to field emission are referred to as nonthermionic or cold cathodes because the overall temperature of the cathode is substantially below that required for thermionic emission. The basic differences between thermionic and nonthermionic cathodes are summarized in Table I (Guile, 1971). The pressure listed in this table refers to the immediate vicinity of the cathode attachment.

As previously mentioned the cathode fall zone is characterized by a net positive space charge producing an electric field which gives rise to an ambipolar charge carrier flow in opposite directions. Assuming that the positive ions in this zone carry at least 5% of the total current, Mackeown

TABLE I

BASIC PROPERTIES OF THERMIONIC AND NONTHERMIONIC CATHODES

Cathode	Surface temperature	Current density	Pressure	Cathode attachment
Thermionic (hot)	>3500 K	10^3–10^4 A/cm^2	$\simeq$Ambient	Fixed or slowly moving
Nonthermionic (cold)	<3000 K	10^6–10^8 A/cm^2	$>$Ambient	Rapidly moving

(1929) found for the field strength in this zone based on the space charge limited current densities of electrons j_e and ions j_i the following expression

$$E^2 = 7.6 \times 10^5 V_c^{1/2}[(1845\, M)^{1/2} j_i - j_e], \tag{30}$$

where E is obtained in volts per centimeter if j_i and j_e are introduced in amperes per square centimeter, and M represents the molecular weight of the ions. This equation is particularly interesting in connection with field emission.

Fowler and Nordheim (1928) considered electron emission from metal surfaces under the influence of very strong electric fields. The electron current density due to the resulting field emission process may be expressed by (Rieder, 1967)

$$j_e = 1.54 \times 10^{-6} \frac{E^2}{e\phi_c} \exp\left[\frac{-6.83 \times 10^9 (e\phi_c)^{3/2}}{E} f\left(\frac{3.79\, E^{1/2}}{e\phi_c} \times 10^{-5}\right)\right]. \tag{31}$$

From this equation the current density is obtained in units of amperes per square meter if the field strength is introduced in volts per meter and $e\phi_c$ in electron volts. The function f decreases from $f(0) = 1$ to $f(1) = 0$. This equation is valid for relatively high field strengths ($>10^7$ V/cm) and low cathode temperatures, i.e., for cathodes with a low boiling point. From Eqs. (30) and (31) it follows that field emission is only possible for current densities $j > 10^7$ A/cm^2 even if unrealistically low work functions are postulated.

For intermediate temperatures and substantial field strengths both thermionic emission (Eq. 29) and field emission (Eq. 31) may be simultaneously involved. The corresponding emission process is known as thermionic field emission or TF-emission (Lee, 1957, 1958, 1959).

In the case of a nonthermionic cathode the arc attachment at the cathode never consists of a single emitting area. The arc root is usually composed of a number of small emitting sites which are close together at higher pressures and which seem to move randomly around with sites disappearing and reappearing simultaneously in spite of the steady-state conditions in the arc.

The number of emitting sites increases with current, which may be an indication that the current per site is fairly constant. It seems that the surface state (oxide layers, impurities, mechanical imperfections, etc.) is an important, if not the governing, parameter for the behavior of the cathode roots on nonthermionic cathodes. Even under extremely clean and macroscopically well defined conditions the definition of the surface state on a microscopic scale remains a formidable problem. Analytical approaches are faced with this dilemma and it appears quite natural that there are so many seemingly conflicting theories because there may be no single theory which applies to all possible conditions.

Guile (1971) summarizes in his review many observations and measurements in the cathode region of nonthermionic arcs, as for example spot size, current density and cathode fall, spot splitting, movement, and evaporation with the resulting cathode tracks and cathode erosion, surface cleaning and oxidation, influence of electrode material including grain size and boundaries, surface state and chemisorbed gases, and the influence of the pressure (vacuum arcs and high pressure arcs).

In addition to the previously discussed electron emission processes at the cathode, possible contributions by other mechanisms cannot be ruled out. Electrons may be liberated from the cathode by individual field components due to the statistical variation of the space charge, by the Auger effect, by the external photoeffect, by excited (metastable) atoms, by positively charged oxide layers on the cathode surface, by lowering of the work function produced by monatomic surface layers (for example Th on W) or by negative space charges inside the cathode. The various possibilities have been summarized by Guile (1971) in conjunction with many pertinent references. In a very recent paper Holmes (1974) shows that emission by excited atom bombardment and thermionic field emission leads to a consistent model for Hg and Cu vapor arcs. Calculated cathode falls, current densities, temperatures, and particle densities in the cathode region are in reasonable agreement with available experimental data.

Secondary effects, such as vapor and plasma jets originating in the cathode region or on the cathode surface itself may exert a strong influence on the cathode region and sometimes on the entire arc.

Cathode jets have been observed in thermionic as well as in nonthermionic arcs, particularly at higher current levels. These cathode jets may be attributed to four different sources.

(1) Electromagnetically induced jets.
(2) Vaporization of cathode material and/or surface impurities.
(3) Ablation and explosive release of cathode material.
(4) Chemical reactions on the cathode surface producing gases.

The interaction of the arc current with its own magnetic field leads, in arc sections of variable cross section, to the phenomena of induced plasma jets. These phenomena are not restricted to the cathode or anode region of an arc; they may also occur in other parts of the arc column where the conditions of variable column cross section are met. If an arc, for example, is forced through a diaphragm which reduces its cross section, plasma jets are induced with the flow directed away from the location of the most severe constriction of the arc column (Maecker, 1955). Such an arrangement (Fig. 26) acts as a pump; gas is continuously moved toward the opening of the diaphragm by suction. Once the gas is ingested and heated by the arc, it is then accelerated away from the orifice in the diaphragm (as shown in Fig. 26).

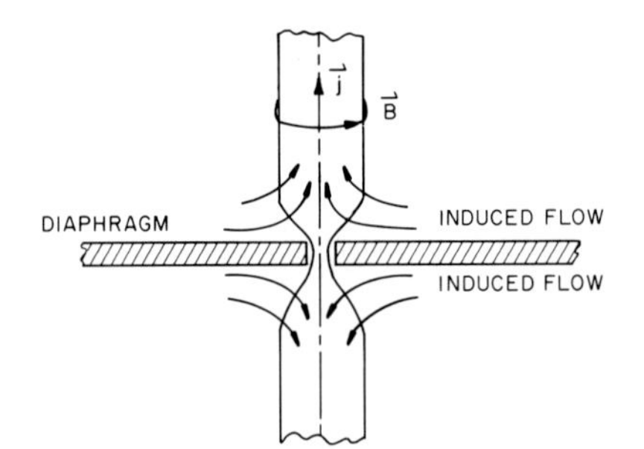

Fig. 26. Schematic of the pumping action induced by arc constriction.

For an analytical description of this phenomenon, momentum and continuity equations are required which, in vector notation, may be written, for a steady arc neglecting viscous effects, as

$$\rho \frac{d\mathbf{v}}{dt} + \operatorname{grad} p = \mathbf{j} \times \mathbf{B}, \tag{32}$$

$$\operatorname{div}(\rho \mathbf{v}) = 0, \tag{33}$$

where ρ is the plasma density, $\mathbf{v}$ the plasma velocity vector, p the pressure, and $\mathbf{B}$ the self-magnetic field vector. The $\mathbf{j} \times \mathbf{B}$ force which, in general, is responsible for the pinch effect in current-carrying plasma columns, may build up a pressure gradient and/or accelerate the plasma. Equation (33) determines which fraction of the magnetic body force is used for plasma acceleration. For a rotationally symmetric arc the radial pressure gradient and the resulting overpressure in the arc may be expressed by

$$\Delta p(r) = \int_r^R j(r)B(r)\, dr, \tag{34}$$

where R is the arc periphery.

With

$$\text{rot}\,\mathbf{B} = \mu_0 \mathbf{j} \tag{35}$$

one obtains

$$B(r) = (\mu_0/r)\int_0^r jr\,dr \tag{36}$$

where $\mu_0 = 1.26 \times 10^{-6}$ H/m is the permeability constant.

If the current density distribution $j(r)$ is known, $\Delta p(r)$ can be calculated. Maecker (1955) calculated this overpressure assuming a uniform current density distribution (one step model) over the arc cross section of magnitude

$$j = I/\pi R^2, \tag{37}$$

where I represents the total arc current. Combining Eqs. (34), (36), and (37) one finds

$$\Delta p(r) = (\mu_0 Ij/4\pi)[1 - (r^2/R^2)], \tag{38}$$

i.e., the overpressure in the arc axis is proportional to the product of total arc current and current density.

As soon as there is a constriction of the arc channel in the cathode region, the current density as well as the self-magnetic field will increase; according to Eq. (38), this will also increase the overpressure in the arc axis, i.e., an axial pressure gradient will be generated pointing towards the cathode. This pressure gradient induces a flow in the opposite direction with a maximum velocity

$$v_{\text{max}} = (\mu_0 Ij/2\pi\rho)^{1/2}, \tag{39}$$

where ρ is the average plasma density. For free-burning 200 A carbon arcs maximum velocities in the order of 100 m/s have been found (Finkelnburg and Maecker, 1956). A more sophisticated but still simplified analysis by Cowley (1973) provides velocity profiles of electrode jets. The flow is carried by a progressively smaller proportion of the arc column as its diameter increases. This holds for cathode as well as for anode jets. The maximum velocity depends critically on the arc constriction in the cathode region, which may be influenced by the cathode shape (Petrie and Pfender, 1970) in the case of thermionically emitting cathodes. In nonthermionic arcs the cathode jet velocities near the surface of Cu and Hg arcs at very low pressures are on the order of 10^4 m/s (Reece, 1963), whereas for higher pressures the velocity drops substantially.

The cathode jets listed under (2), (3), and (4) originate at the cathode surface; these jets contain cathode material and/or impurities either in vapor form or as particular matter, including gases stemming from chemical reactions on the cathode surface. Oxidation of carbon steel, for example, will produce CO and CO_2.

Since metal vapors as well as many impurities have a lower ionization potential than permanent gases, these materials will have a strong influence on the plasma properties including the transport coefficients of the plasma in the cathode region.

The cathode jets may also enhance the "stiffness" of the cathode region and the adjacent arc column. In fact, the electromagnetically induced cathode jet may serve as a stabilizing mechanism for a free-burning arc (see Section E.3). Transverse magnetic fields can influence cathode jets and they also demonstrate that the vapor flow from the cathode oscillates (Hermoch, 1959; Hermoch and Teichmam, 1966).

The interaction of externally applied magnetic fields with arcs seems to be of continuing interest for arc applications as well as for basic research. If, for example, a transverse magnetic field is applied to an arc which may move freely, one finds that the arc including the electrode attachments move in a direction mutually perpendicular to the current flow and the magnetic field. Arcs at pressures $p > 100$ kPa move in the direction given by the magnetic body force $\mathbf{j} \times \mathbf{B}$ (j is the current density in the arc and B the applied magnetic field). At reduced pressures, however, some arcs move in the reverse direction (retrograde motion). Retrograde motion which appears to be associated with the cathode region has been one of the most puzzling problems in arc physics. Although many theories have been advanced and much experimental work has been done to explain this phenomenon [see Mailänder (1973) for references], there is still no generally accepted explanation. It seems that no general explanation can be expected as long as the emission mechanism and other processes in the cathode region cannot be clearly defined because there are strong indications that retrograde motion is intimately linked with processes in the cathode region.

(b) *The Anode Region* There are a number of similarities and features common to both the cathode and anode region in an arc, which have been already mentioned in the first part of this paragraph. In contrast to the cathode region, the anode region plays a more passive role that is reflected in a comparatively small number of investigations and available data.

The anode region, as any other part of an electrical discharge, is governed by the conservation equations including the current equation. Unfortunately, any attempt to solve these equations for the anode region faces three major problems. First, the conventional conservation equations apply only as long as the continuum approach is valid. Since the anode fall spacing is in the order of one mean free path length of the electrons, the continuum approach is no longer valid for that part of the anode region. Secondly, the application of the conservation equations requires that the plasma is in LTE or at least that its thermodynamic state is known. There is serious doubt that LTE exists for the entire anode region. Close to and in the anode fall region deviations from LTE are anticipated due to differences in electron and heavy particle temperatures and due to deviations from chemical composition equilibrium. Finally, the specification of realistic boundary conditions faces similar problems as in the cathode region.

The principal task of the anode fall, namely to provide an electrical connection between the high temperature plasma of the arc and the low temperature anode, embraces several effects imposed by the conservation equations. Conservation of energy requires, for example, that field strength and current density adjust themselves so that the total net energy losses suffered by a volume element are compensated for by the Ohmic heating in the same volume element. Because of the steep gradients of the plasma parameters in front of and normal to the anode surface, losses in the axial direction may be substantially larger than those in the radial direction.

Conservation of charge carriers demands ion production in the anode fall which, however, amounts to only a small fraction (approximately 1%) of the total current. At the anode surface the current is exclusively carried be electrons (ion emitting anodes shall be excluded) that give rise to a net negative space charge. This space charge tapers off with increasing distance normal to the anode due to ion production mentioned before, until the unperturbed state in the plasma column is reached. The potential drop in the anode fall zone is a consequence of the net space charge adjacent to the anode surface (Poisson equation). The specific tasks of the anode fall may be summarized as follows:

(1) Production of ions in order to maintain the ion flux passing through any cross section of the plasma column towards the cathode. Since the ion mobility is very small compared to the electron mobility, the required ion production is rather small.

(2) The directed velocity of the ions due to the high field strength in the anode fall has to be transformed into random motion as the plasma column is approached to match the boundary conditions at the interface between the plasma column and anode fall regions.

(3) The opposite is true for the electrons, which receive a large directed velocity component as they approach the anode. In this way, the anode fall provides the necessary electron collection.

Charge carrier generation in the anode fall zone may occur by two basically different ionization mechanisms, namely *field ionization* (F-ionization) and *thermal ionization* (T-ionization). Details of these processes are described by Bez and Höcker (1954a,b, 1955a,b, 1956a,b) and by Ecker (1961).

Field ionization seems to play an important role in low intensity arcs where anode falls on the order of the ionization potential of the working fluid have been observed. For the high intensity arcs considered in this survey, the anode falls are substantially lower (a few volts; in principle they may even be negative). For such arcs, thermal ionization is the governing ionization mechanism. Electrons in the high energy tail of the Maxwellian distribution are responsible for ionization.

Direct (based on probe measurements) and indirect (based on anode energy balance) measurements of anode falls have been reported. The errors involved in these measurements are usually large, sometimes exceeding 100%. Busz-Peuckert and Finkelnburg (1955, 1956) found that the anode fall in free-burning nitrogen arcs can be expressed by a linear relationship

$$V_a = 22 - 0.1I \qquad \text{for} \quad 50 < I < 200 \quad \text{A}. \tag{40}$$

For arcs in argon atmosphere the situation is more complex because the electrode gap enters as an additional parameter. Measured anode falls in argon for arc currents from 50 to 500 A range from 7 to less than 1 V. Measurements of the anode fall in a coaxial arc configuration resembling an MPD (Magneto-Plasma-Dynamic) arc reveal average anode falls of $1.55\,^{-0.95}_{+0.75}$ V using a direct method and $2.07\,^{-1.7}_{+1.6}$ V derived from an energy balance of the anode (Bose and Pfender, 1969). These anode falls obtained in argon at low pressures (0.01 to 5 kPa) seem to be lower than those reported for atmospheric pressure arcs with exception of high-pressure argon arcs operated with extremely small electrode gaps of 0.1 cm (Busz-Peuckert and Finkelnburg, 1955). Schoeck and Maisenhälder (1966) show that the anode fall of argon arcs depends only slightly on the pressure for pressures $p > 100$ kPa. From anode melting in a 10 A atmospheric pressure argon arc, Sugawara (1967) derived anode falls of 2.4 V. He suggested that the high anode surface temperature may reduce the anode fall. For traveling rail arcs up to 600 A, Kesaev (1965) estimates anode falls in the range from 4 to 7 V in argon and from 2.5 to 10 V in air for $p \geq 100$ kPa.

The attachment of the arc at the anode surface may be diffuse as well as severely constricted (spot) and it is still not entirely clear under which conditions constriction will occur. Chemical reactions on the anode surface as, for example, encountered in arcs operated in atmospheric air or in other oxidizing fluids, seem to favor a constricted anode root that, at the same time, may travel more or less randomly over the anode surface with appreciable velocities. Anode evaporation is another mechanism which leads, in general, to spot formation. In high current arcs with relatively small electrode separation (a few centimeters) the diffuse anode attachment is directly associated with the cathode jet. The well-known bell shape of a free-burning high intensity arc is a typical example. The intense cathode jet impinging on the anode surface pushes hot plasma against the anode, eliminating the need for ionization in the anode fall zone. By increasing the electrode gap under otherwise identical conditions the influence of the cathode jet at the anode is diminished and, finally, at sufficiently large gaps the arc forms a single or several spots at the anode surface. Further evidence that anode spot formation and cathode jet are intimately related is illustrated in Fig. 27. In this configuration the axis of the cathode is parallel to the surface of a plane anode

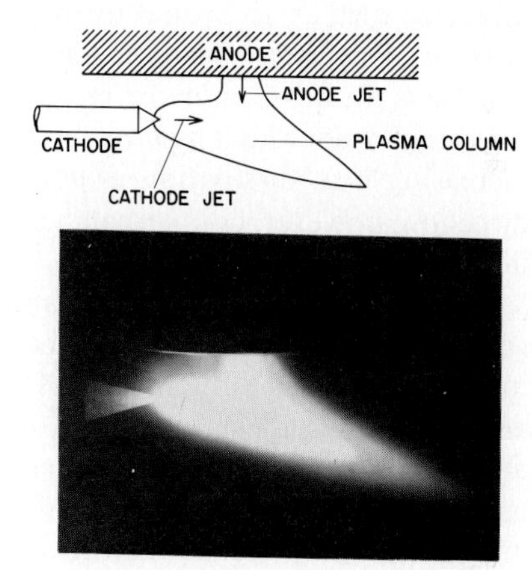

Fig. 27. Interaction of cathode and anode jet in an arc.

so that the cathode jet does not impinge on the anode. Although the photograph reveals a certain moderate constriction of the anode attachment, the deflected cathode jet provides a better indication that there is an appreciable constriction of the anode attachment. Any constriction of the current path leads to the previously described pumping action which results in this case in an anode jet which causes the observed deflection of the cathode jet from the anode surface. The relative strength of the two jets determines the angle of deflection.

These observations suggest that the anode attachment is governed by the thermal conditions at and adjacent to the anode surface. Any effect which has a favorable influence on the energy balance in the anode region, in the sense that internal heat generation by the arc may be decreased, seems to favor a diffuse arc attachment. In a suitable arc configuration the cathode jet is able to provide a continuous flow of hot plasma into the anode region reducing in this way the necessity of heat generation by the arc itself.

It is well known from the behavior of wall-stabilized cylindrical arc columns (which will be discussed in Section II.E.3) that these columns constrict increasingly as the radial heat losses increase. Simultaneously, the heat dissipation in the arc column increases or, at a given arc current, the field strength in the arc column rises. The described effect is analogous to the situation in the anode region if the influence of the cathode jet on this region is strongly reduced or entirely eliminated. The relatively low temperature in the vicinity of the anode induces, as a primary effect, a certain constriction which, however, is always accompanied by the already mentioned pumping effect. The cold gas adjacent to the anode surface is accelerated toward the center of the arc and to a certain degree, ingested into the arc, reducing its diameter further according to this additional heat removal mechanism. A stronger constriction leads to higher velocities of the cold gas which in turn constricts the arc even more, etc. At first glance there seems to be no bound for this process which may be termed a "flow induced thermal pinch." The increasing heat dissipation in conjunction with heat conduction in radial direction, however, counterbalances the constriction due to the induced gas flow establishing a steady-state situation.

Based on the described mechanism a theoretical model for the arc constriction in the anode region has been developed with a number of simplifying assumptions (Chou and Pfender, 1973; Pfender and Schafer, 1975). As an example, Fig. 28 shows the current density distribution for three different argon arcs. According to the arc constriction in front of the anode, the current densities j_B at the anode end of the transition region (contraction region) are substantially higher than in the arc column. This effect is even more pronounced for nitrogen and hydrogen which reveal maximum current densities on the order of 10^4 A/cm^2. At the same time, the field strength and the plasma temperature also increase in the contraction region. There is evidence that the steep gradients of the plasma parameters in this transition zone cause deviations from LTE.

Since the arc attachment at the anode may be diffuse or sharply constricted, the corresponding current densities range from 10^2 to 10^5 A/cm^2 (Sommerville, 1959; Ludwig, 1968; Shih, 1972; Beaudet and Drouet, 1974). These results appear to be consistent with anode heat flux measurements (Cobine and Burger, 1955; Shih and Dethlefsen, 1971; Paulson and Pfender, 1973; Smith and Pfender, 1976). Maximum anode heat fluxes have been found to be in the range from 10^5 to 10^6 W/cm^2. The correlation between current densities and heat fluxes follows from an anode energy balance. Local heat fluxes to the anode may be expressed by (Eckert and Pfender, 1967)

$$q_a = q_{con} + q_r + j_e[\tfrac{5}{2}(kT_e/e) + V_a + \phi_a]. \tag{41}$$

In this equation q_{con} stands for local heat fluxes by conduction and convection, q_r represents the radiative heat flux from the plasma, and the last term describes the energy transferred by the electrons to the anode. T_e is the temperature of the electrons entering the anode fall zone, V_a the anode fall, ϕ_a the work function of the anode material, and j_e the electron current density at the anode surface. In a steady-state situation the energy gain according to Eq. (41) must be balanced by losses which may consist of conduction and convection to the surrounding gas, radiation from a single or several hot anode spots, and energy losses due to ablation and vaporization of anode material.

In most cases of practical importance, heat transfer to the anode is governed by the electron flow [last term in Eq. (41)], i.e.,

$$q_a \simeq j_e[\tfrac{5}{2}(kT_e/e) + V_a + \phi_a] \tag{42}$$

and

$$q_a/j_e = U_I \simeq (\tfrac{5}{2}(kT_e/e) + V_a + \phi_a). \tag{43}$$

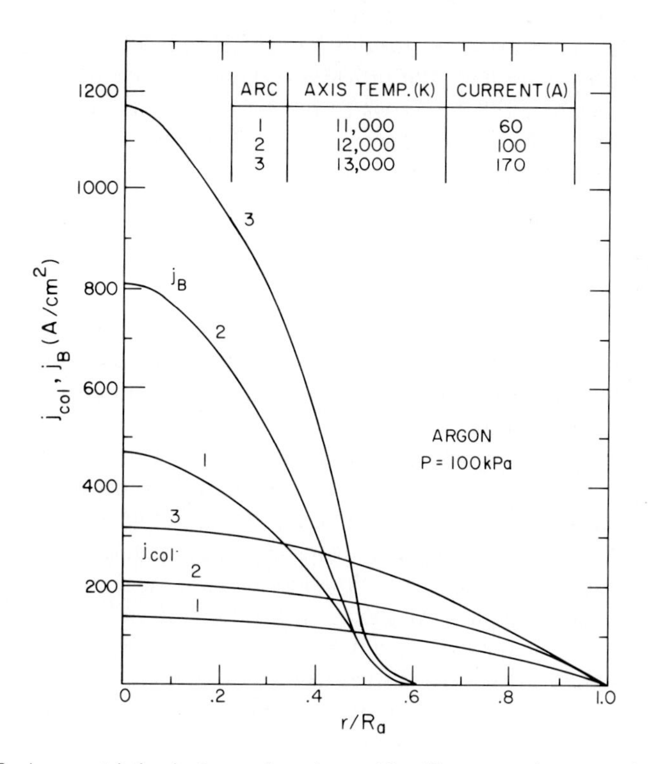

Fig. 28. Arc constriction in the anode region and its effect on anode current densities.

For known values of U_{I} the anode heat flux follows from measured current densities. Experiments in argon, for example, over a wide range of arc parameters show that U_{I} is in the range from 6.2 to 10.4 V with an average around 8 V (Paulson and Pfender, 1973).

A number of peculiar effects in the anode region associated with evaporation of anode material, in particular for carbon anodes, have been described by Finkelnburg and Maecker (1956). Macroscopic flow components in the anode region which may be caused by electromagnetically induced plasma jets, by evaporation of anode material, by chemical reactions on the anode surface, or by a flow through the anode itself (transpiration cooling), may have a significant effect on the space-charge distribution in the anode region, and therefore, on the anode fall. If the previously mentioned flow effects are current dependent, then the anode fall will also become a function of the arc current (anomalous anode fall). Finkelnburg (1948) studied this effect on an evaporating carbon anode in a high-current arc. The low anode falls in high-intensity arcs with relatively small electrode separation seem to be related to the action of the cathode jet impinging on the anode surface as previously mentioned. Since the cathode jet velocity is an increasing function of the arc current, a reduction of the anode fall with increasing current is to be expected. Besides the previously mentioned reduction of the ion flux from the anode fall zone toward the arc column, the dimension of the anode region is reduced by the impinging cathode jet.

D. Arc Characteristics and Electrical Stability

The electric arc, consisting of the arc column and the electrode regions, may in principle be described by solutions of the conservation equations, regardless of the chosen arc configuration and the influence of external effects (flow fields, magnetic fields, etc.). Solutions of the conservation equations for the entire arc would provide useful information on the arc behavior, as for example, the axis temperature in the arc as a function of the power input, or the variation of the arc voltage with current keeping other parameters constant (pressure, electrode gap, electrode material, working fluid, etc.). Unfortunately, this approach is hampered by a number of presently insurmountable problems. As shown in Section II.C solutions of the conservation equations for the arc column alone are already very cumbersome and, at present, limited to the simplest conceivable configuration with reasonably well-defined boundary conditions. The problems associated with the electrode regions discussed in the previous paragraph are even more complex. Although many effects in the electrode regions are qualitatively understood, the state-of-the-art is still far remote from a quantitative description, for example, of the electrode falls.

In spite of this drawback, our knowledge of the behavior of arcs has substantially increased over the past decades by a fruitful combination of mainly qualitative predictions with meaningful experimentation.

One of the most common and also most useful representations of the overall arc behavior is the current–voltage characteristics or, in modified form, the current–field strength characteristics. The latter, of course, apply only for the arc column.

1. Current–Voltage Characteristics

In this paragraph only characteristics of steady (dc) arcs will be considered. For ac and transient characteristics the reader may consult the literature mentioned in the introduction to this chapter.

The measurement of current–voltage characteristics is a straightforward procedure as long as the arc is stable. The characteristics consist of pairs of current and voltage readings which, by definition, represent steady values, although it may in some cases, take rather long to establish a steady-state situation.

There is a wealth of information on measured arc characteristics (low-intensity arcs) in the older literature with numerous attempts to establish empirical relations of the form $V_{\text{arc}} = f(I)$. The relation developed by Ayrton (1902)

$$V_{\text{arc}} = a + bL + [(c + dL)/I] \tag{44}$$

applies only to arcs with falling characteristics since the current appears in the denominator. L represents the arc gap and the constants a, b, c, and d must be determined from the experiment. Many modifications of the Ayrton equation have been proposed for various types of arcs. Without the possibility of attaching physical significance to the various constants, such equations are of little value.

The overall arc voltage is, according to Fig. 24, given by

$$V_{\text{arc}} = V_{\text{c}}' + V_{\text{a}}' + \int_{d_c}^{L-d_a'} E\,dx, \tag{45}$$

where E represents the field strength in the arc column. Since d_{c}' and $d_{\text{a}}' \ll L$, the integration may be taken from $x = 0$ to $x = L$ without introducing a significant error. For long arcs the contribution of the last term dominates whereas for short arcs the current-voltage characteristic reflects essentially the current dependence of $V_{\text{c}}' + V_{\text{a}}'$.

The characteristics of low-current arcs ($I < 50$ A) are usually falling provided that the arc can freely expand with increasing current and that there is no severe influence on the arc by evaporation of electrode material. This applies, for example, to free-burning arcs in air or other gases as long as the arc remains relatively short ($L \leq 20$ cm). Due to free convection effects, the shape of the column of long arcs changes continuously, accompanied by severe voltage fluctuations. Stabilization of the arc in this situation becomes a necessity. Artificial stabilization of the arc column, which will be discussed in the next section, changes the energy balance of the arc drastically so that the resulting characteristic may be rising.

In contrast to low current arcs, high current or high-intensity arcs frequently show rather flat or even rising characteristics. The falling trend of a characteristic is a result of the increasing arc conductance with increasing current caused either by an increase of the electrical conductivity (temperature), or the arc diameter, or by both. In high intensity arcs this general trend may be overbalanced by disproportionally high losses from the arc to which the arc responds with an increase of the field strength.

The foregoing discussion illustrates that no general prediction about the characteristics of arcs can be made because such external conditions as stabilizing walls, magnetic or gas dynamic fields etc. may reverse trends.

Eberhart and Seban (1966) found, for example, that the characteristic of a free-burning high intensity argon arc with a water-cooled anode can be represented by

$$V_{\text{arc}} = 4.3 I^{0.25} L^{0.3} \qquad (I\,[\text{A}],\, L\,[\text{cm}]) \tag{46}$$

for the parameter range $200 \leq I \leq 2300$ A and $0.5 \leq L \leq 3.15$ cm. For this arc the characteristic shows a slightly rising trend whereas the characteristic of a wall-stabilized arc also in argon, shown in Fig. 29, (Maecker, 1960a) indicates at low currents the typical falling trend of low current arcs,

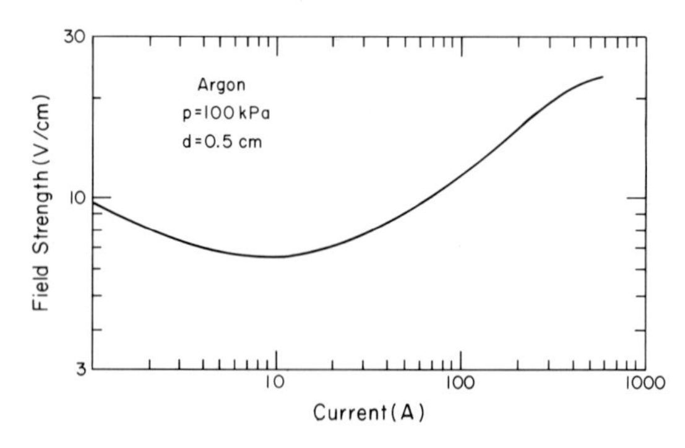

Fig. 29. Measured characteristic of a wall-stabilized argon arc (Maecker, 1960a).

changing to a positive slope at higher currents. The falling trend of a characteristic may continue to relatively high currents in gas vortex-stabilized arcs as shown in Fig. 30 (Pfender, 1963b). The importance of pressure variations on the characteristic is demonstrated in Fig. 31 for a wall-stabilized arc in argon (Kopainsky, 1971b).

Operation of an arc with falling characteristic requires electrical stabilization which will be discussed in the following paragraph.

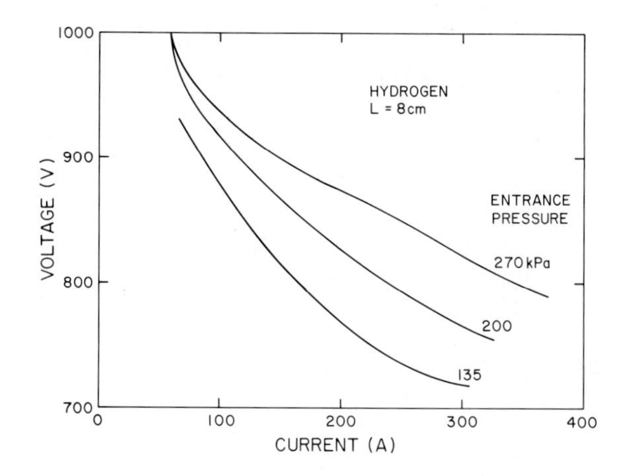

Fig. 30. Measured characteristics of a vortex-stabilized hydrogen arc.

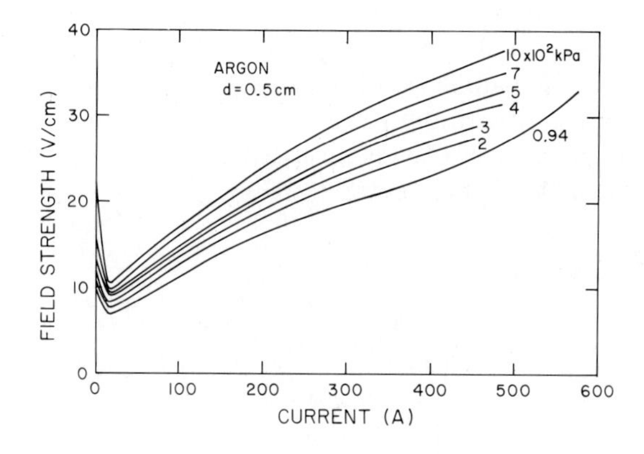

Fig. 31. The influence of pressure on the characteristics of a wall-stabilized argon arc (Kopainsky, 1971b).

2. Electrical Stability

Arcs with rising characteristics do not require any precautions in the electric circuit, in contrast to arcs with falling characteristics. For the following discussion only arcs with falling characteristics will be considered. Figure 32b shows a typical arc characteristic with load line according to the circuit diagram in Fig. 32a. The load line intercepts the characteristic in points A and B but only point A is an electrically stable point of operation. This fact is expressed by the Kaufman stability criterion, which states that for a stable point of operation

$$\frac{dv}{di} + R > 0 \qquad \text{or} \qquad R > \left|\frac{dv}{di}\right|. \tag{47}$$

According to this inequality the load line must intercept the characteristic from above for producing a stable point of operation. In a more detailed electrical stability analysis the arc inductance and capacitance must be taken into account. A calculation based on small disturbances from the equilibrium state leads to two conditions for arc stability

$$\frac{dv}{di} + R > 0 \qquad \text{and} \qquad \frac{1}{L}\frac{dv}{di} + \frac{1}{RC} > 0. \tag{48}$$

The first condition is identical with the Kaufman criterion whereas the second criterion establishes an upper limit for the load resistance

$$R < \frac{L}{C|dv/di|}, \tag{49}$$

where L represents the arc inductance which is assumed to be in series with an induction-free arc and C the capacitance of the arc configuration which is assumed to be in parallel with a capacity-free arc configuration. The second criterion is not critical for high current arcs because L is an increasing function of the current, C is relatively small and $|dv/di|$ does not assume extremely large values in high current arcs.

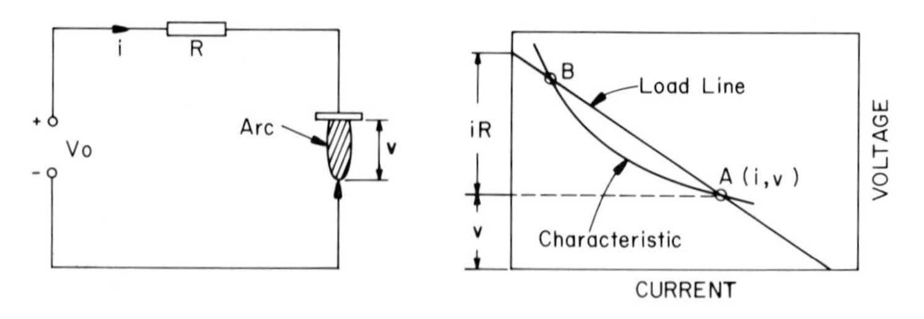

Fig. 32. Electric stability of an arc with falling characteristic.

E. Classification of Arcs

1. Thermal Arcs

The thermal arc is defined as a discharge mode in which the thermodynamic state of the arc column approaches LTE. As outlinèd in Section II.C, LTE is favored by high pressure and temperature levels which bring the electron density close to its maximum in the arc. The latter condition, in turn, is directly associated with the energy balance of the arc column which has been also discussed in Section II.C. In general, high temperatures in the arc column will prevail if the energy dissipation per unit length of the arc IE is high (I = arc current, E = electric field strength), or if the losses are kept very small. Without taking special precautions for reducing energy losses from the arc column, the minimum value of IE must be in the neighborhood of 100 kW/m, a value which may be achieved for example in a typical arc of $I = 100$ A and $E = 1000$ V/m. Such arcs are known as high intensity arcs ($I > 50$ A). For all practical purposes one may equate thermal arcs with high intensity arcs, and both terms are used in the literature. Depending on the arc condition, temperatures in high intensity arcs range from 10^4 to 10^5 °K.

If IE were the only criterion for achieving a certain arc temperature, one might argue that, for example, an arc of $I = 10$ A and $E = 10{,}000$ V/m should yield the same temperature as an arc of $I = 100$ A and $E = 1000$ V/m operated in the same gas at the same pressure. This, of course, is only true if the energy balance for both cases remains the same.

At this point it is interesting to notice that an increase of the heat transfer in the arc fringes (arc constriction) will increase the electric field strength and, therefore, the energy dissipation for a fixed arc current. As previously argued, a higher energy dissipation will result in higher temperatures in the arc core. The net effect of cooling an arc by heat conduction and/or convection is, therefore, an increase of its axis temperature, an effect which has been utilized extensively in the development of arc gas heaters.

It is obvious that there must be a transition region between the extremely high temperatures of the arc column and the relatively low electrode temperatures. In these transition regions, known as the cathode and anode regions (see Fig. 2), there will be steep temperature and potential gradients which clearly distinguish these regions from the arc column with its relatively small potential gradient. A more detailed discussion of the electrode regions has been given in Section II.C.

A brief comment, however, should be made about the charge carrier balance in the electrode regions, in particular in the cathode region. In general, the current in an arc—as in any other gaseous discharge—is carried by electrons and positive ions (negative ions are only of importance in special types of arcs). Under the influence of the applied electric field, electrons are traveling towards the anode and the positive ions are drifting in the opposite direction. Because of the huge mass ratio of these particles and the associated difference in their mobilities, the ion contribution to the total current is negligible in the arc column. In the anode region as well, the current is essentially carried by the electrons, assuming that the anode does not emit ions. In the cathode region, the situation may be entirely different. Since it is implied in this paragraph that the arc is self-sustaining, i.e., no external energy source is required for sustaining a steady arc, the arc must take care of its own electron supply at the cathode. In the case of refractory cathodes, a certain percentage of the current flow in the cathode region is due to ions which impinge on the cathode surface giving off their charge. At the same time the ions release kinetic and potential energy, heating the cathode to temperature levels sufficient for thermionic emission. This is one of the possible electron emission mechanisms at the cathode. A more complete description of cathode effects is given in Section II.C.

2. Nonthermal Arcs

In view of the applications considered in this chapter, nonthermal arcs are of minor importance. Therefore, this paragraph will be rather short.

In contrast to thermal arcs, nonthermal arcs are characterized by strong deviations from LTE. These deviations may manifest themselves in different ways depending on which condition or conditions for the existence of LTE are violated. A well-known violation of LTE conditions appears in deviations from kinetic equilibrium, i.e., the electron temperature T_e may be substantially higher than the heavy particle temperature T_g. This deviation from kinetic equilibrium is mainly due to the poor collisional coupling between electrons and heavy particles at reduced pressures. Since the collision frequency is proportional to the pressure, the difference between T_e and T_g should gradually vanish as the pressure increases. This effect is schematically shown in Fig. 4 indicating that low pressure arcs ($p < 10$ kPa) deviate, in general, from kinetic equilibrium. Another frequently observed deviation from LTE in arcs is associated with the ionization equilibrium. The low intensity carbon arc ($I \leq 20$ A) is a typical example. In such arcs the plasma composition can not be derived from the Saha equation because the electron density is relatively low. In spite of the fact that ionization and other microprocesses in the arc plasma are thermal processes, this type of arc does not qualify as a thermal arc.

From these considerations it follows that low pressure arcs in general and high pressure arcs at low temperature levels (low intensity arcs) must be classified as nonthermal arcs.

Before concluding this paragraph it should be pointed out that thermal arcs as well as nonthermal arcs may be either self-sustaining or nonself-sustaining. The nonself-sustaining arc mode seems to lack practical importance as far as thermal arcs are concerned whereas nonthermal (low pressure) arcs operating in the nonself-sustaining mode have found interesting applications. Thyratrons and gas-filled rectifier tubes (which to a large degree are now replaced by solid state devices) employ externally heated cathodes similar to conventional electron tubes. For lowering the work function, which is crucial for electron emission, oxide cathodes have been used for low voltage and thoriated tungsten cathodes for high voltage applications.

The low voltage arc operating at overall voltages below the first ionization potential of the working fluid represents a special type of nonself-sustaining arc. By proper choice of the work function of cathode and anode, the arc voltage may even become negative resulting in a thermionic converter ("plasma thermocouple").

3. Classification of Arcs According to their Stabilization

In view of arc applications, this particular classification appears to be useful. There is a direct link between the method of stabilizing the arc column and the options available for the design of arc gas heaters.

Most electric arcs require for their stable operation some kind of stabilizing mechanism which must be either provided externally or which may be produced by the arc itself. The term "stabilization" as applied in this paragraph refers to a particular mechanism which keeps the arc column in a given, stable position, i.e., any accidental excursion of the arc from its equilibrium position causes an interaction with the stabilizing mechanism such that the arc column is forced to return to its equilibrium position. This stable position is not necessarily a stationary one; the arc may, for example, rotate or move along rail electrodes with a certain velocity. Stabilization implies in this situation that the arc column can only move in a well defined pattern, controlled by the stabilizing mechanism.

(a) *Free-Burning Arcs* As the name implies no external stabilizing mechanism is imposed on the arc in this case; but this does not exclude the arc generating its own stabilizing mechanism.

Let us assume that a low intensity arc is established in atmospheric air between a pair of horizontal carbon rods. Due to free convection effects, the arc column bends upward forming an "arc" (Rother, 1957). Historically, the arc derives its name from this peculiar shape which is referred to as convection-dominated or convection-stabilized arc. If gravity is eliminated (von Engel, 1965) the arc column becomes rotationally symmetric and wider and, at the same time, the arc voltage drop is somewhat reduced as a consequence of the more favorable energy balance. Free convection effects are not quite as obvious if the arc is operated vertically. The induced free convection flow velocities of laboratory-scale arcs are in the order of 1 m/sec. Thus, free convection effects can be easily eliminated by additional flow fields, a situation which arises, for example, in high intensity arcs. Although high intensity arcs may be operated in the free-burning arc mode, they are frequently classified as self-stabilized arcs [see Part (e) of this section] if the induced gas flow, due to the interaction of the self-magnetic field and the arc current, is the dominating stabilizing mechanism. Therefore, free-burning high intensity arcs to which the described conditions apply will be discussed in Part (e).

Arcs operated at extremely high currents (up to 100 kA) known as ultrahigh current arcs, should also be mentioned in this category. Although most experiments in this current range utilize pulsed discharges, the relatively long duration ($\simeq 10$ msec) of the discharge justifies classifying them as arcs. There is considerable interest in such arcs in connection with melting and steelmaking, utilization in chemical arc furnaces and high power switchgear. Visual observations of ultrahigh current arcs in arc furnaces reveal a rather complex picture of large, grossly turbulent plasma volumes, vapor jets emanating from the electrodes, and parallel current paths with multiple, highly mobile electrode spots. In this situation there is no evidence for any dominating stabilizing mechanism. Induced gas flows and vapor jets exist simultaneously, interacting with each other in a complicated way. Depending on the polarity of the arc and the electrode materials, stable vapor jets have been observed which are able to stabilize the arc column. Thus, the generation of vapor jets by the arc represents another possible mechanism for self-stabilization of arcs.

In a recent survey Edels (1973) gives a comprehensive description of the characteristic features and properties of ultrahigh current arcs, including 120 pertinent references. Continuing studies in this area are mainly concerned with radiation properties (Strachan and Blackburn, 1973) and flow fields (Strachan and Barrault, 1973) in such arcs.

(b) *Wall-Stabilized Arcs* The principle of wall-stabilization of arcs has been known for more than 70 years, having been introduced in connection with arc lamps. A long arc enclosed in a narrow tube with circular cross section will assume a rotationally symmetric, coaxial position within the tube. Any accidental excursion of the arc column towards the wall will be compensated by increasing heat conduction to the wall which reduces the temperature and, therefore, the electrical conductivity at this location. In short, the arc will be forced to return to its equilibrium position. In this situation, increased thermal conduction and the associated secondary effects represents the stabilizing mechanism.

Arcs enclosed in glass or quartz tubes are widely used as light sources. A comprehensive review on arc light sources has been published by Waymouth (1971).

In order to cope with the extremely high wall heat fluxes experienced with high intensity arcs enclosed in small diameter tubes, Maecker (1956) introduced metal tubes as arc vessel, consisting of a stack of insulated, water-cooled disks (usually Cu). This arrangement is known as the wall-stabilized, cascaded arc which has been extensively used as a basic research tool (Maecker, 1959a, 1960a,b, 1961; Schmitz and Uhlenbusch, 1960; Richter, 1961; Shumaker, 1961; Emmons and Land, 1962; Schmitz *et al.*, 1963; Edels and Fenlon, 1965; Patt and Schmitz, 1965a,b; Pfender *et al.*, 1965; Christmann and Hertz, 1965; Kimblin and Edels, 1966; Maecker and Steinberger, 1967; Motschmann, 1967a,b; Hermann, 1968; Seeger, 1968; Morris, 1969; Sauter, 1969; Uhlenbusch and Gieres, 1969; Kruger, 1970; Seeger, 1970; Uhlenbusch *et al.*, 1970a,b; Bauder and Schreiber, 1971; Giannaris and Incropera, 1971; Kopainsky, 1971b; Hackmann *et al.*, 1972; Incropera and Murrer, 1972; Wiese *et al.*, 1972; Cremers *et al.*, 1973; Ernst *et al.*, 1973a,b; Garz, 1973; Hermann, 1973; Incropera, 1973; Srivastava and Weissler, 1973).

The same principle found widespread application in the design of arc gas heaters which will be discussed later on.

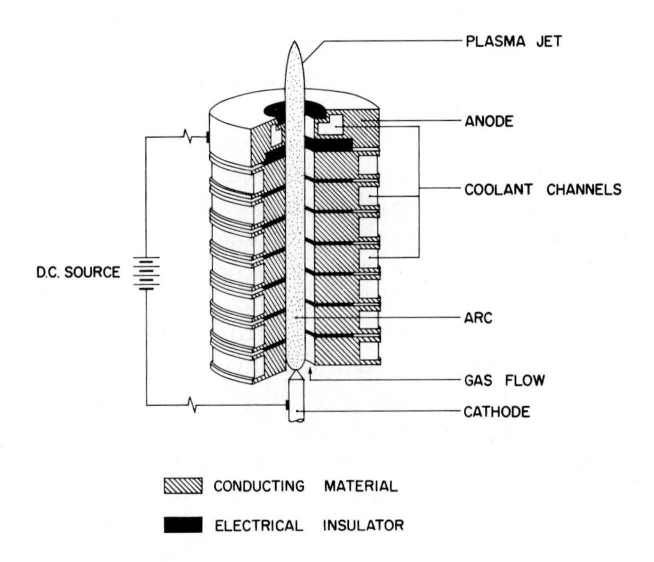

Fig. 33. Cut-away view of a wall-stabilized, cascaded arc.

Figure 33 shows a cutaway view of a typical wall-stabilized arc. Due to the much higher electrical conductivity of metals compared with that of the arc column, segmentation of the tube enclosing the arc is necessary because a continuous metal tube would cause a double arc (arcing from the cathode to the metal tube and from the metal tube to the anode) seeking the path of least resistance. If E is the field strength in the arc column and d the thickness (in field direction) of an individual segment, the following condition has to be met for avoiding double arcing:

$$\int_0^d E\,dx < V_c + V_a.$$

The minimum voltage required for establishing and maintaining an arc is the sum of cathode fall V_c and anode fall V_a. A reduction of the inside diameter of the segments (arc constriction) keeping the other arc parameters the same leads to a marked increase of the field strength according to the enhanced energy losses by heat conduction. In this situation the maximum thickness of a segment may be only 1.2–2 mm (Maecker and Steinberger, 1967; Uhlenbusch and Gieres, 1969).

It is essential for retaining the wall-stabilizing effect that the diameter of the vessel containing the arc is appreciably smaller than the diameter of a free-burning arc would be, operated under the same conditions. If the vessel diameter is too large, the stabilizing effect due to heat conduction is lost and, at the same time, free convection effects may cause serious distortions of the arc column.

The mechanism of wall-stabilization as such does not require any flow in the constrictor tube, although a certain axial flow may be desirable or, under certain conditions, inevitable. The latter is particularly true if cathode jet phenomena are involved; these have been discussed in Section II.C.2. A small axial flow component in wall-stabilized arcs is frequently desirable for removing impurities in the plasma released by the electrodes or the confining walls. Wall-stabilized arcs which operate with strong superimposed axial flow belong in the category of arc gas heaters and they will be discussed in the corresponding section.

Since arc constrictor tubes with circular cross section are able to produce rotationally symmetric arcs, this configuration is usually preferred, although more complex geometrics are also possible.

The maximum possible temperature or enthalpy attainable in a constricted, wall-stabilized arc is limited by the highest permissible heat flux which the wall is able to withstand. Sophisticated water-cooling arrangements permit wall heat fluxes up to 2×10^5 kW/m^2 (Maecker and Steinberger, 1967).

In order to overcome this limitation a modified wall-stabilized arc configuration has been suggested (Pfender *et al.*, 1969). By replacing the water-cooled wall with a transpiration-cooled constrictor, convective/conductive heat fluxes to the wall may be intercepted by the radially oriented gas flow through the porous wall as indicated schematically in Fig. 34. The tempera-

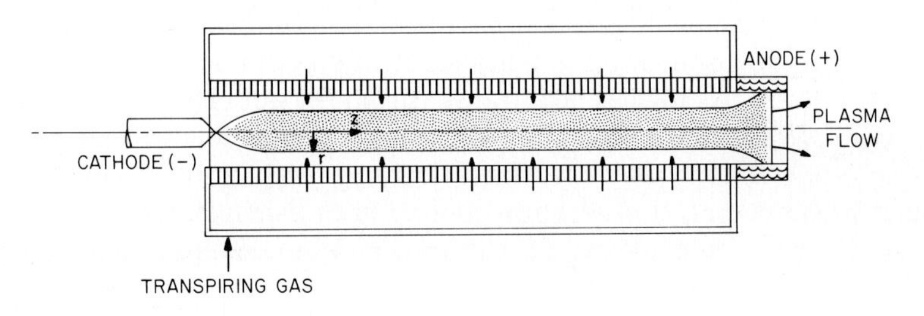

Fig. 34. Schematic of a transpiration-cooled arc.

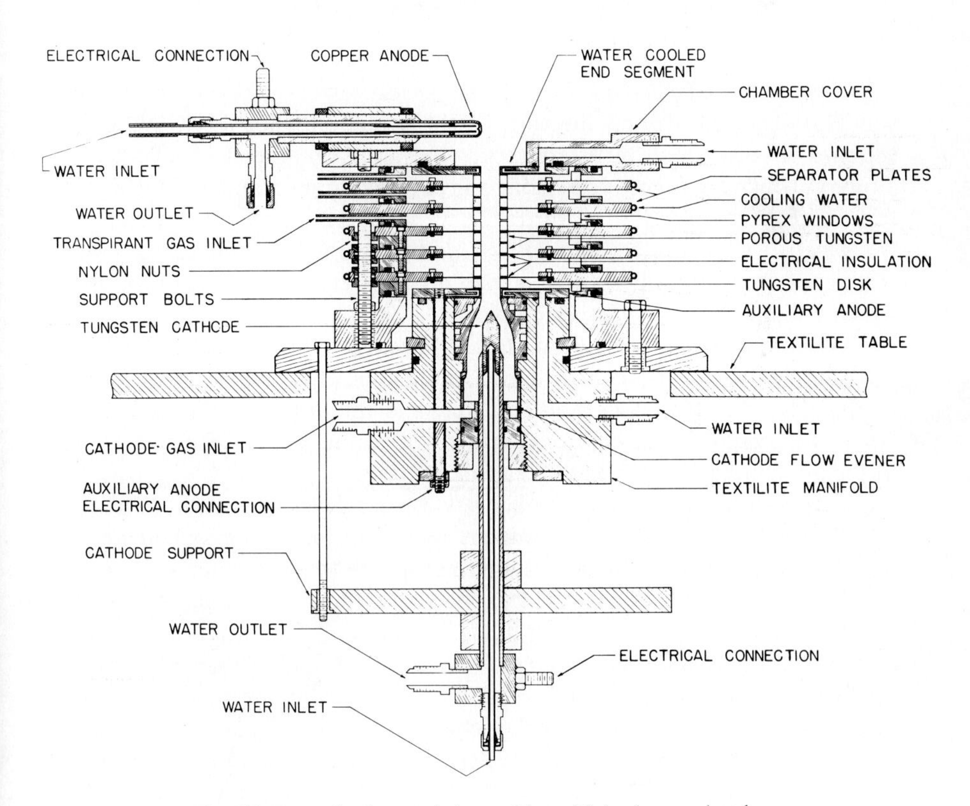

Fig. 35. Transpiration-cooled arc with multiple plenum chambers.

ture gradient at the wall is drastically reduced by this cooling mechanism and, therefore, appreciably higher power inputs per unit length of the arc and correspondingly higher axis temperatures become feasible. Theoretical considerations (Anderson and Eckert, 1967) indicate that axis temperatures up to approximately 60,000°K in hydrogen at atmospheric pressure are possible if limitations due to electrode problems are disregarded. In other gases, as for example argon and nitrogen, the possible peak temperatures are lower owing to the high volumetric radiation of these gases at high temperatures levels. As long as heat transfer from the hot plasma to the wall of the porous constrictor is mainly caused by conduction and convection, transpiration cooling is a very efficient cooling mechanism. If radiation becomes predominant, transpiration cooling loses its effectiveness.

Another problem associated with transpiration cooling is its inherently unstable behavior leading to strong temperature variations along the constrictor wall. A more uniform temperature distribution has been obtained with a segmented plenum chamber as shown in Fig. 35 (Knowles and Heberlein, 1973).

(c) *Vortex-Stabilized Arcs* The principle of vortex-stabilization of arcs was first reported around the turn of the century (Schoenherr, 1909). In the case of vortex- or whirl-stabilization the arc is confined to the center of a tube in which an intense vortex of a gas or liquid is maintained. Centrifugal forces drive the cold fluid towards the walls of the arc chamber which, in this way, is well protected thermally. In addition to the circumferential component of the vortex flow there is also an axial component superimposed which continuously supplies cold fluid.

A well-known example of a vortex-stabilized arc is the so-called Gerdien arc (Gerdien and Lotz, 1922), which is schematically illustrated in Fig. 36. In this case the stabilizing fluid is water, and the arc plasma is generated from water vapor in the core of the vortex. Because of the extreme cooling of the arc fringes, the power dissipation per unit volume and the associated arc temperatures reach much higher values than feasible in wall-stabilized arcs. Burhorn *et al.* (1951) measured arc temperatures in excess of 50,000°K using an arc current of 1450 A, confining the arc to a diameter <2.3 mm.

For actual applications of vortex-stabilized arcs various gases and gas mixtures are utilized as working fluids. In an analytical study by Krichel *et al.* (1968), nitrogen is considered as the working fluid. Figure 37 shows a schematic of a gas vortex-stabilized arc arrangement developed for the generation of fully ionized atmospheric pressure hydrogen plasmas (Pfender, 1963a,b) which may be further heated by high current pulse discharges (Pfender and Bez, 1961). Both electrodes in this arrangement are water cooled. The working fluid enters tangentially at the anode through small orifices. The vortex generated in this way confines the arc to the center of the arc chamber reducing the arc diameter to approximately 2–3 mm. In this particular design the plasma emanates through central holes in the electrodes forming an anode and cathode plasma jet. At sufficiently high mass flow rates the arc cathode root is forced to attach to the rear of the cathode which eliminates contamination of the plasma inside of the arc chamber by in-

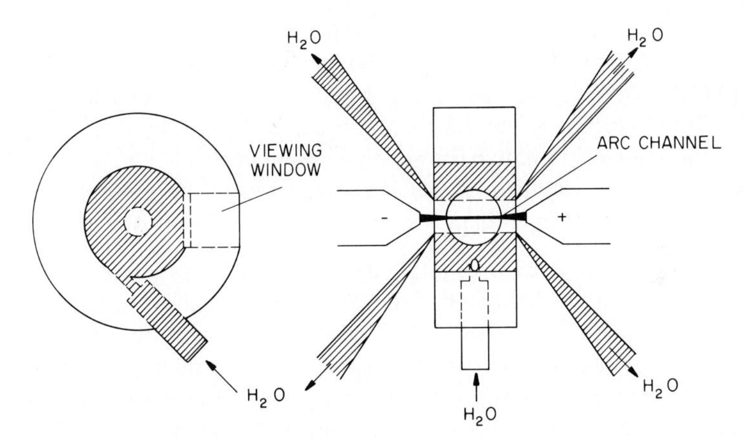

Fig. 36. Schematic of water-vortex-stabilized (Gerdien) arc.

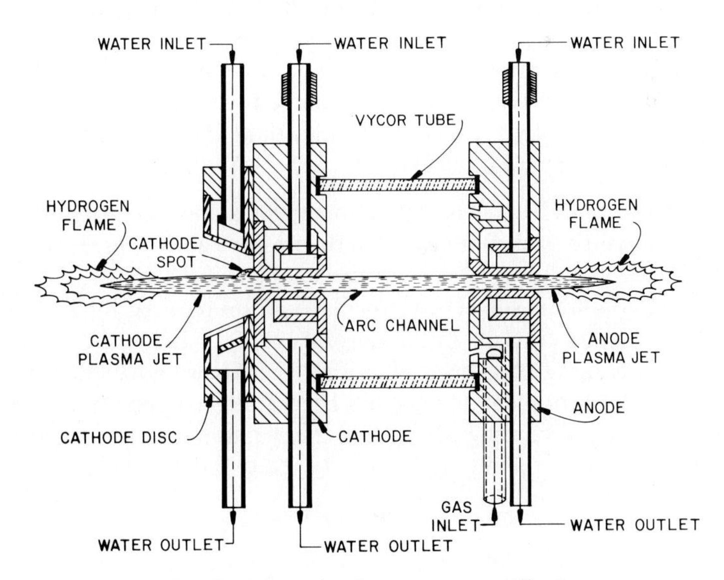

Fig. 37. Schematic of a gas vortex-stabilized arc.

evitable cathode erosion. An insulated water-cooled metal disk attached to the rear of the cathode prevents undesirable excursions of the cathode attachment. The intense convective cooling of the arc fringes due to the vortex flow around the arc enhances the power dissipation per unit length of the arc column, which, in turn, results in high axis temperatures. In the previously mentioned vortex-stabilized hydrogen arc, axis temperatures close to 25,000°K have been reached (Pfender, 1963b).

A modified arrangement for stabilizing an arc by a rotating flow has been reported by Foitzik (1940). In this arrangement the rotating gas flow is generated by rotating the entire discharge vessel ("Wälzbogen"). The vortex flow induced by friction between the wall and the working fluid is strong enough to eliminate natural convection effects.

An interesting experiment has been reported by Polman *et al.* (1973). They operate an arc in a torus geometry stabilized by a toroidal vortex.

(d) *Electrode-Stabilized Arcs* In extremely short arcs (electrode gaps on the order of 1 mm) the behavior of the arc is determined by the vicinity of the electrodes. The arc column as such does not exist any more. The remaining arc consists of the nonuniform electrode regions which, in contrast to a fully developed arc column, reveal strong axial gradients of the plasma properties (Ecker, 1961). As previously mentioned, these regions may be considered as thermal boundary layers which, in this situation, may even partially overlap. The contour of the remaining part of the arc approaches the shape of an ellipse with the electrode roots as focus points (Schmitz, 1952).

The shape of low intensity free-burning arcs is dominated by free convection effects. If free convection is eliminated (zero gravity arcs) the arc behavior is governed by the electrodes and, therefore, such arcs must also be classified as electrode-stabilized arcs.

(e) *Self-Stabilized Arcs* The transition from a low intensity to a high intensity arc, which occurs at atmospheric pressure above currents of 50 A, manifests itself in a drastic change of the stability of the arc column. Below 50 A the arc column is subject to irregular motion induced by free convection effects. For arc currents in the range from 50 to 100 A the column becomes suddenly motionless and stiff with a visually well defined boundary. The cathode jet phenomenon which has been described previously gives rise to this transition. According to Eq. (39) the maximum velocity in the jet is related to total current and current density at the cathode, viz.,

$$v_{max} \sim (Ij)^{1/2}.$$

As soon as this velocity substantially exceeds those induced by free convection effects (in the order of 1 m/sec), the described transition will occur. The current at which this transition takes place depends on the conditions in the cathode region (current density and variation of current density in axial direction). The phenomenon is usually reversible, i.e., by lowering of the current a transition to the free convection dominated low intensity arc occurs (King, 1955, 1956).

It should be emphasized that the arc temperature at which the transition from low to high intensity arc occurs is substantially lower for arcs in molecular gases (for example nitrogen or air). In molecular gases a hot core is formed in the arc as soon as the dissociation temperature in the axis is surpassed which changes the current density distribution and, therefore, influences the jet formation.

(f) *Magnetically Stabilized Arcs* Since an arc is an electrically conducting medium it will interact not only with its own magnetic field as discussed in the previous paragraph, but also with externally applied magnetic fields. Over the past 10 years this interaction attracted increasing interest because of its potential for many arc applications. Magnetically influenced or stabilized arcs are extensively used in the development of arc gas heaters, in circuit breakers, in arc furnaces, in arc welding, in plasma propulsion, etc. According to the existing literature the interaction of arcs with magnetic fields may be divided into the following categories:

(1) Magnetic stabilization of arcs in cross flow.
(2) Magnetically deflected arcs.
(3) Magnetically driven arcs.

The first category refers to arcs exposed to strong cross flows so that the arc column bends in the downstream direction if the electrode roots are fixed. At the same time arc length and arc voltage drop increase and for a sufficiently strong flow the required arc voltage may surpass the available voltage, i.e., the arc extinguishes. In order to stabilize an arc in this situation a magnetic field may be applied so that the drag force exerted on the arc by the flow is balanced by the $\mathbf{j} \times \mathbf{B}$ force. A comprehensive survey on the interaction of arcs with magnetic and flow fields has been published by Roman and Meyers (1966) including 90 references. The same authors (Roman and Myers, 1967) found from an experimental study of a magnetically balanced arc in cross flow that the magnetic field strength required for balancing the arc is proportional to v^2, where v is the gas velocity, i.e., the arc behaves like a solid body as far as aerodynamic drag is concerned. The cross section of the arc assumes the shape of an ellipse with the major axis normal to the flow. Without the balancing $\mathbf{j} \times \mathbf{B}$ force, the major axis of the ellipsoidal cross section of the arc is in flow direction. These findings have been confirmed by a number of other studies including detailed analyses. Many refined experimental studies have been reported including balanced arcs in supersonic flow (Goldstein and Fay, 1967; Benenson and Baker, 1969; Lord, 1969; Winograd and Klein, 1969; Anderson, 1969; Hodnett, 1969; Nicolai and Kuethe, 1969; Bartels and Uhlenbusch, 1970a; Benenson and Baker, 1971; Schrade, 1973; Bose, 1973; Malghan and Benenson, 1973; Grosse-Wilde and Uhlenbusch, 1973).

Ragaller (1974) addressed the question of stability/instability of an arc in a general way and developed a formalism which allows the evaluation of magnetic instabilities in arcs.

The second category deals with magnetically deflected arcs and the secondary effects induced by this deflection. Although the applied magnetic field exerts a strong influence on the behavior of the arcs discussed in this category, the primary stabilizing effect is due to confining walls. In this sense, this type of arc may also be classified as a magnetically influenced, wall-stabilized arc.

Most of the newer work in this area has been reported by Maecker's group (Maecker and Preibisch, 1968; Raeder, 1968; Seeger, 1968; Sauter, 1969; Seeger, 1970; Nathrath, 1970; Rosenbauer, 1971). These papers are primarily concerned with measurements and calculations of the flow field induced by an external, transverse magnetic field in a wall-stabilized arc configuration.

The interaction of curved arcs with their own magnetic field can produce similar effects as observed in magnetically deflected arcs. Wynands *et al.* (1970, 1973) calculated the induced flow field and other properties of a nitrogen arc confined in a torus geometry.

Arcs which derive their shape from the applied magnetic field should also be mentioned in this category. Tiller (1973) and Stäblein (1973) report on experimental and analytical studies of a torus-shaped arc operated between two parallel disks.

The principles of magnetic arc displacement and arc motion have been discussed by Maecker (1971). If the position of the arc is defined in terms of the temperature distribution (location of maximum temperature = arc center) the arc motion may be considered as the vector sum of the mass velocity, produced by external forces or by a pressure gradient, and the relative velocity of the arc with respect to the surrounding gas determined by thermodynamic effects (nonuniform temperature gradients around the temperature maximum). In the case of magnetic arc deflection in a wall-stabilized arc configuration, the arc acts as a pump producing a double-vortex (Fig. 38). The arc, of course, does not follow the flow, whereas an arc between rail electrodes or in a coaxial arc configuration with freely moving arc roots follows the flow. The later arc type falls in the third category which refers to magnetically driven arcs.

Magnetically driven arcs are also classified under magnetically stabilized arcs as previously explained. Such arcs have been extensively studied, from a more basic point of view in connection with the earlier discussed phenomenon of retrograde motion (Mailänder, 1973; Hermoch, 1973). With applications for arc gas heaters in mind, the coaxial, magnetically rotated arc has been of great interest for efficient heating of gases to well-controlled temperature levels in the range from 3×10^3 to 6×10^3°K (Chen and Lawton, 1968; Lawton, 1971; Humphrys and Lawton, 1972; Djakov and Nedelkov, 1974). More details on the performance of such devices as arc gas heaters will be discussed in Section III.

Reduction of electrode erosion by magnetic rotation of an arc is a well established technique (Guile and Hitchcock, 1974). In connection with arc quenching in circuit breakers, magnetically propelled arcs along rail electrodes have been studied for a wide range of parameters (Novak and Fuchs, 1974; Beaudet and Drouet, 1974).

Investigations of arc motion under the influence of oscillating transverse magnetic fields have been reported by Bartels and Uhlenbusch (1970b).

(g) *Classification of Arcs according to the Cathode Emission Mechanism* This paragraph should merely clarify the terms which are sometimes used for classifying arcs. Because of the uncertainty associated with the emission mechanism at arc cathodes, this classification is, in general, of little value. The term thermionic arc (T-emission) implies that the electron emission at the cathode is predominantly thermionic. In the same sense arcs are classified as field emission arcs (F-emission) or arcs with mixed modes of electron emission (TF-emission).

Finally, arcs with thermionically emitting cathode are classified as self-sustaining or nonself-sustaining. In the case of nonself-sustaining arcs the cathode is heated externally from a separate heat source. As mentioned earlier, this type of arc is not of interest in the context of this survey.

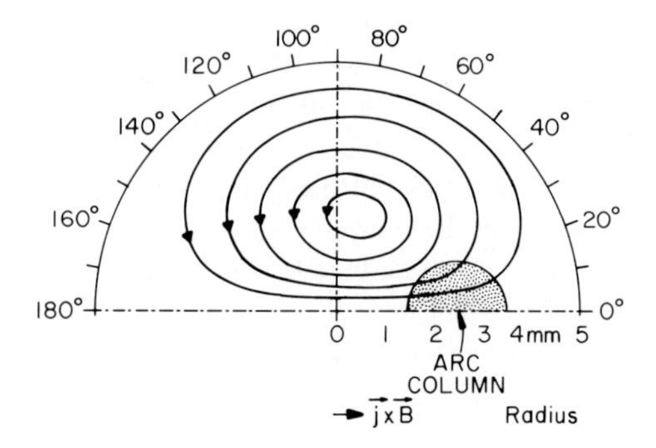

Fig. 38. Flow pattern induced by magnetic deflection of an arc confined in a tube.

III. ARC GAS HEATERS AND PLASMA TORCHES

Since the underlying physical principles of an arc gas heater and of a plasma torch are the same, the term "arc heater" will, for the following considerations, include plasma torches. In an arc gas heater the gas to be heated, which is usually at atmospheric or higher pressure, must be brought into "contact" with the arc. The art of designing an arc heater lies mainly in finding ways of transferring the energy dissipated in the arc in the desired way to the flowing gas within the restraints imposed by physical laws.

The attractiveness of arc heaters for actual and potential applications is due to the ease with which high temperature levels can be produced, far beyond the levels feasible in conventional combustion processes and without the contamination due to combustion products. Although the application of electric arcs for heating gases to high temperature levels has been known for more than 70 years, the development of arc gas heaters is still, to a large degree, an empirical science. With a very few exceptions arc gas heaters have defied a comprehensive analytical treatment. Even in extremely simple, laboratory-type arc configurations, the analytical treatment may be faced with insurmountable problems, as discussed in previous paragraphs. In particular, there are three areas in which basic knowledge is still lacking:

(1) Interaction of arcs with magnetic and/or flow fields.
(2) Effects in the electrode regions and at the electrodes.
(3) Thermodynamic state of the arc plasma (deviations from LTE).

Additional complexities introduced by the "mission oriented" design and secondary effects during operation of actual gas heaters aggravates the situation further. In spite of these problems and the lack of theoretical guidance, experimental ingenuity has produced astonishing results over the past 20 years.

Arc gas heaters have been designed for a wide spectrum of applications including chemical and material processing, nitrogen fixation, coal gasification, and melting and refining of materials. They are successfully employed in extractive metallurgy, for welding, cutting, spraying, surfacing, spherodizing; they are also used in space-related applications, for example, reentry simulation and ablation studies. The latter application provided a strong impetus for research in the field of arc gas heaters, especially during the late fifties and the sixties. The power levels at which arc gas heaters are operated varies from a few kilowatts (typical for certain types of plasma torches) to several megawatts or tens of megawatts. In the case of arc furnaces, power levels in the order of 100 MW are of interest. Chemical processing, for example, frequently requires heating of large gas volumes to modest enthalpy (temperature) levels whereas aerospace applications need extremely high enthalpy levels. It will be shown in the following paragraphs that the first type of arc heater (low enthalpy device) may operate as a highly efficient energy conversion device in contrast to high enthalpy arc heaters which are inherently inefficient.

Thermal arcs, as previously explained, are well suited as the heat source for arc gas heaters although other heat sources as, for example, high frequency plasma generators, have been also proposed. Among the various arc parameters, the plasma temperature and the electron density (or the degree of ionization) may be singled out as those parameters which govern the selection for a specific application. Table II shows a survey of approximate

TABLE II

SURVEY OF ARC TEMPERATURES AND ELECTRON DENSITIES

T[K]			N_E[cm^{-3}]	
100 $\times 10^3$	Magnetically confined arcs ($N_E \approx 10^{15}$ cm^{-3})	10^4 kPa pressure arcs ($T \simeq 12 \times 10^3$ °K)	10^{19}	
50	Gerdien arc (1500 A)			
			10^{18}	
30	Free burning arcs, 500 A (close to cathode) Vortex stabilized arcs		10^{17}	
20	Transpiration cooled arcs Wall stabilized arcs			
			10^{16}	
10	Free burning arcs (main arc body) XE high pressure arc lamps			Thermal arcs
				Nonthermal arcs
	Low intensity arcs in air and in noble gases		10^{15}	
5	Metal vapor arcs (Hg, Ag, Au) Arcs in Na, K			
4			10^{14}	

arc temperatures and electron densities. The numbers in this table refer to the maximum values in the axis of an arc. For arcs which are in LTE or close to LTE the Saha equation provides a straightforward correlation between electron densities and temperatures for a given pressure and a given gas or gas mixture.

Thermal arcs have been discussed in previous paragraphs, except for those listed at the top of this table. Arcs at reduced pressure in hydrogen, confined by strong axial magnetic fields, may reach temperatures up to 10^5°K in the axis (Mahn *et al.*, 1964; Heidrich, 1965; Mahn and Ringler, 1966). For arcs at extremely high pressures (10^4 kPa) the electron density may reach values in the order of 10^{18} or 10^{19} cm^{-3}, depending on the temperature (Bauder and Bartelheimer, 1973).

The continuing interest in arc heater technology and its enormously increasing activities which began in 1955 are reflected by the vast number of papers published during the past 20 years. Developments up to 1960 are summarized by John and Bade (1961) in a comprehensive review, covering arc gas heater applications for chemical synthesis, refractory processing, with special emphasis on reentry simulation and space propulsion. The increasing interest in reentry simulation and space propulsion in the early sixties is documented by a number of technical reports (Eschenbach and Skinner, 1961; Boldman *et al.*, 1962; Cann *et al.*, 1963, 1964; Shepard *et al.*, 1964; Marlotte *et al.*, 1964) and by two AGARD volumes devoted to arc heaters and MHD accelerators (1964). Approximately 20 out of a total of 28 papers in these two volumes are concerned with arc heater developments and related studies. Aerospace related developments in the U.S. continued to dominate the scene in the later part of the sixties (Eschenbach *et al.*, 1965; Watson, 1965; Vorreiter and Shepard, 1965; Shepard *et al.*, 1967, 1968; Marlotte *et al.*, 1968; Beachlèr, 1968; Smith and Folek, 1969; Richter, 1969), leveling off in the early seventies (Harder and Cann, 1970; Richter, 1970; Shepard, 1972; Cann, 1973; Painter, 1974).

The development and application of arc gas heaters and plasma torches in areas which are not aerospace related are difficult to assess because many promising developments in industry are considered as proprietary. The patent literature, however, indicates that there has been increasing activity in this field over the past 15 years. The various designs and claims about the performance of arc gas heaters in the patent literature will not be discussed in this survey.

Most of the available pertinent literature on arc gas heaters and plasma torches in nonaerospace fields has been published abroad. A monograph on "Plasma Technology" (Gross *et al.*, 1969) contains, among other plasma applications, chapters on the principles of plasma torches, the design of torches in various countries, and their application.

In arc furnaces, the primary objective is heating of materials in the condensed phase. Therefore, a discussion of arc furnaces is considered to be beyond the scope of this survey.

Landt (1970), in a survey on "Developments in the area of inorganic plasma chemistry," considers the various types of arc gas heaters suitable for thermal plasma processing, including some applications, particularly in Germany. Sayce (1971) gives a comprehensive review on the state-of-the-art in the development and application of arc gas heaters in extractive metallurgy. Thorpe (1971) reports on actual and potential applications of arc gas heaters for exploitation of metallic as well as nonmetallic minerals.

Two of the most active groups in the USSR involved in research and development of arc gas heaters should also be mentioned, although no attempt can be made to provide a proper assessment of their published research results which is almost exclusively in Russian. The group headed by Zhukov in Novosibirsk (Institute of Thermal Physics, Siberian Division, Academy of Science of the USSR) as well as the group headed by Yas'ko in Minsk (Institute of Heat and Mass Transfer of the BSSR Academy of Sciences) are concerned with all conceivable aspects of arc gas heaters (Anshakov *et al.*, 1973; Shaskhov and Yas'ko, 1973; Shaboltas and Yas'ko, 1974; Sharakhovskii, 1974).

A. Components of Arc Gas Heaters and Plasma Torches

This section will be devoted to a brief discussion of the structural components of arc gas heaters, including magnetic and flow fields. It will be shown that there is a wide variety of possible combinations of available components, but not every conceivable combination leads necessarily to a viable design.

1. Electrodes

The choice of the electrodes as of any other components of an arc heater depends, of course, on the desired performance for a particular application. In certain applications, consummable electrodes (carbon, graphite, or metal) may be acceptable or even desirable, whereas in other applications extreme care must be exercised to avoid contamination from the electrodes. In the case of nonconsumable electrodes, electrode erosion must be minimized to ensure adequate lifetime of the electrodes and to avoid costly shutdown of the plant.

In order to reduce electrode wear, electrodes are frequently protected by efficient cooling systems. This measure, however, may reduce the efficiency of the arc gas heater which is undesirable, especially for the continuous operation of high power devices. An alternate approach for reducing electrode erosion makes use of a rapid movement or rotation of the arc roots by means of flow and/or magnetic fields. Finally, the arc roots may be artificially split into several attachments by dividing the electrodes into segments insulated from each other and connected over small series resistors to the main bus bar of the power supply. In this arrangement unsymmetries of the current distribution to the various segments are eliminated because an accidental increase of the current to one segment, for example, will immediately increase the voltage drop over the corresponding series resistor. Thus, the arc sees at this particular segment a lower driving potential which restores the original current distribution.

(a) *Cathodes* Two types of cathodes are widely used for arc gas heaters, namely the so-called hot cathodes and cold cathodes. The term "hot" implies that thermionic emission is the predominant but not necessarily the only emission mechanism at the cathode, whereas "cold" refers to cathodes at which electron liberation occurs predominantly by field emission. Again, field emission may not be the sole emission mechanism. The TF-emission mechanism, for example, is frequently associated with cold cathodes.

The shape of hot cathodes is mainly determined by the range of arc currents for which the arc gas heater is designed. "Stick" cathodes, consisting of a rod of a few millimeters up to 15 mm diameter, with or without sharp or rounded tip, are useful for currents up to approximately 10^3 A. "Button" type cathodes may be adequate for currents up to around 5×10^3 A. For extremely high current applications ($I > 5 \times 10^3$ A) hollow cathodes may be used where the electron emission can be distributed over a relatively large area. Cathode erosion may become a severe problem if air, oxygen, or other reactive gas mixtures are used or produced in the arc heater. In order to cope with this problem inert gas shielding has been proposed. This method is frequently applied for protection of thoriated tungsten or tungsten cathodes which are probably the most widely used cathode materials in plasma torches. Barium–calcium–aluminate-enriched tungsten has been also proposed as a cathode material (Neurath and Gibbs, 1963; Avco Report 1963, 1967). In gas shielding arrangements, a small flow rate of inert gas (usually argon) is maintained to keep reactive gases away from the cathode. Another approach for reducing excessive cathode erosion by reactive gases employs cathode inserts of special materials which still provide thermionic emission without being sensitive to reactive gases. Zirconium (Weatherly and Anderson, 1965) is, for example, such a material.

In certain applications consumable cathodes, consisting either of graphite or carbon, may be of interest. In arc furnaces the ingot to be remelted may be the cathode itself, although the anode is usually preferred because of the higher total heat transfer rates.

In spite of the usually much higher total heat transfer rates to the anode, the erosion problems at the cathode are frequently more severe because the specific heat fluxes at the cathode are substantially higher than those at the anode if the arc constricts severely in front of the cathode. As a rule, constriction becomes more severe as the pressure increases. Therefore, there must be an upper pressure limit for successful operation of a hot cathode in an arc gas heater. This limit is around 10^3 kPa. In many arc gas heater designs the hot cathode is either imbedded in a watercooled holder or cooled by the entering gas flow which prevents overheating of the cathode and, at the same time, confines the cathode arc attachment to the desired location. Overcooling of the cathode results in erratic behavior of the cathode roots and in excessive erosion.

Cold cathodes are frequently made as hollow cathodes (tubes) or in the shape of wells, protected by highly effective cooling systems. Owing to the high current densities at the cathode and the associated extremely high specific heat fluxes ($>10^6$ W/cm^2), even the most sophisticated cooling system cannot provide adequate protection if the cathode spot remains at the same location for any extended time period. For this reason the natural tendency of the cathode spot to move randomly over the cathode surface is enhanced by a properly shaped gas flow pattern and/or by external magnetic fields. The resulting movement or rotation of the cathode spot distributes the heat load over a substantially larger surface area, reducing cathode erosion.

In contrast to hot cathode devices, arc heaters equipped with cold cathodes are not subject to pressure limitations although cathode erosion may increase with increasing pressure.

Since cold cathodes are usually protected by a cooling system, candidate materials with high thermal conductivity are preferred as, for example, Cu or Cu alloys. Sintered copper–tungsten is also in use as cathode materials, as are probably many more which are within the proprietary domain of industry.

(b) *Anodes* The anode, which plays a more passive role in an arc than the cathode, may consume a large fraction of the total power dissipated in the arc, especially in short arcs (Eckert and Pfender, 1967). This usually requires thermal protection of the anode either by water cooling or by transpiration cooling (Eckert *et al.*, 1962; Pfender *et al.*, 1965; Sheer and Korman, 1973). Some arc heater applications make use of the radiation emitted by uncooled anodes (graphite or carbon) which, at the same time, continuously lose mass due to evaporation or sublimation. Such anodes may be considered as consumable anodes. Although anode erosion of thermally protected anodes is, in general, not quite as severe as cathode erosion, the specific heat fluxes on anodes of high intensity arcs may reach values on the order of 10^5 to 10^6 W/cm^2 (Cobine and Burger, 1955; Smith and Pfender, 1976).

It is customary to use anodes in the shape of tubes, disks, nozzles, or rings depending on the contemplated application of the arc heater. Nozzle-shaped anodes are used in the standard design of an arc plasma torch. The nozzle serves in this case not only as an anode; it acts simultaneously as an arc constrictor increasing the enthalpy level of the emanating plasma jet. For generating supersonic plasma flows, supersonic nozzles may serve at least partially as anodes (Vorreiter and Shepard, 1965). The interaction of the gas flow with the arc in the vicinity of the anode surface may cause a severe constriction of the anode arc attachment which, in turn, gives rise to very high specific heat fluxes and possibly excessive anode erosion. Distribution of the heat flux over a larger anode surface area can be accomplished in the same way as described for the cathode.

The previously mentioned interaction of the gas flow with the arc is responsible for another peculiar phenomenon frequently observed in plasma torches. The constricted portion of the arc column close to the anode surface (anode column) forms the conducting link between the main body of the arc and the anode surface. This link which is under the influence of a cross flow may move in the direction of the flow or remain stationary depending on the balance of aerodynamic drag force and $\mathbf{j} \times \mathbf{B}$ force, where B is the self-magnetic field (Wutzke *et al.*, 1967, 1968). If the anode column moves downstream along with the gas flow, arc length and arc voltage increase until breakdown occurs between cathode and that part of the anode which is close to the cathode. This event causes a sudden reduction of the arc length and of the arc voltage (arc shunting). The new anode column travels again downstream and the process is repeated, i.e., the arc restrikes periodically leading to sawtooth-shaped voltage fluctuations. This mode of operation spreads the heat flux automatically over a larger anode surface area. For certain applications, however, the observed fluctuations of the arc column and of the associated plasma properties in the emanating plasma jet are undesirable (Pfender and Cremers, 1965).

The choice of the proper anode material is not as critical as that of the cathode material. For water-cooled anodes Cu or other metals or metal alloys with high thermal conductivity are usually preferred. Porous graphite seems to perform adequately for transpiration-cooled anodes.

2. Arc Constrictor

For applications which require high enthalpy levels the principle of wall-stabilization may be introduced. By forcing the arc through a constrictor tube located between cathode and anode, the power input per unit length of the constricted arc and the associated enthalpy level increase significantly. This principle may be found in many conventional plasma torches which are equipped with nozzle-shaped anodes. As previously explained, the nozzle serves also as an arc constrictor. In the transferred mode of operation (the arc attaches to an exterior anode—for example, to the work piece in arc welding or cutting) the nozzle is kept at a floating potential and serves only as an arc constrictor.

Arc constriction is of particular importance in arc gas heaters developed for aerospace-related testing (supersonic arc wind tunnels). The constrictor tube consists usually of a stack of insulated, water-cooled Cu segments (Shepard *et al.*, 1964). In later developments gas has been injected between some or all of the segments, providing additional thermal protection of the constrictor tube by film cooling (Shepard, 1972). This principle may be extended to transpiration cooled constrictor tubes consisting of porous materials surrounded by a single or several pressurized plenum chambers (Eckert and Anderson, 1964; Anderson and Eckert, 1967; Pfender *et al.*, 1969).

3. Gas Flow Pattern

The main goal of an arc gas heater is the well-controlled, efficient heating of gases to a certain enthalpy level. An important prerequisite for achieving this goal is the proper choice of the gas flow pattern and the resulting interaction of the arc with the superimposed flow including stabilization of the arc column. There are a number of options available to the designer:

(a) coaxial flow,
(b) cross flow,
(c) radial flow,
(d) vortex flow.

These options refer to the flow pattern with respect to the arc column.

In the case of coaxial flow the gas is usually introduced from the cathode end of the arc heater, flows parallel to the axis of the arc, and forms a shroud of cold gas around the arc column. Some of the cold gas from the shroud permeates gradually into the arc, is heated to the prevailing temperature level and accelerated downstream as shown schematically in Fig. 33 for a constricted arc. The coaxial flow pattern is particularly useful in conjunction with arc constrictors. It is widely used in high enthalpy arc heaters as well as in plasma torches. Sometimes a vortex component is superimposed on the coaxial flow.

Arcs in cross flow have been extensively studied during the past 15 years (references in Section II) and a few applications have been also reported which use this principle for the entire length of the arc column. However, in almost every conceivable arc heater configuration, some parts of the arc (sometimes only those parts in the cathode or anode region) are exposed to cross flow which may be the origin of some peculiar effects as for example discussed previously for plasma torches.

In a few instances, arcs have been exposed to radial flow. The transpiration cooled arc is a typical example in this category. Cremers *et al.* (1973) studied the effect of fluid constriction of an arc by localized radial injection of gas. In actual arc gas heaters, additional gas is sometimes injected along the constrictor tube, but this injection is usually in the tangential, not radial, direction (Shepard, 1972).

Besides coaxial flow, vortex flow is the most important flow pattern found in arc gas heaters. There are three major reasons for this preference of vortex flow. First of all, vortex flow provides an excellent stabilizing mechanism for the arc even in rather wide tubes. Second, vortex flow leads inherently to efficient thermal protection of the tube wall in which the arc is operated. The cold gas shroud at the tube wall eliminates the need for segmentation of the tube. Finally, the intensive interaction of the gas flow with the arc gives rise to high fields strengths, i.e., arc heaters which use vortex flow are usually high voltage devices with high thermal efficiencies. These high efficiencies are a consequence of the lower arc currents for a given power level compared with other types of arc heaters. Electrode losses, in general, increase with increasing current.

Depending on the Reynolds number, the flow in an arc heater may be laminar or turbulent. Analytical work associated with constricted arcs in coaxial flow has been restricted to laminar flow (Watson, 1965; Watson and Pegot, 1967).

4. Magnetic Field Configuration

Externally applied magnetic fields in arc gas heaters may affect the interaction of the gas flow with the arc by driving the arc due to $\mathbf{j} \times \mathbf{B}$ interaction. This implies that the magnetic field itself or at least a component of the field is perpendicular to the axis of the arc. Another frequent application of magnetic fields in arc gas heaters is for the reduction of electrode erosion. As explained earlier, the heat flux at an electrode may be distributed over a larger area by driving the arc attachments with high velocity over the electrode surfaces. Since the desired $\mathbf{j} \times \mathbf{B}$ interaction is confined to the cathode or anode region of the arc, the majority of the arc column is exposed to an axial magnetic field which is expected to have little influence on the arc as long as the arc axis is parallel to the magnetic field lines.

Recent studies of a vortex-stabilized arc in an axial magnetic field show that the apparent arc diameter increases with increasing magnetic field strength (Cann, 1973). Time-resolved measurements indicate that this apparent increase in arc diameter is actually caused by a helical shape of the arc column. A cylindrical arc exposed to an axial magnetic field is obviously not a stable configuration. Unavoidable disturbances of the arc symmetry interact with the magnetic field and drive the arc into a helical shape.

Magnetic balancing of an arc driven by or displaced by aerodynamic forces has been of little importance in the development of arc gas heaters.

B. Classification and Performance of Arc Gas Heaters

In an arc heater electric energy is converted into thermal and kinetic energy as shown schematically in Fig. 39. The prevalent energy conversion path is usually from electric energy over joule heating to thermal energy. Only in cases of strong interaction of the arc with its own or an externally applied magnetic field is an appreciable fraction of the available electric energy used for acceleration of the plasma. By stagnation, the kinetic energy may be also converted into thermal energy.

The principally different requirements of the performance capabilities of arc heaters used for reentry simulation and related tasks and those desired for chemical and material processing suggest dividing arc heaters into two broad categories. Aerospace applications require arc heaters designed for extremely high enthalpy and high impact pressure levels. The efficiency in such applications is of minor importance as long as the necessary investment for the power supply remains within reasonable limits. Testing (for example reentry simulation) lasts usually only for seconds or at most for minutes.

In contrast, applications in chemical engineering and in metallurgy are mainly concerned with relatively low enthalpy levels and pressure levels around atmospheric pressure, but the efficiency is probably the most impor-

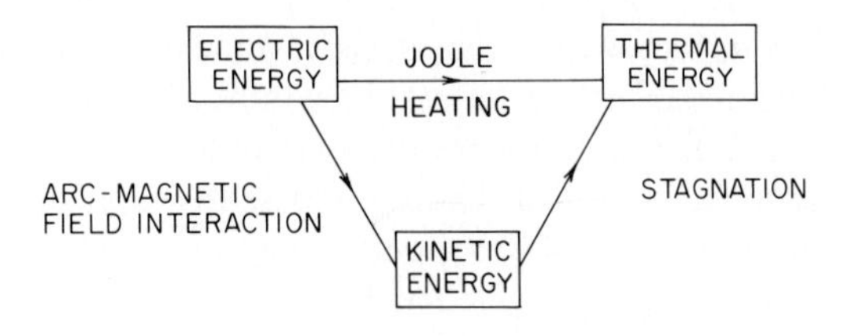

Fig. 39. Schematic of the energy conversion process in an arc gas heater.

tant consideration because such heaters should run continuously for weeks or even months.

The suggested classification does not imply that arc gas heaters which fall between these two extremes are nonexistent or technically not viable.

Sometimes arc heaters are classified according to the cathode temperature, or more precisely, according to the dominating electron emission mechanism at the cathode ("hot" cathode and "cold" cathode arc heaters). This classification will not be used explicitly in this survey.

1. High Enthalpy Arc Gas Heaters

As mentioned earlier arc gas heaters needed for reentry studies and aerodynamic testing should be capable of generating high enthalpy gas flows (in extreme cases up to 10^8 kJ/kg) with impact pressures in the order of 10^4 kPa. Reaching this goal still remains a challenge for arc heater designers.

There are two basically different designs which approach these extreme values, but fail in certain aspects of the required performance. The constricted arc heater with coaxial flow, developed by a group at NASA-Ames and also at AEDC produces relatively high enthalpy levels but cannot operate reliably at high pressures. The vortex-stabilized arc heater (known as the Linde design in the U.S.) can operate at pressure levels in the order of 10^4 kPa, but its enthalpy level is, in general, lower than in the other type. These two designs form the basis of many developments in the arc gas heater field and, therefore, they will be briefly discussed in the following subsections.

(a) *Constricted Arc Heater with Coaxial Flow* The basic outlay of a constricted arc heater designed for a hyperthermal supersonic tunnel (Stine, 1963; Shepard *et al.*, 1964) is shown in Fig. 40. A hot cathode (stick type) protected by an inert gas atmosphere (not shown in the sketch) represents the upstream electrode. The main gas flow (for example air) enters a plenum chamber downstream of the cathode chamber. The constrictor diameter and length is chosen so that a given gas flow is choked at a desired pressure, and that the enthalpy profile becomes fully developed before the flow enters the diverging nozzle. The smoothly contoured nozzle produces the supersonic flow. The constrictor tube consists of a stack of insulated water-cooled copper disks. The maximum enthalpy that can be achieved in this configuration without burnout is limited by the maximum heat flux which can be absorbed by the constrictor wall. The arc is maintained throughout the length of the constrictor and most of the supersonic nozzle. The arc attaches to the multiple anode (downstream electrode) in a diffuse manner because the anodes are in a low pressure environment. The multiple anode arrangement promotes axial symmetry and limits the current to an individual anode.

With this design, average enthalpies of 10^5 kJ/kg have been obtained at pressures upstream of the constrictor tube from 100 to 250 kPa (Vorreiter and Shepard, 1965). For this particular design the performance has been limited by the current-carrying capability of the electrodes. A later version of this facility produced a mass-average enthalpy of 3×10^5 kJ/kg at 100 kPa and 3.5×10^4 kJ/kg at 700 kPa. The air stream leaving the supersonic nozzle carries more than 2 MW of power (Shepard *et al.*, 1968).

For simulating entries into the atmospheres of the outer planets a 20 MW arc heater has been developed (Shepard, 1972) which produces stagnation enthalpies of approximately 9×10^5 kJ/kg. The device relies on heat capacity cooling rather than watercooling which reduces the operating periods to approximately 1 sec. A schematic of the arrangement is shown in Fig. 41.

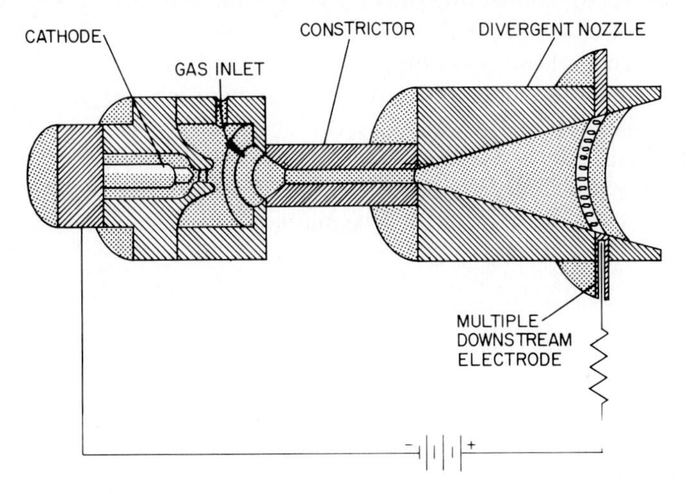

Fig. 40. Schematic of a constricted-arc gas heater for reentry simulation (Shepard *et al.*, 1964).

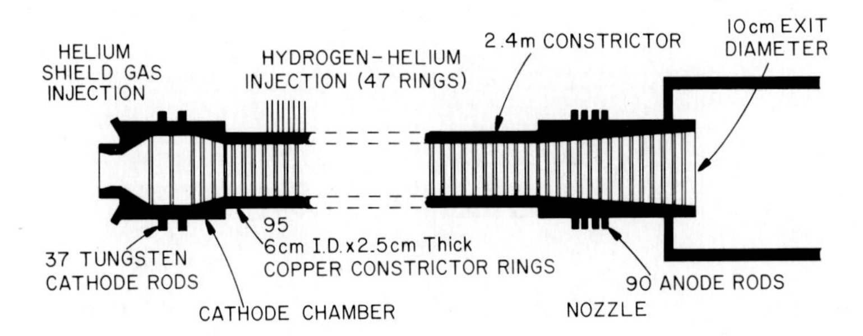

Fig. 41. Schematic of a 20 MW heat-capacity cooled, constricted-arc gas heater for simulation of entries into the atmospheres of the outer planets (Shepard, 1972).

The highest enthalpies which have been reached so far with constricted arc heaters seem to be around 10^7 kJ/kg (Richter, 1969, 1970). According to the scaling laws of constricted arc heaters the optimum performance is expected to be around 2×10^7 kJ/kg at stagnation pressures of approximately 1.5×10^4 kPa (Harder and Cann, 1970). As the pressure level in an arc heater of this type is increased, the radiation losses increase also (the plasma approaches the properties of a blackbody radiator) and it becomes increasingly difficult to transfer more energy to the arc plasma without exceeding the maximum permissible heat load to the walls of the constrictor, which is on the order of 10 kW/cm^2.

For a given average enthalpy level the temperature distribution within the constrictor tube should be kept as flat as possible to avoid excessive radiation losses due to temperature peaks. By superimposing an rf discharge onto an arc, more uniform heating of the plasma may be achieved (Schreiber *et al.*, 1973) since an arc heats predominantly the region close to the axis in a rotationally symmetric configuration whereas an rf discharge heats predominantly the fringes.

Constricted arc heaters in which a transition to turbulent flow occurs offer another possibility to realize flatter temperature profiles. An arrangement in which an ac magnetic field induces turbulence is shown in Fig. 42.

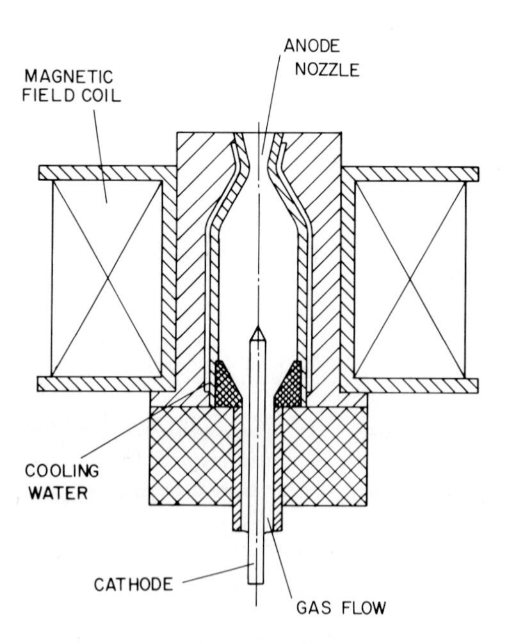

Fig. 42. Arc gas heater for inducing turbulence (Peters, 1964).

Another configuration in which transition to turbulence occurs is shown in Fig. 43 (courtesy of M. I. Sazonov, 1974, Inst. of Thermal Physics, Novosibirsk, USSR). The transition to turbulent flow is indicated by a sudden increase of the field strength E and a gradual increase of the wall heat flux Q. This heater uses a hot cathode in the form of a button, protected by a small flow (g_1) of nitrogen. Anode erosion is reduced by a magnetic field coil around the anode. Unfortunately, arc stability seems to be a severe problem in such arc gas heaters (Anshakov *et al.*, 1973).

In contrast to the previously discussed high power arc heaters, plasma torches, that belong in the category of constricted arc heaters, usually operate at lower power levels, typically in the range from 1 kW to 1 MW, although there is no well-defined border line. In spite of the lower power levels, plasma torches are capable of producing relatively high peak enthalpies (up to 10^5 kJ/kg) by matching the constrictor diameter with the power level and the desired mass flow rate. In conventional plasma torches, designed for operation around atmospheric pressure, no attempt is made to optimize the enthalpy level. This would require substantially longer constrictor tubes with correspondingly high heat losses and lower efficiencies.

As far as the mass-average enthalpy is concerned, most plasma torches must be classified under low-enthalpy devices which will be discussed in Section III.B.2.

(b) *Vortex Stabilized Arc Gas Heaters* Arc heaters in which a strong vortex confines the arc in the center of a tube are particularly suited for

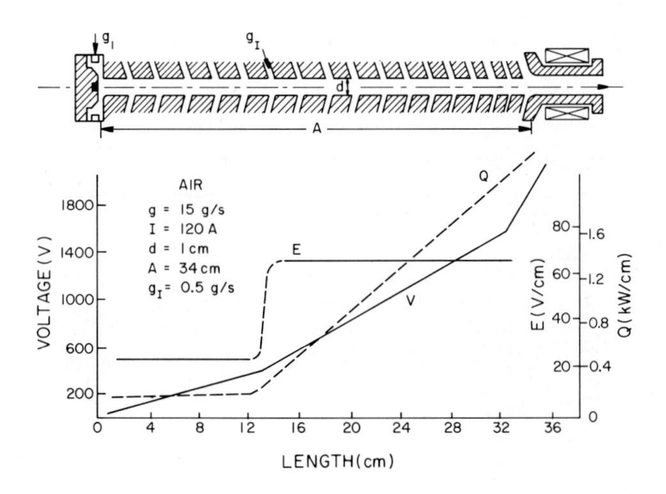

Fig. 43. Characteristics of a constricted-arc gas heater with transition to turbulence (courtesy of M. Sazonov).

operating at high pressures and relatively high enthalpy levels. Figure 44 shows the principle of a vortex stabilized arc gas heater with supersonic nozzle and Fig. 45 depicts a set of typical performance characteristics of such an arc heater (Eschenbach and Skinner, 1961). The voltage drop in this type of arc heater may exceed 15 kV, and power levels up to 50 MW are feasible (Beachler, 1970). This arc heater is capable of, for example, heating air flowing at a rate of 3.6 kg/sec to stagnation enthalpies of approximately 1.5×10^4 kJ/kg at 1.25×10^4 kPa using 37 MW of electrical power.

For arc heaters in which the energy balance is dominated by radiation (high-pressure devices) Harder and Cann (1970) derived correlations among the various arc heater parameters which should be useful for the design of such devices. Their derivation is based on a linearization of the governing equations. The derived correlations hold for laminar as well as for turbulent flow.

For vortex-stabilized arc heaters the decay of the vortex in the downstream direction, especially for long heaters (several meters length), may cause a loss of arc stability. Therefore, in some designs a vigorous vortex is maintained by introducing gas through tangential slots along the confining tube.

Another approach for correlating data for various types of vortex-stabilized and other arcs has been discussed by Shashkov and Yas'ko (1973). Although arcs do not rigorously obey similarity laws, the authors obtained encouraging results by applying approximate similarity considerations.

There are a number of different configurations to which this approximate similarity analysis can be applied. Figure 46 shows as an example three configurations in which the arc is vortex-stabilized. Figure 47 shows the usual and the generalized current–voltage characteristic of an arc heater

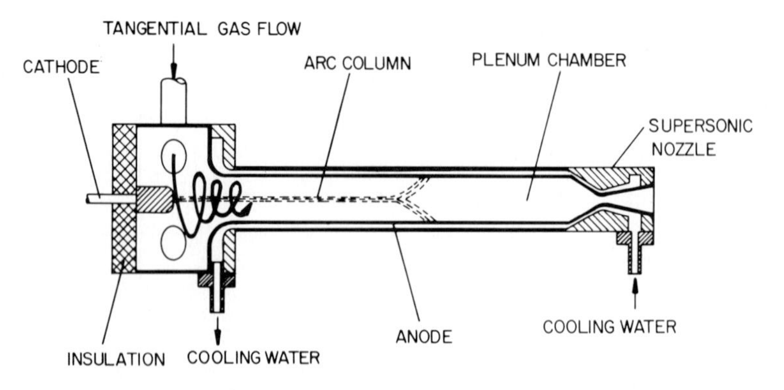

Fig. 44. Schematic arrangement of a "Linde-type" arc gas heater.

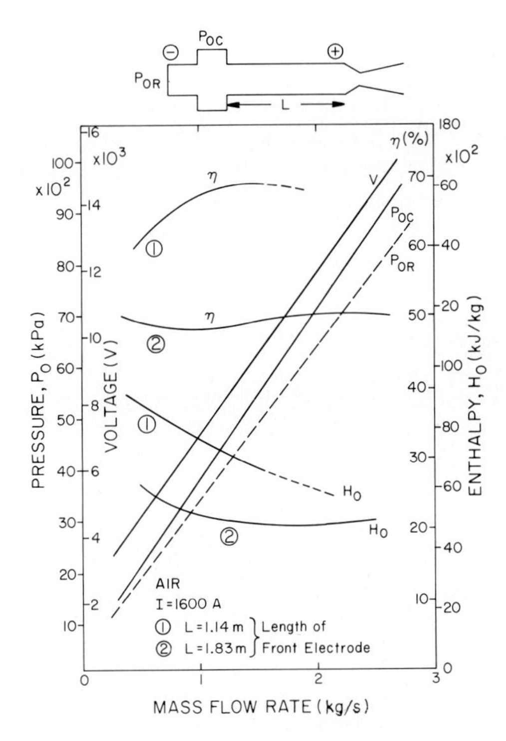

Fig. 45. Performance characteristics of a "Linde-type" arc gas heater (Eschenbach and Skinner, 1961).

operating in air and designed according to Fig. 46a (Shashkov and Yas'ko, 1973); I represents the arc current, V the arc voltage, g the mass flow rate, σ the electrical conductivity and h the enthalpy. The subscript zero refers to reference values. Both electrodes are water cooled. The data in Fig. 47 can be represented by the following semiempirical relationship:

$$Vd\sigma_0/I = 0.71(I^2/gd\sigma_0 h_0)^{-0.84}. \tag{50}$$

The deviation of the data points from this generalized expression is less than 18%. Another example of a semiempirical correlation for air arc heaters is shown in Fig. 48 (courtesy of M. I. Sazonov, 1974, Inst. of Thermal Physics, Novosibirsk, USSR). In this design magnetic field coils are used for electrode protection. By correlating the arc voltage V with I^2/gd, g/d, and pd, a wide parameter range can be covered with reasonable scatter of the data points. These representations provide useful guidelines as long as a new design falls within the parameter range covered by such semiempirical relationships. Extrapolations beyond this parameter range, however, need careful justification.

Since the arc in a vortex-stabilized arc heater is almost entirely surrounded by the electrodes, the efficiency will be strongly influenced by heat transfer to the electrodes. Painter (1974) reported results of electrode heat transfer studies for an air heater with a maximum power capability of 12 MW covering arc currents up to 800 A and pressures up to 2×10^4 kPa. Electrode losses in this heater increase linearly with current and with the square

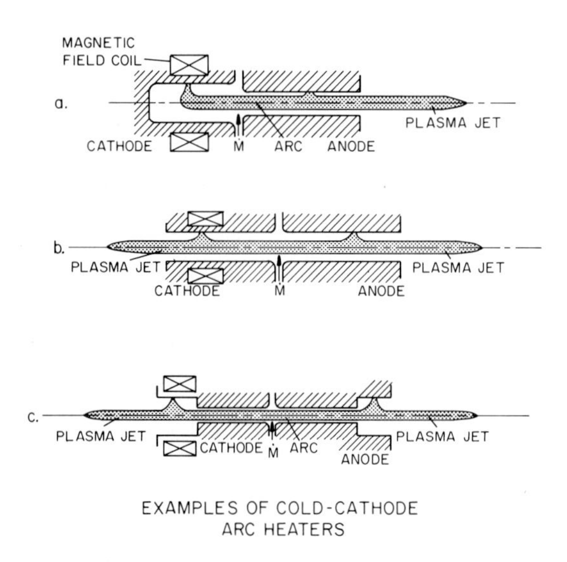

Fig. 46. Design options for vortex-stabilized, cold-cathode arc gas heaters.

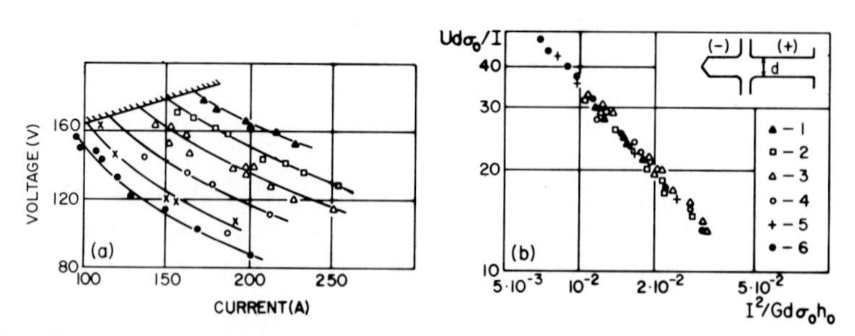

Fig. 47. Characteristics of a vortex-stabilized, cold-cathode arc gas (air) heater (Shashkov and Yas'ko, 1973). (a) Current-voltage characteristics. (b) generalized characteristic.

Symbol:	1	2	3	4	5	6	
Mass flow rate:	2.05	1.83	1.4 to 1.6	1.26	1.15	1.01	g/s

root of the pressure. As expected, heat transfer by radiation becomes the dominating mechanism as the pressure approaches 2×10^4 kPa. As mentioned earlier, temperature peaks in the arc plasma are detrimental for achieving high efficiencies, because radiation losses increase exponentially with temperature. Ideally, the temperature distribution in an arc heater should be flat; the entire mass passing through the heater should reach a uniform temperature according to the desired enthalpy level. An attempt to approach this goal has been reported by Cann (1973). By exposing a vortex-stabilized arc to an axial magnetic field, the arc is forced into a helical path which enlarges the apparent arc diameter and produces a fairly uniform temperature distribution which is neither sensitive to the arc current nor to the pressure. The arrangement used for such experiments is shown in Fig. 49 (Cann, 1973).

2. Low Enthalpy Arc Gas Heaters

Many applications, especially in the growing field of chemical processing, require arc heaters which produce relatively low enthalpy gas flows compared with those used, for example, for reentry testing. The efficiency of devices designed for extremely high enthalpy levels is usually of minor im-

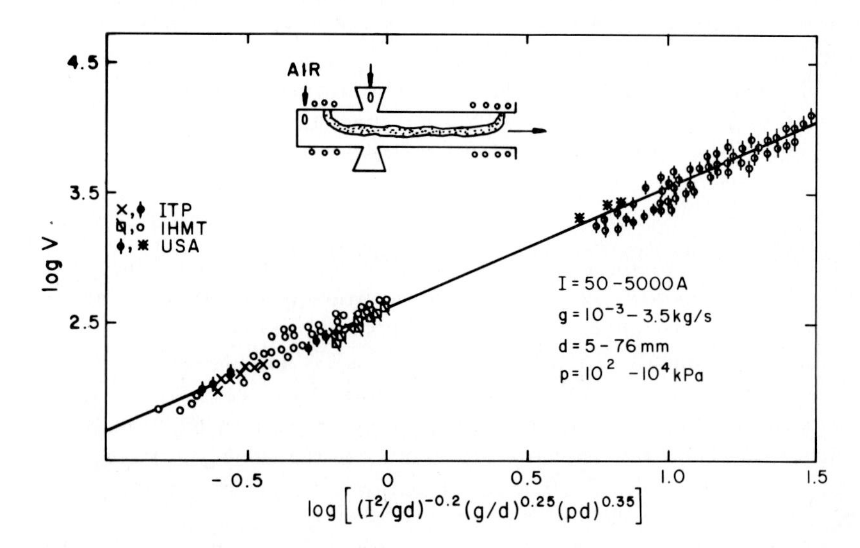

Fig. 48. Semiempirical correlation for a vortex-stabilized cold-cathode, air arc heater (courtesy of M. Sazonov). ITP = Institute of Thermal Physics, Siberian Branch of the USSR Academy of Sciences, Novosibirsk, USSR. IHMT = Institute of Heat and Mass Transfer of the BSSR Academy of Sciences, Minsk, USSR.

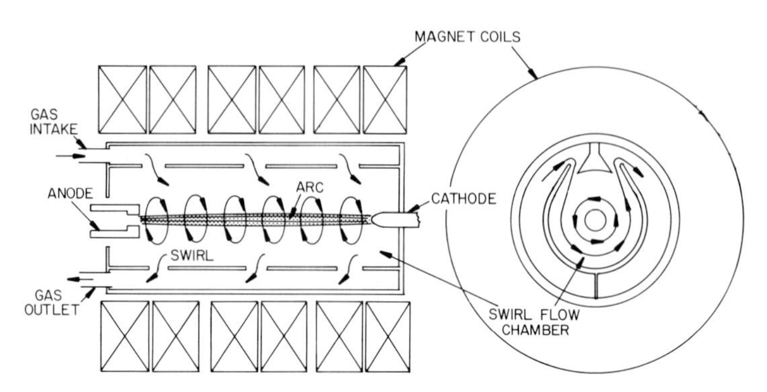

Fig. 49. Schematic of a vortex-stabilized arc gas heater with superimposed axial magnetic field (Cann, 1973).

portance because of the short duration of the tests. In contrast, the efficiency of the arc gas heaters to be covered in this paragraph is one of the most important design parameters. This aspect may receive further emphasis in the future because of the increasing cost of energy.

As indicated earlier, there is no sharp boundary between high and low enthalpy arc heaters. In fact, a number of existing heaters fall somewhere between these two extremes. Plasma torches, for example, in which the plasma column is forced through a constrictor either attached or separate from the anode, belong in this category.

The relatively modest, but uniform enthalpies needed in certain applications (for example in chemical processing) suggest a distribution of the heat dissipation over larger volumes than provided by the natural size of an arc. This goal may be reached in two different ways, either by rotating an arc or by expanding the column of an arc. Both approaches have been pursued and will be briefly discussed at the end of this paragraph.

(a) *Plasma Torches* A plasma torch is probably the simplest tool for generating a high temperature plasma. Figure 50 shows schematically the essential components of a plasma torch. The arc is initiated between the tip of the cathode (typically thoriated tungsten) and the water-cooled anode. The working gas is introduced either axially or with an additional swirl component. The latter improves arc stability in the vicinity of the cathode and rotates the anode root. The gas heated by the arc emanates as a plasma jet from the torch orifice. This jet may be used in a variety of applications such as welding, cutting, spraying, sphereodizing, surface treating, chemical processing, etc. In these applications the properties of the plasma jet are of immediate interest.

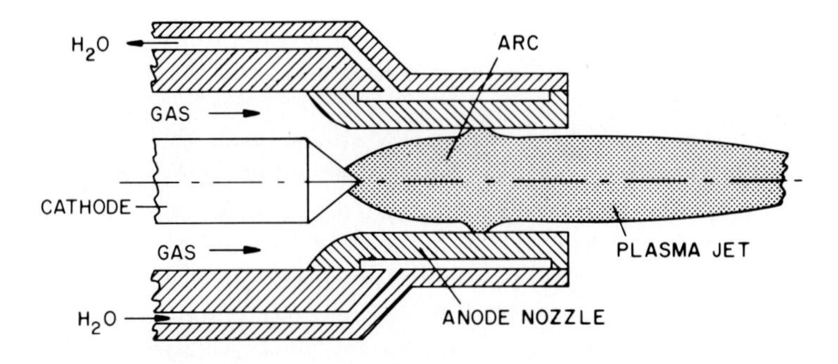

Fig. 50. Typical electrode configuration of an arc plasma torch.

The plasma jet may be laminar or turbulent depending on the chosen arc parameters. In the laminar regime (relatively small mass flow rates) the torch operates quietly and the highly luminous plasma jet may reach lengths up to 30 cm in atmospheric air at sufficiently high power levels. The transition to turbulence (relatively high mass flow rates) is characterized by increasing audible noise and decreasing length of the plasma jet. The maximum temperature in the plasma jet is a function of the operating parameters and may vary from 8000 to 20,000°K close to the nozzle orifice. Since the plasma jet is a field-free plasma, the plasma temperature decays rapidly with increasing distance from the nozzle orifice, especially when turbulent mixing enhances the energy exchange between the plasma jet and the surrounding atmosphere. A reduction of the energy exchange with the ambient atmosphere (for example by lowering the ambient pressure) produces substantially longer plasma jets.

The flow velocities in the jet may range from almost zero to sonic velocities. If the conventional anode nozzle is replaced by a supersonic nozzle, a supersonic plasma flow may be produced.

Plasma torches have been designed for power levels from a few KW up to approximately 1 MW. Thermal efficiencies are in the range from 30 to 90%. Usually inert gases or their mixtures are used as working fluids. Hydrogen, ammonia, hydrocarbons, oxygen and other corrosive gases can also be used as working fluids if the necessary precaution is taken to protect the electrodes. Frequently, corrosive gases are introduced into the plasma flow downstream of the anode orifice to circumvent electrode problems.

For certain applications a magnetic field coil surrounds the anodė which distributes the anode heat load over a large area; in this way the lifetime of the anode is increased and the level of contamination of the plasma is reduced.

For applications which require high specific heat fluxes (for example welding or cutting) it is customary to transfer the arc to the work piece. In this transferred mode the work piece is the anode and the nozzle serves as arc constrictor.

Plasma torches have been extensively studied over the past 20 years. The water-vortex-stabilized arc in combination with a ring-shaped electrode served as one of the first arrangements for producing a plasma jet (Weiss, 1954). Temperatures up to 13,000°K in the axis of this plasma jet were found. Similar temperatures have been measured in the axis of atmospheric pressure argon plasma jets (Jahn, 1961, 1963; Brzozowski and Celinski, 1962; Cremers and Pfender, 1964; Grey and Jacobs, 1964; Ahlborn, 1965). These temperatures refer to field-free plasma jets. If parts of the plasma jet carry electric currents, these parts will be substantially hotter (Bruckner, 1963). Various options for generating plasma jets with high temperatures and velocities including supersonic velocities have been discussed by Peters (1961). Besides the temperature distribution in plasma jets, other properties (for example velocity and enthalpy distributions, fluctuations etc.) are also of interest. Corresponding measurements and calculations are reported in some of the previously mentioned references and also by Freeman *et al.* (1962), Freeman (1962), Tatento and Saito (1963), Au and Sprengel (1966), and Adcock (1967).

The possible application of plasma torches for space propulsion was extensively studied during the early sixties. These studies culminated in the development of so-called thermal arc-jet engines shown as cutaway views in Figs. 51 and 52. Both engines are designed for a power level of 30 kW. Hydrogen or ammonia may be used as working fluids (propellants). In the radiation-cooled engine, the anode is made of pure tungsten and also serves

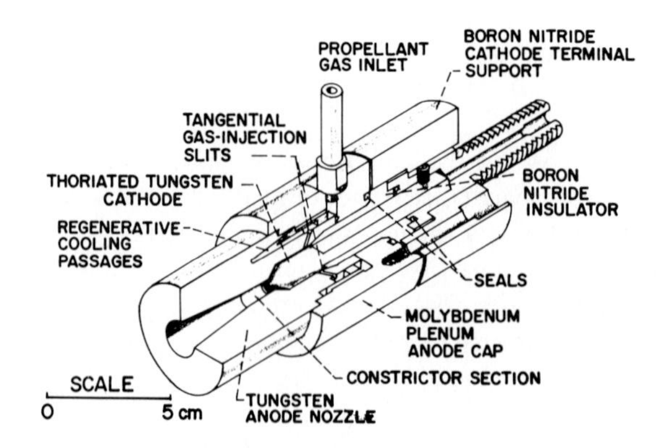

Fig. 51. Cutaway view of a 30 kW dc radiation-cooled arc-jet engine (John, 1964).

as an exhaust nozzle. This nozzle has the usual convergent-divergent shape to produce the desired supersonic flow. Part of the energy losses to the electrodes, especially to the anode, may be recovered with regenerative cooling by circulating the propellant through passages in the anode before injecting it into the arc (Fig. 52). Such engines produce specific impulses of

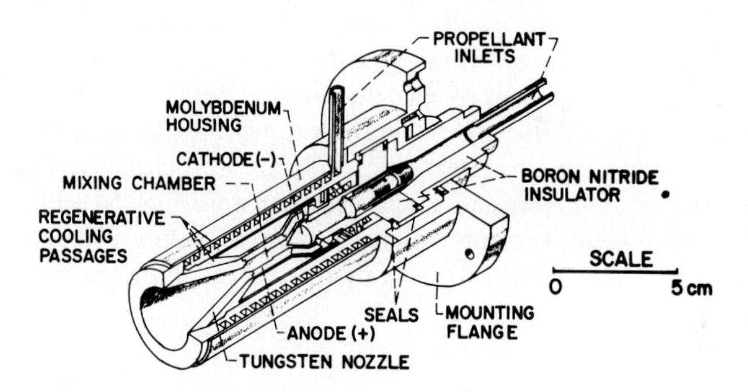

Fig. 52. Cutaway view of a 30 kW dc regeneratively-cooled arc-jet engine (Todd, 1964).

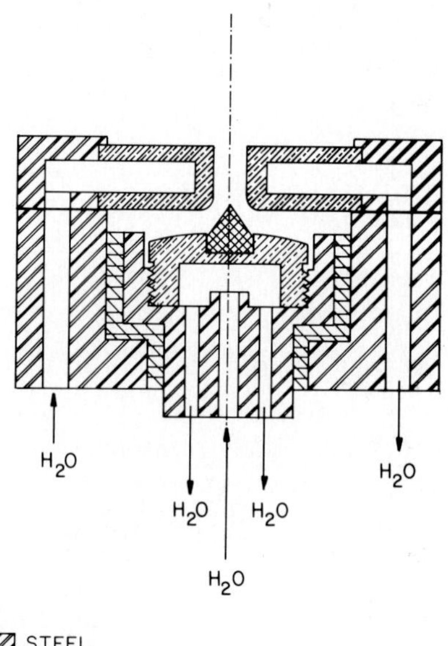

Fig. 53. Actual design of an arc plasma torch (schematically).

approximately 10^3 sec, and thrusts of 250 to 300 g with an overall efficiency of approximately 40%. In general, thermal arc-jet engines may be considered as well developed, reliable engines. Applications for space missions will, however, not arise until low specific-weight power supplies are available. However, the wealth of information accumulated and the improvements in design and materials made during this development period, are reflected today in plasma torch hardware designed for industrial applications.

In the following figures, some typical data will be shown obtained with a small, modified, commercial plasma torch (Boffa and Pfender, 1968). Figure 53 shows a cross-section of the torch without the gas flow passage. The inside nozzle diameter is 6.3 mm. Velocity and enthalpy distributions at a distance of 5 mm from the nozzle are shown in Figs. 54 and 55, respectively. For these measurements an enthalpy probe (Grey and Jacobs, 1967) was used. The vertical bars in these figures reflect the experimental error.

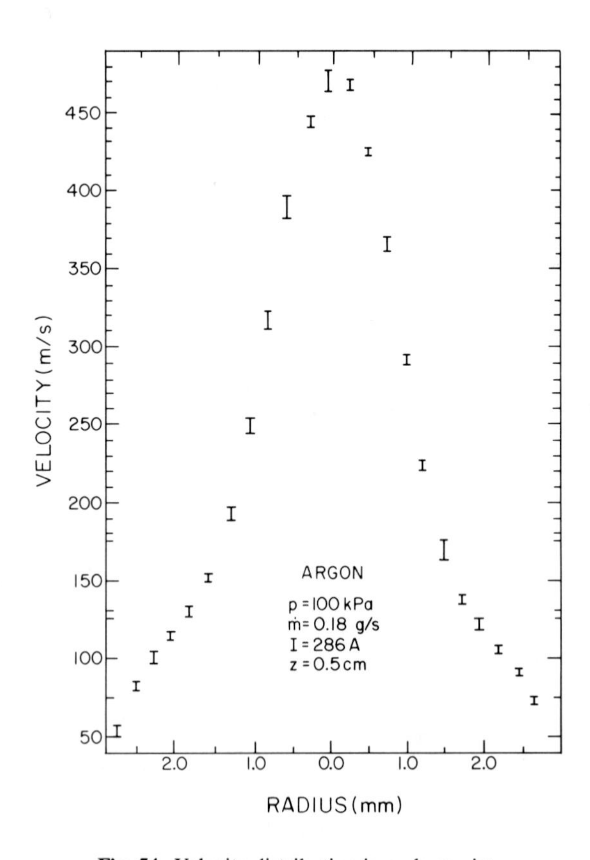

Fig. 54. Velocity distribution in a plasma jet.

The corresponding temperature distributions, measured spectrometrically, are shown in Fig. 56. For comparison, temperature distributions at lower currents are also included in this figure. For the highest current (286 A) the power input to this torch is approximately 5.5 kW and the efficiency is close to 50%. Due to the relatively small mass flow rate the plasma jet is laminar at this current level.

(b) *Magnetically Spun Arcs* In the late fifties and early sixties the magnetically spun arc operated in a coaxial electrode configuration attracted great interest as a possible candidate for reentry simulation. A sketch of such a heater is shown in Fig. 57 (Boldman *et al.*, 1962). The magnetic field coil wrapped around the anode produces a magnetic field with a component perpendicular to the current flow in the vicinity of the anode. The **j** × **B** force in this region drives the arc in azimuthal direction and a constant velocity of this magnetically spun arc is established by a balance between

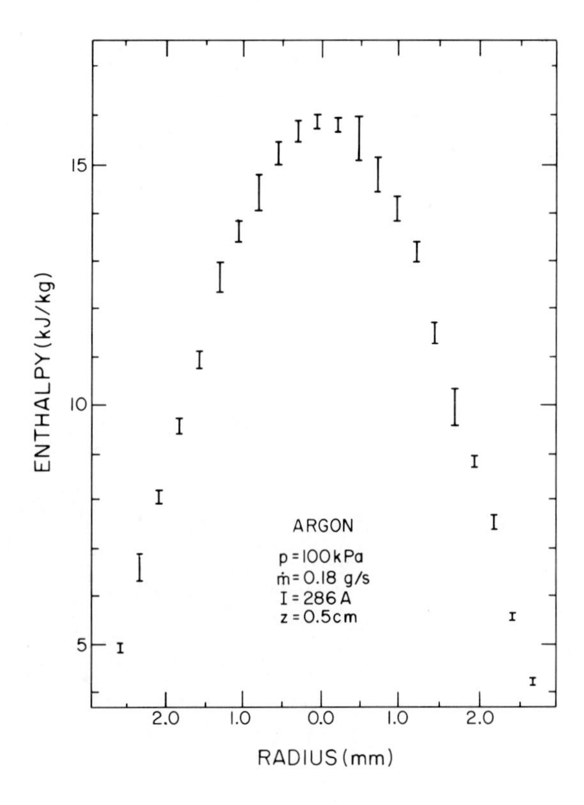

Fig. 55. Enthalpy distribution in a plasma jet.

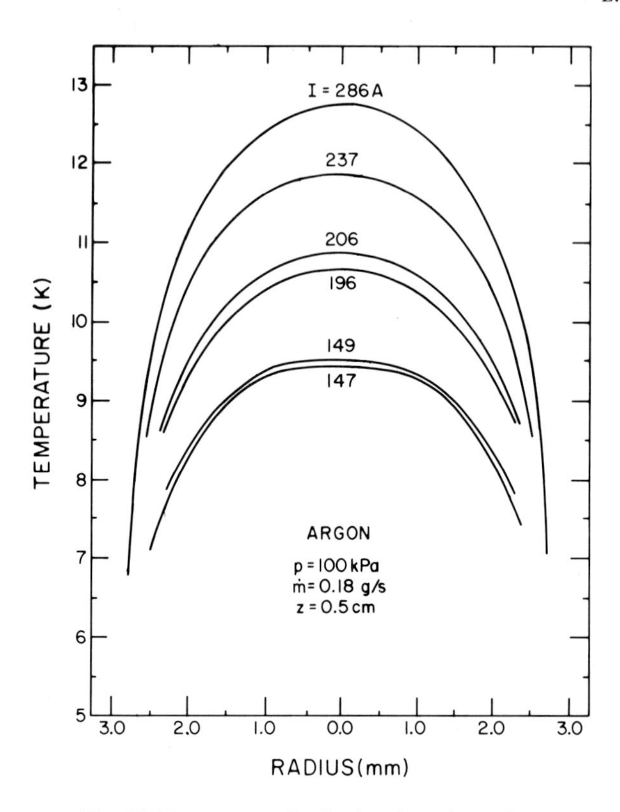

Fig. 56. Temperature distributions in a plasma jet.

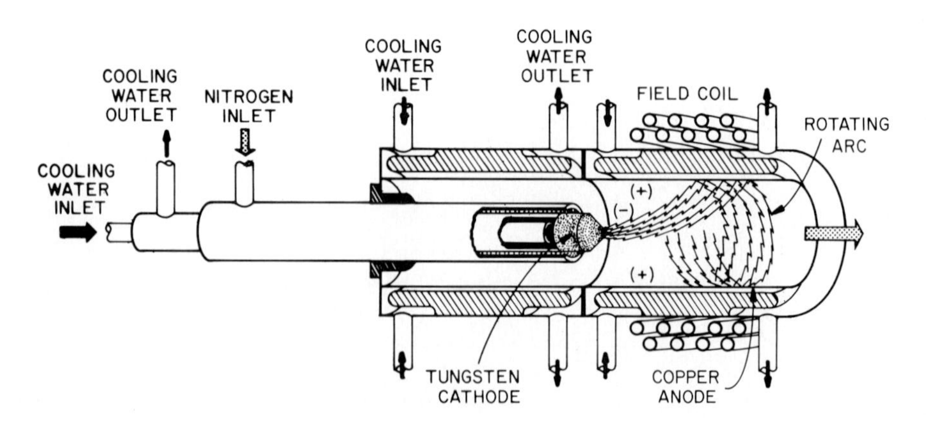

Fig. 57. Schematic of a magnetically spun arc gas heater.

magnetic driving force and drag force acting on the arc column due to its motion relative to the surrounding gas. It was anticipated that this device would provide an efficient and almost uniform heating of large gas volumes blown through the coaxial gap between the electrodes by spinning the arc with sufficiently high velocities. It was further argued that with increasing spinning velocities an increasing fraction of the annular gas volume would be ionized and eventually the arc would fill the entire gap between the electrodes optimizing the energy exchange between plasma and cold gas. Unfortunately, the expected high enthalpy levels could not be confirmed because the arc continues to rotate as a rather constricted spoke, resulting in relatively poor energy exchange between arc and cold gas.

Nevertheless, this arrangement offers four features which render it very attractive as arc heater for chemical processing:

(1) Rotation of the arc distributes the anode heat load and reduces anode erosion in this way.

(2) The enthalpy level of the heated gas can be easily controlled by monitoring arc current and spinning velocity.

(3) Relatively large volumes of gas can be heated to fairly uniform temperatures.

(4) Rotation of the arc increases the potential drop so that more power can be delivered to the gas at a given current.

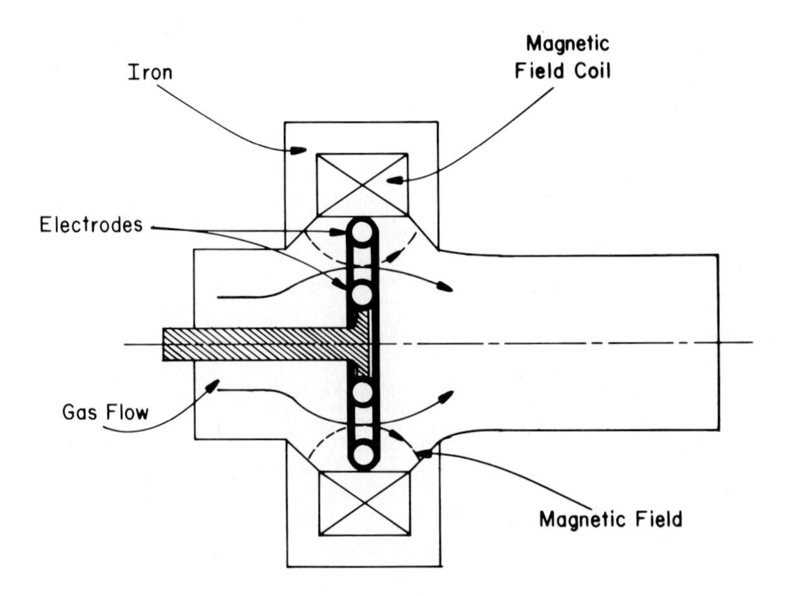

Fig. 58. Concentric electrode arrangement for a magnetically spun arc gas heater (Peters, 1964).

Another possible arrangement of a rotating arc gas heater is shown in Fig. 58. In this case the electrodes consist of two concentric metal rings so that the arc current flows essentially in radial direction providing a strong interaction with the applied magnetic field.

Studies of the influence of arc current, magnetic field, air flow, and gap size on the arc velocity in a similar configuration indicate that experimental data can be correlated with semiempirical relationships (Shaboltas and Yas'ko, 1974; Sharakhovskii, 1974; Shashkov and Yas'ko, 1973).

(c) *Expanded Arcs* It is a well-known fact that an arc operated in a wide tube will not fill the entire tube unless convection effects are eliminated or at least reduced. The gravity-free arc and experiments with rotating arc vessels (Wälzbogen) described in Section II demonstrate this effect. The principle of expanding an arc by rotating the arc enclosure has been successfully exploited in the design of expanded arc furnaces (Sayce, 1971). It is conceivable to use the same principle for producing expanded plasma flows with rather flat enthalpy profiles.

REFERENCES

Adcock, B. D. (1967). *J. Quant. Spectrosc. Radiat. Transfer* **7,** 385.

AGARDograph 84 (1964). Arc Heaters and MHD Accelerators for Aerodynamic Purposes, Part 1, September.

AGARDograph 84 (1964). Arc Heaters and MHD Accelerators for Aerodynamic Purposes, Part 2, September.

Ahlborn, B. (1965). *Z. Naturforsch.* **20a,** Heft 3, 466.

Amdur, I., and Mason, E. A. (1958). *Phys. Fluids* **1,** No. 5, 370.

Anderson, J. E. (1969). *Progr. Heat Mass Transfer* **2,** 419.

Anderson, J. E., and Eckert, E. R. G. (1967). *AIAA J.* **5,** No. 4, 699.

Anshakov, A. S., Zhukov, M. F., and Timoshevsky, A. N. (1973). Arc dynamics in arc heater tunnel, *Proc. Int. Conf. Phenomen. Ionized Gases, 11th, Prague, Czechoslovakia* Contributed Papers, p. 225. Czech. Acad. Sci., Inst. of Phys., 18040 Prague 8, Na Slovance 2, CSSR.

Au, G. F., and Sprengel, U. (1966). *Z. Flugwiss* **14,** Heft 4, 188.

AVCO Report (1963). Theoretical and Experimental Investigation of Arc Plasma Generation Technology. AVCO RAD-TR-63-12, Part 2, Vol. 2.

AVCO Report (1967). Experiments to Establish Current-Carrying Capacity of Thermionically-Emitting Cathodes, AVSSD-0043-67-RR.

Ayrton, H. (1902). The Electric Arc. The Electrician Co., London.

Bartels, K., and Uhlenbusch, J. (1970a). *Atomkernegie (ATKE)* **16,** 263.

Bartels, K., and Uhlenbusch, J. (1970b). *Z. Angew. Phys.* **29,** Heft 2, 122.

Bauder, U. H., and Bartelheimer, D. L. (1973). *IEEE Trans. Plasma Sci.* **PS-1,** No. 4, 23.

Bauder, U. H., and Maecker, H. (1971). *Proc. IEEE* **59,** No. 4, 588.

Bauder, U. H., and Schreiber, P. W. (1971). *Proc. IEEE* **59,** No. 4, 633.

Beachler, J. C. (1968). Design and Shakedown Operation of the Air Force Flight Dynamics Laboratory's 2 Ft (4 Megawatt) Electro-Gasdynamic Facility, Air Force Flight Dynam. Lab. Rep. 68-3, July. Wright-Patterson Air Force Base, Ohio.

Beachler, J. C. (1970). Operating Characteristics of the Air Force Flight Dynamics Laboratory Reentry Nose Tip (RENT) Facility, NBS SP 336, Paper No. 55 893.

Beaudet, R., and Drouet, M. G. (1974). *IEEE Tr. Power Appl. Syst.* **93,** No. 4, 1054.

Behringer, K., Kollmar, W., and Mentel, J. (1968). *Z. Phys.* **215,** 127.

Benenson, D. M., and Baker, A. J. (1969). *AIAA J.* **7,** No. 12, 2335.

Benenson, D. M., and Baker, A. J. (1971). *AIAA J.* **9,** No. 8, 1441.

Bez, W., and Höcker, K. H. (1954a). *Z. Naturforsch.* **9a,** 64.

Bez, W., and Höcker, K. H. (1954b). *Z. Naturforsch.* **9a,** 72.

Bez, W., and Höcker, K. H. (1955a). *Z. Naturforsch.* **10a,** 706.

Bez, W., and Höcker, K. H. (1955b). *Z. Naturforsch.* **10a,** 714.

Bez, W., and Höcker, K. H. (1956a). *Z. Naturforsch.* **11a,** 118.

Bez, W., and Höcker, K. H. (1956b). *Z. Naturforsch.* **11a,** 192.

Bober, L., and Tankin, R. S. (1970). *J.Q.S.R.T.* **10,** 991.

Boffa, C., and Pfender, E. (1968). Enthalpy Probe and Spectrometric Studies in an Argon Plasma Jet, HTL TR No. 73, Univ. of Minnesota.

Boldman, D. R., Shepard, C. E., and Fakan, J. C. (1962). Electrode Configurations for a Wind-Tunnel Heater Incorporating the Magnetically Spun Electric Arc, NASA TN D-1222.

Bose, T. K. (1973). *Plasma Phys.* **15,** 819.

Bose, T. K., and Pfender, E. (1969). *AIAA J.* **8,** 1643.

Bower, W. W., and Incropera, F. P. (1969). *Wärme-Stoffübertrag.* **2,** 150.

Brückner, V. R. (1963). *Ber. Deut. Keramisch. Ges.* **40,** Heft 11, 603.

Brzozowski, W. S., and Celinski, Z. (1962). *Bull. Acad. Polon. Sci.* **10,** No. 5, 293.

Burhorn, F. (1959). *Z. Phys.* **155,** 42.

Burhorn, F., and Wienecke, R. (1960a). *Z. Phys. Chem.* **215,** 269.

Burnhorn, F., and Wienecke, R. (1960b). *Z. Phys. Chem.* **215,** 285.

Burnhorn, F., and Wienecke, R. (1960c). *Z. Phys. Chem.* **213,** 37.

Burhorn, F., Maecker, H., and Peters, Th. (1951). *Z. Phys.* **131,** 28.

Busz-Peuckert, G., and Finkelnburg, W. (1955). *Z. Phys.* **140,** 540.

Busz-Peuckert, G., and Finkelnburg, W. (1956). *Z. Phys.* **144,** 244.

Cambel, A. B. (1963). "Plasma Physics and Magneto-Fluid Mechanics," Series in Missile and Space Technol. McGraw-Hill, New York.

Cann, G. L. (1973). An Experimental Investigation of a Vortex Stabilized Arc in an Axial Magnetic Field, ARL 73-0043.

Cann, G. L. *et al.* (1963). Thermal Arc Jet Research. Aeronaut. Syst. Div., Tech. Documentary Rep. No. ASD-TDR-63-632.

Cann, G. L., Buhler, R. D., Harder, R. L., and Moore, R. A. (1964). Basic Research on Gas Flows Through Electric Arcs-Hot Gas Containment Limits, ARL 64-49.

Capitelli, M., and Devoto, R. S. (1973). *Phys. Fluids* **16,** No. 11, 1835.

Capitelli, M., and Ficocelli, E. (1970). Transport Properties of Mixed Plasmas: He-N_2, Ar–N_2 and Xe–N_2 plasmas at one atm, between 5000 and 35000 K. Centro di Studio per la Chimica dei Plasmi del Consiglio Nazionale delle Ricerche—Instituto di Chimica Generale e Inorganica—Univ. degli Studi, Bari, Italy.

Capitelli, M., Ficocelli, E., and Molinari, E. (1970a). Equilibrium Compositions and Thermodynamic Properties of Mixed Plasmas, I: He–N_2, Ar–N_2 and Xe–N_2 plasmas at 1 atm, between 5000 and 35000 K. Centro di Studio per la Chimica dei Plasmi del Consiglio Nazionale delle Ricerche—Instituto di Chimica Generale e Inorganica—Univ. degli Studi—Bari, Italy.

Capitelli, M., Ficocelli, E., and Molinari, E. (1970b). II: Argon-Oxygen Plasmas at $10^{-2} - 10$ atm between 2000 and 35000 K.

Capitelli, M., Ficocelli, E., and Molinari, E. (1970c). III: Argon-Hydrogen Plasmas at 10^{-2} – 10^3 atm between 2000 and 35000 K.
Chen, D. C., and Lawton, J. (1968). *Trans. Inst. Chem. Eng.* **46,** T270.
Chou, T. S., and Pfender, E. (1973). *Wärme-Stoffübertrag.* **6,** 69.
Christmann, H., and Hertz, W. (1965). Das Abklingen des Elektrischen Leitwertes in der Säule eines N_2-Kaskadenbogens nach Unterbrechung des Stromes, *Proc. Int. Conf. Ionizat. Phen. Ionized Gases, 7th, Beograd* **1,** 769. Gradevinska Knjiga Publ. House, Beograd, Yugoslavia, 1966.
Clark, K. J., and Incropera, F. P. (1972). *AIAA J.* **10,** No. 1, 17.
Cobine, J. D., and Burger, E. E. (1955). *J. Appl. Phys.* **26,** No. 7, 895.
Cowley, M. D. (1973). On Electrode Jets, *Conf. Phen. Ionized Gases, 11th, Prague, Czechoslovakia* Contributed Papers, p. 249. Czech. Acad. of Sci., Inst. of Phys., 18040 Prague 8, Na Slovance 2, CSSR.
Cremers, C. J., and Pfender, E. (1964). Thermal Characteristics of a High and Low Mass Flux Argon Plasma Jet, ARL 64-191.
Cremers, C. J., Hsia, H. S., and Mahan, J. R. (1973). *IEEE Trans. Plasma Sci.* **PS-1,** No. 3, 10.
Devoto, R. S. (1966). *Phys. Fluids* **9,** No. 6, 1230.
Devoto, R. S. (1967). *Phys. Fluids* **10,** No. 2, 354.
Devoto, R. S. (1973). *Phys. Fluids* **16,** No. 5, 616.
Devoto, R. S., and Mukherjee, D. (1973). *J. Plasma Phys.* **9,** Part 1, 65.
Djakov, B. E., and Nedelkov, I. (1974). *J. Phys. D: Appl. Phys.* **7,** No. 18, L199.
Drawin, H. W. (1970a). *High Temp.–High Pressures* **2,** 359.
Drawin, H. W. (1970b). *J. Quant. Spectrosc. Radiat. Transfer* **10,** 33.
Drawin, H. W. (1973). *Phys. Lett.* **42A,** No. 6, 423.
Drawin, H. W., and Emard, F. (1973). *Z. Naturforsch.* **28a,** 1289.
Drawin, H. W., and Emard, F. (1974). *Z. Phys.* **270,** 59.
Drawin, H. W., Emard, F., and Katsonis, K. (1973). *Z. Naturforsch.* **28a,** 1422.
Drellishak, K. S., Aeschliman, D. P., and Cambel, A. B. (1965). *Phys. Fluids* **8,** No. 9, 1590.
Eberhart, R. C., and Seban, R. A. (1966). *Int. J. Heat Mass Transfer* **9,** 939.
Ecker, G. (1961). *Ergebnisse exakten Naturwissensch.* **33,** 1.
Eckert, E. R. G., and Anderson, J. E. (1964). Performance Characteristics of a Fully-Developed Constricted Transpiration-Cooled Arc, AGARDograph 84, Part 2, p. 751.
Eckert, E. R. G., and Pfender, E. (1967). *Adv. Heat Transfer* **4,** 229.
Eckert, E. R. G., Schoeck, P. A., and Winter, E. R. F. (1962). *Int. J. Heat Mass Transfer* **5,** 295.
Eddy, T. L., Pfender, E., and Eckert, E. R. G. (1973). *IEEE Trans. Plasma Sci.* **PS-1,** No. 4, 31.
Edels, H. (1973). *Proc. Int. Conf. Phen. Ionized Gases, 11th, Prague, Czechoslovakia* Invited Papers, p. 9. Czech. Acad. of Sci., Inst. of Phys., 18040 Prague 8, Na Slovance 2, CSSR.
Edels, H., and Fenlon, F. H. (1965). *Brit. J. Appl. Phys.* **16,** 219.
Elenbaas, W. (1935). *Physica* **2,** 169.
Elwert, G. (1952a). *Z. Naturforsch.* **7a,** 432.
Elwert, G. (1952b). *Z. Naturforsch.* **7a,** 703.
Emmons, H. W. (1967). *Phys. Fluids* **10,** No. 6, 1125.
Emmons, H. W., and Land, R. I. (1962). *Phys. Fluids* **5,** No. 12, 1489.
Ernst, K. A., Kopainsky, J. G., and Maecker, H. (1973a). *IEEE Trans. Plasma Sci.* **PS-1,** No. 4, 3.
Ernst, K. A., Kopainsky, J., and Mentel, J. (1973b). *Z. Phys.* **265,** 253.
Eschenbach, R. C., and Skinner, G. M. (1961). Development of Stable High Power, High Pressure Arc Air Heaters for a Hypersonic Wind Tunnel, WADD-TR-61-100. Linde Co., Indianapolis, Indiana, July.

Eschenbach, R. C. *et al.* (1965). Performance Improvement of Air Heaters for Aerodynamic Wind Tunnels, Air Force Flight Dynamics Lab. Rep. 65-87. Linde Co., Div. of Union Carbide.

Evans, D. L., Marchand, J. M., Braun, W. G., and Oss, J. P. (1970). Radiation Property Measurements on an Arc Heated Argon Plasma, AIAA Paper No. 70-42, 8th Aerospace Sci. Meeting, New York.

Finkelnburg, W. (1948). "Hochstromkohlebogen." Springer-Verlag, Berlin and New York.

Finkelnburg, W., and Maecker, H. (1956). *Encyclo. Phys.* **22,** 254.

Finkelnburg, W., and Peters, Th. (1957). *Encyclo. Phys.* **28** (Spektrokopie II), 79.

Fisher, E., Hackmann, J., and Uhlenbusch, J. (1969). *Z. Naturforsch.* **24a,** Heft 9, 1427.

Foitzik, R. (1940). *Wiss. Veröff. Siemens-Konz.* **19,** 28.

Fowler, R. H., and Nordheim, L. (1928). *Proc. Roy. Soc. London Ser. A* **119,** 173.

Freeman, M. P. (1962). *Temp. Its Measurement Contr. Sci. Ind.* **3,** Part 2, 969.

Freeman, M. P., Li, S. U., and Jaskowsky, W. V. (1962). *J. Appl. Phys.* **33,** No. 9, 2845.

Garz, D. (1973). *Z. Naturforsch.* **28a,** 1459.

Gerdien, H., and Lotz, A. (1922). *Wiss. Veröff. Siemens-Konz.* **2,** 489.

Giannaris, R. J., and Incropera, F. P. (1971). *J. Quant. Spectrosc. Radiat. Transfer* **11,** 291.

Giannaris, R. J., and Incropera, F. P. (1973). *J. Quant. Spectrosc. Radiat. Transfer* **13,** 162.

Goldenberg, H. (1959). *Brit. J. Appl. Phys.* **10,** 47.

Goldstein, M. E., and Fay, J. A. (1967). *AIAA J.* **5,** No. 8, 1510.

Grey, J., and Jacobs, P. F. (1964). *AIAA J.* **2,** No. 3, 433.

Grey, J., and Jacobs, P. F. (1967). *AIAA J.* **5,** No. 1, 84.

Griem, H. R. (1964). "Plasma Spectroscopy." McGraw-Hill, New York.

Griem, H. R. (1974). "Spectral Line Broadening by Plasmas." Academic Press, New York.

Gross, B., Grycz, G., and Miklossy, K. (1969). "Plasma Technology." Iliffe, London and STNL, Prague.

Grosse-Wilde, H., and Uhlenbusch, J. (1973). *IEEE Trans. Plasma Sci.* **PS-1,** No. 3, 55.

Guile, A. E. (1971). *IEE Rev.* **118,** 1131.

Guile, A. E., and Hitchcock, A. H. (1974). *J. Phys. D: Appl. Phys.* **7,** No. 4, 597.

Hackmann, J., Michael, H., and Uhlenbusch, J. (1972). *Z. Phys.* **250,** 207.

Harder, R. L., and Cann, G. L. (1970). *AIAA J.* **8,** No. 12, 2220.

Heidrich, U. (1965). *Z. Naturforsch.* **20a,** Heft 3, 475.

Heller, G. (1935). *Physica* **6,** 389.

Hermann, W. (1968). *Z. Phys.* **216,** 33.

Hermann, W. (1973). *Z. Naturforsch.* **20a,** Heft 3/4, 443.

Hermann, W., and Schade, E. (1970). *Z. Phys.* **233,** 333.

Hermoch, V. (1959). *Czech. J. Phys.* **9,** 221.

Hermoch, V. (1973). *IEEE Trans. Plasma Sci.* **PS-1,** No. 3, 62.

Hermoch, V., and Teichmann, J. (1966). *Czech. J. Phys.* **195,** 125.

Hodnett, P. F. (1969). *Phys. Fluids* **12,** No. 7, 1441.

Holmes, A. J. (1974). *J. Phys. D: Appl. Phys.* **10,** No. 7, 1412.

Hoyaux, M. F. (1968). "Arc Physics." Springer-Verlag, Berlin and New York.

Huddlestone, R. H., and Leonard, S. L. (eds.) (1965). "Plasma Diagnostic Techniques." Academic Press, New York.

Humphreys, J. F., and Lawton, J. (1972). *J. Phys. D: Appl. Phys.* **5,** 701.

Incropera, F. P. (1973). *IEEE Trans. Plasma Sci.* **PS-1,** No. 3, 3.

Incropera, F. P., and Murrer, E. S. (1972). *J.Q.S.R.T.* **12,** 1369.

Jahn, R. E. (1961). *Proc. Int. Conf. Ionizat. Phen. Gases, 5th, Munich* p. 955. North-Holland Publ., Amsterdam.

Jahn, R. E. (1963). *Brit. J. Appl. Phys.* **14,** 585.
John, R. R. (1964). Thirty kW Plasmajet Rocket-Engine Development, Rep. No. RAD TR-64-6, AVCO Corp.
John, R. R., and Bade, W. L. (1961). *ARS J.* No. 31, 4.
Jordan, D. L., and Swift, J. D. (1973). *Int. J. Electron.* **35,** No. 5, 595.
Kesaev, I. G. (1965). *Sov. Phys.–Tech. Phys.* **9,** 1146.
Kimblin, C. W. (1974). *IEEE Trans. Plasma Sci.* **PS-2,** No. 4, 310.
Kimblin, C. W., and Edels, H. (1966). *Brit. J. Appl. Phys.* **17,** 1607.
King, L. A. (1955). *Appl. Sci. Res.* **5,** 189.
King, L. A. (1965). Theoretical Calculation of Arc Temperatures in Different Gases, Coll. Spectr. Int. G, 152.
Knoche, K. F. (1968). Enthalpie-Dichte-Diagramme für Hochtemperaturplasmen und Anwendungsbeispiele, VDI-Forschungsheft 526, VDI-Verlag, Düsseldorf/Germany.
Knopp, C. F., and Cambel, A. B. (1966). *Phys. Fluids* **9,** No. 5, 989.
Knowles, G. R., and Heberlein, J. V. (1973). *IEEE Trans. Plasma Sci.* **PS-1,** No. 3, 15.
Kopainsky, J. (1971a). *Z. Phys.* **248,** 417.
Kopainsky, J. (1971b). *Z. Phys.* **248,** 405.
Krichel, R., Druxes, H., and Schmitz, G. (1968). *Z. Phys.* **217,** 336.
Kruger, C. H. (1970). *Phys. Fluids* **13,** No. 7, 1737.
Kuhn, W. E. (1956). "Arcs in Inert Atmospheres and Vacuum." Wiley, New York.
Kuhn, V., and Motschmann, H. (1964). *Z. Naturforsch.* **19a,** Heft 5, 658.
Landt, U. (1970). *Chem. Ing.-Tech.* **42,** Jahrg. Nr. 9/10, 617.
Lawton, J. (1971). *J. Phys. D: Appl. Phys.* **4,** 1946.
Lee, T. H. (1957). *J. Appl. Phys.* **28,** 920.
Lee, T. H. (1958). *J. Appl. Phys.* **29,** 734.
Lee, T. H. (1959). *J. Appl. Phys.* **30,** 166.
Lee, T. H., and Greenwood, A. (1963). Space Charge and Ionization Regions Near the Arc Cathode, ARL Rep. 63-163.
Lee, T. H., Greenwood, A., Breingau, W. D., and Fullerton, H. P. (1964). Voltage Distribution, Ionization and Energy Balance in the Cathode Region of an Arc, ARL Rep. 64-152.
Lochte-Holtgreven, W., Editor. (1968). "Plasma Diagnostics." North-Holland Publ., Amsterdam.
Lord, W. T. (1969). *J. Fluid Mech.* **35,** Part 4, 689.
Ludwig, H. C. (1968). *Weld. Res. Suppl.* **244 s,** May.
Lukens, L. A., and Incropera, F. P. (1972). *Int. J. Heat Mass Transfer* **15,** No. 5, 935.
Mackeown, S. S. (1929). *Phys. Rev.* **34,** 611.
Maecker, H. (1955). *Z. Phys.* **141,** 198.
Maecker, H. (1956). *Z. Naturforsch.* **11a,** 457.
Maecker, H. (1959a). *Z. Phys.* **157,** 1.
Maecker, H. (1959b). The Properties of Nitrogen Up to 15,000 K, Rep. 324, AGARDograph.
Maecker, H. (1960a). *Z. Phys.* **158,** 392.
Maecker, H. (1960b). The properties of nitrogen up to 15,000 K, *Proc. Int. Conf. Ionizat. Phen. Gases, 4th, Uppsala, Sweden, 1959* **IIB,** 378. North-Holland Publ., Amsterdam.
Maecker, H. (1961). Fortschritte in der Bogenphysik, *Proc. Int. Conf. Ionizat. Phen. Gases, 5th, Munich* **II,** 1793. North-Holland Publ., Amsterdam, 1962.
Maecker, H. (1961). *Proc. IEEE* **59,** No. 4, 439.
Maecker, H., and Preibisch, H. (1968). *Z. Angew. Phys.* **25,** Heft 1, 29.
Maecker, H., and Steinberger, S. (1967). *Z. Angew. Phys.* **23,** Heft 6, 456.

Mahn, C., and Ringler, H. (1966). Measurements on a stationary DC hydrogen arc with flow in an axial magnetic flow, *Proc. Int. Conf. Phen. Ionized Gases, 7th, Beograd* **I,** 408.

Mahn, C., Ringler, H., Wienecke, R., Witkowski, S., and Zankl, G. (1964). *Z. Naturforsch.* **19a,** Heft 10, 1202.

Mailänder, M. (1973). A new experiment on retrograde motion, *Proc. Int. Conf. Phen. Ionized Gases, 11th, Prague, Czechoslovakia,* Contributed Papers, p. 81. Czech. Acad. of Sci., Inst. of Phys., 18040 Prague 8, Na Slovance 2, CSSR.

Malghan, V. R., and Benenson, D. M. (1973). *IEEE Trans. Plasma Sci.* **PS-1,** No. 3, 38.

Marlotte, G. L., Harder, R. L., and Pritchard, R. W. (1964). The Radiating Arc Column, AGARDograph 84, Part 2, 633.

Marlotte, G. L., Cann, G. L., and Harder, R. L. (1968). A Study of Interactions Between Electric Arcs and Gas Flows, ARL Rep. 68-0049. Thermomech. Res. Lab., Aerospace Res. Lab., Wright-Patterson Air Force Base, Ohio (AD 673368).

Marr, G. V. (1968). "Plasma Spectroscopy." Elsevier, Amsterdam.

Mason, E. A., Munn, R. J., and Smith, F. J. (1967). *Phys. Fluids* **10,** No. 8, 1827.

Meador, E. W., Jr., and Staton, L. D. (1965). *Phys. Fluids* **8,** No. 9, 1694.

Morris, J. C. (1969). *J. Quant. Spectrosc. Radiat. Transfer* **9,** 1629.

Morris, J. C., Rudis, R. P., and Yos, J. M. (1970). *Phys. Fluids* **13,** No. 3, 608.

Motschmann, H. (1967a). *Z. Phys.* **205,** 235.

Motschmann, H. (1967b). *Z. Phys.* **200,** 93.

NASA Contractor Report (1966). CR-575, prepared by AVCO Corp., Wilmington, Massachusetts.

Nathrath, N. (1970). *Z. Naturforsch.* **25a,** Heft 11, 1609.

Nelson, H. F. (1972). *J. Spacecr. Rockets* **9,** No. 3, 177.

Nelson, H. F., and Goulard, R. (1968). *J. Quant. Spectrosc. Radiat. Transfer* **8,** 1351.

Neurath, P. W., and Gibbs, T. W. (1963). *J. Appl. Phys.* **34,** 277.

Nicolai, L. M., and Kuethe, A. M. (1969). *Phys. Fluids* **12,** No. 10, 2072.

Nizovsky, V. L., Khodakov, K. A., and Shabashov, V. I. (1973). Non-equilibrium in a stabilized deuterium arc, *Proc. Int. Conf. Phen. Ionized Gases, 11th, Prague, Czechoslovakia* Contributed Papers, p. 205. Czech. Acad. of Sci., Inst. of Phys. 18040 Prague, Na Slovance 2, CSSR.

Novak, J., and Fuchs, V. (1974). *Proc. IEE* **1,** 81.

Olsen, H. N. (1959). *Phys. Fluids* **2,** No. 6, 614.

Painter, J. H. (1974). High-Pressure Arc Heater Electrode Heat Transfer Study, *AIAA/ASME Thermophys. Heat Transfer Conf., July 15* AIAA Paper No. 74-731.

Patt, H. J., and Schmitz, G. (1965a). *Z. Phys.* **185,** 1.

Patt, H. J., and Schmitz, G. (1965b). *Z. Phys.* **188,** 1.

Paulson, R., and Pfender, E. (1973). *IEEE Trans. Plasma Sci.* **PS-1,** No. 3, 65.

Peters, Th. (1956). *Z. Phys.* **144,** 621.

Peters, Th. (1961). *Astronaut. Acta* **7,** 150.

Peters, Th. (1964). Arc Heaters Producing Flat Temperature Distributions, AGARDograph 84, Part 2, 883.

Petrie, T. W., and Pfender, E. (1970). *Weld. J. Res. Suppl.* **49,** 588-S.

Pfender, E. (1963a). Generation of an almost fully ionized, spectrally clean, high density hydrogen plasma, *Proc. Int. Conf. Ionizat. Phen. Gases, 6th, Paris, France* **34,** 369. North–Holland Publ., Amsterdam, 1964.

Pfender, E. (1963b). Die Aufheizung von Wasserstoff in einer wirbelstabilisierten Hochstromentladung, Jahrbuch der WGLR, 389, F. Vieweg and Sohn, Braunschweig, Germany.

Pfender, E., and Bez, W. (1961). Ein wirbelstabilisierten Bogen in reiner Wasserstoffatmosphäre als Vorionisierungsstrecke für eine stromstarke Impulsentladung, *Proc. Int. Conf. Ionizat. Phen. Gases, 5th, Munich, Germany* **1,** 897. North–Holland Publ., Amsterdam, 1962.
Pfender, E., and Cremers, C. J. (1964). Thermal Characteristics of a High and Low Mass Flux Argon Plasma Jet, ARL 64-191, Aerospace Res. Lab.
Pfender, E., and Cremers, C. J. (1965). *AIAA J.* **3,** 1345.
Pfender, E., and Schafer, J. (1975). *J. Heat Transfer Ser. C.* **97,** No. 1, 41.
Pfender, E., Eckert, E. R. G., and Raithby, G. D. (1965). Energy transfer studies in a wall-stabilized, cascaded arc, *Proc. Int. Conf. Ionizat. Phen. Gases, 7th, Beograd* **I,** 691. Gradevinska, Knjiga Publ. House, Beograd, Yugosalvia, 1966.
Pfender, E., Gruber, G., and Eckert, E. R. G. (1969). Experimental investigation of a transpiration-cooled, constricted arc, *Proc. Int. Symp. High Temp. Tech., 3rd* p. 593. I.U.P.A.C., Butterworths, Washington, D.C.
Plantikow, U. (1969). *Z. Phys.* **227,** 271.
Plantikow, U. (1970). *Z. Phys.* **237,** 388.
Plantikow, U., and Steinberger, S. (1970). *Z. Physik.* **231,** 109.
Polman, R. W., Busser, J. J., and Hellingman, P. (1973). Arc stabilization with a toroidal vortex, *Proc. Int. Conf. Phen. Ionized Gases, 11th, Prague, Czechoslovakia* Contributed Papers, p. 234. Czech. Acad. of Sci., Inst. of Phys., 18040 Prague 8, Na Slovance 2, CSSR.
Raeder, J. (1968). *Z. Naturforsch.* **23a,** Heft 3, 424.
Ragaller, K. (1974). *Z. Naturforsch.* **29a,** 556.
Reece, M. P. (1963). *IEE* **110,** 793.
Richter, J. (1961). *Z. Astrophys.* **53,** 262.
Richter, R. (1969). Ultra-High Pressure Arc Heater Studies, AECD TR 69-180. Arnold Eng. Develop. Center, Arnold Air Force Station, Tennessee, September.
Richter, R. (1970). "Ultra-High Pressure Arc Heater Studies, (Phase III)," AEDC TR 70-106, Arnold Engineering Develop. Center, Arnold Air Force Station, Tennessee, March.
Rieder, W. (λ967). "Plasma und Lichtbogen." Friedr. Vieweg, Braunschweig, Germany.
Roman, W. C., and Meyers, T. W. (1966). Survey of Investigations of Electric Arc Interactions with Magnetic and Aerodynamic Fields, Aerospace Res. Lab. TR 66-0184.
Roman, W. C., and Myers, T. W. (1967). *AIAA J.* **5,** No. 11, 2011.
Rosenbauer, H. (1971). *Z. Phys.* **245,** 295.
Rother, H. (1957). *Ann. Phys.* **20,** 230.
Runstadler, P. W. (1965). Laminar and Turbulent Flow of an Argon Arc Plasma, T.R. No. 22. Eng. Sci. Lab., Harvard Univ., Cambridge, Massachusetts.
Sauter, K. (1969). *Z. Naturforsch.* **24a,** Heft 11, 1694.
Sayce, I. G. (1971). Plasma Processes in Extractive Metallurgy, *Adv. Extr. Met. Refining, Proc. Int. Symp.* p. 241.
Schmitz, G. (1952). *Z. Phys.* **132,** 23.
Schmitz, G., and Uhlenbusch, J. (1960). *Z. Phys.* **159,** 554.
Schmitz, G., Patt, H. J., and Uhlenbusch, J. (1963). *Z. Phys.* **173,** 552.
Schoeck, P., and Maisenhälder, F. (1966). *Beitr. Plasmaphys.* Heft 5, 345.
Schoenherr, O. (1909). *Elektrotech. Z.* **30,** 365.
Schottky, W. (1914). *Phys. Z.* **15,** 872.
Schrade, H. O. (1973). *IEEE Trans. Plasma Sci.* **PS-1,** No. 3, 47.
Schreiber, P. W., Hunter, A. M., II, and Benadetto, K. R. (1973). *IEEE Trans. Plasma Sci.* **PS-1,** No. 4, 60.
Seeger, G. (1968). *Z. Angew. Phys.* **25,** Heft 1, 23.
Seeger, G. (1970). *Z. Angew. Phys.* **29,** Heft 6, 357.
Shaboltas, A. S., and Yas'ko, O. I. (1974). *J. Eng. Phys.* **19/6,** 1529.

Sharakhovskii, L. I. (1974). *J. Eng. Phys.* **20/2,** 222.
Shashkov, A. G., and Yas'ko, O. I. (1973). *IEEE Trans. Plasma Sci.* **PS-1,** No. 3, 21.
Sheer, C., and Korman, S. (1973). *IEEE Trans. Plasma Sci.* **PS-1,** No. 3, 76.
Shepard, C. E. (1972). *AIAA J.* **10,** No. 2, 117.
Shepard, C. E., Boldman, D. R., and Fakan, J. C. (1962). Electrode Configurations for a Wind-Tunnel Heater Incorporating the Magnetically Spun Electric Arc, NASA TN D-1222.
Shepard, C. E., Watson, V. R., and Stine, H. A. (1946). Evaluation of a Constricted-Arc Supersonic Jet, NASA TN D-2066.
Shepard, C. E., Vorreiter, J. W., Stine, H. A., and Winovich, W. (1967). A Study of Artificial Meteors as Ablators, NASA TN D-3740.
Shepard, C. E., Ketner, D. M., and Vorreiter, J. W. (1968). A High Enthalpy Plasma Generator for Entry Heating Simulation, NASA TN D-4583.
Shih, K. T. (1972). *J. Appl. Phys.* **43/12,** 5002.
Shih, K. T., and Dethlefsen, R. (1971). *J. Heat Transfer Ser. C* **93,** No. 1, 119.
Shih, K. T., and Pfender, E. (1970). *AIAA J.* **8,** No. 2, 211.
Shumaker, J. B., Jr. (1961). *Rev. Sci. Instrum.* **32,** No. 1, 65.
Smith, J., and Pfender, E. (1976). *IEEE Trans. Power Apparatus Syst.* **PAS-95,** No. 2, 704.
Smith, R. T., and Folek, J. L. (1969). Operating characteristics of a multi-megawatt arc heater used with the AFFDL fifty megawatt facility, *Proc. Ann. Tech. Meeting, 15* p. 281. Inst. of Environ. Sci.
Sommerville, J. M. (1959). "The Electric Arc." Wiley, New York.
Srivastava, S., and Weissler, G. L. (1973). *IEEE Trans. Plasma Sci.* **PS-1,** No. 4, 17.
Stäblein, H. G. (1973). Theoretical investigation of the full circle arc, *Proc. Int. Conf. Phen. Ionized Gases, 11th, Prague, Czechoslovakia* Contributed Papers, p. 241. Czech. Acad. Sci., Inst. of Phys., 18040 Prague 8, Na Slovance 2, CSSR.
Stine, H. A. (1963). The hyperthermal supersonic aerodynamic tunnel, *NASA Proc. Int. Symp. High Temp. Tech., September.* Butterworths, Washington, D.C., 1964.
Stine, H. A., and Watson, V. R. (1962). The Theoretical Enthalpy Distribution of Air in Steady Flow Along the Axis of a Direct-Current Electric Arc, NASA, TN D-1331.
Stine, H. A., Watson, V. R., and Shepard, C. E. (1964). Effect of Axial Flow on the Behavior of the Wall-Constricted Arc, AGARDograph 84, Part 1, 451.
Strachan, D. C., and Barrault, M. R. (1973). Axial velocities in the high current free burning arc, *Proc. Int. Conf. Phen. Ionized Gases, 11th, Prague, Czechoslovakia* Contributed Papers, p. 239. Czech. Acad. Sci., Inst. of Phys. 18040 Prague, Na Slovance 2, CSSR.
Strachan, D. C., and Blackburn, T. R. (1973). Radiation losses from high current free burning arcs, *Proc. Int. Conf. Phen. Ionized Gases, 11th, Prague, Choslovakia* Contributed Papers, p. 238, Czech. Acad. of Sci., Inst. of Phys., 18040 Prague, Na Slovance 2, CSSR.
Sugawara, M. (1967). *Brit. J. Appl. Phys.* **18,** 1777.
Tatento, H., and Saito, K. (1963). *Jpn. J. Appl. Phys.* **2,** 192.
Taussig, R. T. (1970). *J. Quant. Spectrosc. Radiat. Transfer* **10,** 449.
Thorpe, M. L. (1971). High-temperature technology and its relationship to mineral exploitation, *Adv. Extr. Met. Refining, Proc. Int. Symp.* p. 275.
Tiller, W. (1973). Experimental investigations on the radially free full circle arc, *Proc. Int. Conf. Phen. Ionized Gases, 11th, Prague, Czechoslovakia* Contributed Papers, p. 240. Czech. Acad. of Sci., Inst. of Phys., 18040 Prague 8, Na Slovance 2, CSSR.
Todd, J. P. (1964). Thirty kW Arc-Jet Thruster Research, Rep. No. FRO 24-10338 (APL-TDR-64-58), Giannini Scientific Corp.
Uhlenbusch, J. (1972a). Determination of transport coefficients from arc measurements, *Proc. Yugoslav Symp. Summer School. Phys. Ionized Gases, 6th,* p. 479. Inst. of Phys., Beograd, Yugoslavia.

Uhlenbusch, J. (1972b). Calculation of arc properties using balance equations, *Proc. Yugoslav Symp. Summer School Phys. Ionized Gases, 6th* p. 441. Inst. of Physics, Beograd, Yugoslavia.
Uhlenbusch, J., and Detloff, L. (1966). *Z. Naturforsch.* **21a,** Heft 6, 843.
Uhlenbusch, J., and Fischer, E. (1971). *Proc. IEEE* **59,** No. 4, 578.
Uhlenbusch, J., and Gieres, G. (1969). *Z. Angew. Phys.* **27,** 66.
Uhlenbusch, J., Fischer, E., and Hackman, J. (1970a). *Z. Phys.* **238,** 404.
Uhlenbusch, J., Fischer, E., and Hackmann, J. (1970b). *Z. Phys.* **239,** 120.
von Engel, A. (1965). "Ionized Gases." Oxford Univ. Press (Clarendon), London and New York.
Vorreiter, J. W., and Shepard, C. E. (1965). Performance characteristics of the constricted-arc supersonic jet, *Proc. Heat Transfer Fluid Mech. Inst.* p. 42.
Watson, V. R. (1965). Comparison of detailed numerical solutions with simplified theories of the constricted-arc plasma generator, *Proc. Heat Transfer Fluid Mech. Inst.* p. 24. Stanford Univ. Press, Stanford, California.
Watson, V. R., and Pegot, E. B. (1967). Numerical Calculations for the Characteristics of a Gas Flowing Axially through a Constricted Arc, NASA, TN, D-4042.
Waymouth, J. F. (1971). "Electric Discharge Lamps." M.I.T. Press, Cambridge, Massachusetts.
Weatherly, M. H., and Anderson, J. E. (1965). *Electrochem. Technol.* **3,** No. 3–4, 80.
Weiss, R. (1954). *Z. Phys.* **138,** 170.
Wiese, W. L., Kelleher, D. E., and Paquette, D. R. (1972). *Phys. Rev. A* **6,** No. 3, 1132.
Winograd, Y. Y., and Klein, J. F. (1969). *AIAA J.* **7,** No. 9, 1699.
Wutzke, S. A., Pfender, E., and Eckert, E. R. G. (1967). *AIAA J.* **5,** No. 4, 707.
Wutzke, S. A., Pfender, E., and Eckert, E. R. G. (1968). *AIAA J.* **6,** No. 8, 1474.
Wynands, A., Druxes, H., and Schmitz, G. (1970). *Z. Phys.* **239,** 306.
Wynands, A., Klein, U., and Schmitz, G. (1973). *IEEE Trans. Plasma Sci.* **PS-1,** No. 4, 53.

Chapter 6

Relativistic Electron Beam Produced Plasmas

GEROLD YONAS and ALAN J. TOEPFER
FUSION RESEARCH DEPARTMENT
SANDIA LABORATORIES
ALBUQUERQUE, NEW MEXICO

I. INTRODUCTION

In this chapter we will discuss in fairly general terms the properties and applications of intense relativistic electron beams (REBs). At the outset we wish to differentiate this area of interest from the use of nonrelativistic beams generated in a cw mode using thermionic cathodes which produce typically less than 1 A/cm^2 at the cathode. We will instead emphasize beams generated using 10^5 to 10^7 V potentials, produced by nanosecond pulse-forming techniques such as pulse-charged transmission lines, to form $\lesssim$100-ns pulses with $\sim 10^3$ A/cm^2 at the cathode surface.

ISBN 0-12-349701-9

The very existence of such high-power beams of electrons, capable of delivering in excess of 10 TW from the largest REB accelerator, is not generally known since the development of this technology arose entirely from military applications and, although unclassified, has not been widely publicized. On the other hand, the technical basis for REB accelerators is relatively simple in concept; they are far less costly to obtain than other forms of high-power beams and the applications are sufficiently broad in scope that a wider knowledge of this technology is certain to arise in the future.

This article is intended to serve as an introduction to newcomers to the field and as such will emphasize present-day applications fairly qualitatively and without a great deal of theoretical detail. First, we should give a general introduction to the concept of an REB accelerator and its uses before proceeding to specific applications in greater depth.

As defined above, REB accelerators generally consist of either a capacitively or inductively stored energy supply which charges a pulse-forming line (a cable of sorts) which then feeds into a vacuum diode through a fast-closing switch. The electrons are accelerated from a cold cathode surface which becomes covered with a layer of plasma early in the voltage pulse. The concept is shown conceptually in Fig. 1. If the anode is thick compared with the electron stopping distance (for instance, 90% of the energy contained by a 1-MeV beam will stop in a thickness of material corresponding to an areal density of roughly 0.3 gm/cm^2), the beam energy will be converted to thick target bremsstrahlung and thermal energy of the target. This technique of classical stopping of an REB at the anode has been a major application of REBs for the generation of intense bursts of x rays, and has more recently come under consideration for heating of solid targets of thermonuclear fuel for inertially confined thermonuclear fusion. On the other hand, a thin, low atomic number anode foil can be employed, thus permitting the beam to enter a target volume just beyond the anode with negligible loss in energy. The beam can then be drifted over some distance in a neutral gas or preformed plasma, and external magnetic fields can be applied to assist in either beam transport or beam confinement. In this beam mode, REB accelerators have been used to heat magnetically confined plasmas, to produce intense microwave bursts, to collectively accelerate ions to many times the beam kinetic energy, to excite high-pressure gas lasers, and to ignite chemical mixtures to produce chemical lasers.

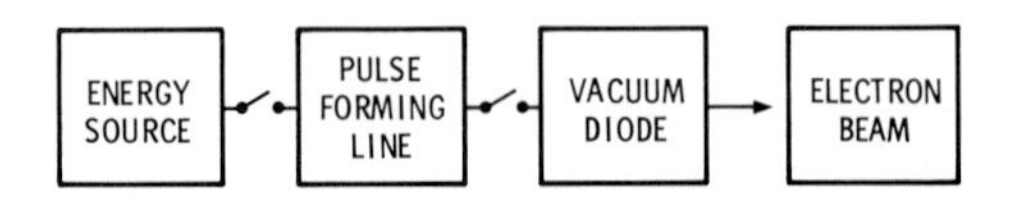

Fig. 1. Block diagram showing major REB accelerator components.

We will discuss these applications after first presenting the technological basis for REB accelerators (generally called pulsed power), the behavior and properties of high-current diodes, and the basic ideas of beam dynamics and interaction with first a neutral gas and then preformed plasma.

II. PULSED POWER TECHNOLOGY AS APPLIED TO REB ACCELERATORS

As mentioned above, the basic initial element of an REB accelerator is a high-voltage, single-pulse power supply which can deliver from a few kilojoules to over 100-kJ of electrical energy to a pulse-forming line. In the majority of present-day applications, this is accomplished using a Marx generator, which is simply a group of capacitors charged in parallel and then connected in a fraction of a microsecond by switches into a series configuration. These switches either cascade naturally or are triggered from an outside source (see Fig. 2). Each capacitor is charged to V_0 which gives a Marx output voltage of nV_0 delivered from a total parallel capacitance of nC. The Marx generator is typically insulated with transformer oil or high-breakdown-strength gas, and each capacitor is nominally charged to between 50 and 100 kV. When the output voltage rises in less than 1 μs at an untriggered switch S_A and exceeds the strength of the dielectric, breakdown takes place, delivering the pulse to the transmission line with an impedance Z_0. When the switch S_B closes, a voltage wave $nV_0/2$ will be delivered to a matched load and a discharge wave $-nV_0/2$ will propagate back toward the charging supply. After the wave makes two transits of the pulse line, which occurs in a time equal to twice the line length divided by the speed of light in the dielectric, the transmission line will be discharged. During the pulse, the output power will be given by $P = (nV_0/2)^2/Z_0$.

In order to increase the voltage and power which can be delivered with such a system, a Blumlein circuit is typically used to replace a simple transmission line. The Blumlein configuration is essentially a double transmission line version of a Marx generator in which two lines are charged in

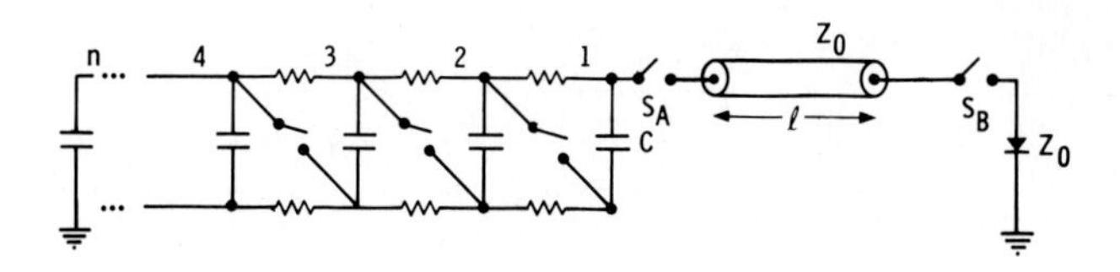

Fig. 2. REB accelerator schematic shown as a Marx generator charged cable feeding a diode.

parallel and discharged in series by means of a single switch (Fig. 3). In this case the intermediate conductor (2) is initially charged by the capacitor bank to nV_0; when S_A closes, a wave propagates to the point of termination of this line. After the wave is partially reflected and partially transmitted, $-nV_0$ is applied between the output conductor (1) and ground. This voltage will be applied to the load until the wave can make two additional transists of the line. The output power is then $P_{(\mathrm{Blumlein})} = (nV_0)^2/Z_0$. Multiple lines of this kind employing triggered switching can also be used to feed one or more diodes as shown in Fig. 4 (Prestwich, 1975).

The combination of a high-voltage Marx generator/Blumlein concept, as well as many of the present-day pulsed power techniques used in REB accelerators were pioneered in England by J. C. Martin and his group at the Atomic Weapons Research Establishment beginning in the early 1960s. REB development began later in the United States at Physics International Company, Ion Physics Company, Maxwell Laboratories, Field Emission Company, the Naval Research Laboratory, Sandia Laboratories, and at Cornell University.

One of the first Marx–Blumlein accelerators in the terawatt range was Hermes II at Sandia Laboratories developed by Martin (1969). The largest such accelerator presently in operation was developed at Physics International Company and consists of four separate pulse-forming lines delivering four beams of 12 MV and 400 kA each, with a total energy of ~2 MJ in a single 160-ns pulse (Fig. 5) (Bernstein and Smith, 1973).

As stated already, the energy content of the beam will be limited by the capacitance of the pulse-forming line, which depends on the dielectric constant and allowable electric field. These parameters are listed below for typical materials used in pulse-forming lines for microsecond charging pulses under ideal conditions (Martin and Smith, 1965–1970); see Table I. The apparent advantages of Mylar or other solid dielectrics have not been generally useful in practice because a volume breakdown results in irreparable damage to the insulating material, whereas a breakdown in a liquid dielectric is easily cured if the liquid is circulated and the bubbles or the debris particles formed are removed. Another important feature of liquid

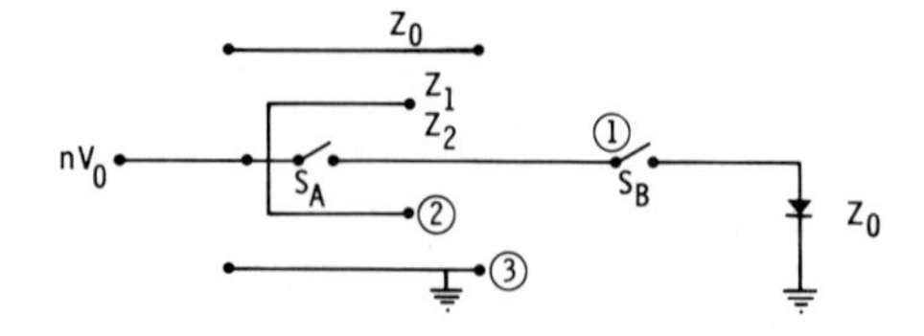

Fig. 3. Schematic of Blumlein pulse forming line driving an electron beam diode.

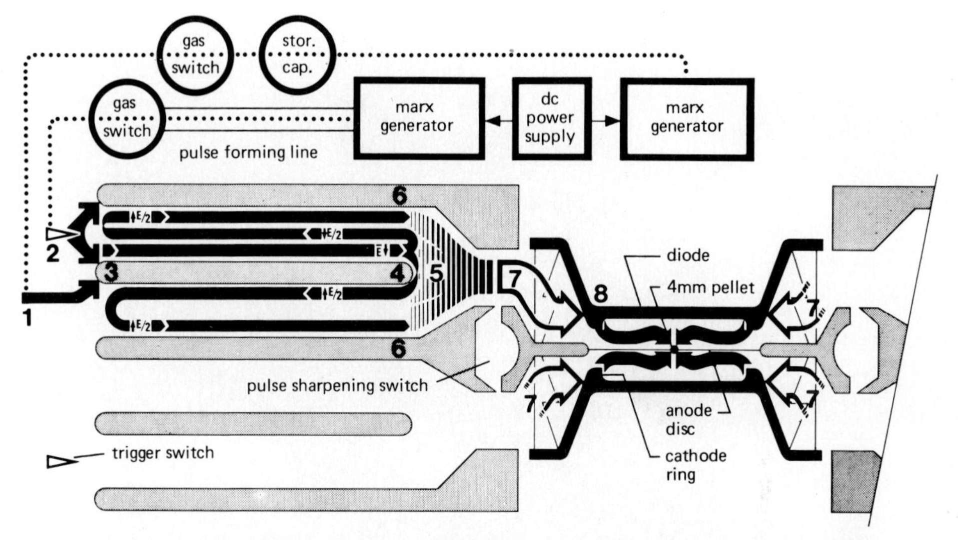

Fig. 4. A multiple Blumlein pulse forming line configuration illustrating the method of pulse generation: 1, The center transmission line is charged; 2, a high voltage pulse is applied to the trigger switch which closes the circuit from the charged line to ground; 3, an electromagnetic wave is initiated which traverses the length of the transmission line three times; 4, as the wave begins the second traverse, the energy pulse begins to flow into the diode (5); 6, as the wave completes the third traverse, all of the stored energy has been released; 7, the pulse of energy is generated simultaneously by the multiple Blumlein transmission lines; 8, the electron beam is generated at the two cathodes delivering energy to a common anode.

dielectrics is that the critical electric field for breakdown is time dependent and obeys the following relationship due to the dependence of streamer propagation velocity on local electric field:

$$E_B t^{1/3} A^{1/10} = K, \tag{1}$$

where E_B is in megavolts per centimeter, t is in microseconds, and A is the electrode area in square centimeters. This expression is for parallel plane electrodes; however, where the electric field diverges, as for instance in coaxial systems, water shows a polarity effect as shown in Table II (Martin and Smith, 1965–1970).

If one can then charge the pulse-forming line in, say, one-tenth the time, then because of the time dependence of breakdown, the stored energy density will increase roughly by a factor of five, thus permitting smaller and generally lower inductance components. To accomplish the required rapid charging of pulse-forming lines, either a low-inductance Marx can be used or a transfer capacitor of large enough interelectrode dimensions to avoid

Fig. 5. The Aurora REB accelerator.

breakdown can be placed between the slower Marx and the transmission line to be charged.

In order to deliver the energy rapidly to either the line or transfer capacitor, the switch S_A must itself have a fast risetime. The risetime will depend on the switch inductance and the time for the discharge channels in the dielectric medium to lose their resistivity (resistive risetime). Techniques have been developed to distribute the current in liquid dielectric switches so that the inductive effect is negligible and, by enhancing the electric field in the switch, the resistive phase can be as short as a few nanoseconds in water or oil, even at voltages of several megavolts (Prestwich, 1974; VanDevender and Martin, 1975). A high-pressure gas such as SF_6 can also be used with additional complications of gas supply and reliability of the high-pressure solid dielectric housing, but in a gas discharge the resistive phase is essentially negligible. The total switch risetime is the sum of the inductive risetime T_L and the resistive risetime T_R. T_L is equal to L/Z, where L is the switch inductance and Z is the impedance of the line feeding the switch. Useful empirical formulas for T_R are given in Table III (Martin, 1970).

TABLE I

PROPERTIES OF DIELECTRICS USED IN PULSED POWER APPLICATIONS

DIELECTRIC	ϵ	BREAKDOWN FIELD E_B MV/m	ENERGY DENSITY $= \frac{1}{2} \epsilon E_B^2$ MJ/m^3
OIL[a]	2.4	41	0.018
GLYCERIN[a]	44.0	10	0.021
WATER[a]	80.0	22	0.16
MYLAR[b]	2.8	320	1.3

[a] Effective pulse duration of $\sim 0.3\ \mu s$; electrode area of 1000 cm^2.

[b] DC breakdown volume of 800 cm^3.

TABLE II

BREAKDOWN CONSTANT FOR COAXIAL PULSE CHARGED LINES FOR POSITIVE (K_+) AND NEGATIVE (K_-) CENTER CONDUCTOR

	OIL	WATER
K_+	0.5	0.3
K_-	0.5	0.6

TABLE III

RESISTIVE RISE TIMES FOR MATERIALS USED IN SWITCHES[a]

	T_R (ns)
GASES	$\frac{0.19}{Z^{1/3} E^{4/3}} (\rho/\rho_0)^{1/2}$
SOLIDS	$\frac{5}{Z^{1/3} E^{4/3}}$
LIQUIDS	$\frac{5}{Z^{1/3} E^{4/3}}$

[a] E is in megavolts per centimeter and ρ/ρ_0 is the ratio of the gas density to air at STP.

The above concepts have been used to produce accelerators in the 10^{12} W range; however, the problem becomes more difficult for higher powers. In order to produce an output power in the 10-TW range with a diode voltage of a few megavolts or less, the diode and line impedance must be less than 1 Ω. Such a low impedance cannot be conveniently achieved with a single line because of the excessively large dimensions required, and two other approaches are normally taken. Either a transformer which is typically a tapered coaxial section is inserted between the pulse-forming line and the diode (Block *et al.*, 1971; Martin, 1973) or a number of higher impedance lines are connected in parallel, using synchronized low-jitter switching, to the diode (Yonas *et al.*, 1971; Martin and Prestwich, 1974; Prestwich, 1975) (see Fig. 6).

If these techniques are to permit accelerators to be extended to achieve short pulses at 10^{14} W as needed for some of the applications discussed later, then the inductive limitation of switches will require that literally thousands of switch channels be generated in order to achieve fast-rising pulses with less than 0.1-Ω transmission lines. One approach under consideration is to use separate gas switches, but questions of cost and reliability may make this prohibitive. Fortunately, recent studies based on the earlier work of Martin (1970) indicate that rapidly rising electric fields in liquid switches can be used to produce discharge channels which are closely spaced. Prestwich (1974) and Smith (1974) have reported on triggered multichannel oil and water switching, and Johnson (1974) and VanDevender and Martin (1975) have investigated untriggered oil and water, respectively (see Fig. 7).

Once the propagating wave in the transmission line reaches the diode, it then passes into the vacuum region through a solid dielectric interface

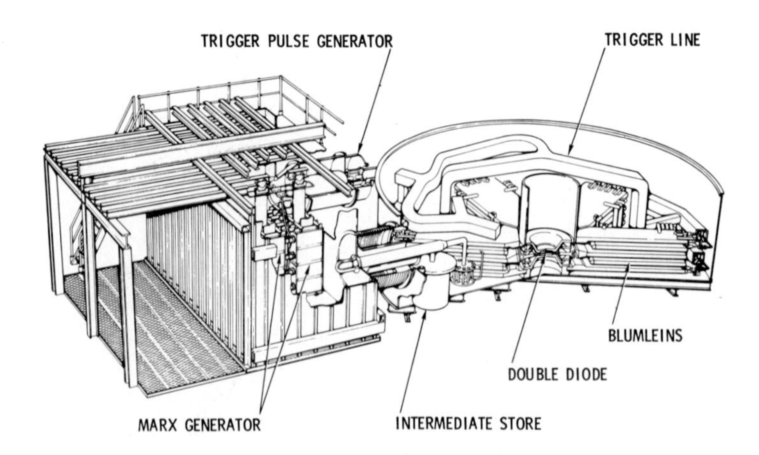

Fig. 6. Artist's conception of Sandia Laboratories' Proto-I, showing eight of twelve strip Blumlein pulse forming lines.

between the liquid dielectric and the vacuum. The breakdown strength on the vacuum side limits the energy density and here, as in much of the work cited thus far, the basis for insulator design has been largely empirical. Two types of diodes commonly employed are shown in Fig. 8 (Martin, 1969, 1973).

Finally the electric field is applied to the diode and a beam is formed.

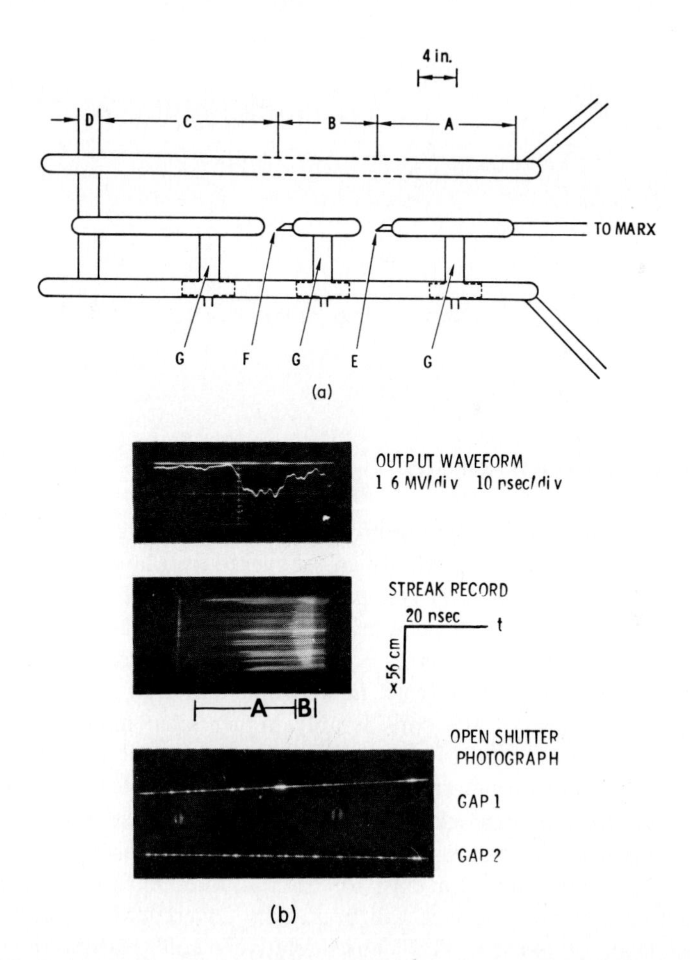

Fig. 7. (a) Apparatus used to demonstrate untriggered water switches with fast rise time and low jitter. The lines extend 120 cm into the paper. A is the transfer capacitor section; B is the pulse-forming line; C is the output line terminated in a load resistor D. E and F are edge-plane gaps and G represents resistive voltage monitors that are uniformly graded in the high field region between the lines. The gaps are viewed through screens or slits in the upper line. (b) The output waveform at 1.6 MV/division corresponds to 800 kA/division in the 2 Ω line. The streak record first records the streamers in the second gap (B). The open shutter photograph of the slits is also shown for comparison with the streak record.

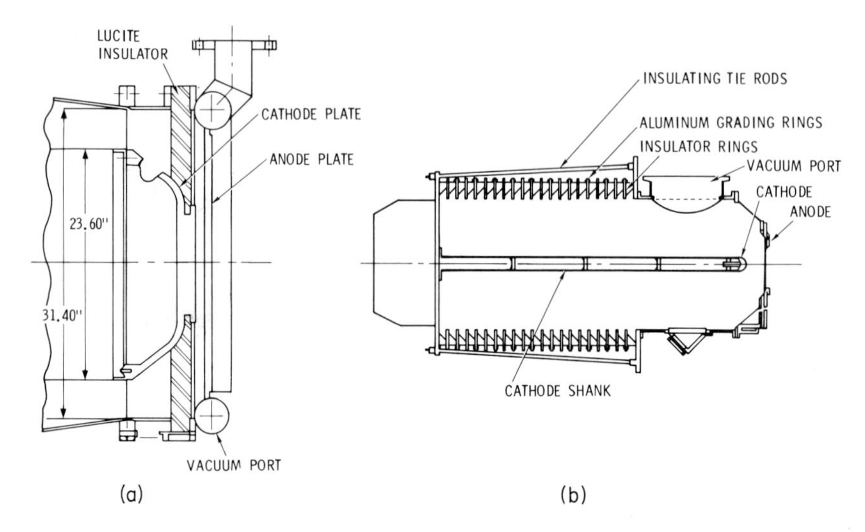

Fig. 8. Diode configurations employed on (a) low impedance and (b) high impedance accelerators.

III. DIODE PHENOMENOLOGY

As mentioned earlier, the electrons in an intense beam vacuum diode are emitted from a cold cathode which is covered by a layer of plasma. The existence of this plasma layer is a naturally occurring phenomenon which is critical to the concept of generating beams of high currents. Whenever macroscopic electric fields of $\geq 2 \times 10^5$ V/cm are applied to any solid surface, field emission begins from microscopic material imperfections or "whiskers," because of the considerably higher local microscopic fields. Messyats *et al.* (1970) and Messyats (1975) have shown that these whiskers can supply field emission current densities of 10^9 A/cm^2, leading to resistive heating. Subsequent explosion of the whiskers produces plasmas having thermal expansion velocities of 2 to 4 cm/μs. Since thousands of such microscopic emission sites can exist for each square centimeter of cathode surface area, any such surface will be rapidly covered with a uniform expanding layer of plasma. This dense, relatively cold plasma then serves as a copious electron supply, or effective cathode, and the emission current density will be governed by space charge limitations as described by the Child–Langmuir relation. For a right circular cathode of radius R and anode–cathode gap d the impedance will therefore be given by

$$Z\,(\Omega) = (k(V)/V^{1/2})(d(t)/R)^2 \quad (V \text{ in MV}), \tag{2}$$

where $k(V)$ is a slowly increasing function of voltage (Boers and Kelleher, 1969) and is equal to 136 for nonrelativistic beams. Parker *et al.* (1974) have correlated the measured diode impedances for smooth-surfaced cathodes and found an initial field emission phase early in the pulse followed by a space-charge-limited behavior from a cathode plasma which expands at a constant velocity of 2–3 cm/μs. He also observed that when the energy density at the surface of a carbon anode exceeded 400 J/gm, the diode impedance began to drop more rapidly, indicating the existence of a virtual anode also formed by a layer of plasma expanding into the gap. Since the anode temperature was far too low to account for carbon varporization, Parker concluded that an impurity layer of absorbed gas was probably responsible for this effect.

Bradley and Kuswa (1972), as well as Parker, found evidence of plasma motion using streak photography. They observed luminous fronts emerging from both cathode and anode with velocities of 10^6–10^7 cm/s. The first quantitative measurement of diode plasma behavior resulted from holo-

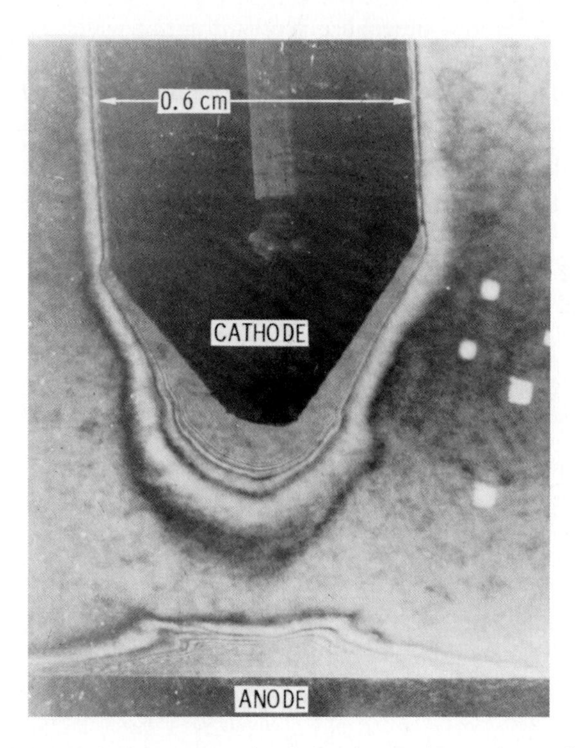

Fig. 9. Holographic interferrogram of a diode taken during an electron beam pulse and showing formation of anode and cathode plasmas.

graphic interferometry used by Mix *et al.* (1973) to observe the evolution of plasmas in the 10^{17}–$10^{19}/\text{cm}^3$ range with a pointed cathode. In Fig. 9 we see an interferogram taken at roughly 50 ns into a pulse indicating that the most tenuous cathode plasma observed ($\sim 10^{17}/\text{cm}^3$) moved at $\sim 10^7$ cm/s.

The anode plasma shown in Fig. 9 can also supply a space-charge-limited ion current that will serve to depress the electron-produced potential barrier near the cathode and further increase the electron current density. For a one-dimensional flow, the ion current density J_i and the electron current density J_e will be related to the nonrelativistic, space-charge-limited value without ions J_c, by the following relations (Poukey, 1975a,b; Wheeler, 1974):

$$J_i/J_c = (m/m_i)^{1/2}(1 + \alpha)^{1/2} \tag{3}$$

and

$$J_e/J_c = f(\alpha), \tag{4}$$

where $2\alpha = eV/mc^2$ and $f(\alpha)$ is given in Fig. 10.

As the voltage and current rise during a given pulse, a critical current can be reached for low-impedance accelerators for which magnetic effects start to be important, and a simple estimate can be made for this critical current (Friedlander *et al.*, 1968). This is accomplished by determining the critical current for which an electron emitted at the cathode edge (outer

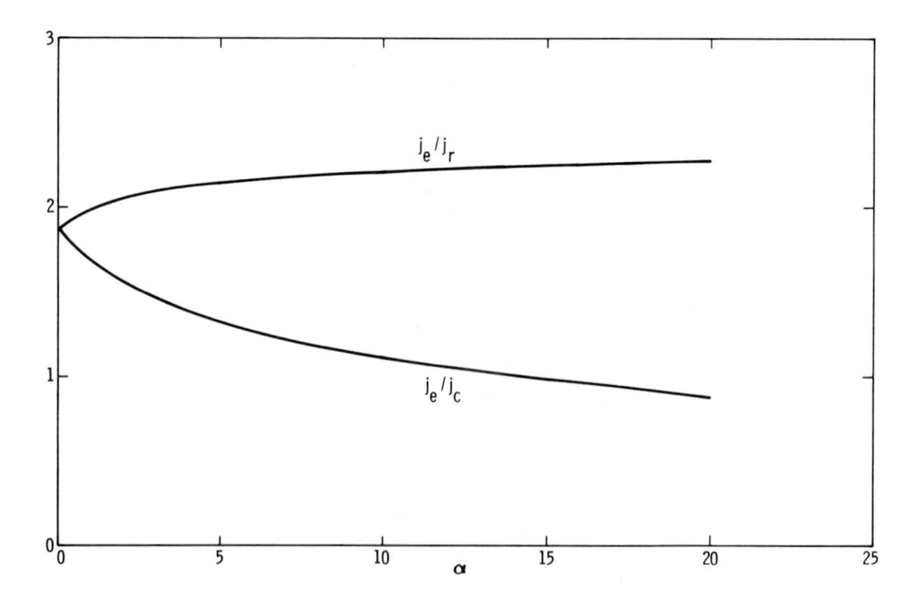

Fig. 10. Electron current density J_e versus $\alpha = eV/2mc^2$ for one dimensional bipolar-flow diode, J_e is normalized to the nonrelativistic J_c and relativistic J_r Child's-law current densities.

radius R) would reach the anode (at a distance d from the cathode) with zero component of velocity normal to the anode, neglecting the change in the azimuthal field with radius (a reasonable estimate for $R/d \gg 1$).

By integrating the electron radial equation of motion and setting the radial velocity at the anode equal to βc, we obtain

$$I_c \text{ (A)} = 8500\beta\gamma(R/d), \tag{5}$$

where $\gamma = 1 + 2\alpha$ and $\beta = (1 - \gamma^{-2})^{1/2}$.

Until 1967 the high-impedance oil-insulated Blumlein systems generally available could not exceed this critical current. The first studies of such very high current diodes were begun by Yonas and Spence (1968) using an accelerator developed by I. Smith at Physics International Company. This accelerator employed an 8-Ω coaxial oil-insulated line and, by mismatching to a lower impedance load, currents of up to $\sim$200 kA could be obtained at diode voltages of $\sim$200 kV.

A convenient measure of the relative importance of magnetic effects in accelerated beams has been pointed out by Lawson (1958). He showed that beam properties could be classified according to the value of the parameter ν/γ where ν, the so-called Budker (1956) parameter, is defined as

$$\nu = N(e^2/mc^2) = Nr_0 = I \text{ (A)}/17{,}000\beta, \tag{6}$$

where $N = 2\pi \int_0^a n_b r \, dr$ is the number of electrons per unit length of beam and r_0 the classical electron radius. Equation (5) can then be written

$$I/I_c = (\nu/\gamma)(2d/R), \tag{7}$$

and, for a 200-kV, 200-kA diode using Eqs. (2) and (5), we find $d/R \sim 0.1$ and that $I/I_c > 1$. A self-pinched beam was therefore expected and in fact was observed. The high ν/γ beams thus produced were employed in beam transport experiments discussed in the next chapter.

The two-dimensional flow resulting from beam pinching was not amenable to previous one-dimensional analyses, and the first of several analytical models to describe the impedance of this self-pinched flow was developed by de Packh (1968). He considered a simplified problem assuming that the electron trajectories followed conical equipotentials, i.e., by requiring a distribution of current such that the electrostatic force was just balanced by the Lorentz force, the electrons could flow along diode equipotentials, thus leading to the name "parapotential flow" (see Fig. 11). This model was later simplified by Creedon (1972) who provided relativistic generalizations of parapotential flow and gave the following impedance prediction for one special subclass of such solutions:

$$Z \,(\Omega) = \frac{V \text{ (V)}}{8500(R/d)\gamma \ln\{\gamma + (\gamma^2 - 1)^{1/2}\}}. \tag{8}$$

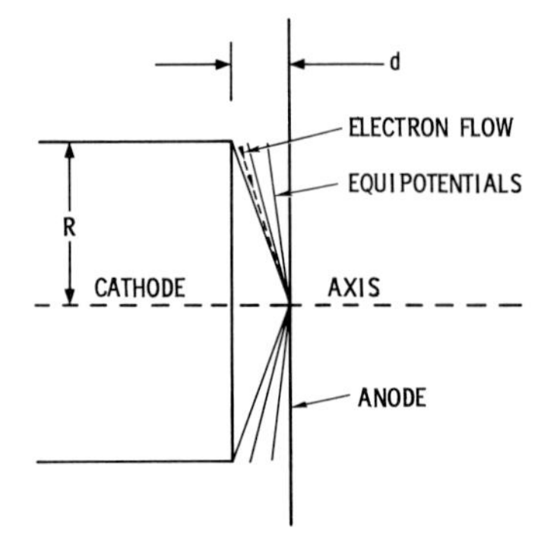

Fig. 11. Schematic of equipotential distribution and electron trajectory for parapotential flow.

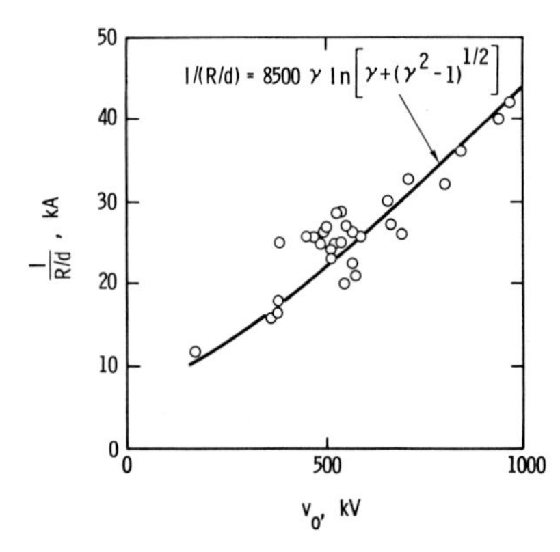

Fig. 12. Comparison of theoretical and experimental results for impedance of diode with pinched flow (from Physics International Co., San Leandro, California, Rep. DNA 3787F, May 1973).

Although there is no general agreement as to the validity of this model, a remarkable amount of the data collected thus far is in general agreement with Eqs. (7) and (8); see Fig. 12. In order to rationalize an apparent lack of plasma motion as evidenced by diode impedance behavior, Creedon *et al.* (1973) also postulated that the self-magnetic field of the beam restricted the expansion of cathode and anode plasmas.

In an attempt to treat the problem self-consistently, Poukey *et al.* (1973) have attempted to include the effects of two-dimensionality, as well as anode and cathode plasmas. Recently Poukey (1975a,b) has considered ions emitted from an anode plasma. In these studies a self-consistent, two-dimensional, particle-in-cell numerical simulation is used to solve the equations of motion for as many as 10,000 test electrons. In this numerical approach, the radiation field is neglected and the code is run until a steady solution is reached, indicating the current density distribution at the anode

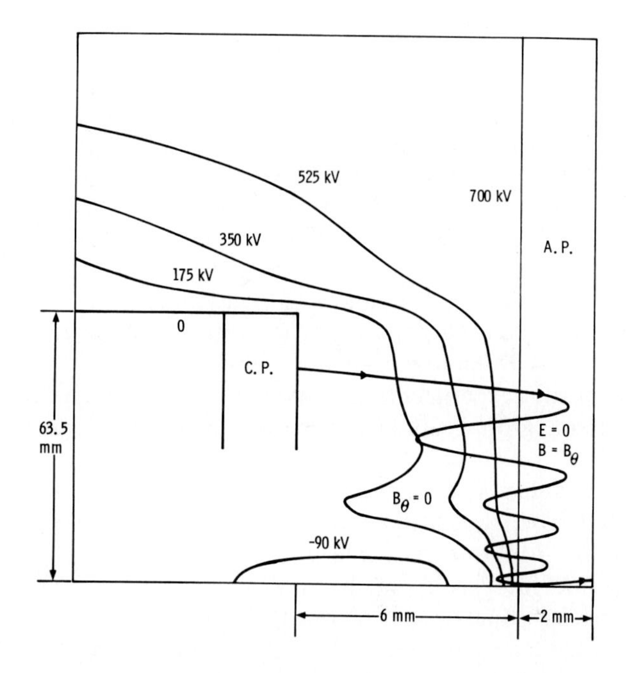

Fig. 13. Equipotential lines and a typical electron trajectory as calculated for a 700 kV diode and employing a hollow cathode. C.P. and A.P. denote layers of conducting cathode and anode plasmas respectively.

and the diode impedance. Beam self-pinching consistent with experiment has been shown with this numerical technique by assuming that the anode plasma layer excludes the diode electric field but not the magnetic field. The resulting electron motion is a result of a combination of **E** and **B** fields as shown in Fig. 13. A critical feature in these solutions is the treatment of the anode plasma and a rather different approach to the electron motion in the anode plasma has been used in another analytical solution to the problem by Goldstein *et al.* (1974). In that model the anode plasma is assumed to be perfectly conducting, and it is found that the trajectories in the vacuum gap determine the beam focusing properties. Measurements of the beam density distribution at the anode (Chang *et al.*, 1974) have indicated agreement with the numerical simulations, but the question is still an open one. The critical questions that remain relate to the distribution and properties of plasmas in these diodes and a variety of techniques are being employed to provide a more complete understanding of high-current, self-focusing diodes.

IV. INTENSE BEAM INJECTION INTO NEUTRAL GASES

A. Introduction

In this section we consider the interaction and propagation of electron beams in initially neutral gases. This is accomplished in practice by injecting the beam through a thin anode which separates the diode from a drift chamber filled with a gas at a preset density. As shown in Fig. 14 (Yonas and Spence, 1969), diagnostics allow the determination of injected beam parameters (voltage and current), beam net current as a function of position, transmitted beam current and beam radial distribution.

The first of such experiments involved the use of open shutter photography to determine gross beam behavior from the self-luminosity of the gas (see Fig. 15).

In order to understand this behavior, we refer first to a formalism, which was actually derived almost 10 years before any of the data to be discussed were taken. As mentioned previously, the paramter v/γ serves as a convenient measure for classifying the properties of intense beams. Alfvén (1939) first realized that $v/\gamma = 1$ represented a limit to the propagation of a charge-neutralized electron beam. This is most easily shown following the work of Lawson (1958).

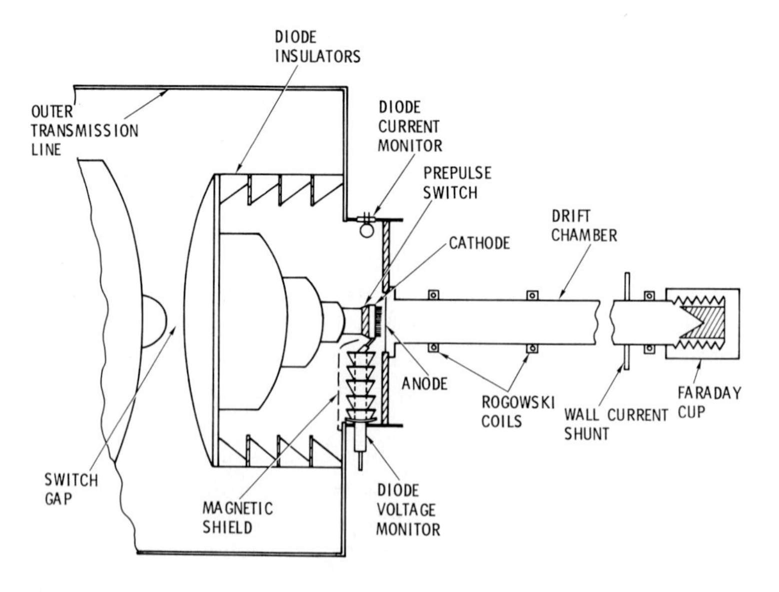

Fig. 14. Typical experimental apparatus employed in neutral gas experiments.

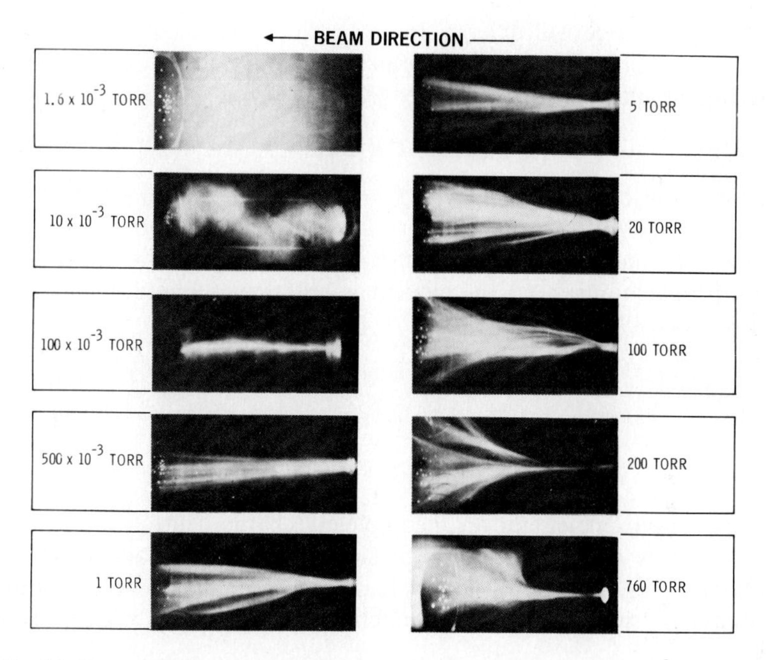

Fig. 15. Beam behavior at various neutral gas pressures as shown from gas luminosity using open shutter photography (from Physics International Co., San Leandro, California, Rep. [unpublished]).

B. Lawson Formalism

Under the restrictive assumption that electron motion is almost parallel to the beam axis (paraxial approximation), the radial equation of motion for an electron with the beam radius r_b is

$$\frac{d^2r}{dz^2} = 2\frac{v}{\gamma}\frac{r}{r_b{}^2}\frac{(1 - f_e - \beta^2)}{\beta^2}. \tag{9}$$

$f_e = n_i/n_e$ is the fractional charge neutralization by a background of stationary ions. Charge neutralization can be achieved by beam injection into a neutral gas whereupon electron collisional ionization will produce ions at the rate $dn_i/dt = n_g n_b \sigma \beta c$, where n_g is the background density, n_b is the beam density, and σ is the ionization cross section. For linearly rising beam current in time, the charge neutralization time is given by $t_{cn} = 2/(\beta c n_g \sigma)$, which is of the order of 10 ns for 1 Torr H_2, for example. Olson (1975) has shown that this time can be further reduced if ion avalanche effects are considered. Nevertheless, prior to radial force neutralization $f_e = 1 - \beta^2$, the beam envelope will diverge. In particular, from this formalism we can see that for no space-charge neutralization, Eq. (9) can be rearranged to describe the blowup of the beam envelope ($r = r_b$):

$$r_b(d^2r_b/dz^2) = 2\nu/\beta^2\gamma^3. \tag{10}$$

We should also note that even if $f_e = (1 - \beta^2)$, there will still be a longitudinal electrostatic potential barrier since the beam will not be fully charge neutralized, and the resultant axial longitudinal electric field can prohibit beam propagation unless the current is less than a limiting value (Olson and Poukey, 1974).

$$I_L = \beta(\gamma - 1)(mc^3/e)[1 + 2\ln(R/r_b]^{-1}(1 - f_e)^{-1} \tag{11}$$

or

$$I/I_L = [\nu/(\gamma - 1)][1 + 2\ln(R/r_b)](1 - f_e).$$

This potential barrier, for an incompletely charge neutralized beam with $I > I_L$, can result in substantial transient fields near the anode which can accelerate ions to energies larger than the beam energy. This example of collective acceleration of ions by intense electron beams will be discussed later.

If the beam is fully charge neutralized, $f_e = 1$ and Eq. (1) becomes

$$\frac{d^2r}{dz^2} + 2\left(\frac{\nu}{\gamma}\right)\left(\frac{r}{r_b{}^2}\right) = 0 \tag{12}$$

with a sinusoidal solution of wave length

$$\lambda = 2\pi r_b/(2\nu/\gamma)^{1/2}. \tag{13}$$

The paraxial assumption is not valid when $\lambda/r_b \lesssim 1$, i.e., $\nu/\gamma \gtrsim 1$. Even for $\nu/\gamma \ll 1$, the model is not valid in a region of pinching where particle orbit crossing, a finite temperature effect, occurs (Poukey *et al.*, 1971). Lawson recognized this and pointed out that $\nu/\gamma = 1$ can be considered as the demarcation between a beam of electrons and a plasma stream with a high transverse temperature.

The first calculations valid for nonparaxial beam flow were carried out by Poukey and Toepfer (1974). The transition from cold beam to particle orbit crossing and finally to hot beam is illustrated in Fig. 16 for a beam with $\nu/\gamma \sim 1$ propagating in an axial electric field. In the figure, the streamline solution, based on a model which assumes an isotropic pressure tensor for the beam, is seen to be valid in the limit of a cold beam and the fully pinched hot beam, but breaks down in the intermediate region where orbit crossing begins to occur. To this date, no analytic model has been found which adequately treates the intermediate region.

C. Current Neutralization

When discussing the properties of high v/γ beams, we must first introduce another vital concept, namely current neutralization. Ford *et al.* (1966) and Link (1967), after first experiments by Graybill and Nablo (1966), presented a modification of Lawson's model to account for observations of beam defocusing in a background of 1 Torr N_2 where space-charge effects should not have been important. Ford *et al.* (1966) suggested that a plasma current I_p driven by the inductively generated longitudinal electric field could reduce the net current I_{net} and thus the magnetic force on the beam electrons. He defined a magnetic neutralization term $f_m = I_p/I_b$ such that $I_{net} = I_b - I_p = I_b(1 - f_m)$. Then Eq. (9) becomes

$$\frac{d^2r}{dz^2} = \left(\frac{2v}{\gamma}\right)\frac{r}{{r_b}^2\beta^2}[1 - f_e - \beta^2(1 - f_m)] \tag{14}$$

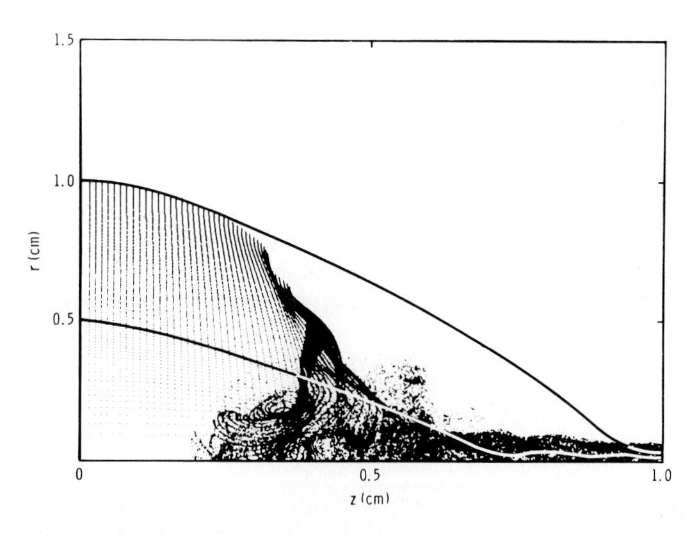

Fig. 16. Axially dependent REB equilibrium for beam with $f_m = 0, f_e = 1$, propagating in an axial electric field, after Poukey and Toepfer (1974). Dot density is proportional to electron density X radius. Solid curves indicate streamlines from finite temperature fluid theory. Transition from paraxial flow ($Z < 0.2$) to particle orbit crossing ($0.2 < Z < 0.6$) to hot beam ($Z > 0.6$) is illustrated.

and, in the case of complete charge neutralization ($f_e = 1$), we find

$$\lambda = 2\pi r_b/[(2\nu/\gamma)_{net}]^{1/2},$$

where $(\nu/\gamma)_{net} = (\nu/\gamma)(1 - f_m)$. Return current can therefore effectively reduce the importance of self-field effects. Creedon (1967) first pointed out that the conductivity of the beam-generated plasma would control the degree of current neutralization, and Yonas and Spence (1969) correlated detailed measurements of the current neutralization phenomenon with a semiempirical model based on zero-order solutions for beam-generated fields.

This model assumed that, for beams over the space-charge limiting current [Eq. (11)], beam blowup prior to charge neutralization would prevent propagation. After $f_e = 1$ is reached and the beam begins to propagate, the induced longitudinal electric field due to the rapid rise of the beam current would result in avalanche ionization of the background gas. As a result of the avalanche breakdown of the gas, the conductivity would rise essentially instantaneously after a delay time τ_B tending to hold the net current at the value of beam current at the time of avalanche; (see Fig. 17). If finite conductivity effects are added in a zero-order approximation by assuming that all beam properties have only a slight radial variation, then the solution for I_{net} is modified as shown in the figure.

τ_B can be correlated with measurements of the formative time of pulsed breakdown from the work of Felsenthal and Proud (1965) for a variety of gases supporting the model that the avalanche gas breakdown is the dominant ionization process responsible for production of sufficient conductivity to provide current neutralization. The conductivity after breakdown can also be determined empirically from the measured decay time of the plasma current after the termination of the beam pulse τ_D (see Fig. 18) (Yonas and Spence, 1968). These simple ideas can be extended further by considering the implications to beam propagation of beam-induced avalanche and current neutralization.

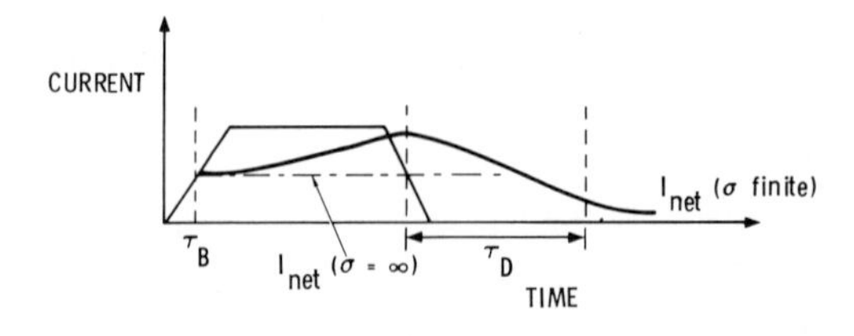

Fig. 17. Beam current and net current waveforms shown for cases of finite and infinite conductivity. τ_B is the gas breakdown time, and τ_D is the net current decay time.

As a result of avalanche breakdown of the background gas, the equations describing the beam-generated electromagnetic fields in the drift chamber change their character from wavelike to diffusion dominated. By combining Ampere's and Ohm's laws, we find

$$\nabla \times \mathbf{B} = \mu_0 \mathbf{J}_{\text{net}} + \mu_0 \varepsilon_0 \frac{\partial \mathbf{E}}{\partial t} = \mu_0(\mathbf{J}_\text{b} + \sigma \mathbf{E}) + \mu_0 \varepsilon_0 \frac{\partial \mathbf{E}}{\partial t}. \tag{15}$$

Combining this with $\nabla \times \mathbf{E} = -\partial \mathbf{B}/\partial t$, we obtain

$$\nabla^2 \mathbf{B} - \mu_0 \sigma \frac{\partial \mathbf{B}}{\partial t} + \frac{\varepsilon_0}{\sigma} \frac{\partial^2 \mathbf{B}}{\partial t^2} = \mu_0 \nabla \times \mathbf{J}_\text{b}. \tag{16}$$

For time scales of interest, the displacement current is found to be negligible, leading to a diffusion equation for the magnetic field:

$$\nabla^2 \mathbf{B} - \mu_0 \sigma \frac{\partial \mathbf{B}}{\partial t} = \mu_0 \nabla \times \mathbf{J}_\text{b}. \tag{17}$$

After solving for **B**, the longitudinal electric field can be calculated. General solutions to these equations have been found by several authors under various assumptions and by McArthur and Poukey (1973) including effects of collisional and avalanche ionization, as well as electron–neutral and

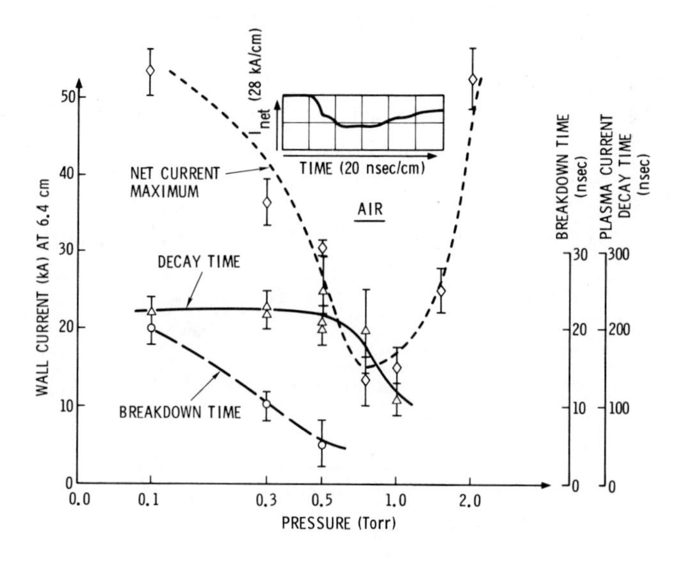

Fig. 18. Net current, plasma current decay time, and breakdown time versus pressure in air. ○, breakdown time; △, decay time; ◇, net current maximum.

electron–ion collisions in calculating the background gas conductivity. Based on this complete model, the last authors found reasonable agreement with measurements of net current over a fairly wide range of pressures.

An important thing to note here is that, prior to τ_B, substantial longitudinal electric fields can exist which can erode the beam front as it propagates, and again a simplified treatment will exemplify this fact. If a radially uniform beam fills the drift chamber, then E_z on axis is given by

$$E_z = -\frac{\partial}{\partial t} \int_0^{r_b} B_\theta(r', t)\, dr', \tag{18}$$

where

$$B_\theta(r', t) = (\mu_0/2\pi r') \int_0^{r'} (\mathbf{J}_{\text{net}})_z 2\pi r''\, dr'', \tag{19}$$

thus giving (in rationalized mks units)

$$E_z = 10^{-7} \frac{dI_{\text{net}}}{dt}. \tag{20}$$

This expression will also hold near the axis for any beam profile with a weak radial dependence.

After breakdown, this longitudinal electric field is greatly reduced if the plasma conductivity is sufficiently large to provide for substantial current neutralization of the beam.

For typical values of dI/dt of 10^{12}–10^{13} A/s prior to breakdown, the longitudinal field is 10^5–10^6 V/m, which can obviously contribute to important energy losses of the head of a 1-MeV beam as it propagates. After break-

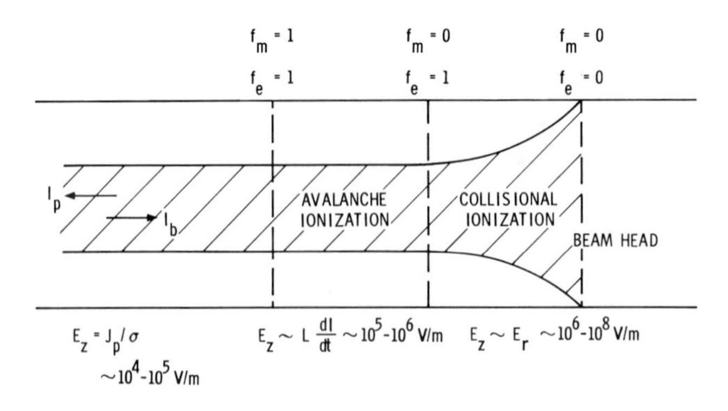

Fig. 19. Schematic of beam front penetration into neutral gas showing regions of collisional and avalanche ionization.

down, the plasma conductivity rises to the order of 2500 mho/m. Knowing this conductivity, we can then determine the longitudinal electric field, $E_z = J_p/\sigma$. If current neutralization is almost complete and the beam current density is less than 10^8 A/m^2 as in past experiments, then $E_z < 4 \times 10^4$ V/m. Electrons will therefore propagate with almost their longitudinal velocity achieved at the diode until they reach the head of the beam region where current neutralization is incomplete. In this region they will then be decelerated in the inductively generated electric field. From breakdown data we learn that, at 10^5–10^6 V/m, breakdown will occur in 3–8 ns in 1-Torr air, so that for 10^{12} A/s the net current at breakdown and during most of the pulse will be $(dI/dt)\tau_B \lesssim 10^4$ A (see Figs. 18 and 19 for a schematic representation of these concepts). Even if the beam current continues to rise for 50 ns, giving a peak current of 5×10^4 A for $dI/dt = 10^{12}$ A/s, the net current will remain at $\sim 10^4$ A and f_m as defined above will be close to one. As we shall see, such low net currents are an important feature of the efficient propagation of high ν/γ beams.

D. Beam Behavior

As can be seen from Fig. 15, REB equilibria are not in general axially uniform (Poukey *et al.*, 1971). The beam propagation behavior is determined by the charge and magnetic neutralization capabilities of the background medium and the magnitude of the beam current as shown in Table IV. At pressures >100 Torr, the beam propagation is also modified by scattering of primary electrons from ions in the background medium (Mosher, 1974). For $\nu/\gamma \geq 1$ and $f_e = f_m = 1$, the beam will not propagate efficiently because the beam electrons acquire a large transverse energy in the diode, and free-streaming motion will cause the beam to spread rapidly (Yonas and Spence, 1968).

TABLE IV

EFFECT OF BACKGROUND GAS PRESSURE ON BEAM BEHAVIOR

ν/γ	f_e	f_m	BEAM BEHAVIOR	TYPICAL GAS PRESSURE
$\ll 1$	0	0	PARAXIAL MOTION	ALL
≥ 1	0	0	BLOWS UP	10^{-3} TORR (AIR)
≥ 1	1	0	PINCHES	10^{-1} TORR AND 760 TORR (AIR)
≥ 1	1	1	ELECTRON MOTION FORCE FREE	1 TORR (AIR)

For $v/\gamma \geq 1$, $f_e = 1$, and $f_m < 1$, the transverse energy of the beam electrons can be confined by self-magnetic fields. For this case, there is an infinite number of axially independent equilibria that can be calculated using both fluid and Vlasov models. First work was carried out by Bennett (1934, 1955) using a fluid model, and he obtained the well-known relation between beam transverse energy $W_\perp$ and net current I:

$$I^2 = 2NW_\perp, \tag{21}$$

where N is the number of electrons per unit length. An expression similar to Eq. (21) can be derived, including current neutralization, thus leading to a relationship between the longitudinal and transverse beam energy for any v/γ (Yonas *et al.*, 1969):

$$\frac{E_{\text{trans}}}{E_{\text{longit}}} = (v/\gamma)(1 - f_m)^2. \tag{22}$$

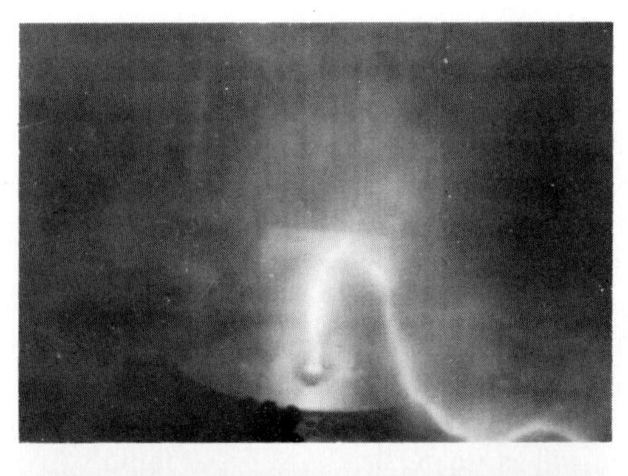

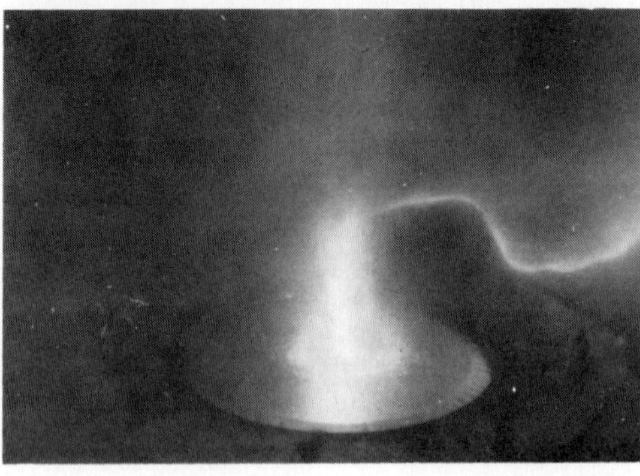

Fig. 20. Open shutter photographs (taken at right angles) showing unstable high v/γ pinched beam behavior.

Since this work was done, generalized fluid models have been presented by Benford and Book (1971) and Toepfer (1971). Axially uniform Vlasov equilibria have been worked out by Hammer and Rostoker (1970), Yoshikama (1971), and Lawson (1973). To date, only the equilibria of Bennett (1934) and Toepfer (1971) have been compared with low v/γ experiments (Briggs *et al.*, 1974; Poukey *et al.*, 1973), and the agreement has been to within the (rather large) experimental error. High v/γ beam propagation experiments (Yonas and Spence, 1968) indicate that such beams may propagate in a pinched mode although in an obviously unstable manner (Fig. 20), and no detailed experimental studies of such beams have been carried out.

Axially dependent equilibria of propagating beams have been calculated by Poukey and Toepfer (1974) using a "particle-in-cell" numerical model and an analytical fluid theory for the beam envelope. For $v/\gamma > 1$, the PIC calculation did not lead to propagating equilibria for initial conditions corresponding to those existent in beams produced from conventional diodes, although envelope calculations (Fig. 21) predict a stable propagating self-pinched equilibrium for high v/γ beams.

Even if a high $v/\gamma, f_m = 0$ beam can be created and stabilized, it will still suffer large energy losses due to the inductively generated electric field [Eq. (20)]. The solution to this problem of beam transport is to use an external

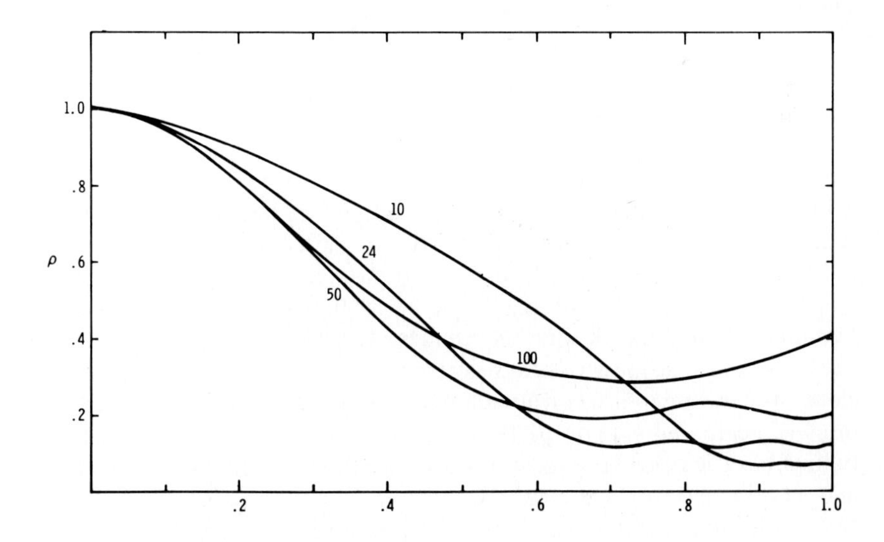

Fig. 21. Z-dependent fluid equilibria for $v/\gamma = 10$ beam after Poukey and Toepfer (1974). Curves are parametrized by initial beam temperature (keV).

longitudinal magnetic field to prevent radial beam spreading and to either preionize a sufficiently dense gas or to choose a background gas such that τ_B is much less than the beam risetime. Stallings *et al.*, (1972) have shown that a beam with $\nu/\gamma \gg 1$ can be transported with little loss over 1 meter in a longitudinal magnetic field of ~ 1 T at an optimum pressure of about 1 Torr in air in order to minimize self-field effects. Prono *et al.* (1975) have extended these results at a variety of pressures in hydrogen and found similar results. It appears that there is a linear loss in the magnitude of the peak beam current with distance which has not been fully explained as yet. This loss is still not large, with as much as 90% of the current transported over 1 m under optimum conditions.

Kolb *et al.* (1973) have used these techniques to propagate beams in the 20–30-kJ range with currents of several hundred kiloamperes over 130 cm with negligible energy loss and have correlated their experimental observations with an adiabatic, high β, fluid model, i.e., they assume that the beam transverse pressure can be balanced by the pressure of the externally applied magnetic field. Using this model and assuming even that a sizable fraction of the beam energy is in transverse motion, fields in the 1–2-T range are found to be sufficient for efficient beam control.

We have seen that beams of arbitrarily high currents can be transported over sizable distances but, to be useful particularly for fusion research, they must be coupled in some way to either a low-density plasma or to a solid. These subjects are covered in the next two sections.

V. BEAM INTERACTION WITH PLASMA

A major application for intense pulsed REBs is the heating of magnetically confined plasmas to fusion temperatures. Theoretical studies of the REB-plasma interaction show that collective processes should dominate the beam-to-plasma energy transfer, and stopping powers have been calculated which are orders of magnitude larger than would be expected from two-body Coulomb collisions between beam electrons and plasma electrons and ions. The streaming of beam electrons through the plasma, and plasma electrons carrying return current relative to ions, will excite collective oscillations of beam and plasma particles which couple to transverse and longitudinal electric fields. The most unstable modes grow and modify the beam and plasma electron distribution functions, at rates comparable to the electron plasma frequency ω_p. In the nonlinear regime, trapping, parametric instabilities, and nonlinear Landau damping lead to saturation of the instabilities and thermalization of the energy.

When an REB is injected into a fully ionized plasma, the return current carried by plasma electrons will essentially cancel the beam current, resulting in $I_{net} = 0$. For beam plasma density ratios,

$$n_b/n_p < (kT_e/Mc^2)^{1/2},$$

where T_e is the plasma electron temperature (assumed $\geq T_i$, the ion temperature), M the ion mass, and c the velocity of light, the plasma electron drift velocity will be below the threshold for ion-acoustic instability, and the dominant instability will be driven by the interaction between beam and plasma electrons. This regime corresponds to the weak beam case ($n_b/n_p \ll 1$). The linear theory for this interaction has been worked out by Bludman *et al.* (1960) in the electrostatic approximation and more recently by Godfrey *et al.* (1975) and Wright and Hadley (1975) for the full electromagnetic dispersion relation.

For electrostatic modes in a cold-beam, cold-plasma system, maximum growth rates are

$$\gamma_{es} = \omega_p \frac{\sqrt{3}}{2}\left(\frac{n_b}{2n_p\gamma_b}\right)^{1/3} F(\theta, \gamma_b), \tag{23}$$

where ω_p is the ambient electron plasma frequency, $n_b(n_p)$ the beam (plasma) electron density, θ the angle of propagation of the wave relative to the beam drift velocity $\mathbf{V}$, and $\gamma_b = (1 - \mathbf{V}\cdot\mathbf{V}/c^2)^{-1/2}$. The function $F(\theta, \gamma_b)$ has a maximum value of order 1 and, for a cold beam with no external magnetic fields, is maximum near $\theta = \pi/2$. This is a result of the relativistic dependence of beam electron mass on energy. If the beam has a finite spread in perpendicular velocity corresponding to an rms velocity $\langle v_\perp{}^2\rangle$ (Fainberg *et al.*, 1969), modes propagating at large θ are Landau damped, and the most unstable mode propagates at an angle θ_m, where

$$\tan^2\theta_m = \frac{2^{7/6}}{3}\frac{c}{\langle v_\perp{}^2\rangle}\left(\frac{n_b}{n_p\gamma_b}\right)^{1/3} \tag{24}$$

and $F(\theta_m, \gamma_b) = 1$. In the presence of a magnetic field $\mathbf{B}$, parallel to $\mathbf{V}$, electron motion perpendicular to $\mathbf{B}$ is restricted and if $\omega_c \gg \omega_p$, where $\omega_c = eB/m_e c$, then $\theta_m = 0$ and $F(\theta_m, \gamma_b) = 1/\gamma_b{}^{2/3}$.

The electromagnetic modes grow perpendicular to the direction of beam motion. For $\mathbf{B} = 0$ and $\theta = 90°$ (Weibel, 1959),

$$\gamma_{em} = \beta_b\left(\frac{n_b}{n_p\gamma_b}\right)^{1/2}\left(\frac{k^2c^2}{k^2c^2 + \omega_p{}^2}\right)^{1/2}\omega_p, \tag{25}$$

where $\beta_b = (1 - 1/\gamma_b{}^2)^{1/2}$. For $|\mathbf{B}| > 0$ (Godfrey *et al.*, 1975), the Weibel mode is stabilized when $\omega_c > \beta_b \omega_p (n_b \gamma_b / n_p)^{1/2}$, and for larger magnetic fields, the electromagnetic modes have growth rates

$$\gamma_{em} = (\omega_p/2)(n_b \omega_p / n_p \omega_c)^{1/2} G(\theta, \omega_c/\omega_p), \tag{26}$$

where $G \rightarrow 0$ as $\theta \rightarrow 0$ and has a maximum value $\sim O(1)$.

The nonlinear phase of the electrostatic beam-plasma electron instability has been studied in the kinetic limit by Fainberg *et al.* (1969), Rudakov (1970), Breizman and Ryutov (1970, 1971), and Breizman *et al.* (1972). In this limit, the spread in longitudinal velocity of the beam electrons is sufficiently large that quasilinear relaxation of the beam electron distribution function occurs. Calculations by Rudakov (1970) and Breizman and Ryutov (1971) also consider effects of nonlinear Landau damping on beam relaxation. Numerical simulation calculations of the nonlinear saturation level of electrostatic modes in one-dimensional systems corresponding to $\omega_c \rightarrow \infty$ have been carried out by Matsiborko *et al.* (1972) in the limit $n_b/n_p \ll 1$ and by Toepfer and Poukey (1973) and Thode and Sudan (1973) for arbitrary values of n_b/n_p. Thode and Sudan also calculated the growth of ion density fluctuations driven by the nonlinear electrostatic waves, and Toepfer and Poukey (1973) have considered the effect of ion density fluctuations and beam temperature on the two-stream saturation levels. Beam stopping lengths predicted by Rudakov (1970) and Breizman and Ryutov (1971) differ qualitatively and quantitatively because of assumptions regarding the dominant nonlinear processes. Beam energy loss and instability saturation levels from the one-dimensional simulation all agree as to the dominant mechanism of saturation, i.e., trapping of beam electrons in the unstable waves to form a hot electron tail, and also agree quantitatively in the parameter and space–time regions where the calculations overlap. In Fig. 22, these results will be compared with experiments to be discussed below.

Nonlinear calculations for the saturation of beam-plasma electromagnetic modes have been carried out by Lee and Lampe (1973). These calculations indicate that the instability causes a charge and current neutralized beam to split up into filaments, each of which self-pinches to a density $\simeq n_p$, the background plasma density, on time scales of orders $\leq 10^2 \omega_p^{-1}$. On a longer time scale, the individual filaments coalesce to form a beam of density $n_b \simeq n_p$ which is not current neutralized. More recent results (Papadopoulos *et al.*, 1974) in which both electrostatic and electromagnetic modes were calculated shows that the electrostatic component of the instability dominates in the interaction, and disperses the filaments which are initially formed.

For $(kT_e/mc^2)^{1/2} > n_b/n_p > (kT_e/Mc^2)^{1/2}$, the return current driven by the beam exceeds the ion acoustic velocity in the plasma and, for $T_e \gg T_i$,

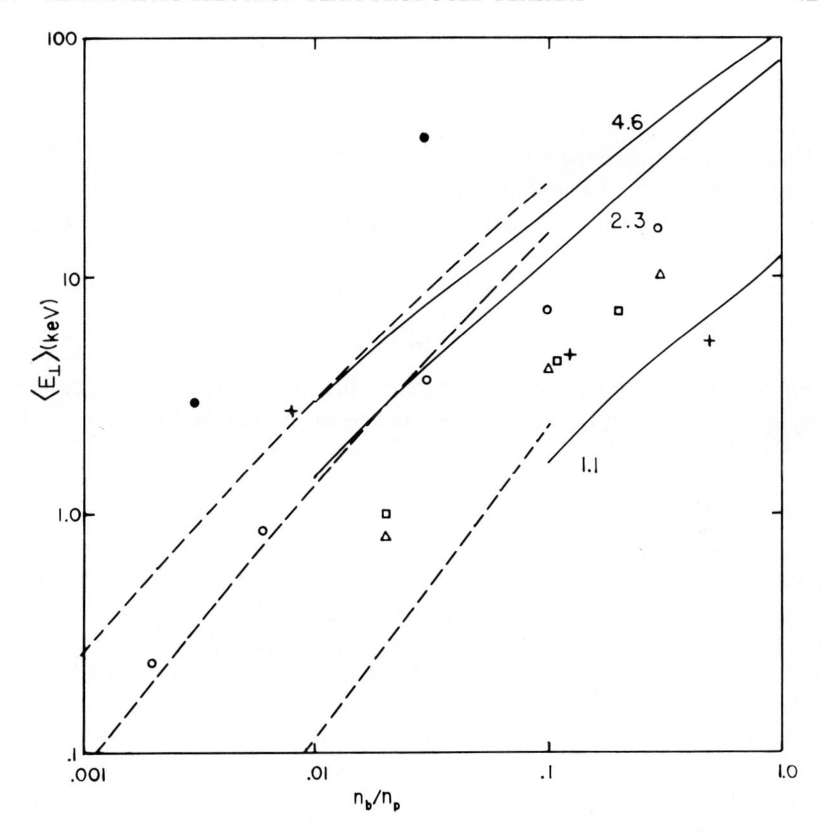

Fig. 22. Average plasma electron energy as inferred from diamagnetic loop measurements in the REB-plasma heating experiments of Altyntsev (1971) (●), Abrashitov (1973) (○), Miller (1973) (□), Ekdahl (1974) (△), and Kapetenakos (1973) (+). Curves parametrized by γ_b indicate two-stream saturation energy as calculated by Toepfer and Poukey (1973), and Thode and Sudan (1973).

drives electrostatic fluctuations in the plasma which couple to unstable ion acoustic waves. For $n_b/n_p > (kT_e/mc^2)^{1/2}$, unstable electrostatic modes of the type first studied by Buneman (1959) are excited. In either case, if the turbulence becomes strong enough to alter the plasma resistivity, then the resulting finite electric field supported by the plasma will act to slow down the beam (Swain, 1973). Also, the kinetic energy of streaming of plasma electrons will be converted to thermal energy through joule heating at a rate

$$\frac{dW_p}{dt} \simeq \frac{1}{\sigma_{\text{eff}}} j_p^2, \tag{27}$$

where j_p is the plasma current density, and

$$\sigma_{\text{eff}} = \omega_p{}^2/4\pi\nu^*, \tag{28}$$

where ν^* is the effective plasma electron momentum transfer collision frequency due to turbulence. Calculations by Thode and Sudan (1975) indicate that $\nu^* \simeq 0.3\Omega_p$, where Ω_p is the ion plasma frequency. The total energy per unit length available to the plasma from return-current-driven instabilities has been shown by Lovelace and Sudan (1971) to be

$$E = \mathscr{L} I_A I_b, \tag{29}$$

with $\mathscr{L}$ the inductance per unit length of the system, I_b the beam current, and I_A the Alfvén critical current for the beam, corresponding to $\nu/\gamma = 1$.

First experiments in which an intense relativistic electron beam was injected into a preionized gas were carried out by Roberts (1965, 1969) and by Benford *et al.* (1969). These experiments investigated the efficiency of propagation of beams in plasmas using witness plates, open-shutter photography, and calorimeters to record energy transport, and Rogowski loops to determine net current in the drift tube. Efficient transport was observed for $\nu/\gamma < 1$ beams in plasmas with $n_p \gg n_b$ and poor transport for $\nu/\gamma > 1$. Almost complete cancellation of the beam current by return currents in the plasma was observed.

Following these initial experiments, a number of experiments have been carried out to investigate the efficiency of coupling between beam and plasma in linear geometries (Levine *et al.*, 1971; Altyntsev *et al.*, 1971; Abrashitov *et al.*, 1973; Miller and Kuswa, 1973; Kapetanakos and Hammer, 1973; Goldenbaum *et al.*, 1974; Ekdahl *et al.*, 1974). Table V is a summary of beam and plasma parameters in these experiments, along with the diagnostics employed. Other experiments have been carried out but have not been listed in the table because they entailed few diagnostics of the plasma (Smith, 1972; Friewald *et al.*, 1971; Mather *et al.*, 1973) or of the beam (Korn *et al.*, 1973).

The experiments of Levine *et al.* (1971) studied principally the transport of a beam in neutral and preionized gas. The results for propagation in neutral gas confirmed those previously obtained by Yonas and Spence (1968). In the case of injection into a plasma, it was found that a high degree of current neutralization was obtained so that $(\nu/\gamma)_{\text{net}} < 1$. Results were consistent with calculations of Hammer and Rostoker (1970) which predicted $I_{\text{net}} \simeq [c/\omega_p r_b] I_b$, where the quantity in brackets is the ratio of plasma electromagnetic skin depth to beam radius. An anomalously low value was obtained for the velocity of propagation of the beam front through the plasma. Even if electrons with high transverse velocities were lost to the walls, the beam-front velocity was 40% lower than expected. In order to explain the anomalously low beam-front velocity, the authors suggested that collective effects occurring via the two-stream instability resulted in some resistance to the flow of beam electrons through the plasma. They did not attempt, however, to measure the resultant heating of the plasma which would have occurred due to the two-stream interaction.

The experiments by Altyntsev *et al.* (1971) with the RIUS-5 accelerator were the first which measured the transfer of beam energy to plasma. The

TABLE V

PARAMETERS FOR RELATIVISTIC BEAM PLASMA EXPERIMENTS

REFERENCE	BENFORD (1969)	LEVINE (1971)	ALTYNTSEV (1971)	ABRASHITOV (1973)	GOLDENBAUM (1974)	MILLER & KUSWA (1973)	EKDAHL (1974)	KAPETENAKOS & HAMMER (1973)
γ	1.4	1.8	6	3	2.2	1.7	2	1.8
ν/γ	12	2.5	.1	.1	1	2	1.5	1.5
$n_b (cm^{-3})$	10^{12}	$2 \cdot 10^{11}$	$5 \cdot 10^{11} (3 \cdot 10^{10})$	$\leq 5 \cdot 10^{11}$	$\leq 10^{11}$	10^{12}	10^{12}	$5 \cdot 10^{11}$
r_b (cm)	1	5	1.25(3.5-5)	1.25-1.5	2.5	1.25	1.5-3	2.5
E(kJ)[a]	1.2	.8	1.3	.07-.13	1.5	.63	.8	.4
CHAMBER								
L (m)	.25	3	3	2.3	5	.3	1.8	.4
R (cm)	2.5	7	10	6.5	≥ 2.5	5	20	7.5
PLASMA								
$n_p (cm^{-3})$	—	$>2 \cdot 10^{13}$	$10^{10}-10^{14}$	$3-4 \cdot 10^{14}$	$5 \cdot 10^{13}-7 \cdot 10^{14}$	$10^{12}-10^{15}$	$10^{12}-6 \cdot 10^{13}$	$10^{11}-3 \cdot 10^{14}$
T_0 (eV)	—	.53	—	2-3	5	1	—	1
r_p (cm)	<2.5	<7	3.5-5	4	≥ 2.5	1.5	3-5	—
$\hat{j}_p$	$\hat{Z}$	$\hat{\theta}$	$\hat{\theta}$	$\hat{A}$ (PENNING)	$\hat{Z}$ & $\hat{\theta}$	$\hat{Z}$	NONE	$\hat{Z}$
GAS [PRESSURE (μ)]	AIR (50, 190)	METHANE (5-370)	H_2, Ar	H_2	He(100)	H_2(2-15)	—	He(10)
EXTERNAL $\underset{\sim}{B}$(kg)[b]	0	$.5\hat{Z}$	$(0-2.5)\hat{Z}$[b]	$15\hat{Z}$[b]	$5\hat{Z}$	$1.5\hat{Z}$[b]	$2.6\hat{Z}$[b]	$(1-5)\hat{Z}$[b]
DIAGNOSTICS								
WITNESS PLATE	X			X				
CALORIMETER		X	X	X				
MAGNETIC PROBES ($\hat{\theta}$)		X				X	X	X
DIAMAGNETIC LOOPS ($\hat{Z}$)			X	X		X	X	X
SPECTROSCOPY		STARK BROADENING			TIME DEPENDENT EMISSION			
μ-WAVE (cm)			.8, 3, 6	.2, .4, .8. 1.3, 3	3	.8	.4, .8	
X-RAY						X	X	
ROGOWSKI LOOP	X		X	X		X		
FARADAY CUP			X			X		
PIEZO-ELECTRIC PROBE							X	
WALL SCINTILLATORS		PILOT B						
ELECTROSTATIC PROBE				X				
THOMSON SCATTERING					X			
NEUTRAL ATOM DETECTORS			X				X	

[a] E (kJ) $= 0.6 IV\tau$, where I, V, τ are nominal beam current, voltage, and pulse length.
[b] Refers to mirror field configuration.

degree of heating of the plasma was measured by using diamagnetic loops to infer the increase in plasma pressure transverse to the axial magnetic field after beam passage. Net current, beam energy transport, initial plasma density, microwave radiation, and backscattered high-energy electrons were measured. Most efficient energy transfer was observed to occur at $n_b \approx 0.1\ n_p$, and plasma heating decreased with decreasing n_b/n_p. Under conditions of optimum coupling, a radial broadening of the beam density profile was observed and the total energy deposited in the plasma was ≤ 180 J, corresponding to $\sim 10\%$ of the total beam energy in a distance of 3 m. On the other hand, the efficiency of beam energy transport was decreased by 80%, consistent with the observed radial scattering of beam electrons. A result unique to this experiment was the observation of an electron thermal wave which propagated at velocities ranging up to 10^{10} cm/s which was apparently caused by enhanced beam energy deposition near the anode.

The experiments of Altyntsev *et al.* (1971) stimulated further work by Abrashitov *et al.* (1973). Abrashitov injected a slightly lower voltage beam with less total energy but higher density than Altyntsev into an H_2 plasma produced by a pulsed Penning-type discharge. A 50–70% loss in beam energy delivered to the calorimeter at the end of the drift tube was observed when $n_b/n_p \gtrsim 0.1$, and was attributed to enhanced radial scattering of the beam to the chamber walls. Good current neutralization was observed only when $n_b/n_p < 0.01$. The plasma heating was observed to peak at $n_b/n_p = 0.006$, where $W_p/W_b \simeq 0.14$, with $W_p(W_b)$ the plasma (beam) energy density per unit length in the drift tube. At a given density, plasma diamagnetism was observed to increase monotonically with B_z. At densities $>10^{14}$ cm^{-3} and axial magnetic fields <3 kG, the beam was observed to excite magneto-acoustic oscillations in the plasma column.

In contrast to the experiments of Altyntsev *et al.* (1971) and Abrashitov *et al.* (1973), those of Miller and Kuswa (1973), Kapetanakos and Hammer (1973), Goldenbaum *et al.* (1974), and Ekdahl *et al.* (1974) were carried out with beams having $\nu/\gamma \geq 1$. In all these experiments the plasma perpendicular energy was found to increase with increasing beam to plasma density ratio, and the plasma heating as inferred from diamagnetic loop measurements was consistent with saturation levels calculated for the one-dimensional relativistic, electron–electron two-stream instability.

Figure 22 is a plot of plasma perpendicular energy as determined from diamagnetic loop measurements in the experiments of Altyntsev *et al.* (1971), Abrashitov *et al.* (1973), Miller and Kuswa (1973), Ekdahl *et al.* (1974), and Kapetanakos and Hammer (1973). $\langle E_\perp \rangle$ is the average perpendicular energy per electron. Also shown are calculations of Toepfer and Poukey and of Thode and Sudan for $\langle E_\perp \rangle$ due to the one-dimensional, beam-plasma electron two-stream instability. In this case, the energy per unit volume deposited by the beam in the electrostatic field E at saturation is

$$\varepsilon_w = (1/8\pi L) \int_0^L E^2(z)\, dz, \tag{30}$$

where L is the system length. An equal amount of energy is given up to the plasma electrons. After thermalization the average perpendicular energy available per plasma electron is $\langle E_\perp \rangle = \frac{4}{3}\varepsilon_w$. A series of curves for $\langle E_\perp \rangle$, parametrized by the initial beam energy γ_0 are shown in Fig. 22 for numerical simulations assuming initially cold beams. Also shown are estimates for $\langle E_\perp \rangle$ according to Thode and Sudan (1973):

$$\langle E_\perp \rangle = \tfrac{2}{3}(n_b/n_p)[S/(1 + S)^\alpha](mc^2\gamma_b), \tag{31}$$

where $S = \gamma_b \beta_b{}^2 (n_b/2n_p)^{1/3}$, $\alpha = \frac{5}{2}$ for $S \gtrsim 1$, and $\alpha = \frac{3}{2}$ for $S \ll 1$.

The disagreement between the numerical calculations and Eq. (31) for $n_b/n_p \gtrsim 0.03$ is caused by the breakdown of the single-mode approximation in the analytic model of Thode and Sudan for large beam–plasma density ratios.

With the exception of the data of Kapetanakos and Hammer (1973), the qualitative variation of $\langle E_\perp \rangle$ with n_b/n_p is consistent with that expected from the nonlinear, one-dimensional, two-stream calculations. Absolute comparison with experiment is not valid because of the assumption made in the numerical simulation of infinite, one-dimensional beam–plasma systems, and uncertainties in the radial extent of the heated plasma volume in the experiments. There appears to be good agreement between the experiments of Ekdahl *et al.* (1974) and Miller and Kuswa (1973), as would be expected from the similarity of parameters indicated in Table V.

The results of the linear beam–plasma interaction experiments indicate that, for present beams ($n_b \sim 10^{10}$–10^{12} cm^{-3}) injected into plasmas of interest for fusion application ($n_p \sim 10^{14}$–10^{16} cm^{-3}), the primary two-stream electron–electron instability dominates, leading to modest plasma heating.

The numerical simulations indicate that the bulk of the plasma electrons remains cold, with most of the absorbed energy going into the tail of the distribution function. In order to take advantage of nonlinear processes occurring on longer time scales, injection of beams into very long plasmas, such as a multiple mirror system (Budker, 1973) or a toroidal system, has been proposed. Preliminary experiments on REB injection into a toroidal system have been carried out by Benford *et al.* (1974) and Gilad *et al.* (1974). Extension of these experiments to tokamak systems has been proposed by Ikuta (1972) and Swain *et al.* (1975).

Magnetohydrodynamic coupling of beam energy to plasma ions and neutrals was observed by Vandevender, *et al.* (1974) for a 36-kA, 350-kV, 100-nsec electron beam injected into neutral hydrogen gas. Energy transfer resulting in 100-eV ion energies resulted from an inverse pinch effect induced by plasma return currents. In these experiments $n_b/n_p \ll 1$ and electron heating was significantly less than the observed ion heating.

VI. BEAM INTERACTION WITH SOLID TARGETS

The interaction of pulsed electron beams with solid targets is of interest for the determination of material properties under conditions of high pressure and temperature, and has applications to the heating of inertially confined thermonuclear targets.

One of the first uses of intense, pulsed REBs was in studies of the dynamic response of materials to rapid heating. Because relativistic electrons deposit their energy in depth, such experiments utilizing short pulse beams characterize the material response under conditions of constant-volume heating. Also, to the extent that the equation-of-state and phase transition characteristics of a material are known in the pressure and temperature regimes of the experiment, the electron beam energy deposition profile can be determined.

Theoretical models for the rate of energy loss by relativistic electrons in solids have been developed over the past four decades by such authors as Bethe, Fermi, and Landau. Present models involve the use of Monte Carlo numerical techniques (Berger and Selzer, 1964) to follow individual electron and photon transport in a solid. The codes usually include transport for secondary knock-on electrons and continuous bremsstrahlung and characteristic x rays; and photons are allowed to produce photoelectrons, Compton electrons, and electron–positron pairs. Annihilation quanta, fluorescence radiation, and Auger electrons are also computed. More recent applications of Monte Carlo codes to calculate electron transport in gas laser devices and advanced x-ray converters include the effect of externally applied magnetic fields on the charged particle transport.

The Monte Carlo codes do not calculate the response of the material to energy absorbtion. For highly focused finite-pulse-length electron beams, propagation of a relief wave into the deposition region from the front surface of the target can alter the coupling of the beam to the target. To calculate this phenomenon, two-dimensional hydrodynamic computer codes employing equation-of-state models with phase transitions and including models for elastic–plastic flow, fracture, and a single-temperature model for ionization assuming Saha equilibrium have been utilized. Figure 23 shows the

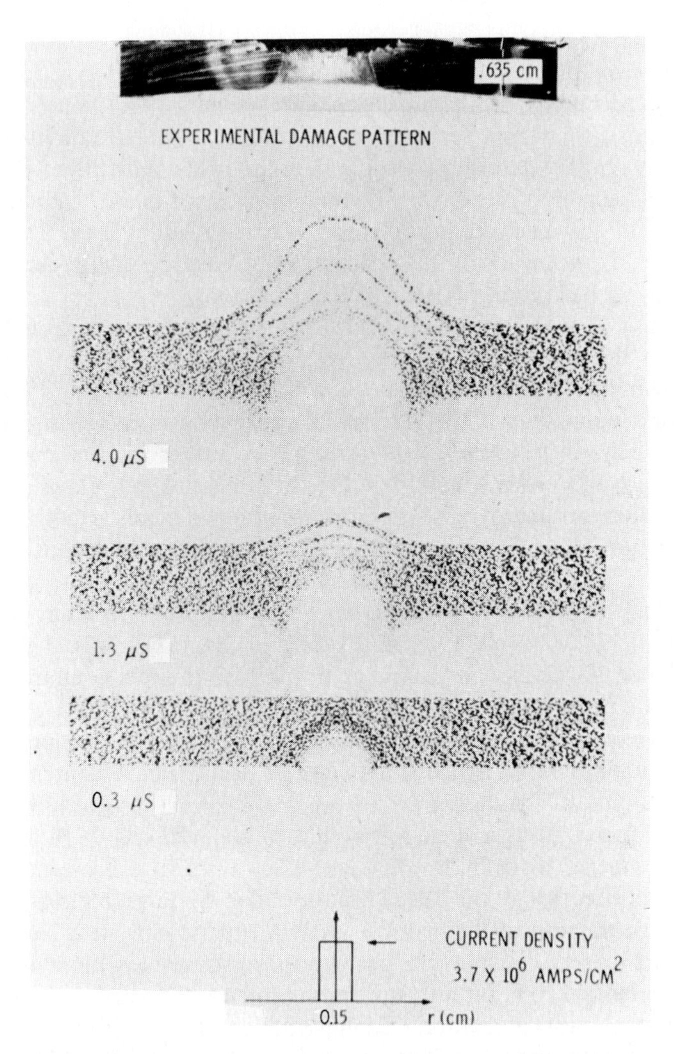

Fig. 23. Comparison between observed crater formed in aluminum target and results of two-dimensional hydrodynamic calculation.

results of a cratering calculation for a 500-keV, 50-ns, 1-mm radius, 4-kJ electron beam, incident on a 0.95-cm-thick al plate (Widner and Thompson, 1974) compared with the experimental results. The electron beam energy desposition profile was assumed to be consistent with that obtained from Monte Carlo calculations. Comparison of experimental and calculated damage patterns in witness plates of varying materials and thicknesses indicates that crater size is most sensitive to total beam energy, and is weakly dependent on pinch radius, pulse length, and energy deposition profile. Recent experiments by Perry (Perry and Widner, 1976) which measured shock transit times and rear surface velocities of flat Al targets are more sensitive to energy deposition profile than are measurements of crater size. Results indicate that pressures >1 Mbar are attained in the beam deposition volume and for electron beam energy fluxes $\lesssim 10^{12}$ W/cm^2, within the estimated 15% accuracy of the measurements, electron energy deposition is consistent with classical Monte Carlo calculations.

The use of intense relativistic electron beams to heat fusionable solid targets to thermonuclear temperatures has been proposed by Winterberg (1968), Linhart (1970), Babykin *et al.* (1971), Rudakov and Samarsky (1973), and Yonas *et al.* (1974). The initial approach considered direct heating of the fuel by electron energy deposition. Because of the low mass density of the fuel (solid or gaseous D_2 or DT), it is necessary to have other than classical electron energy absorption dominate the beam target interaction in order for a sizable fraction of the beam energy to be deposited in the target. The possibility of using a high atomic number liner to *confine* the heated fuel was also considered by several authors (Winterberg, 1972; Linhart, 1972; Maxwell, 1967; Babykin and Starykh, 1972). The effect of such a liner, also called a "tamp," is to increase the fuel confinement time by the ratio (liner density/fuel density)$^{1/2}$.

Initial experiments on solid target radiation in which neutron production was measured were carried out by Clark *et al.* (1972). A deuterated polyethylene target was irradiated by a 2.5-MV, 25-kA, 40-ns-duration electron beam produced from a dielectric-rod cathode, yielding $\geq 10^7$ neutrons. However, further work by Bradley and Kuswa (1972), Kerns and Johnson (1974), and Freeman *et al.* (1973), using similar diode geometries, indicates that a large fraction of the neutron yield in experiments with bare D_2 and deuterated targets results from beam target interactions involving collectively accelerated ions rather than thermonuclear processes. These results are consistent with calculations (Linhart, 1973; Clauser, 1975) which imply that $\sim 10^8$ J, delivered to a millimeter-size target in <10 ns would be required to ignite a bare target.

A second approach to inertially confined fusion is similar to pellet implosion schemes first considered for laser-driven fusion (Nuckolls *et al.*, 1972). In this case it has been proposed to utilize a high-Z layer to more efficiently absorb the electron beam energy and compress the fuel to high density. If the layer is thick compared with the incident electron range (Clauser, 1975), ablation of the outer surface drives the relatively cold inner layer (pusher) radially inward, resulting in efficient compression of the fuel. On the other hand, if the high-Z layer is thin compared with an electron range

(Rudakov and Samarsky, 1973), the shell expands symmetrically, resulting in less efficient coupling to the fuel. For targets with thick pushers, recent calculations indicate that "break-even" can be obtained for ~3 MJ delivered in 4 ns to a 1-mm-radius target surrounded by a 0.2-mm-thick Au shell (Fig. 24). For such targets, questions of stability of the ablatively driven pusher and two-dimensional effects produced by asymmetries in loading and the electron beam magnetic fields still need to be answered.

In order to achieve high thermonuclear yields from targets using compressional heating, it is necessary to compress the fuel several orders of magnitude which requires a high degree of spherical symmetry. This requirement is complicated by the fact that the target must be connected to a ground plane, such as an anode foil or plasma, to prevent the target from charging up and repelling the beam electrons.

Experiments to investigate the symmetry of loading of spherical targets by self-focused electron beams have been carried out using two beams to radiate opposite hemispheres of a spherical target, and a single radially focused beam to irradiate the entire surface of a spherical target (Fig. 25).

Two beam irradiation experiments (Chang *et al.*, 1974) involved a 4-mm-diameter, 0.15-mm-thick, spherical Au shell mounted in a 0.38-mm-thick Ta anode which was common to two 13-cm-diameter opposing cathodes. The two beams were obtained by triggering SF_6 gas switches in the two pulse-

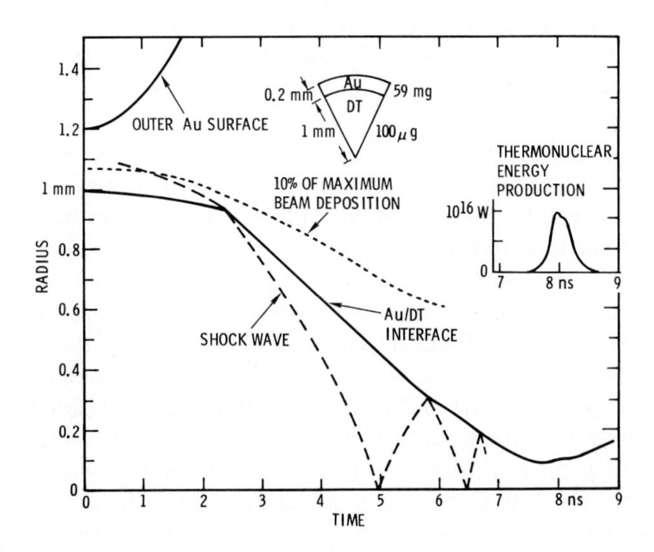

Fig. 24. Radial implosion calculation of electron beam driven target after Clauser (1975). Calculation assumes 1 MeV electrons incident on spherical target (cf. pie diagram) at 8×10^{14} W. Total thermonuclear energy released is 5 MJ.

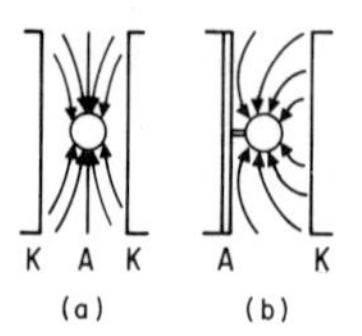

Fig. 25. Diode geometry for electron beam driven targets using: (a) two-beam irradiation, and (b) single-beam irradiation.

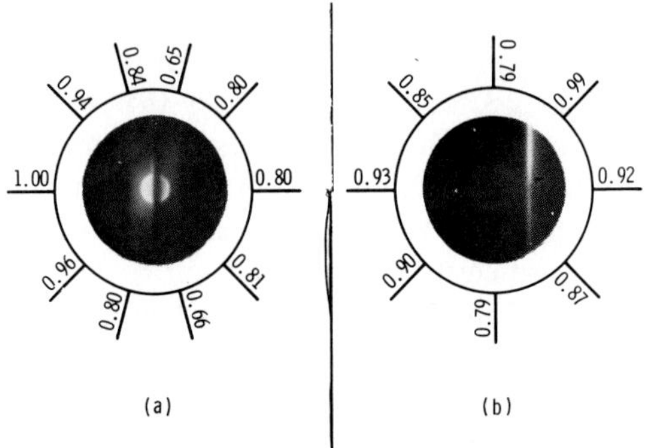

Fig. 26. Time integrated x-ray pinhole photographs for (a) two-beam irradiated target, and (b) single-beam irradiated target. Numbers indicate azimuthal variation of film optical density over target surface image.

forming lines of the Hydra accelerator with a jitter of 6 nsec and feeding the energy to the diode through two 11-Ω curved vacuum coaxial transmission lines. In the experiments, two 600-kV, 300-kA electron beams were generated which had a total energy of up to 25 kJ. It is estimated that $\sim$8 kJ of beam energy was delivered to the target. A time-integrated x-ray pinhole photograph of an irradiated target looking parallel to the anode plane is shown in Fig. 26a. The film exposure is due primarily to the 80-keV characteristic radiation from the Au target, and net optical density varies by $\lesssim 20\%$ over the circumference of the target.

In the single beam experiments (Chang *et al.*, 1975), a 4-mm-diameter spherical target was mounted on a 1-mm-long stalk protruding from the anode. About 5 kJ of 700-keV electrons was deposited in the target in an 80-ns-long pulse. Comparison of anode damage with computer calculation to determine the symmetry of loading indicates that the beam energy was deposited over the entire surface of the spherical target. Time-integrated x-ray pinhole pictures as shown in Fig. 26b are consistent with better than 2:1 symmetry of loading over the entire surface of the spherical target.

The use of electron beams to deposit large amounts of energy instantaneously in solid targets has provided a unique tool for studying the physics of matter at high energy densities. In order to approach the conditions required for producing net energy gain from fusion reactions in an inertially confined plasma, significant advances in pulse power technology and electron beam focusing are required and may be realized on a time scale comparable to the solution of corresponding problems in other approaches to controlled fusion.

VII. OTHER APPLICATIONS OF INTENSE PULSED REBS

A. Production of Electromagnetic Radiation

Historically, the main reason for the development of pulsed REB accelerators was for their use as intense x-ray sources. More recently the applications of beams to the pumping of high-power gas lasers, ignition of chemical lasers, and the production of microwave and infrared radiation have been studied.

1. Generation of X Rays

The conversion of electron beam energy to x rays is typically carried out by the experimental arrangement shown in Fig. 27. The interaction of electrons with a high-Z target produces a flux of x rays via bremsstrahlung production. For a thick target and fixed beam energy, the efficiency of conversion of electron beam energy to x-ray energy is proportional to the atomic number of the target (Evans, 1955). In order to maximize the efficiency of conversion of electron to photon energy, the target is usually less than one electron range thick, so there are some electrons which pass through the target in addition to the transmitted photons. The transmitted electrons are then absorbed in a low-Z material, behind which may also be placed a debris stopper for material from the target and absorber. Because of photon absorption in the target, absorber, and debris stopper, the softer components of the bremsstrahlung spectrum are filtered. For some applications in which

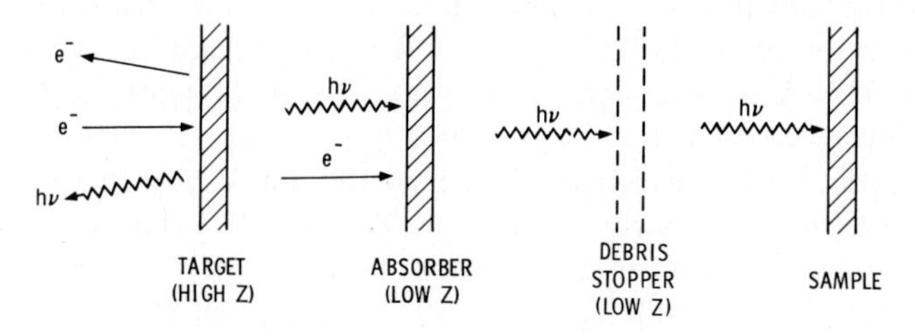

Fig. 27. Experimental configuration employed in production of intense bremsstrahlung sources.

one wishes to deposit a maximum amount of x-ray energy per unit mass in the sample, this is an undesirable result. Recent studies (Ecker, 1972) devoted to minimizing these loss mechanisms have utilized an externally applied uniform magnetic field at an angle to a thin target to cause the beam electrons to make multiple passes through the target. Such an "advanced converter" allows an enhancement of the softer portion of the photon spectrum without significant loss in conversion efficiency (Halbleib, 1974).

2. Direct Pumping of Gas and Chemical Lasers

The high-power levels and large total energies available from pulsed REB accelerators have led to their use for direct pumping of uv and vuv radiation in H_2, N_2, and Xe gas lasers, and as a "match" to initiate chemical pumping reactions in HF chemical lasers.

Using a 10^4-A, 400-keV, 3-ns electron pulse propagating parallel to a 2–10-kG magnetic field in 10–100-Torr H_2, Hodgson and Dreyfus (1972a) have observed stimulated emission from Lyman-band transitions near 1600 Å. In further work, with somewhat higher beam power levels, stimulated emission was observed from H_2 Werner bands near 1200 Å (Hodgson and Dreyfus, 1972b). Laser power levels up to 400 W/cm^2 were inferred from the exposure of film in a uv grating spectrograph. Attempts to pump the gas at higher pressures failed, apparently due to the development of instabilities in the electron beam.

Experiments on electron beam excitation of N_2 molecular lasing levels have been carried out by Clerc and Schmidt (1971), Dreyfus and Hodgson (1972), and Patterson *et al.* (1972). Clerc and Schmidt utilized a 560-keV, 10-J, 3-ns electron beam injected into a 45-cm-long, 8-cm-diameter cylindrical cavity that was equipped with concave mirrors to increase the optical path to 15 m. Stimulated emission from the B $^3\Pi_g \rightarrow$ A $^3\Sigma_u$ transition at 6322 Å and the C $^3\Pi_u \rightarrow$ B $^3\Pi_g$ transition at 3370 Å were observed at pressures between 0.3 and 40 Torr. Dreyfus and Hodgson observed superradiance from the 3370 Å upper laser state in a 1.75-m-long, 2-cm-diameter drift tube without mirrors. In experiments using a more powerful electron beam, Patterson *et al.* (1972) have observed superradiant power outputs of 24 MW at 3371 Å in a 6-ns-long pulse, for a 10^{10} W, 300-keV electron beam injected into N_2 at 20 Torr.

A cascade excitation process in which the 400-keV primary electrons generate intermediate energy (35 eV to 200 keV) electrons, which in turn cause multiple ionization and excitation of the N_2 molecular levels, was proposed by Dreyfus and Hodgson (1972) to explain their experimental results. A theoretical model of electron-beam-excited nitrogen lasers (McArthur and Poukey, 1974) which includes a more complete model for the lasing transitions and the effect of molecular excitation by low-energy plasma electrons indicates that plasma electrons, as opposed to primary and cascade electrons, cause nearly all the excitation. The model also predicts a pressure dependence of laser power as observed by Patterson (1973) and a decrease of laser power at pressures between 5 and 50 Torr as laser diameter is decreased.

A third gas in which optical gain in the vuv has been produced by electron beam pumping is xenon. Koehler *et al.* (1972) have observed stimulated emission from electron-beam pumped Xe at 1730 Å. Gerardo and Johnson (1973) injected a 1.5-MeV, 250-A, 50-nsec pulse into Xe at pressures from 1.3×10^3 to 2×10^4 Torr. Stimulated emission at 1730 Å was observed, and an effective stimulated gain cross section of 7×10^{-19} cm^2 was estimated.

Electron beams have been used to ignite chemically reacting mixtures of H_2 and F_2 (Zharov *et al.*, 1972) and SF_6 and C_2H_6 (Robinson *et al.*, 1973; Gerber and Patterson, 1974, 1975). Experiments utilizing the most powerful electron beam generator have been carried out by Gerber and Patterson, who injected a 55-kA, 2-MeV, 70-nsec FWHM beam of electrons into a 15-cm-diameter, 2-m-long drift tube filled with 20–500 Torr SF_6 with an 0–20% concentration of C_2H_6. Superradiance was observed with a maximum laser energy of 228 J (one direction axially) in the spectral range between 2.65 and 2.95 μm. More recent experiments (Gerber *et al.*, 1974) carried out with a 360-Torr F_2, 140-Torr O_2, 100-Torr SF_6, 100-Torr H_2 mixture have produced 2.3 kJ of laser energy in a 35-ns (FWHM) pulse.

3. Generation of Microwave and Infrared Radiation

Intense bursts of microwaves have been produced by the injection of relativistic electron beams into drift tubes filled with neutral gas (Nation and Gardner, 1971), evacuated drift tubes containing a strong axial magnetic field (Carmel and Nation, 1973; Friedman *et al.*, 1973), a rippled magnetic field (Friedman and Herndon, 1972), and guide tubes having structured walls (Nation, 1970; Friedman and Hammer, 1972; Friedman, 1974; Granatstein *et al.*, 1974a; Carmel *et al.*, 1974a). Intense radiation in the far-infrared (>50 μm) has been produced by beams injected into an evacuated drift tube containing a rippled magnetic field (Friedman and Herndon, 1973). In most recent work, power levels ranging from $\sim 10^5$ W in the 0.4–0.5-mm band (Granatstein *et al.*, 1974b) to 500 MW in the 3-cm band (Carmel *et al.*, 1974b) have been reported.

B. Collective Ion Acceleration

Since Veksler (1956) first publicly introduced the idea of using collective fields of electron beams to accelerate ions, research on the subject has expanded and sometimes contracted at laboratories throughout the world. Work is going on today in many countries, although Veksler's most enthusiastic expectations (1958) have not as yet been realized. Collective fields of the order of 10^6 V/cm have been documented but it has yet to be shown that such fields can be extended over considerable lengths.

Collective methods can be divided into two basic categories: (1) those methods employing a macroscopic space-charge-produced potential well, created by a localized enhancement of the electron beam density, and (2) methods that involve the production of regions of space-charge enhancement within a beam or plasma as a result of growing waves excited in a controlled manner.

In both cases it is necessary to first inject and trap ions in a stationary or slowly moving well and then to accelerate the well such that the acceleration balances the holding power of the well. This is in some ways similar to trying to first catch a marble that has been thrown at you in a round-bottom dish and then to accelerate the dish with the marble staying in it.

The effort on macroscopic space charge wells has proceeded along two lines: (1) using partially neutralized electron rings which contain just enough ions so that the self-magnetic field can hold the ring together, or (2) employing linearly propagating beams. In the ring approach, the same electrons recirculate and are accelerated together with the ions. With linear beams new electrons continually flow into the well to replenish those which give up their energy to the ions. Acceleration by accelerating wave-produced space-charge bunches within beams has also followed two directions, either employing a beam–plasma interaction or waves excited within a beam propagating in a vacuum.

1. Electron Ring Acceleration (ERA)

The electron ring method of Veksler *et al.* (1967) is based on the following:

(a) creating a ring of recirculating electrons relatively slowly using a low current injector,

(b) compressing the ring at a fixed location in space by an increasing magnetic field, loading the ring with ions, and

(c) accelerating the ring longitudinally as the ring is allowed to expand in a decreasing magnetic field.

In another variation of this method, a longitudinal electric field is used to accelerate the ring. This method is being pursued at Dubna (USSR) by Sarentsev and his group (1972), at Garching (Andelfinger, 1973; Andelfinger *et al.*, 1974), and until recently a major effort was also under way at the

Lawrence Berkeley Laboratory (Lambertson *et al.*, 1973; Lambertson, 1974). The major difficulty with the ring approach has been the growth of collective electron and ion instabilities which are found to have sufficient time to grow during the ring formation process. Although using this acceleration method, Sarentsev first reported acceleration of alpha particles to 30 MeV (Sarentsev *et al.*, 1971a,b); their original results have not been duplicated since then. The limitation on this approach seems to be that the number of electrons and ions are limited because of observed instabilities and, as a result, one can expect accelerating fields which are limited to <1 MV/cm using the magnetic approach (Schumacher, 1973; Lambertson, 1974). In the Garching approach, the ring is compressed rapidly to avoid ion-related instabilities, and this group has recently announced acceleration of the ions (Schumacher *et al.*, 1975).

An entirely different approach to ring formation is being followed at the University of Maryland (Destler *et al.*, 1975). Instead of forming the ring by using cross-field injection followed by compression on a microsecond time scale, a hollow cylindrical beam is injected through a static cusp field. The cusp field converts the beam longitudinal velocity into rotational motion in nanoseconds, hopefully avoiding some of the instability problems. On the other hand, the higher currents needed for this approach present limitations in beam quality which tend to restrict the electron number and number density in the ring (See Fig. 28).

The ERA approach has had substantial support for a number of years, and critical experiments in the near future should define whether further effort in this area will be productive.

2. Linear Collective Acceleration (LCA)

Until very recently the LCA approach has been characterized largely by empirical observations of acceleration of ions in the direction of propagating beams. Since there has been no comprehensive research program toward this approach as in the case of the ERA, the true significance of these observations is still a matter of question. The first observation of LCA was made by Plyutto (1961) in which he observed protons accelerated in the direction of extraction of an electron beam from a plasma-filled diode. Later Plyutto *et al.* (1967) reported observing proton energies as high as 4–5 MeV from a 200–399-kV discharge. In a detailed study of diodes prefilled with plasma, Mkheidze *et al.* (1971) clarified these experimental results to some extent. The very high ion energies (up to 30 times the applied voltage) were correlated with a drop in the diode current in a few nanoseconds, and the formation of a pinched electron beam on axis was reported (see Fig. 29). These results have not been explained, and no method has been proposed for

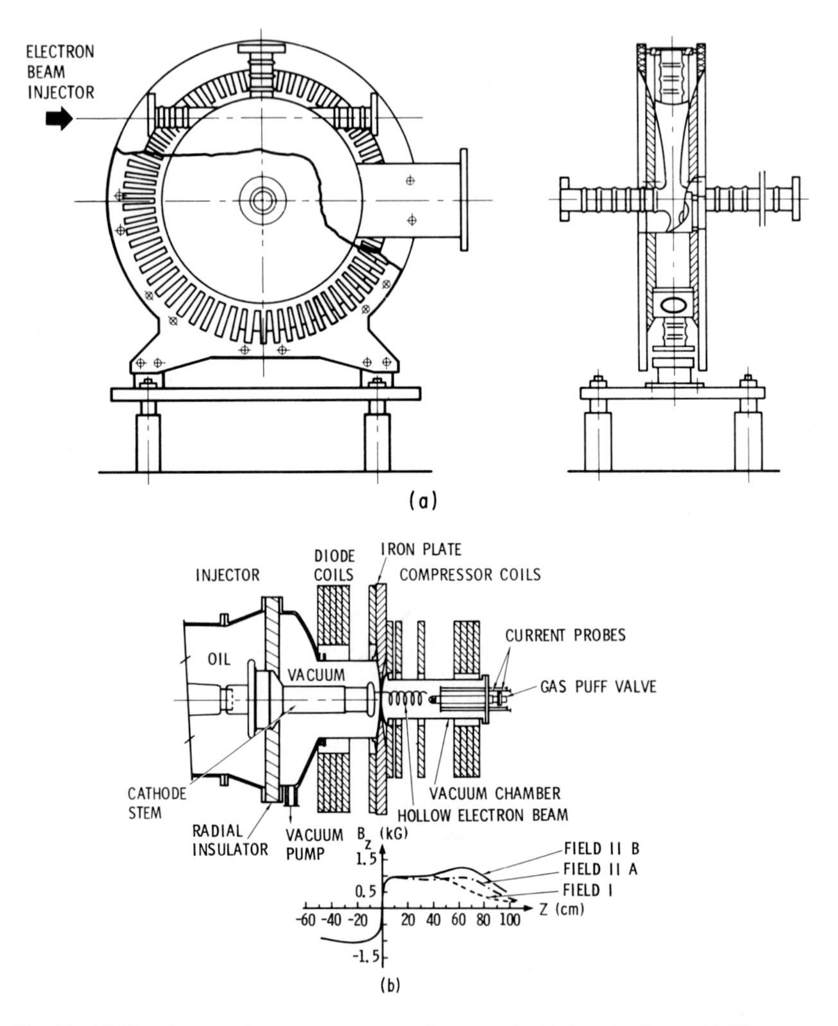

Fig. 28. (a) The electron ring compressor employed at the Dubna Institute. (b) The cusp field approach to ring formation as employed at the University of Maryland.

further extensions, although these data represent the highest multiplication of applied voltage thus observed. Unfortunately, since the experiments were carried out in the 30-kV range, the final ion energy was limited and the accelerating field was less than 1 MV/cm.

An apparently simpler experiment is to use a vacuum diode with a single-

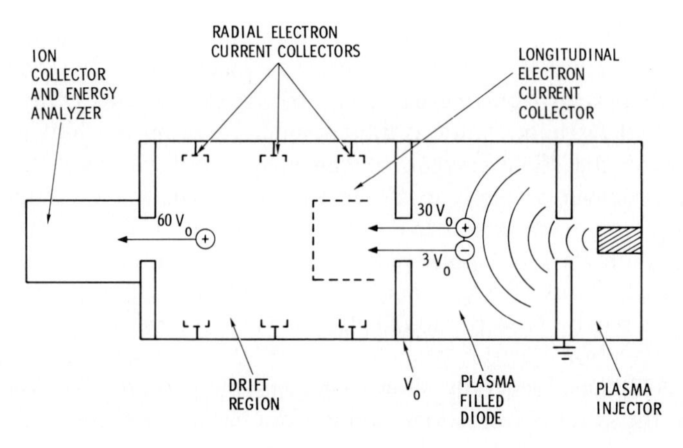

Fig. 29. Schematic of plasma diode apparatus studying collective ion acceleration (Mkheidze *et al.*, 1971).

TABLE VI

TYPICAL DATA FOR PROTONS ACCELERATED IN DIODES[a]

PROTONS		ELECTRONS		DIODE			REFERENCES
$\mathscr{E}_i$ (MeV)	N	$\mathscr{E}_e$ (MeV)	I_0 (kA)	U_0 (MV)	A-K GAP (cm)	CONFIGURATION	
4-5	10^{11} - 10^{12}	–	–	0.2-0.3	–	A	(PLYUTTO, et al., 1967)
0-2.5	10^{11} - 10^{12}	0-0.25	1-2	0-0.1	1-10	A	(PLYUTTO, et al., 1969)
0.7	10^{11} - 10^{12}	0.06	1-3	0.02	1-5	A	(MKHEIDZE AND KOROP, 1971)
2-7	–	–	–	0.2-1.0	2-7	A	(PLYUTTO, et al., 1973)
2-3	–	–	5	0.2-0.3	1-2	C	(KOROP AND PLYUTTO, 1970)
0.08-3	$\sim 10^{15}$	–	50	5	1.8	B	(JOHNSEN & KERNS, 1974)
0.1-3	–	–	100	2	0.6	C	(BRADLEY & KUSWA, 1972)
5-13	$\sim 10^{14}$	–	30	2.5	–	D	(LUCE, et al.(1973), LUCE (1975), SAHLIN (1975)

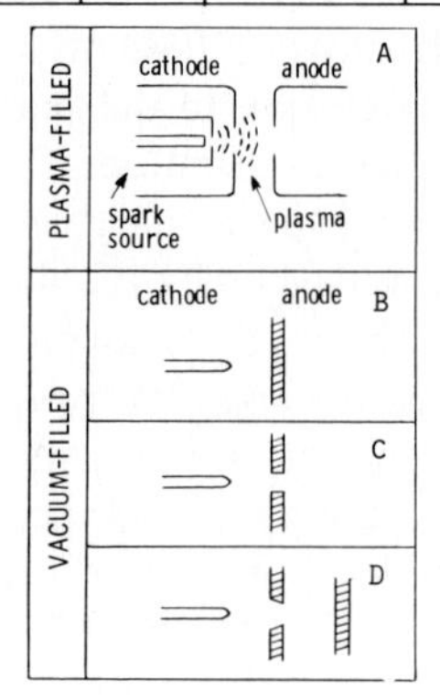

[a] Olson (1975).

pointed cathode and a plane anode as shown in Fig. 9. Such vacuum diodes also show evidence of LCA but, as we saw in Section III, such diodes can be filled with plasma from the electrons. In the experiments of this type it appears that ion acceleration is accompanied by beam pinching and rapid decreases in the diode current, but again no definitive understanding of these phenomena presently exists. In the most remarkable vacuum diode experiment reported thus far, Korop and Plyutto (1971a) found that an 80-kV and only a 100-A pulse produced 400-keV protons. In other experiments using beams of 5 kA and 300 keV, proton energies up to 2.0 MeV were observed by Korop and Plyutto (1971b). Extensions to the tens of kA and 2–5-MV electron beam range (Bradley and Kuswa, 1972; Luce *et al.*, 1973) have generally shown smaller voltage multiplication factors than for the lower current cases, and it is not clear how these results can be extended. A brief summary of LCA results in diodes is given in Table VI.

One explanation has been proposed (Plyutto *et al.*, 1969, 1972) for the vacuum diode results that, as we shall see later, may also have relevance to propagating beams. Plyutto has argued that a space-charge-produced potential well can exist near the moving front of the expanding cathode plasma. The electric field that accelerates the ions is then the electron energy divided by the Debye length at the edge of the cathode plasma. If this model is true, then accelerating fields of about 10^6 V/cm are possible, but of course cannot be extended to large distances. Obviously one would like to create similar fields in drifting beams.

This was first accomplished by Graybill and Uglum (1970) who injected a 1.5-MeV, 40-kA beam into a low-pressure gas ($\sim$0.1 Torr in hydrogen) and observed acceleration of protons with energies up to 5 MeV. These results have been extended by various authors including Rander *et al.* (1970), Ecker and Putnam (1973), Kuswa *et al.* (1973), and Miller and Straw (1975) (see summary Table VII). Rostoker (1969) offered an explanation similar to that of Plyutto, namely acceleration in the space-charge wave caused by a beam. Olson (1975) developed the first definitive explanation considering self-consistent two-dimensional space-charge effects and detailed ionization processes, obtaining good agreement with the various parameter ranges investigated by several authors. Subsequent numerical calculations by Poukey and Olson (1975) confirmed this explanation. A schematic of the two-dimensional potential well formation is shown in Fig. 30 (Olson, 1975). The limitation of this natural process is that the ionization mechanism which defines the width of the acceleration region is relatively slow and results in strong fields only near the anode. Olson (1975) has proposed to provide an

TABLE VII

TYPICAL DATA FOR PROTONS ACCELERATED BY INTENSE RELATIVISTIC ELECTRON BEAM INJECTION INTO NEUTRAL GAS (H_2)

PROTONS				IREB					GUIDE TUBE			REFERENCES
$\mathscr{E}_i$ (MeV)	N	T (nsec)	I_i (A)	$\mathscr{E}_e$ (MeV)	I_0 (kA)	t_r (nsec)	t_b (nsec)	r_b (cm)	p (Torr)	R (cm)	L (cm)	
4-7	$\sim 4 \times 10^{12}$	3-10	~200	1.5	30	10	50	1.25	0.05-0.15	7.6	50	(GRAYBILL, 1972)
2-10	$\sim 2 \times 10^{12}$	3-5	–	1	110	10	60	2.5	0.15-0.65	3.8	73	(ECKER & PUTNAM, 1973)
1-5	–	–	–	1.8	80	60	80	0.5	0.015-0.15	2.5	70	(KUSWA, et al., 1973; KUSWA, 1975; SWAIN, et al., 1974; OLSON, et al., 1974)
5-16	$\sim 10^{13}$	–	–	5	40	25	125	2[a]	0.05-0.35	32	122	(MILLER & STRAW (1975)
1-3, 8	$\sim 10^{12}$	5-15	26	0.65	15-20	15	50	–	0.005-0.4	5	20-50	(KOLOMENSKY, et al. (1974), KOLOMENSKY, 1974)

[a] Annular beam, 2-mm thick.

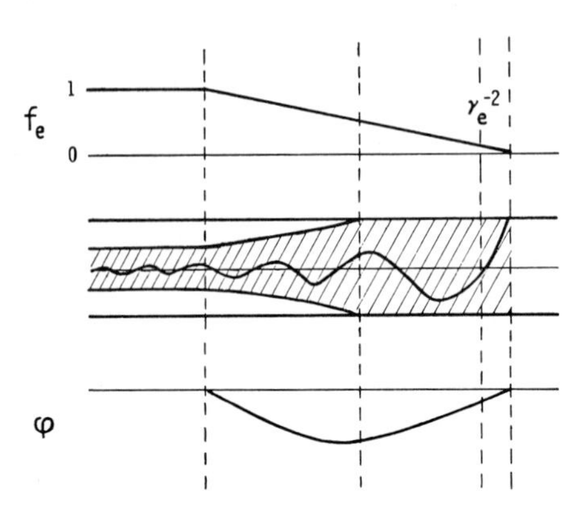

Fig. 30. Moving, self-consistent, two-dimensional, beam front equilibria, showing the fractional space charge neutralization f_e, the beam profile, and the resultant potential φ.

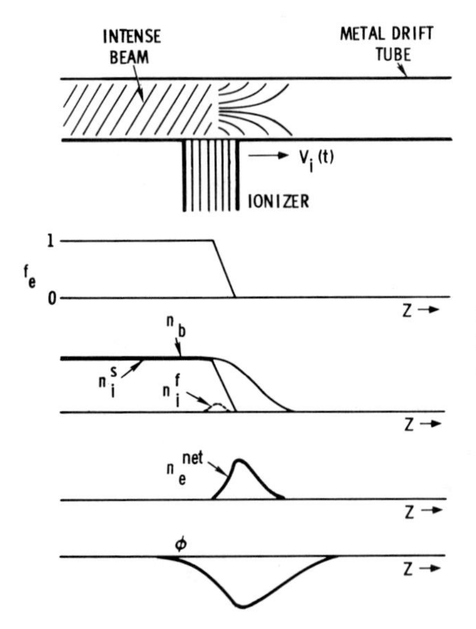

Fig. 31. Schematic of linear collective ion acceleration scheme with external photo ionizer to control potential well acceleration (Olson, 1974; Olson and Poukey, 1974).

independent photoionization source within a background gas of too low a pressure to be affected by collisional ionization. By sweeping this externally driven ionization source at a preprogrammed rate of acceleration, it may be possible to extend the most intense space-charge fields of approximately 1 MV/cm over distances of one or more meters (see Fig. 31).

Other active means of controlling the ion accelerating field have been proposed for wave-related methods. Research on acceleration of ions in localized microscopic regions of charge enhancement created by nonlinear wave growth in plasmas was first proposed by Fainberg (1956). Although this work has been carried out for many years at PhTI, Kharkov (USSR), it has not succeeded in producing accelerated ions thus far. The major difficulty with this approach seems to be in producing and controlling single modes; recent results indicate that premodulation of the beam should aid in producing coherent waves, and work is continuing (Fainberg, 1975).

More recently, Sloan and Drummond (1973) have proposed employing a single cyclotron mode excited in a beam propagating in a vacuum with a strong applied longitudinal magnetic field. By decreasing the field with distance, they propose to increase the wave phase velocity and thus accelerate trapped ions. In any such scheme which accelerates ions starting initially with a high phase velocity, ion injection and trapping is a serious question. In addition, as in the plasma wave methods or any finite amplitude wave processes, it is not clear that a single mode can be excited in a controlled manner.

Of the various applications discussed in this chapter, a successful collective ion accelerator would possibly have the most immediate application. Such an accelerator could be used in such diverse fields as nuclear chemistry, medium energy physics, and cancer therapy. The fact that collective acceleration has already been demonstrated with fields of about 1 MeV/cm should provide incentive that this application will someday turn into a practical reality.

VIII. CONCLUSION

Intense relativistic electron beams have developed rapidly since the introduction of the technology in the mid-sixties. Available power from such accelerators has increased by three orders of magnitude in this time, and a comprehensive knowledge of beam generation and transport has evolved. At the present, potentially significant applications to fusion research, laser excitation, intense radiation sources, and collective ion acceleration are at hand. The richness of the physical phenomena associated with this relatively simple technology continues to add to the level of interest it has aroused and should provide unexpected areas of payoff as research proceeds.

REFERENCES

Abrashitov, Yu. I., *et al.* (1973). Interaction of a High-Power Relativistic Electron Beam with a Plasma in a Magnetic Field. Inst. of Nuclear Physics, Novosobirsk, USSR, Preprint E. Ya. F., pp. 60–73.

Alfvén, H. (1939). *Phys. Rev.* **55,** 425.

Altyntsev, A. T., *et al.* (1971). *ZhETF Pis. Red.* **13,** 197.

Andelfinger, C. (1973). *Particle Accel.* **5,** 105.

Andelfinger, C., Dommaschk, W., Hoffmann, I., Merkel, P., Schumacher, U., Ulrich, M. (1974). *Proc. Int. Conf. High Energy Acclerators, 9th, SLAC,* pp. 218–222.

Babykin, M. V., and Starykh, V. V. (1972). *Sov. Phys.–Tech. Phys.* **16,** 1273.

Babykin, M. V., Zavoiskii, E. K., Ivanov, A. A., Rudakov, L. I. (1971). *In* "Plasma Physics and Controlled Thermonuclear Fusion Research" (*Proc. Int. Conf., 4th, Madison*) Vol. 1, p. 635. IAEA, Vienna.

Benford, G., and Book, D. (1971). Relativistic beam equilibria, *Adv. Plasma Phys.* **4,** 125.

Benford, J., Ecker, B., and Yonas, G. (1969). Beam Propagation in Pre-ionized Media, Physics International Co., San Leandro, California, Rep. No. PIIR-10-70.

Benford, J., Ecker, B., and Bailey, V. (1974). *Phys. Rev. Lett.* **33,** 574.

Bennett, W. (1934). *Phys. Rev.* **45,** 890.

Bennett, W. (1955). *Phys. Rev.* **98,** 6.

Berger, M. J., and Seltzer, S. M. (1964). NASA Publ. SP-3012.

Bernstein, B., and Smith, I. D. (1973). *IEEE Trans. Nucl. Sci.* **NS-20,** 294.

Block, J., *et al.* (1971). *Rec. Symp. Electron, Ion, and Laser Beam Technol. 11th, May,* p. 513.

Bludman, S. A., Watson, K. M., and Rosenbluth, M. N. (1960). *Phys. Fluids* **3,** 747.

Boers, J. E., and Kelleher, D. (1969). *J. Appl. Phys.* **40,** 2409.

Bradley, L. P., and Kuswa, G. W. (1972). *Phys. Rev. Lett.* **29,** 1441.

Breizman, B. N., and Ryutov, D. D. (1970). *Zh. Eksp. Teor. Fiz. Pis. Red.* **11,** 421.

Breizman, B. N., and Ryutov, D. D. (1971). *Zh. Eksp. Teor. Fiz.* **60,** 408.

Breizman, B. N., Ryutov, D. D., and Chebotaev, D. Z. (1972). *Zh. Eksp. Teor. Fiz.* **62,** 1409.

Briggs, R. J., Hestor, R. E., Lamb, W. A., Lauer, E. J., and Spoerlein, R. L. (1974). *Bull. Am. Phys. Soc.* **19,** 902.

Budker, G. I. (1956). *Proc. CERN Symp. High Energy Accelerators PION Phys., Geneva* **1,** 68.

Budker, G. I. (1973). *Eur. Conf. Controlled Fusion Plasma Stud., 6th* **2,** 36.

Buneman, O. (1959). *Phys. Rev.* **115,** 503.

Carmel, Y., and Nation, J. A. (1973). *Phys. Rev. Lett.* **31,** 806.

Carmel, Y., Ivers, J., Kribel, R. E., and Nation, J. (1974a). *Phys. Rev. Lett.* **33,** 1278.

Carmel, Y., Ivers, J., Kribel, R. E., and Nation, J. (1974b). *Bull. Am. Phys. Soc.* **19,** 965.

Chang, J., *et al.* (1974). *Proc. Conf. Plasma Phys. Controlled Nucl. Fusion Res., 5th, Tokyo, Japan, November 11–15.*

Chang, J., Widner, M. M., Kuswa, G. W., and Yonas, G. (1975). *Phys. Rev. Lett.* **34,** 1266.

Clark, J. G., Kerns, J. R., and McCann, T. E. (1972). *Bull. Am. Phys. Soc.* **17,** 1031.

Clauser, M. J. (1975). *Phys. Rev. Lett.* **34,** 570.

Clerc, M., and Schmidt, M. (1971). *C. R. Acad. Sci. Paris* **B272,** 668.

Creedon, J. (1967). Physics International Co., San Leandro, California, Rep., PIIR-17-67.

Creedon, J. (1972). Physics International Co., San Leandro, California, Rep. PIIR-17-72A, October.

Creedon, J., Spence, P., and Huff, R. (1973). *Bull. Am. Phys. Soc.* **18,** 1310.

de Packh, D. (1968). Naval Res. Lab., Washington, D.C., Rep. RPIR 7.

Destler, W. W., Hudgings, D. W., Misra, P. K., and Rhee, M. J. (1975). *Bull. Am. Phys. Soc.* **20,** 181.

Dreyfus, R. W., and Hodgson, R. T. (1972). *Appl. Phys. Lett.* **20,** 195.
Ecker, B. (1972). Physics International Co., San Leandro, California, Rep. No. PIFR-366/371.
Ecker, B., and Putnam, S. (1973). *IEEE Trans. Nucl. Sci.* **NS-20,** 301.
Ekdahl, C., Greenspan, M., Kribel, R. E., Sethian, J., and Wharton, C. B. (1974). *Phys. Rev. Lett.* **33,** 346.
Evans, R. D. (1955). "The Atomic Nucleus," pp. 614–617. McGraw-Hill, New York.
Fainberg, Ya. B. (1956). *Proc. CERN Symp. High Energy Accelerators, Geneva, Switzerland* **1,** 84.
Fainberg, Ya. B. (1975). *Particle Accelerators* **6,** 95.
Fainberg, Ya. B., Shapiro, V. D., and Shevehenko, V. I. (1969). *Zh. Eksp. Teor. Fiz.* **57,** 966.
Felsenthal, P., and Proud, J. M. (1965). *Phys. Rev.* **139A,** 1796.
Ford, F. C., Link, W. T., and Creedon, J. (1966). Physics International Co., San Leandro, California, Rep. PIPB-9, August.
Freeman, B., Gullickson, R., Zucker, O., Bostick, W., and Klapper, H. (1973). *Bull. Am. Phys. Soc.* **18,** No. 10, 1351.
Friedlander, F., Hecktel, R., Jory, J., and Mosher, C. (1968). Varian Associates, San Carlos, California, DASA 2173.
Friedman, M. (1974). *Phys. Rev. Lett.* **32,** 92.
Friedman, M., and Hammer, D. A. (1972). *Appl. Phys. Lett.* **21,** 174.
Friedman, M., and Herndon, M. (1972). *Phys. Rev. Lett.* **28,** 210.
Friedman, M., Hammer, D. A., Manheimer, W. M., and Sprangle, P. (1973). *Phys. Rev. Lett.* **31,** 752.
Friewald, D. A., Prestwich, K. R., Kuswa, G. W., and Beckner, E. H. (1971). *Phys. Lett.* **36A,** 297.
Gerardo, J. B., and Johnson, A. W. (1973). *IEEE J. Quantum Electron.* **QE-9,** 748.
Gerber, R. A., and Patterson, E. L. (1974). *IEEE J. Quantum Electron.* **QE-10,** 333.
Gerber, R. A., and Patterson, E. L. (1975). *IEEE J. Quantum Electron.* **QE-11,** 642.
Gerber, R. A., Patterson, E. L., Blair, L. S., and Greiner, N. R. (1974). *Appl. Phys. Lett.* **25,** 281.
Gilad, P., Kusse, B. R., and Lockner, T. R. (1974). *Phys. Rev. Lett.* **33,** 1275.
Godfrey, B. B., Newberger, B. S., and Taggert, K. A. (1975). *IEEE Plasma Sci.* **PS3,** 60.
Goldenbaum, G. C., Dove, W. F., Gerber, K. A., and Logan, B. G. (1974). *Phys. Rev. Lett.* **32,** 830.
Goldstein, S. A., Davidson, R. C., and Siambis, L. R. (1974). *Phys. Rev. Lett.* **33,** 1471.
Granatstein, V. L., Herndon, M., Parker, R. K., and Sprangle, P. (1974a). *IEEE J. Quantum Electron.* **QE-10,** 651.
Granatstein, V. L., Sprangle, P., and Schlesinger, S. P. (1974b). *Bull. Am. Phys. Soc.* **19,** 965.
Graybill, S. E. (1972). *IEEE Trans. Nucl. Sci.* **NS-19,** 292.
Graybill, S. E., and Nablo, S. V. (1966). *Appl. Phys. Lett.* **8,** 18.
Graybill, S. E., and Uglum, J. R. (1970). *J. Appl. Phys.* **41,** 236.
Halbleib, J. A., Sr. (1974). *J. Appl. Phys.* **45,** 4103.
Hammer, D. A., and Rostoker, N. (1970). *Phys. Fluids* **13,** 1831.
Hodgson, R. T., and Dreyfus, R. W. (1972a). *Phys. Lett.* **31A,** 213.
Hodgson, R. T., and Dreyfus, R. W. (1972b). *Phys. Rev. Lett.* **28,** 536.
Ikuta, K. (1972). *Jpn. J. Appl. Phys.* **11,** 1684.
Johnson, D. J., and Kerns, J. R. (1974). *Appl. Phys. Lett.* **25,** 191.
Johnson, D. L. (1974). *Proc. Int. Conf. Energy Storage, Compression and Switching, Torino, Italy, November 5–7*, p. 515.
Kapetanakos, C. A., and Hammer, D. A. (1973). *Appl. Phys. Lett.* **23,** 17.
Kerns, J. R., and Johnson, D. J. (1974). *J. Appl. Phys.* **45,** 5225.
Koehler, H. A., Ferderber, L., Redhead, D. L., and Ebert, P. J. (1972). *Appl. Phys. Lett.* **21,** 198.

Kolb, A. (1973). Maxwell Laboratories, San Diego, California, Rep., DNA 2959F, MCR 140.
Kolomensky, A. A. (1974). *Int. Conf. High Energy Accelerators, 9th, SLAC, Stanford, California*, p. 254.
Kolomensky, A. A., and Zozulya, Yu. T. (1974). *Nat. USSR Accelerator Conf., 4th, Moscow.*
Korn, P., Sandel, F., and Wharton, C. B. (1973). *Phys. Rev. Lett.* **31,** 579.
Korop, E. D., and Plyutto, A. A. (1971a). *Sov. Phys.–Tech. Phys.* **10,** 830.
Korop, E. D., and Plyutto, A. A. (1971b). *Sov. Phys.–Tech. Phys.* **15,** 1986.
Kuswa, G. W. (1975). *Ann. N. Y. Acad. Sci.* **251,** 514.
Kuswa, G. W., Bradley, L. P., and Yonas, G. (1973). *IEEE Trans. Nucl. Sci.* **NS-20,** 305.
Lambertson, G. R. (1974). *Proc. Int. Conf. High Energy Accelerators, 9th, SLAC, Stanford, California*, pp. 214–217.
Lambertson, G. R., *et al.* (1973). *Particle Accelerators* **5,** 113.
Lawson, J. D. (1958). *J. Elec. Contr.* **5,** 146.
Lawson, J. D. (1973). *Phys. Fluids* **16,** 1298.
Lee, R., and Lampe, M. (1973). *Phys. Rev. Lett.* **31,** 1390.
Levine, L. S., Vitkovitsky, I. M., and Hammer, D. A. (1971). *J. Appl. Phys.* **42,** 1863.
Linhart, J. G. (1970). *Nucl. Fusion* **10,** 211.
Linhart, J. G. (1973). *Nucl. Fusion* **13,** 321.
Link, W. T. (1967). *IEEE Trans. Nucl. Sci.* **14,** 777.
Lovelace, R. V., and Sudan, R. N. (1971). *Phys. Rev. Lett.* **27,** 1256.
Luce, J. S. (1975). *Ann. N. Y. Acad. Sci.* **251,** 217.
Luce, T., Sahlin, H., and Crites, T. (1973). *IEEE Trans. Nucl. Sci.* **NS-20,** 336.
McArthur, D. A., and Poukey, J. W. (1973). *Phys. Fluids* **16,** 1996.
McArthur, D. A., and Poukey, J. W. (1974). *Phys. Rev. Lett.* **32,** 89.
Martin, J. C. (1970). Atomic Weapons Res. Establishment, Aldermaston, England, Rep. SSWA/JCM/70327, March 5.
Martin, J. C., and Smith, I. D. (1965–1970). Series of Unpublished Notes from the Atomic Weapons Res. Establishment. Obtainable from AFWL(EL) Kirtland Air Force Base, New Mexico, 87117.
Martin, T. H. (1969). *IEEE Trans. Nucl. Sci.* **NS-16,** 59.
Martin, T. H. (1973). *IEEE Trans. Nucl. Sci.* **NS-20,** 289.
Martin, T. H., and Prestwick, K. D. (1974). *Proc. Int. Conf. Energy Storage, Compression Switch., Torino, Italy, November 5–7*, p. 57.
Mather, J. W., Carpenter, J. P., Friewald, D. A., Ware, K. D., and Williams, A. H. (1973). *J. Appl. Phys.* **44,** 4913.
Matsiborko, N. G., Onishchenko, I. N., Shapiro, V. D., and Shevchenko, V. I. (1972). *Plasma Phys.* **14,** 591.
Maxwell, D. (1967). Physics International Co., San Leandro, California, Rep. PIIR-41-67, July.
Messyats, G. A. (1975). *Sov. Phys.–Tech. Phys.* **19,** 948.
Messyats, G. A., Littvinov, E. A., and Proshvrovsky, D. I. (1970). *In* "Discharges and Electrical Insulator in Vacuum" (*Proc. Int. Symp., 4th*), p. 82. Waterloo, Ontario.
Miller, P. A., and Kuswa, G. W. (1973). *Phys. Rev. Lett.* **30,** 958.
Mix, L. P., Kelly, J. G., Kuswa, G. W., Swain, D. W., and Olsen, J. N. (1973). *J. Vac. Sci. Technol.* **10,** 951.
Mkheidze, G. P., and Korop, E. D. (1971). *Zh. Tekh. Fiz.* **41, 873** [*English transl.: Sov. Phys.–Tech. Phys.* **16,** 690].
Mkheidze, G. P., Plyutto, A. A., and Korop, E. D. (1971). *Sov. Phys.–Tech. Phys.* **16,** 749.
Mosher, D. (1974). Naval Research Laboratory, Washington, D.C., Rep. 2959, December.
Miller, R. B., and Straw, D. C. (1975). *Bull. Am. Phys. Soc.* **20,** 182.

Nation, J. A. (1970). *Appl. Phys. Lett.* **17,** 491.

Nation, J. A., and Gardner, W. L. (1971). *Nucl. Fusion* **11,** 5.

Nuckolls, J., Wood, L., Thiessen, H., Zimmerman, G. (1972). *Nature* (*London*) **239,** 139.

Olson, C. L. (1974). *Proc. Int. Conf. High Energy Accelerators, 9th, SLAC, Stanford, California,* p. 272.

Olson, C. L. (1975). *Phys. Rev. A* **11,** 288.

Olson, C. L., and Poukey, J. W. (1974). *Phys. Rev. A* **9,** 2631.

Olson, C. L., Kuswa, G. W., Swain, D. W., and Poukey, J. W. (1974). *Proc. Symp. Ion Sources, Berkeley, California, October,* p. III-3-1.

Papadopoulos, K., *et al.* (1974). *Proc. Conf. Plasma Phys. Controlled Nucl. Fusion Res., 5th, Tokyo, November 11–15.*

Parker, R. K., Anderson, R. E., and Duncan, C. V. (1974). *J. Appl. Phys.* **45,** 2463.

Patterson, E. L. (1973). *J. Appl. Phys.* **44,** 3193.

Patterson, E. L., Gerardo, J. B., and Johnson, A. W. (1972). *Appl. Phys. Lett.* **21,** 293.

Perry, F. C., and Widner, M. M. (1976). *J. Appl. Phys.* **47,** 127.

Plyutto, A. A. (1961). *Sov. Phys. JETP* **12,** 1106.

Plyutto, A. A., *et al.* (1967). *Zh. Eksp. Teor. Fiz. Pis. Red.* **6,** 540 [*English transl.: JETP Lett.* **6,** 61].

Plyutto, A. A., Suladze, K. V., Temchin, S. M., and Korop, E. D. (1969). *At. Energ.* **27,** 418.

Plyutto, A. A., *et al.* (1970). *Zh. Eksp. Teor. Fiz. Pisma* **6,** 540.

Plyutto, A. A., Suladze, K. V., Korop, E. D., and Ryzhkov, V. N. (1971). *Int. Symp. Discharges Electrical Insulation Vacuum, 5th, Poznan, Poland,* p. 145.

Plyutto, A. A., *et al.* (1974). *Zh. Tekh. Fiz.* **43,** 1627 [*English transl.: Sov. Phys.–Tech. Phys.* **18,** 1026].

Poukey, J. W. (1975a). *Proc. Symp. Electron, Ion, Photon Beam Technol., 13th, Colorado Springs, Colorado, May.*

Poukey, J. W. (1975b). *Appl. Phys. Lett.* **26,** 145.

Poukey, J. W., and Olson, C. L. (1975). *Phys. Rev. A* **11,** 691.

Poukey, J. W., and Toepfer, A. J. (1974). *Phys. Fluids* **17,** 1582.

Poukey, J. W., Toepfer, A. J., and Kelly, J. G. (1971). *Phys. Rev. Lett.* **26,** 1620.

Poukey, J. W., Freeman, J. R., and Yonas. G. (1973). *J. Vac. Sci. Technol.* **10,** 954.

Prestwich, K. R. (1974). *Proc. Int. Conf. Energy Storage, Compression Switching, Torino, Italy, November 5–7,* p. 451.

Prestwich, K. R. (1975). *Proc. IEEE Particle Accelerator Conf., Washington, D. C., March 12–14,* p. 975.

Prono, D., Ecker, B., Bergstrom, N., Benford, J., and Putnam, S. (1975). Physics International Co., San Leandro, California, Rep. PIFR-557, February.

Rander, J., Ecker, B., Yonas, G., and Drickey, D. (1970). *Phys. Rev. Lett.* **24,** 283.

Roberts, T. G. (1965). The Experimental Verification of Self-Focusing in Intense, Relativistic, Electron Beam. U. S. Army Missile Command, Redstone Arsenal, Alabama, Rep. No. RR-TR-65-17.

Roberts, T. G. (1968). On the Propagation of High Intensity, High Voltage Electron Beams and the Maximum Current Which Such Beams May Possess. U. S. Army Missile Command, Redstone Arsenal, Alabama, Rep. No. RR-TR-69-7.

Robinson, C. P., Jensen, R. J., and Kolb, A. (1973). *IEEE J. Quantum Electron.* **QE-9,** 963.

Rostoker, N. (1969). *Proc. Int. Conf. High Energy Accelerator, 7th, Yerevan,* p. 509.

Rudakov, L. I. (1970). *Zh. Eksp. Teor. Fiz.* **59,** 2091.

Rudakov, L. I., and Samarsky, A. A. (1973). *Proc. Eur. Conf. Controlled Fusion Plasma Phys., 6th, Moscow,* p. 487.

Sahlin, H. L. (1975). *Ann. N. Y. Acad. Sci.* **251,** 238.

Sarentsev, V. P. (1972). *Symp. Collective Methods Acceleration, Dubna.*
Sarentsev, V. P., *et al.* (1971a). Joint Institute for Nuclear Research, P9-5558. Dubna, January.
Sarentsev, V. P., *et al.* (1971b). *Sov. Phys. JETP* **33,** 1067.
Schumacher, U. (1973). Max Planck Inst. fur Plasma Phys. IPP 0/20, Garching.
Schumarcher, U., Andelfinger, C., and Ulrich, M. (1975). *Bull. Am. Phys. Soc.* **20,** 181.
Sloan, M. L., and Drummond, W. E. (1973). *Phys. Rev. Lett.* **31,** 1234.
Smith, D. R. (1972). *Phys. Lett.* **42A,** 211.
Smith, I. D. (1974). *Proc. Int. Conf. Energy Storage, Compression Switching, Torino, Italy, November 5–7.*
Stallings, C. Shope, S., and Guillory, J. (1972). *Phys. Rev. Lett.* **28,** 653.
Swain, D. W. (1973). *Phys. Fluids* **16,** 569.
Swain, D. W., Kuswa, G. W., Poukey, J. W., and Olson, C. L. (1974). *Proc. Int. Conf. High Energy Accelerators, 9th, SLAC, Stanford,* p. 268.
Swain, D. W., Miller, P. A., and Widner, M. M. (1975). Sandia Lab. Albuquerque, New Mexico, Rep. SAND-75-0214.
Thode, L. E., and Sudan, R. N. (1973). *Phys. Rev. Lett.* **30,** 732.
Thode, L. E., and Sudan, R. N. (1975). *Phys. Fluids* **18,** 1564.
Toepfer, A. J. (1971). *Phys. Rev. A* **3,** 1444.
Toepfer, A. J., and Poukey, J. W. (1973). *Phys. Lett.* **42A,** 383.
VanDevender, J. P., and Martin, T. H. (1975). *IEEE Trans. Nuc. Sci.* **NS-22,** 979.
VanDevender, J. P., Kilkenny, J. D., and Dangor, A. E. (1974). *Phys. Rev. Lett.* **33,** 689.
Veksler, V. I. (1956). *Proc. CERN Symp. High Energy Accelerators Ion Phys.* **I,** 80.
Veksler, V. I. (1958). *Sov. Phys. Usp.* **66** (1), 54.
Veksler, V. I., *et al.* (1967). Cambridge Electron Accelerator Rep. No. CEAL-2000, p. 289, Cambridge, England (unpublished).
Weibel, E. S. (1959). *Phys. Rev. Lett.* **2,** 83.
Wheeler, C. B. (1974). *J. Phys. D. Appl. Phys.* **7,** 1597.
Widner, M. M., and Thompson, S. L. (1975). Sandia Lab. Albuquerque, New Mexico, Rep., SAND-74-351.
Winterberg, F. (1968). *Phys. Rev.* **174,** 212.
Winterberg, F. (1972). *Nucl. Fusion* **12,** 353.
Wright, T. P., and Hadley, G. R. (1975). *Phys. Rev.* A, **12,** 686.
Yonas, G., and Spence, P. W. (1968). Physics International Co., San Leandro, California, Rept., DASA 2175, October.
Yonas, G., and Spence, P. (1969). *Symp. Electron, Ion, Laser Beam Technol., 10th,* pp. 143–154. San Francisco Press, San Francisco, California.
Yonas, G., Spence, P., Ecker, B., and Rander, J. (1969). Physics International Co., San Leandro, California, Rep. PIFR 106-2, August.
Yonas, G., Smith, I., Spence, P., Putnam, S., and Champney, P. (1971). *Rec. Symp. Electron, Ion, Laser Beam Technol., 11th, May,* p. 421.
Yonas, G. Poukey, J. W., Prestwich, K. R., Freeman, J. R., Toepfer, A. J., and Clauser, M. J. (1974). *Nucl. Fusion* **14,** 731.
Yoshikama, S. (1971). *Phys. Rev. Lett.* **26,** 295.
Zharov, V. F., Malinovskii, V. K., Neganov, Yu. S., and Chumak, G. M. (1972). *JETP Lett.* **16,** 154.

Chapter 7

Shock Induced Plasmas

P. BOGEN and E. HINTZ
INSTITUT FÜR PLASMAPHYSIK DER
KERNFORSCHUNGSANLAGE JÜLICH GMBH
ASSOCIATION EURATOM-KFA
JÜLICH, FEDERAL REPUBLIC OF GERMANY

I. INTRODUCTION

Plasmas induced by shock waves show the same extraordinary variability as the plasma state in general. Using the electron density as one possible means to achieve some ordering of observed shock phenomena, we find at

ISBN 0-12-349701-9

the lower end of the density scale shocks in the galactic halo with typical densities of the order 10^{-3} cm^{-3} and at the opposite end, laser induced shocks in solid hydrogen pellets at densities around 10^{23} cm^{-3}.

In Table I we have compiled some of the better known laboratory shock devices and cosmic shock phenomena, respectively, together with values of the density, the temperature, and the magnetic field in the preshock plasma. Apparently plasmas within a density range comprising 26 orders of magnitude with temperatures varying between a few electron volts and a few kiloelectron volts are accessible to the probing investigation of scientists, and a large number of new and interesting physical phenomena as well as applications useful in science and technology may be expected. It is beyond the scope of this article to discuss the whole range of plasmas induced by stationary or nonstationary shock waves, which can be prepared in the laboratory or which can be investigated by means of space vehicles. The latter will not be considered at all in this article. With regard to laboratory shocks we will restrict the following discussion to a few shock wave types and applications. We have tried to choose examples that are relatively well understood and which are of practical interest. Concerning the many shock wave types and shock wave phenomena which have been discussed in the literature but cannot be dealt with here, we refer the reader to available

TABLE I

CHARACTERISTIC DATA OF THE INITIAL CONDITIONS OF SOME LABORATORY SHOCK DEVICES AND OF COSMIC SHOCK PHENOMENA

References[a]	Initial state	u_1 (cm/s)	n_1 (cm^{-3})	B_1 (G)	T_1 (°K)
(1) Laser driven shock	Solid	$5.8 \cdot 10^6$	$5 \cdot 10^{22}$		
(2) Coaxial electromagnetic shock tube	Neutral	$4 \cdot 10^8$	10^{15}	$\leq 10^4$	300
(3) Theta pinch *A*	Plasma	$2 \cdot 10^7$	10^{16}	$2 \cdot 10^3$	20,000
(4) Theta pinch *B*	Plasma	10^8	10^{13}		20,000
(5) Plasma focus	Neutral gas	10^8	10^{17}	$\approx 10^6$	300
(6) Bow shock	Plasma	$4 \cdot 10^7$	5	$5 \cdot 10^{-5}$	$T_i = 5 \cdot 10^4$ $T_e = 1.5 \cdot 10^5$
(7) Shocks in the corona (type II radio bursts)	Plasma	10^8	10^8	1	$2 \cdot 10^6$
(8) Collision of extragalactic gas clouds with the galactic halo	Neutral	$5 \cdot 10^7$	10^{-3}	10^{-6}	100

[a] (1) van Kessel and Sigel (1974); (2) Patrick (1959), Gross (1973); (3)–(5) Plasma Physics and Controlled Nuclear Fusion Research, IAEA, Vienna, 1971; (6) Hundhausen (1970); (7) McLean (1973); (8) Savedoff *et al.* (1967).

books and review articles (Pai, 1959; Kantrowitz and Petschek, 1966; Sagdeev, 1966; Chu and Gross, 1969; Hintz, 1970; Gross, 1971; Tidmann and Krall, 1971; Kidder, 1971; Biskamp, 1973).

One class of experiments makes use of shock waves because they are a relatively cheap and convenient means to provide high temperatures which otherwise require more costly apparatus or cannot be achieved at all. In this case, the extremely high heating rates occurring in high Mach number shocks are utilized. Well known examples are shock heated plasmas in controlled thermonuclear research, and one of the preferred experimental methods is the theta pinch. With this type of experiment high temperature plasmas have been produced in a density range from 10^{13} cm^{-3} to about 10^{18} cm^{-3}. Most of the physical processes have been thoroughly explored, and one promising concept for a fusion reactor is based on the theta pinch (Ribe *et al.*, 1974; Kaufmann and Köppendörfer, 1974). Theta pinches have also been used successfully as light sources for the uv- and x-ray part of the spectrum. Their techniques and macroscopic properties will be discussed later in detail.

Another category of experiments uses shock waves to produce controlled deviations from thermal equilibrium in a thermal plasma to study the resulting transport and relaxation phenomena. We will discuss two limiting cases here: Shock structure and relaxation processes in a nonmagnetized, collision-dominated plasma and shock structure and resistive dissipation in a magnetized, "collision-free" plasma.

We have given a short survey on the objectives of this article. The material will be organized in the following way. In Section II those results of magnetohydrodynamic (mhd) shock wave theory will be summarized which are important for the rest of this chapter. Some general features of shock structures will also be described. Experimental methods applied in pinch experiments will be described in Section III, in particular, typical electrical circuits, electrotechnical components, and ways to produce the initial plasma. A survey of the studies of plasma behavior and plasma heating in theta pinches will be given in Section IV. With the aid of rather simple examples the behavior of nonthermal plasmas in the front of shock waves will be discussed in Section V.A for the case of a collision-dominated plasma and in Section V.B for that of a collision-free plasma.

II. SOME CHARACTERISTIC PROPERTIES OF SHOCK WAVES

A. General Remarks

In this section those results of shock wave theory are compiled which are necessary for understanding the physics of shock waves and which are useful for a discussion of shock wave applications. From the great variety of shock waves which are possible in plasmas, we are going to discuss only two rather simple cases, namely, plane stationary shocks in

(a) collision-dominated, magnetic field free plasmas, i.e., $\omega_{ce}\tau_{ee} \ll 1$, $\beta \gg 1$;

(b) magnetized plasmas with the direction of wave propagation perpendicular to the magnetic field and $\omega_{ce}\tau_{ee} \gg 1$, $\beta \ll 1$.

Here ω_{ce} is the electron cyclotron frequency, τ_{ee} is the electron–electron collision time, $\beta = 8\pi nk(T_e + T_i)/B^2$ is the ratio of plasma pressure to magnetic field pressure, $n = n_e = n_i$ is the particle density, T the particle temperature, and B the magnetic field; the subscripts *e* and *i* denote electrons and ions, respectively.

In case (a) plasma flow can be described by the Euler equations; for case (b) the plasma pressure in the Euler equations has to be replaced by the magnetic field pressure $B^2/8\pi$. For the range of plasma parameters and wavelengths for which this description of the plasma is correct, we can consequently make use of the results of hydrodynamic theory. With regard to nonlinear waves this means that the Riemann solution is applicable, i.e., for case (a), the propagation velocity of a finite amplitude perturbation is equal to $u + c_s(u)$, where u is the flow velocity and $c_s = [(\gamma_e kT_e + \gamma_i kT_i)/m_i]^{1/2}$ is the sound velocity. Simple nonlinear waves with a positive density gradient in the direction of wave propagation steepen in the course of time and develop into a discontinuity. In gasdynamic waves steepening of the wave front at sufficiently strong gradients is limited by dissipation processes; this should also be true for case (a). In the case of mhd waves the equations have to be modified for wavelengths smaller than the ion cyclotron radius. Taking inertia effects into account, the waves become dispersive. In addition to dissipation dispersive effects may now limit the growth of the gradient; as a result shock fronts with an oscillatory structure may occur (Adlam and Allen, 1958; Davis *et al.*, 1958; Gardner *et al.*, 1958; Sagdeev, 1960, 1962).

B. Shock Relations

We consider a plane stationary shock normal to the direction of plasma flow in a frame of reference where the shock front is at rest. We denote the equilibrium state ahead of the shock, where all gradients are zero and $u > c_s$, by the subscript 1 and the corresponding one behind the shock by the subscript 2. The region where the gradients do not disappear we call the shock front (Fig. 1). The properties of the post shock state can be calculated from those of the preshock state by means of the shock relations. These equations are derived from the conditions for particle, momentum, and energy con-

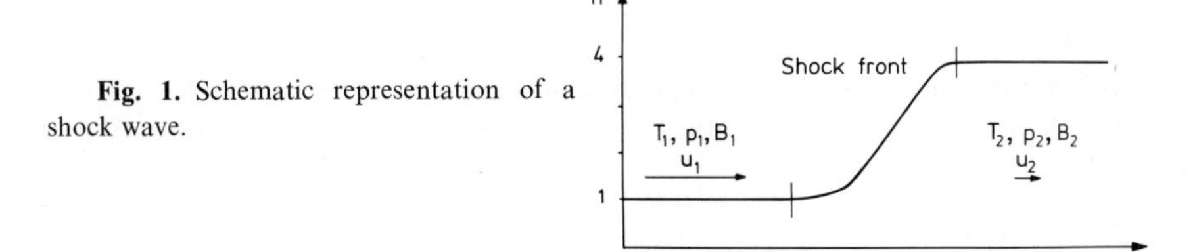

Fig. 1. Schematic representation of a shock wave.

servation. For a fully ionized plasma with radiation being negligible, the shock relations are identical with those for an ideal gas with a constant ratio of specific heats:

$$\eta = \frac{n_2}{n_1} = \frac{u_1}{u_2} = \frac{(\gamma + 1)M^2}{(\gamma - 1)M^2 + 2}, \tag{1}$$

$$\frac{p_2}{p_1} = \frac{2\gamma M^2}{\gamma + 1} - \frac{\gamma - 1}{\gamma + 1}, \tag{2}$$

$$\frac{T_2}{T_1} = \frac{[2\gamma M^2 - (\gamma - 1)][(\gamma - 1)M^2 + 2]}{(\gamma + 1)^2 M^2}. \tag{3}$$

Here η is the compression ratio, p is the plasma pressure, and $M = u_1/c_{s1}$ is the Mach number. For γ the relation $\gamma = (f + 2)/f$ holds, where f is the number of degrees of freedom of the particles. We want to point out that $T = T_e + T_i$ in the shock relations.

The change in plasma properties across a shock wave is an irreversible process and the corresponding entropy change is given by

$$S_2 - S_1 = \frac{R}{(\gamma - 1)} \ln\left[\frac{p_2}{p_1}\left(\frac{n_1}{n_2}\right)^{\gamma}\right], \tag{4}$$

where R is the universal gas constant. For weak shocks the logarithmic terms may be expanded into powers of $(M^2 - 1)$:

$$S_2 - S_1 = R \cdot \tfrac{2}{3}[\gamma/(\gamma + 1)^2](M^2 - 1)^3 + \cdots. \tag{5}$$

The entropy can only increase; this requires $M > 1$. For weak shocks the entropy increase is small and may often be neglected. In this case the equations for isentropic processes can be used.

For strong shock waves, i.e., for $M \gg 1$ the shock relations simplify considerably

$$\eta = (\gamma + 1)/(\gamma - 1) = f + 1, \tag{6}$$

$$p_2/p_1 = [2/(\gamma + 1)]M^2, \tag{7}$$

$$T_2/T_1 = [(\gamma - 1)/(\gamma + 1)]p_2/p_1. \tag{8}$$

The compression ratio tends to a limiting value as $M \to \infty$. The pressure and the temperature ratio, however, increase with M without limit.

Switching now to the laboratory frame of reference, where the shock front is moving with a velocity u_1, one obtains in the strong shock limit:

$$m_i u_p{}^2/2 = (m_i \cdot u_2^{*2})/2 = (f/2)kT_2, \tag{9}$$

where $u_p = u_1 - u_2$ is the velocity of the piston, and $u_2{}^*$ is the flow velocity in the post shock plasma. Equipartition of energy has taken place between flow energy and thermal energy.

In order to convert the flow energy entirely into thermal energy one often lets the plasma flow impinge on a solid wall or introduces a plane of symmetry where the shock is reflected and the plasma brought to rest (Fig. 2). If we denote the plasma state behind the reflected shock with the subscript 3, we obtain the equations connecting state 3 and state 2

$$\frac{n_3}{n_2} = \frac{1 + [(\gamma + 1)/(\gamma - 1)](p_3/p_2)}{[(\gamma + 1)/(\gamma - 1)] + (p_3/p_2)}, \tag{10}$$

$$\frac{p_3}{p_2} = \frac{\{2[(\gamma - 1)/(\gamma + 1)] + 1\}(p_2/p_1) - [(\gamma - 1)/(\gamma + 1)]}{[(\gamma - 1)/(\gamma + 1)](p_2/p_1) + 1}, \tag{11}$$

and in the strong shock limit

$$n_3/n_2 = \gamma/(\gamma - 1), \qquad p_3/p_2 = (3\gamma - 1)/(\gamma - 1). \tag{12}$$

We now consider shock waves in a magnetized plasma. We assume that plasma and magnetic field are uniform and that the direction of shock wave propagation is perpendicular to the magnetic field. The shock relations then contain two dimensionless parameters instead of one, e.g., the

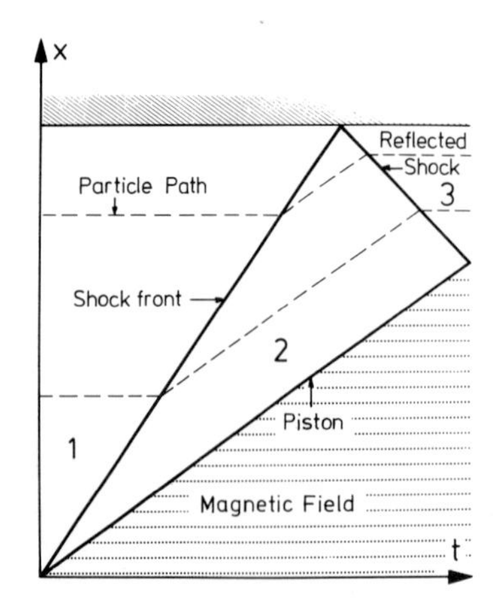

Fig. 2. Incident and reflected shock wave.

Alfvén Mach number $M_A = (4\pi u_1^2 n_1 m_i / B_1^2)^{1/2}$ and β_1:

$$\frac{1}{\eta} = \frac{1}{2}\left[\frac{\gamma-1}{\gamma+1} + \frac{\gamma}{\gamma+1}\,\frac{\beta_1+1}{M_A^2}\right] + \frac{1}{2}\left\{\left[\frac{\gamma-1}{\gamma+1} + \frac{\gamma}{\gamma+1}\,\frac{\beta_1+1}{M_A^2}\right]^2 + \frac{4\cdot(2-\gamma)}{(\gamma+1)M_A^2}\right\}^{1/2}, \tag{13}$$

$$\frac{p_2}{p_1} = 1 + \frac{2(\eta-1)}{\beta_1}\,\frac{\gamma\beta_1 + \frac{1}{2}(\gamma-1)(\eta-1)^2}{(\gamma+1)-\eta(\gamma-1)}, \tag{14}$$

$$n_2/n_1 = B_2/B_1 = u_1/u_2. \tag{15}$$

These equations have been evaluated numerically (Zeyer, 1975). The result is shown in Fig. 3. Going to the strong shock limit in which $B_1^2/8\pi$ and $\frac{3}{2}n_1 kT_1$ are negligible compared to $(n_1 m_i/2)u_1^2$, the relations get independent of the magnetic field and are identical with Eqs. (6)–(8). For β_2 one obtains

$$\beta_2 = [2f/(f+1)^3]M_A^2 \qquad (f = 3;\ \beta_2 = \tfrac{3}{32}M_A^2). \tag{16}$$

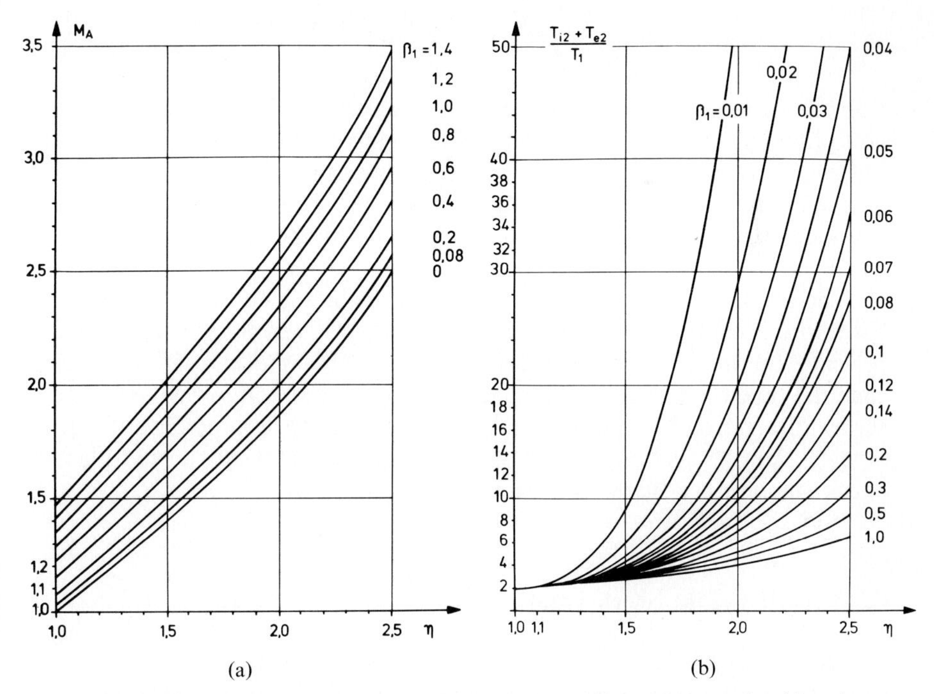

Fig. 3. Numerical evaluation of mhd shock relations. Alfvén Mach number (a) and post shock temperatures (b) as function of the compression ratio with β_1 as parameter (Zeyer, 1975).

So far we have considered stationary plane shocks. For plasma shock wave experiments a strong magnetic field is often applied at the boundary of a cylindrical plasma, and thereby cylindrically converging shocks are generated which are not stationary. At sufficiently large radii the shock wave will behave like a plane wave. In the vicinity of the axis, however, one expects that the shock strength increases. For these applications it would be useful to have at least an approximate understanding of the effects of cylindrical convergence. Most calculations on converging cylindrical or spherical shocks start with the assumption that the jump conditions derived for plane stationary shocks are locally valid for converging shocks. This assumption is certainly justified if the shock radius is large compared to the shock thickness. For an ideal fluid a similarity solution (Gouderley, 1942), which is valid near the center of convergence, predicts for $M \gg 1$ that M increases as

$$M = M_0(r_0/r)^{i\alpha} \tag{17}$$

with $i = 1$ for cylindrical and $i = 2$ for spherical shocks. Witham (1958) gives the approximation

$$\alpha = \left\{1 + \left[\frac{2}{\gamma(\gamma - 1)}\right]^{1/2}\right\}^{-1} \cdot \left\{1 + \left[\frac{2(\gamma - 1)}{\gamma}\right]^{1/2}\right\}^{-1}. \tag{18}$$

For $\gamma = \frac{5}{3}$, we obtain $\alpha = 0.225$. With $r \to 0$, when M as well as T and p go to infinity, an approximate value of the temperature on the axis may be obtained if a minimum radius $r_{\min}$ of about the shock width is introduced and then $M_{\max} = M_0(r_0/r_{\min})^\alpha$ is used in Eq. (8). The density stays finite; a maximum value of $23n_{e1}$ is calculated for $\gamma = \frac{5}{3}$ in the reflected cylindrical shock.

The similarity solution does not provide pressure balance at the piston. Therefore it can be valid only for radii $r < R^*$ for which the time needed for a perturbation to travel from the piston to the shock front is longer than the time needed for the shock to reach the center, or, in mathematical terms, the piston should not intersect the limiting characteristic. This gives a limit $r < R^* \approx R_0/2$ (R_0 = tube radius). For larger r, the plane shock solution is a better approximation.

C. Properties of the Shock Front

The nonequilibrium region between the plasma in front of the shock (state 1) and that region behind the shock where thermal equilibrium can again be assumed (state 2) is called the shock front. In this layer the plasma is accelerated, compressed, and heated. From Eq. (5) we know that an increase of entropy occurs. This indicates that dissipation processes will be important for a description of the shock front. The amount of entropy increase across the shock is independent of the special way in which energy is dissipated and depends only on the one or two parameters necessary for describing the shock. The dissipation processes will, however, determine the length scales over which the entropy change occurs. For the applications that we are considering it will be important to have an estimate of the thickness of the shock front. It would also be useful to have some idea under which conditions the various dissipation mechanisms do occur and how energy will be distributed between electrons and ions.

Intuitively one would expect that for medium and high Mach number shocks in nonmagnetized plasmas viscous dissipation as a result of ion–ion collisions will be the dominating process in the shock front. For these conditions it is the only process by which the required amount of flow energy can be converted into heat. Electron viscosity is unimportant since the electrons carry only a negligible amount of the energy, and heat conduction can only be of secondary significance since it transports heat only from one point to another after it has first been generated by some other process. A similar statement is true for energy exchange between electrons and ions. Neglecting relaxation processes one might expect that the shock is about one mean free path λ thick.

For low and medium strength shocks in a magnetized low β plasma the situation is different: In the magnetic field gradient of the shock front a significant fraction of the flow energy of the ions can be converted into energy of the drift motion of the electrons. In this case there will be high relative velocities between electrons and ions which can be dissipated by electron–ion collisions; as a result, resistive dissipation may be dominating.

One result of a rather complete theory of low Mach number plasma shock waves (Grad and Hu, 1967) is that the weak shock profile, just as in ordinary gases, is given by $\tanh(\varepsilon X/L)$ where $\varepsilon = \Delta p/2\gamma p_1$ is the shock strength and L is a scale length which is a complicated function of $\omega_{ce}\tau_{ei}$ and β. For $\beta \gg 1$, $\omega_{ce}\tau_{ei} \ll 1$, L is essentially the mean free path; for $\beta \ll 1$, $\omega_{ce}\tau_{ei} \gg 1$, it is the resistive shock thickness $c^2/4\pi\sigma u_1$; where σ is the Spitzer conductivity. Concerning the thickness of resistive shocks one has to bear in mind that resistive dissipation requires flow energy first to be transferred from ions to electrons by electric fields in the shock. Furthermore, the drift motion of electrons is predominantly in the direction perpendicular to that of shock propagation; one would expect therefore that L may be smaller than λ.

In any case, the shock thickness $\Delta = L/\varepsilon$ decreases with increasing shock strength. This means that the rate of change of certain flow variables and therefore, the deviation from thermal equilibrium can be controlled by the shock strength ε.

On the basis of this discussion one would expect that shocks do occur only on a length scale that permits collisions to take place. Observations in space as well as in the laboratory plasmas, however, have shown that this condition is not always satisfied. There are shocks with a front width too thin for binary collisions to account for the observed dissipation. Such shocks are often called collisionless. A great variety of "collisionless" shocks has been identified and investigated (Biskamp, 1973) and a number of mechanisms have been proposed which can replace binary collisions and, on the basis of collective interactions, explain the entropy increase required. Here we will discuss only one example, namely, turbulent resistive shocks, where turbulent electric fields develop as a result of the high electron drift velocity and accomplish the randomization of drift energy.

III. EXPERIMENTAL METHODS

Shock waves have been generated and systematically investigated in various experimental devices of which the better known are listed in Table I. Their merits and disadvantages depend on their respective aims and cannot be discussed here. Although the geometry and the dimensions of the shock tubes may be different, the electrical circuits employed in most cases are very similar. Some of the more customary circuits and their basic components will be described in the following. Since the experimental part of this chapter will mainly deal with investigations on theta pinches, special attention will be given to their technical needs.

Apart from electrotechnical requirements, the successful operation of shock devices often depends critically on the possibility to prepare plasma with the required parameters in a controlled and reproducible way. Some established methods of plasma production will be discussed at the end of this section.

A. Requirements on the Design of Shock Wave Experiments

We want to analyze the possibilities for producing shock waves by means of the fast compression of a plasma by an external magnetic field. Our considerations will be primarily related to theta pinches, i.e., the compression of a cylindrical plasma by an axial magnetic field. Most of the results, however, can be adapted to other geometries. We consider a long, single winding, cylindrical coil of circular cross section filled with a uniform, fully ionized plasma of high electrical conductivity. For the compression one would like to apply a step function like magnetic field pulse. Its amplitude B_p will be related in some way (which we have yet to find), to the desired shock strength and to the plasma parameters. The rise time of the pulse should be short compared to the transit time of a pressure perturbation across the radius R_0 of the plasma. In Section III.B we will discuss how to realize such pulses technically. Here we will simplify the situation. We consider the discharge of a capacitor bank (capacitance C, charging voltage U, internal inductance L_S) through the coil. L_S must be small compared to the inductance of the coil L_C for the rise time of the magnetic field to be small compared to the transit time of the shock. We want to know what voltage is needed to heat the plasma by means of a shock wave to a temperature T_2. In the strong shock limit Eq. (9) relates the piston velocity u_p to the pressure of the post shock plasma which in turn is equal to the pressure of the magnetic field of the piston, i.e.,

$$B_p{}^2/8\pi = p_2 = n_2 kT_2 = [m_i u_p{}^2 \cdot (f + 1)n_1]/f. \tag{19}$$

After a transient phase, after which the radius of the piston R_p according to our assumption will still be of the order of R_0, the voltage at the coil (of length l) is given by

$$U_c = J \cdot \frac{dL_c}{dt}. \tag{20}$$

Using $U_c = 2\pi R_p E_p$ and $L_c = (4\pi^2/c^2)(R_0{}^2 - R_p{}^2)/l$ one obtains

$$cE_p = u_p \cdot B_p. \tag{21}$$

Together with Eq. (19) this leads to

$$E_p = (1/c)f^{1/2}(f + 1)^{1/2}(8\pi n_1/m_i)^{1/2} kT_2, \tag{22}$$

$$E_p = [(f + 1)/f]^{1/2}(u_p{}^2/c)(8\pi n_1 m_i)^{1/2}. \tag{23}$$

For deuterium and with $f = 3$ this gives $E_p = 1.1 \cdot 10^{-19} u_p{}^2 n_1{}^{1/2}$ and $E_p = 1.5 \cdot 10^{-11} n_1{}^{1/2} T_2$, where n_1 is measured in ions per cubic centimeter, T_2 in degrees Kelvin, and E_p in volts per centimeter. It must be kept in mind that the validity of this formula is restricted to strong shocks, i.e., $u_1{}^2 \gg c_s{}^2 + V_A{}^2$ and to conditions which guarantee that $\Delta \ll R_0$.

For the generation of medium strength collisional plasma shocks electric field strengths of about 50 V/cm are needed. In case of shocks in low β plasmas (e.g., $\beta_1 = 10^{-2}$) the Alfvén velocity $V_A \approx \beta^{-1/2} c_s$ is the characteristic velocity for the propagation of pressure perturbations and E_p values of about 500 V/cm are required to generate resistive shocks ($M_A \lesssim 3$). For the applications of interest here diameters of the compression coil between 20 and 50 cm may be appropriate. Such diameters would allow a shock thickness up to 1 cm, a length which would still permit a reasonable spatial resolution. To generate the electric fields mentioned above, coil voltages, respectively charging voltages of the capacitor bank up to 75 kV are needed. Considering the radii of the discharge tube and the shock velocities envisaged, the voltage rise time at the coil should be below 10^{-8} s. In order that the total charging voltage may appear at the coil, it is necessary that the internal inductance of the capacitor bank is small compared to the coil inductance $L_c \approx (4\pi^2/c^2)R_0{}^2/l$, where l is the length of the coil. According to Hintz (1970) $l = 4R_0$ should be sufficient to provide a magnetic field with negligible curvature inside the coil. In this case $L \approx 10^{-8} \cdot R_0[H]$. The only process by which the coil ends can influence the central part of the shock wave, on which most of the measurements are performed, is by thermal conduction of the electrons [cf. Eq. (39)]. It may be necessary to increase the coil length to a value that excludes such interference from the end.

Typically, a coil inductance of the order 100 nH may be expected; accordingly, the internal inductance of the capacitor bank (including switches and connections to the coil) should not exceed the order of 10 nH.

B. Electrical Circuits and Components

A systematic discussion of the electrical circuits and of the electrotechnical components, which have been used for plasma shock devices, is not possible within the framework of this article. Here we can only explain some general principles, describe some established procedures and give some selection criteria. Concerning more special and advanced techniques, we will refer to the literature. In case of components which are commercially available, we will restrict ourselves to giving the names of companies.

Figure 4 shows a schematic diagram of a standard LC circuit used for magnetically driven shock waves. According to Section III.A it should have the following features: The capacitor bank should have a charging voltage of about 50 kV and an internal inductance that is small compared to that of the compression coil, i.e., typically of the order 10 nH. A lower limit for the capacitance C can be obtained by considering that a nearly constant

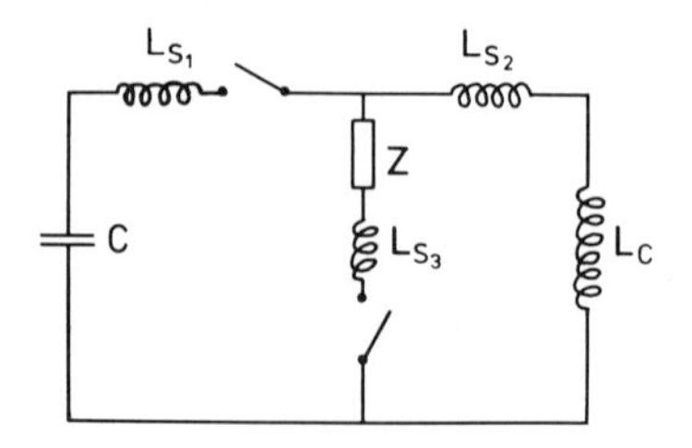

Fig. 4. Electrical circuit diagram for shock devices, schematic.

voltage is needed at the coil for at least one transit time of the piston across the radius. Then there is sufficient time for the shock to be reflected at the axis, and all flow energy can be converted into thermal energy. Since the changing inductance of the coil acts like a resistive impedance, the characteristic time for the discharge of C is $C\cdot(dL_c/dt)$. In order to obtain a voltage drop of less than 20%, one needs $C\cdot(dL_c/dt) > 5\cdot(R_0/u_p)$. From this follows $C/l \approx [(3\cdot 10^7)/u_p{}^2]$ [F/cm], e.g., with $u_p = 2\cdot 10^7$ cm s^{-1} and $l = 50$ cm, $C = 4\cdot 10^{-6}$ F.

Capacitor banks used as energy supply for shock waves will typically have energies of up to 10 kJ/m. The energy content can be much larger if for specific purposes the shock compression is to be followed by an adiabatic compression. The bank energy is then determined by the final magnetic field: $(C/2)U^2 = (B^2/8)R_0{}^2\cdot l$.

The observation period of the shock heated plasma can be increased by introducing a so-called crowbar switch into the electrical circuit. As indicated in Fig. 4, it provides the possibility for establishing a short circuit across the compression coil. By closing the switch at the time of current maximum, the current in the coil will decay monotonically with a time constant L_c/Z, where Z is the total ohmic resistance of the circuit consisting of compression coil and crowbar switch. The inductance and the ohmic resistance of the crowbar switch and of the leads connecting it to the collector plate must be small compared to the corresponding quantities of the rest of the circuit. If this can be achieved, larger voltage drops across the capacitor bank do not appear and oscillations of the current are avoided.

We will now discuss practical solutions; how to realize capacitor banks with the specified characteristics. Apart from the specifications of commercially available capacitors, switches etc., the actual design depends on further boundary conditions imposed on the experimental installation, e.g., available personnel and space. With 10 nH being set as the upper limit for the internal inductance of the capacitor banks that are to serve as energy supply for plasma shock devices, the internal inductance of the various components of the circuit should be low and of comparable magnitude. By using low energy storage capacitors and by connecting a number N of them in parallel, it is always possible for a given energy to achieve very low total inductance values $L_{tot} = (1/N)L_i$. However, in order to keep the complexity of the apparatus low, N should be a small number. For example, if P is the probability that one spark gap, employed as a switch, prefires, then $(1 - P)^N$ is the probability that a bank with N switches connected in parallel does not prefire. For crowbar switches in addition to low inductance a low effective ohmic resistance is asked for.

Typically one would consider a life expectation of 10^4 discharges for both capacitors and switches to be sufficient. In most applications repetition rates need not be higher than 1 per minute. With 10^{-8} s very high standards are set with regard to the accuracy required for the closing of the switches. At the same time the probability of a switch to prefire must be low.

The electrical leads connecting the various components with each other and with the compression coil must also be of low inductance. The electric strength of dielectric materials commercially available, e.g., polythene or Mylar foil, permits the construction of suitable transmission lines. Difficulties do arise with the construction of connecting terminals.

The properties of commercially available capacitors* and spark gap switches may be summarized as follows:

capacitors: energy 1–10 kJ; voltages up to 75 kV; inductance 20–30 nH; current per unit 180–500 kA;

switches: working voltage 30–55 kV; inductance 20–30 nH; life expectation 10^3–10^5 discharges.

These data indicate that capacitor banks for the applications considered here will as a rule consist of several capacitors and switches connected in parallel to obtain the desired low inductance. This may also be appropriate from a safety point of view.

In order to achieve simultaneous firing of a number of spark gaps connected in parallel, some simple rules have to be followed. Switches have to be separated electrically for a longer time than the jitter time of the spark gaps. This can be effected, e.g., by cables of suitable length or by the capacitance of the collector plate together with the inductance of switches and capacitors connected to it (Friedrich and Hintz, 1966).

Spark gaps are fired as a rule by the application of steep high voltage pulses. Usually the required pulses are generated by discharging a low inductance capacitor across high voltage cables of suitable impedance (see, e.g., Beerwald and Hintz, 1960).

Concerning detachable low inductance connections between different parts of a high voltage circuit, remarkable progress has been achieved by introducing high voltage gaskets (Weniger, 1970). If the high voltage insulation of two conductors at some place changes from solid dielectric to a gas, e.g., to air, then the width of the gap at the interface can be reduced by pressing a gasket of elastic insulating material (e.g., silicon rubber) with sufficiently high pressure against the solid dielectric surface. At higher pressures the breakdown voltage across the gasket approaches that of the gasket material. Special applications may require voltages much higher than the 50 kV which so far have been considered as a kind of standard or they may

* E.g., Maxwell Laboratories Inc., San Diego, California; BICC, Helsby Warrington, Wa6 Odj England; Culham Laboratory, Abingdon, England.

suggest the use of special pulse shapes for the magnetic field. Concerning the demand for voltages up to about 100 kV one may attempt to take advantage of existing technology at 50 kV and use either voltage doubling techniques or low stray inductance transformers (Cosler *et al.*, 1974; Dokopolous *et al.*, 1973).

An example for a capacitor bank where voltage doubling has been used with good success (Dietz, 1969) is shown in Fig. 5. Here the compression coil has a length of 80 cm and a diameter of 40 cm giving a coil inductance of 165 nH. The basic circuit consists of a series–parallel connection of four capacitors with a charging voltage of 60 kV, a capacitance of 1.9 μF, and an internal inductance of 18 nH. The capacitors are connected to a low inductance collector plate by means of four many-channel spark gaps (Friedrich and Hintz, 1966). The maximum electric field attained at the plasma surface was about 750 V/cm corresponding to a voltage of 90 kV around the plasma.

Damage of the capacitors because of overvoltages in case one switch prefires is prevented by special safety measures: (a) electrical bypass leads which switch the nonconducting spark gaps in parallel to the one which prefired, and (b) a trigger circuit which automatically triggers all spark gaps within the first quarter cycle of the discharge after one spark gap has prefired.

For the construction of circuits providing very high voltages new concepts have been applied successfully, e.g., high voltage, low impedance delay lines (Vitkovitzky, 1966; De Silva *et al.*, 1969; Herppich, 1969) and high voltage, pulse-charged capacitor banks (Dokopoulos, 1970).

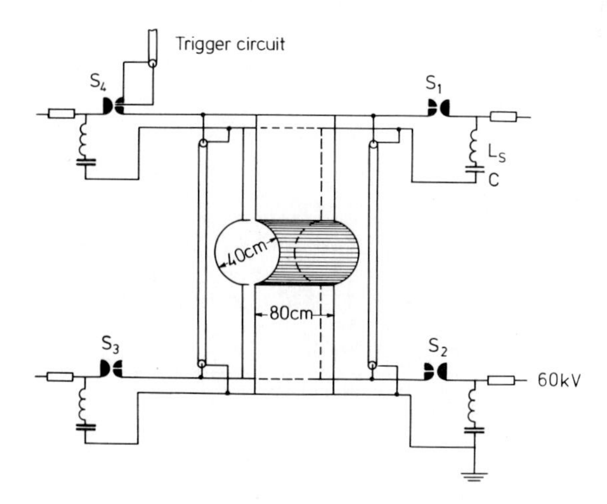

Fig. 5. Circuit diagram for voltage doubling, schematic (Dietz, 1969).

C. Plasma Formation

For the investigation of shocks in plasmas, one has to provide a fully ionized, uniform initial plasma, free of impurities and, possibly, in a uniform magnetic field. Unfortunately, these conditions can normally not be fulfilled simultaneously.

Usually the plasma is formed in two steps: first, the gas is preionized, i.e., electron densities of $n_e \approx 10^{10}$ cm^{-3} are provided, then the preionized gas is heated to about 20,000°K, which means that full ionization is obtained. The preionization is achieved by photoionization (Chodura *et al.*, 1969; Lie, 1973) or by rf-discharges. A high current discharge along the axis or an electrodeless ring discharge (Beerwald *et al.*, 1962; Lupton *et al.*, 1961) is used for plasma heating. Sometimes the plasma is produced outside the compression coil and then injected into it.

Of the various possible techniques of plasma formation we will describe only two. One often applied technique works as follows (see Fig. 6): Breakdown of the gas is achieved by applying an rf electric field between the coil (at ground potential) and copper electrodes outside the discharge tube, a few centimeters from the coil end. Frequencies of about 10 MHz, powers of about 1 kW, and voltages $\gtrsim 1$ kV are typical values of the rf transmitter that is usually operated in a pulsed mode. For the preheating phase, an electrodeless ring discharge is excited by discharging a single condenser or a condenser bank through the main coil. The voltage at the coil is normally limited to values below 10 kV to avoid prefiring of the spark gaps of the main bank. Furthermore, one has to consider that for the main discharge the preheater bank acts as a parallel load to the coil. Concerning the choice of the amplitude of the oscillating magnetic field only the rather general rule can be given that the higher densities certainly require higher fields. It

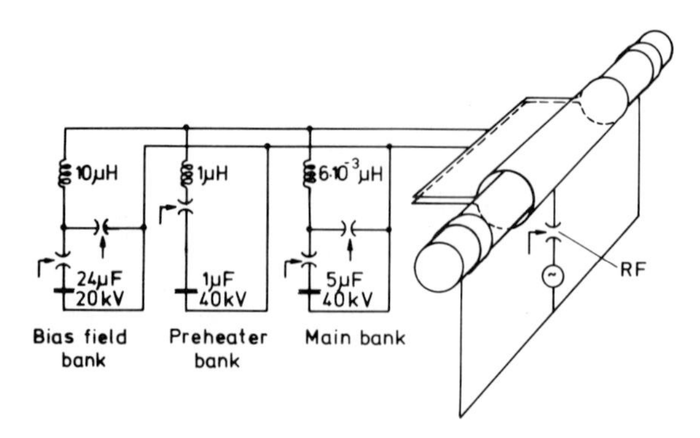

Fig. 6. Circuit diagram of a theta pinch with capacitor banks for the magnetic bias field, for the preheater, and for the main discharge (Hintz, 1969).

must be considered, however, that the magnetic field may get trapped in the plasma.

Electron densities produced by preionization should be so high that diamagnetic effects get noticeable during the early stage of the preheating discharge. The largest ionization rates are observed during the time at which the magnetic field B_z goes through zero (Beerwald, 1965, 1967). Then the velocity of the ionizing electrons is not limited by E/B drifts. Since magnetic fields of opposite polarity diffuse into the plasma during successive half-cycles, strong heating occurs in a later phase by the dissipation of the anti-parallel fields (Bogen and Hintz, 1961). The main bank is fired when the plasma is sufficiently ionized. If the experiment requires a low amount of internal magnetic field, however, one has to wait until the trapped fields are dissipated. This may lead to incomplete ionization, and only a compromise between the demands for a high degree of ionization and low trapped fields can be found. The residual fields in the afterglow plasma depend on the plasma parameters and the exact history of the discharge and are difficult to predict.

The preparation of a low-β plasma with a uniform superimposed magnetic field presents special difficulties. A really satisfactory method is still missing. In some experiments an afterglow plasma is generated as described above and then a slowly rising magnetic field is generated in the coil which diffuses into the plasma. The rise time of the magnetic field is chosen equal to its penetration time into the plasma. The radial density distributions achieved this way are shown in Fig. 7. For details see (Lupton *et al.*, 1961; Hintz, 1969).

At low gas densities photoionization is an attractive method for plasma production. It offers the advantage that the gas can be ionized independent of the magnetic field and of the pressure, provided the optical thickness for the ionizing radiation is small compared to one. Although the degree of

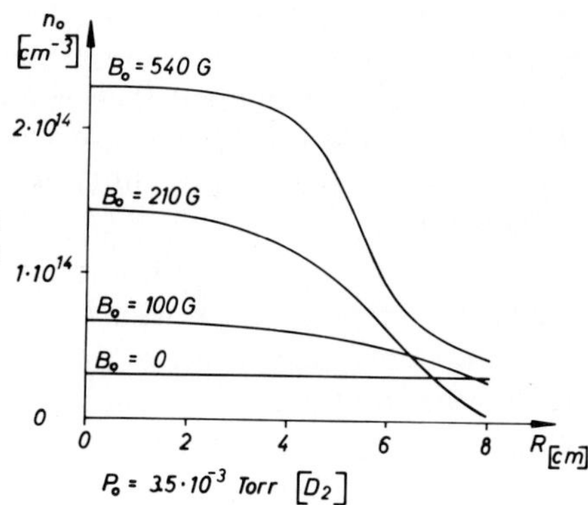

Fig. 7. Density in the preheated plasma as function of radius with internal magnetic field as parameter (Hintz, 1969).

ionization usually is low ($\approx 10\%$), the neutrals need not disturb shock experiments because of their long ionization times. The method is most effective for heavy gases (e.g., Xe) because of their large photoionization cross sections. The light source is usually a high current gas discharge. Since it is normally used without a window, a part of the current may flow through the gas volume which is irradiated and may produce nonuniform ionization and disturbing magnetic fields.

IV. PLASMA HEATING AND PLASMA BEHAVIOR IN THETA PINCHES

A. Fast Compression

Plasma heating in theta pinches generally occurs in two phases: a fast compression phase, where irreversible heating takes place, and an adiabatic compression phase. For a description of the first phase, the shock relations together with the equations for converging cylindrical shocks, given in Section II, can be used and give a good estimate of plasma heating provided the mean free path in the post shock plasma is short. Numerical methods (see e.g., Hain and Kolb, 1962) may give a better description of cylindrical shocks. Figure 9 shows a smear camera photograph of the radially converging shock waves and demonstrates the acceleration of the shock front near the axis which is typical for such shocks. At small trapped magnetic fields and sufficiently long mean free paths, however, a shock front cannot develop since there are no collisions among the particles, and the plasma cannot be accelerated before it reaches the piston. If microinstabilities are absent, the plasma is reflected by the combined action of the magnetic field and a radial electric space-charge field. Since the electron orbits are more easily bent by the magnetic field than those of the ions, the ions tend to get ahead of the electrons. This separation causes a space-charge field $E_r = (1/ed)mu_p{}^2/2$ by which the ions are reflected. The characteristic width of the reflecting layer d is given by $d = (m_i m_e)^{1/2} cu_p/eB$, the geometric mean of ion and electron Larmor radius (Longmire, 1963). With the pressure balance of Eq. (24), given below, the width is equal to $d = \frac{1}{2}c/\omega_{pe}$.

This reflection at the piston results in a plasma flow which may appear similar to that of a shock with $f = 1$ ($\gamma = 3$). The reflected particles with $u_1 = 2u_p$ move ahead like a shock front. However, the particle energy behind the front is not thermal but remains in a beam of particles with a velocity of $2u_p$ moving through a plasma at rest.

A great advantage of this "free particle model" is that analytical expressions can be found for the piston velocity and the ion energies (the electrons remain cold) in the cylindrical geometry of theta pinches (Kever, 1960, 1962). The pressure balance at the boundary yields

$$B^2/8\pi = 2\rho_1(dr/dt)^2 = 2n_{i1}m_i u_p{}^2 \tag{24}$$

with

$$B = (4\pi/c)(J/l), \tag{25}$$

where J is the current in the coil and l the coil length. Assuming a constant voltage U_0 at the capacitors during the fast plasma compression and neglecting ohmic resistance, the circuit equation for J is given by

$$\frac{d(LJ)}{dt} = J\frac{dL}{dt} + L\frac{dJ}{dt} = U_0, \tag{26}$$

where L is the total inductance of the circuit and consists of a variable part $L_c[1 - (r/R_0)^2]$, which is initially zero because the coil is filled with a well conducting plasma, and a constant part L_s, the external inductance, which also includes the nonconducting space between coil and plasma.

Equation 26 can be integrated and yields J as a function of r and t. Substituting J into Eq. (24), one obtains a differential equation, the integration of which gives t as function of r. From this one obtains the compression time t_c and the mean piston velocity $u_p = R_0/t_c$:

$$t_c \approx 2R_0(\pi\rho_1)^{1/4}\left(\frac{3 + 2\lambda}{3\lambda}\right)^{1/2}\left(\frac{2\pi R_0}{cU_0}\right)^{1/2} \tag{27}$$

$$= (16\pi\rho_1)^{1/4}\left(\frac{1}{cE_p}\right)^{1/2}\left(\frac{3 + 2\lambda}{3 + 3\lambda}\right)^{1/2} R_0,$$

$$u_p = \left(\frac{c^2E_p{}^2}{16\pi\rho_1}\right)^{1/4}\left(\frac{3 + 3\lambda}{3 + 2\lambda}\right)^{1/2} \tag{28}$$

with $\lambda = L_c/L_s$, $E_p = (U_0/2\pi R_0)\lambda/(1 + \lambda)$, R_0 = initial plasma radius, $\rho_1 = n_{i1} \cdot m_i$.

The energy $\mathscr{E}_K$ of the plasma at the time t_c can be derived from the work done by the Lorentz forces

$$\mathscr{E}_K = 2\pi\int_0^{t_c}[B^2(t)/8\pi]r\dot{r}\,dt = \tfrac{2}{3}R_0(\rho_1/4\pi)^{1/2}U_0c[\lambda/(1 + \lambda)]. \tag{29}$$

A two-dimensional "temperature" is defined by dividing $\mathscr{E}_K$ by the line density $N = \pi R_0{}^2 n_{i1}$

$$kT_{i\perp} = \frac{2}{3}\left(\frac{m_ic^2}{4\pi^2}\right)^{1/2}\frac{U_0}{N^{1/2}}\frac{\lambda}{1 + \lambda} \tag{30}$$

or in practical units

$$T_{i\perp}(^\circ\mathrm{K}) = 1.0 \cdot 10^{11} \frac{U_E\,(\mathrm{V})}{[N \quad (\mathrm{cm}^{-1})]^{1/2}} A^{1/2},$$

where $U_E = U_0[\lambda/(1 + \lambda)]$ is the voltage applied to the plasma in volts, and A is the atomic weight. High temperatures are favored by high voltages and low line densities.

The free particle model gives simple formulas; its applicability to the experiment, however, is limited. Figure 8 shows the Doppler-broadened D_α line observed side on during the fast compression of a low density plasma at about 50% ionization. In the wings of this line the radiation is emitted by charge exchange neutrals. The intensity distribution reflects the velocity distribution of the ions in radial direction, the wavelength shift of the first peak corresponds to a velocity of $4.1 \cdot 10^7$ cm/s which agrees with the piston velocity derived from magnetic probe measurements and that derived from Eq. (28). The second peak corresponds to a shift of twice the piston velocity

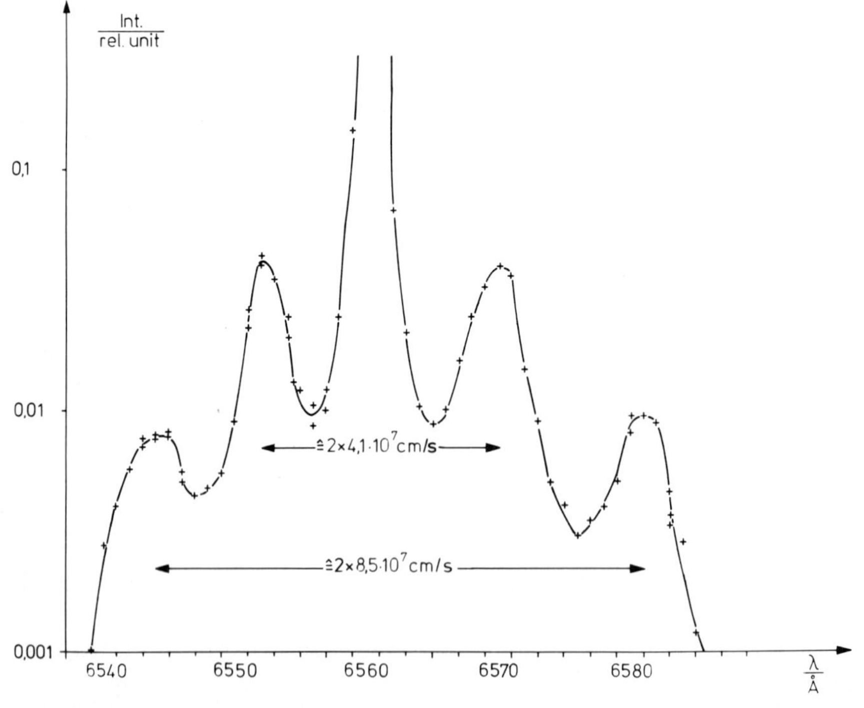

Fig. 8. Profile of the D_α line 0.3 μs after the start of the compression pulse. Tube diameter 40 cm, $n_{e1} = 5.10^{12}$ cm^{-3} (Bogen *et al.*, 1975a).

which is expected for reflected particles. Detailed investigations of the radial density distribution with Thomson scattering of laser light have shown that for this case only 20% of the particles are reflected while the rest are trapped in the piston and that 50% of the energy is carried by the reflected particles. Thus the free particle model is only partially justified. The trapping of the particles in the piston is presumably caused by microinstabilities (Bogen *et al.*, 1975a; Höthker, 1976). Similar results have been reported for higher densities by McKenna *et al.* (1974) who applied magnetic probes and interferometric techniques.

Observations show that after the particle flow has crossed the axis most of the flow energy is converted into thermal energy perpendicular to the magnetic field (Bogen *et al.*, 1969, 1975b, 1976).

B. Adiabatic Phase

The fast heating is followed by an adiabatic phase for which the laws of adiabatic compression can be applied, i.e., $pn^{-\gamma} = \text{const.}$ p is given by the pressure balance equation:

$$p = (B_a{}^2 - B_{int}{}^2)/8\pi = \tilde{\beta} B_a{}^2/8\pi \tag{31}$$

If the internal magnetic field B_{int} can be neglected, one obtains

$$T/T_0 = (B/B_0)^{2(\gamma-1)/\gamma}, \tag{32}$$

$$n/n_0 = (B/B_0)^{2/\gamma}. \tag{33}$$

Which γ values have to be chosen depends on the ratio of the self-collision time to the adiabatic compression time [see Eq. (37)]. If this ratio is small compared to one, a three-dimensional Maxwellian velocity distribution can be established and γ is equal to $\frac{5}{3}$.

After the fast compression, the plasma oscillates around an equilibrium state (Green and Niblett, 1959). These oscillations can be approximately described by (Kever, 1962)

$$m_i N\ddot{r} = -2\pi r\left\{B_a{}^2/8\pi - \left[\bar{p}\left(\frac{\bar{r}}{r}\right)^{2\gamma} + \frac{\bar{B}_{int}{}^2}{8\pi}\left(\frac{\bar{r}}{r}\right)^4\right]\right\}, \tag{34}$$

where $\bar{p}$, $\bar{r}$, and $\bar{B}_{int}$ are mean values during the oscillation. The period τ is

$$\tau \approx 2\pi\,\frac{(m_i N)^{1/2}}{B_a}\cdot\frac{B_a{}^2}{B_{int}{}^2 + (\gamma/2)(B_a{}^2 - B_{int}{}^2)} \approx \frac{(m_i N)^{1/2}}{B_a}\cdot 2\pi. \tag{35}$$

If the external field increases, the oscillation period decreases during the compression. The smear picture of a theta pinch discharge (see Fig. 9)

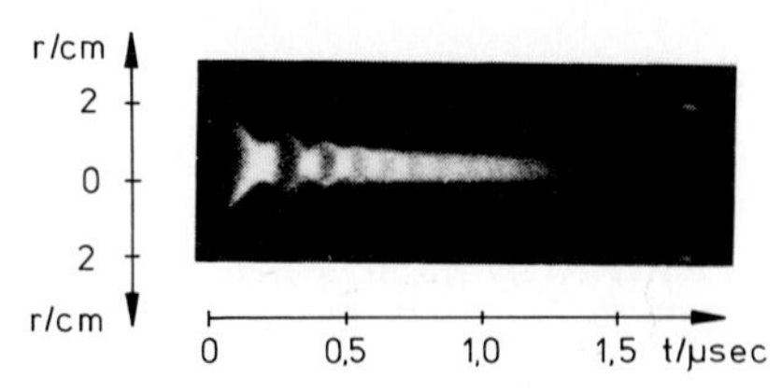

Fig. 9. Smear camera picture of a theta pinch with low trapped magnetic field. Filling pressure 0.3 Torr D_2, maximum magnetic field 45 kG, half period 1.7 μs.

illustrates the hydromagnetic oscillations; the period is as expected from Eq. (35). Therefore by a measurement of this oscillation period and of the external magnetic field the line density N can be determined. Since in the direction parallel to the magnetic field lines the plasma is not confined, a rarefaction wave propagates into the plasma with a speed which in a simplified model is determined by the sound velocity and the $\tilde{\beta}$ value:

$$v = c_s \cdot (1 - \tilde{\beta})^{1/2} \tag{36}$$

(Taylor and Wesson, 1965). Here it is assumed that the transverse plasma pressure can be neglected after the arrival of the wave, but the flux trapped in the plasma cross section, $B_i A$, is conserved. This means $B_{i1} A_1 = B_{i2} A_2$. With $B_{i1} = B_a \cdot (1 - \tilde{\beta})^{1/2}$ and $B_{i2} = B_a$, we obtain $A_2/A_1 = (1 - \tilde{\beta})^{1/2}$. This reduction in cross section which is also observed experimentally (Fig. 9) together with the continuity equation gives the reduction in the velocity of the rarefaction wave. In the neighborhood of $\tilde{\beta} = 1$, the method has to be substituted by a kinetic calculation.

After the fast compression most of the plasma energy is in the radial kinetic energy of the ions and $T_e \ll T_i$. The time for establishing an isotropic thermal distribution of ions and electrons with $T_{i\perp} = T_{i\parallel}$ and $T_i = T_e$ can be estimated by the relaxation times given by Spitzer (1956) for small deviations from thermal equilibrium. Isotropic velocity distribution of one kind of particles, electrons or ions, can be expected after the self-collision time of about

$$\tau = 11.4 \cdot (A^{1/2} T^{3/2} / n Z^4 \ln \Lambda), \tag{37}$$

where A is the atomic weight, T (°K), Z is the electric charge, and $\ln \Lambda$ is the Coulomb logarithm ≈ 10.

The self-collision time is shortest for the electrons. Their equilibration time is normally short compared to the other characteristic times of the pinch. The ion–ion collision time and even more the ion–electron equilibration time

$$\tau_{eq} = 250 \cdot (A / n Z^2 \ln \Lambda) T_e^{3/2} \tag{38}$$

are often long compared to the quarter period of the external field. This is clearly demonstrated by experiments (Andelfinger *et al.*, 1967; Bogen *et al.*, 1969, 1975b). In Fig. 10, the variation of $T_{i\perp}$ and $T_{i\parallel}$ after the fast compression

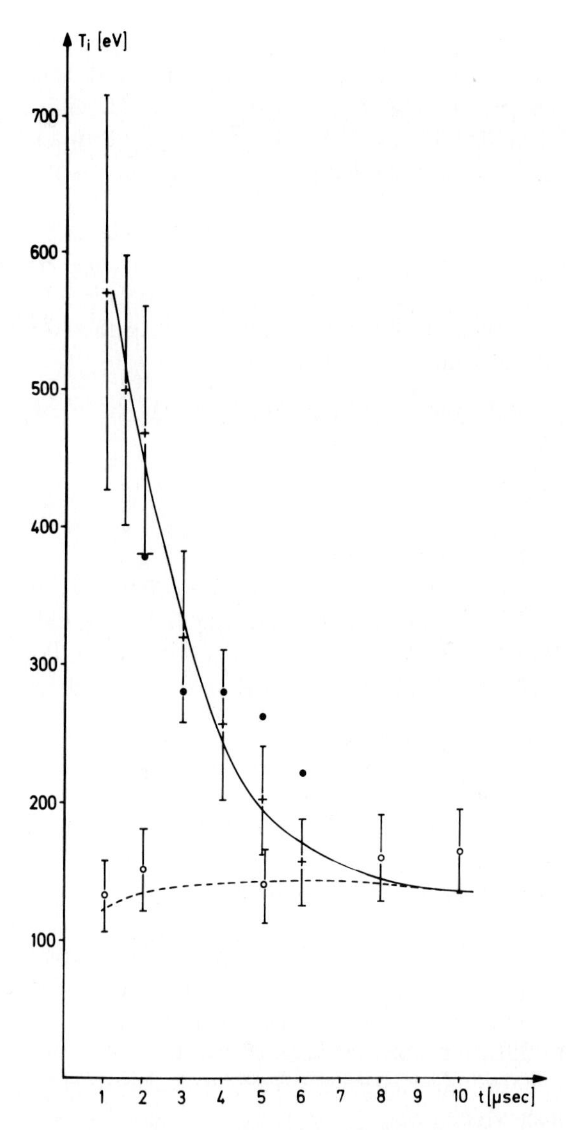

Fig. 10. Temperatures parallel (○) and perpendicular (+) to the magnetic field from D_α profiles and temperatures from pressure balance (●), $n_e \approx 1.5 \cdot 10^{14}$ cm^{-3} (Bogen *et al.*, 1975b).

is shown. With $T_{i\perp} \gg T_{i\parallel}$ initially, an isotropic ion velocity distribution is observed somewhat earlier than expected from Eq. (37). The reasons presumably are that because of large anisotropy Eq. (37) is not applicable and improved formulas (Lehner, 1967) have to be used. In addition, energy loss processes (e.g., charge exchange, ionization) have to be taken into account.

An important limitation of T_e is due to thermal conduction losses to the coil ends (Green *et al.*, 1967; Morse, 1972). One expects an equilibrium determined by electron heating due to collisions with the ions and by electron energy losses because of thermal conduction to the coil ends. Morse gives a simple formula for T_e:

$$T_e[\text{eV}] = \{n_e^2 \,(\text{cm}^{-3}) l_1 l_2 \,(\text{cm}) T_i \,(\text{eV})[(\ln \Lambda)^2 \cdot 10^{-29}/A]\}^{1/5}. \tag{39}$$

Here l_2 is half the coil length, l_1 a thermal sheath thickness of about $0.4l_2$. T_e is not very sensitive to the exact choice of l_1 and l_2 because of the fifth root dependence. This formula often gives good agreement with experiments, but some exceptions are observed mainly at very high β.

Energy losses by line radiation of impurities can impose another limitation on T_e. For the general case an exact calculation is difficult (see e.g., Kolb and McWhirter, 1964). A crude estimate of the radiated power for T_e in the range of 100 eV and for ions with three or more electrons gives (Meade, 1974)

$$P_L = 2 \cdot 10^{-26} n_e^2 f \quad (\text{W/cm}^3); \tag{40}$$

here $f = n_{iz}/n_e$ and n_{iz} is the impurity number density. From $P_L \cdot \tau = \frac{3}{2} n_e k T_e$, one obtains the time τ at which the electron energy is lost by radiation. Accordingly radiation losses become important if

$$n_e \tau \quad (\text{cm}^{-3}\, s) \gtrsim 1 \cdot 10^7 T_e(\text{eV})(1/f). \tag{41}$$

In addition to the discussed energy losses, the diffusion of the magnetic field into the plasma may influence the plasma parameters. A rough estimate for the classical diffusion time, determined by binary collisions, is given by Spitzer (1956). For cylindrical geometry (see e.g., Knoepfel, 1970) a correction has to be applied, which gives

$$\tau_D = 0.174 \cdot 4\pi r^2 \sigma = 3.5 \cdot 10^{-14} T_e^{3/2} r^2 \quad (\text{sec}), \tag{42}$$

where T_e is in degrees Kelvin, r is in centimeters, and $\ln \Lambda = 10$.

Anomalously fast diffusion has been found in some experiments to result in a penetration depth of the external field into the plasma of $\Delta \approx c/\omega_{pi}$. This diffusion may be explained by an enhanced resistivity due to current driven instabilities. If the drift velocity

$$v_d = \frac{c}{4\pi n \cdot e}\frac{dB}{dx} \approx \frac{c}{4\pi n \cdot e}\frac{B_a - B_i}{\Delta}$$

is higher than the critical velocity v_c for an instability to be excited, a penetration depth

$$\Delta = (c/\omega_{pi})(2/\tilde{\beta})^{1/2}\,[1 - (1 - \tilde{\beta})^{1/2}](c_s/v_c) \tag{43}$$

is expected (Krall *et al.*, 1974). For $T_e \gg T_i$ as observed during fast compression (Bogen *et al.*, 1975a) the ion acoustic instability may be excited with $v_c \approx c_s = (kT_e/m_i)^{1/2}$ (= ion sound velocity), leading to the observed value $\Delta \approx c/\omega_{pi}$. If the radius of the plasma is smaller than Δ, the β on the axes is considerably reduced. This gives a critical line density

$$N_c = \pi\Delta^2 n_e = c^2 m_i/4e^2 = 1.63 \cdot 10^{15}\ \mathrm{A\ cm^{-1}},$$

below which a high β plasma cannot be achieved (A = atomic weight).

At sufficiently high line densities and electron temperatures, the decrease of the β value is not very important as measured by compensated loops (Green, 1962), magnetic probes (Hintz, 1962), and Faraday rotation (Bogen and Rusbüldt, 1966; Gribble *et al.*, 1968). $\tilde{\beta}$ values up to more than 99% have been observed in the center of the plasma. Bodin *et al.* (1969) made detailed measurements of the diffusion rates by observing the radial density profile of an 8 m long theta pinch as a function of time. Since end effects were unimportant, the variation of the profile could be attributed to diffusion only. At their parameter set ($T_e \approx 200$ eV, $n_e \approx 5 \cdot 10^{16}\ \mathrm{cm^{-3}}$) diffusion was found to be in agreement with classical binary collision theory, except for a short initial anomalous diffusion phase.

C. Applications

Most experiments on theta pinches have been performed with hydrogen or deuterium as working gases, because these gases are easily fully ionized and because the experiments have been related to thermonuclear research. It has been demonstrated that in theta pinches the temperatures ($T_i \approx 10$ keV) and densities ($n_e \approx 10^{14}$), necessary for a future fusion reactor can be reached. However, in these experiments the energy release from nuclear reactions is very small since the plasma is lost in a short time due to the ends. To avoid end losses, toroidal devices, e.g., high-β stellarators have been built (Ribe *et al.*, 1974; Kaufmann and Köppendörfer, 1974; Fünfer *et al.*, 1975), but up to now the plasma life time is still very short due to instabilities.

Stabilization by conducting walls seems to be possible if a low compression ratio of the plasma is achieved, i.e., if the usual adiabatic compression is abandoned.

Theta pinches in gases different from hydrogen, e.g., He (Bogen, 1970, 1972), A, Xe (Dietz and Hintz, 1970) have also been used, but have not been investigated as extensively because of the complications arising from energy losses by ionization and radiation. For many applications, mixtures of hydrogen with small amounts of heavier gases (impurities) have been of great advantage. They are prepared in a vacuum vessel which is first filled at low pressure with the heavy gas, then the hydrogen is added. Long time intervals are needed for the gases to mix well. Then a linear increase of spectral line intensity versus impurity concentration has often been found (below about 1%) (see Fig. 11) indicating that the plasma has the same impurity concentration as the admitted gas mixture, except for the impurities desorbed from the walls by plasma wall interaction. The latter concentration can be found by extrapolating the straight line to zero intensities (see Fig. 11).

Calculations of the degree of ionization and excitation in shock induced plasmas are often difficult because of the lack of complete thermal equilibrium. Thermal equilibrium can only be expected for the free electrons and for excited states with quantum numbers higher than m^*, which for hydrogen-like ions and $m^* > 2$ is approximately (Griem, 1964)

$$m^* = 126Z^{14/17}n_e^{-2/17}(kT/Z^2\chi_H)^{1/17}\exp(4Z^2\chi_H/17m^{*3}kT), \tag{44}$$

where Z is the nuclear charge and χ_H is the ionization energy of hydrogen.

For $m > m^*$, the number n_{zm} of ions per cubic centimeter in the excited state m can be calculated from the Saha equation:

$$\frac{n_{z+1,1}n_e}{n_{zm}} = \frac{g_{z+1,1}}{g_{zm}} \cdot \frac{2(2\pi m_e kT)^{3/2}}{h^3} \exp\left(-\frac{\chi_{zm}}{kT}\right), \tag{45}$$

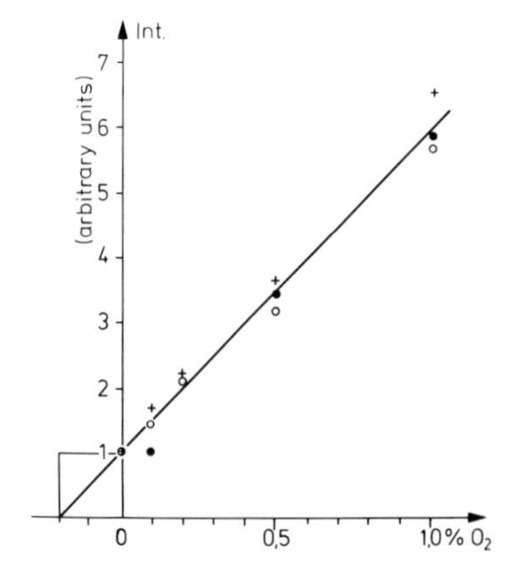

Fig. 11. Intensities of OV (+), OVII (○), and OVIII (●) lines as function of oxygen concentration in deuterium (Schlüter, 1966).

where χ_{zm} is the ionization energy of the ion z from the quantum level m, n_{zm} is the ion density in this quantum level, and g_{zm} is the statistical weight.

If the Saha equation is not applicable, the ionization equilibrium has to be calculated by means of the elementary ionization and recombination rates (Griem, 1964; McWhirter, 1965). In the limit of low densities, the corona equation (Elwert, 1952) has often been applied. For rough estimates it may be used in the form

$$\frac{n_{z+1}}{n_z} \approx 10^8 \frac{\zeta}{m_0} \frac{1}{\chi_z^2} \cdot \frac{kT}{\chi_z} \exp\left(\frac{-\chi_z}{kT}\right), \tag{46}$$

where χ_z is the ionization energy (in electron volts), ζ is the number of electrons in the outer shell, and m_0 is the quantum number of the ground state.

The corona equation takes into account only ionization by electron impact and recombination by the emission of free-bound continua. Several authors (Burgess, 1964; Shore, 1969; Allen and Duprée, 1969) pointed out that the ionization equilibrium can be changed considerably by the dielectronic recombination. This process, which is the inverse of autoionization, proceeds in two steps. First a free electron is caught by an ion $X^{(z+1)}$ which is simultaneously excited to a high energy level (autoionizing level) of $X^{(z)}$. In the second step the excited ion $X^{(z)}$ is "stabilized" by a transition to the ground state and simultaneous emission of a photon. Simple formulas are not available. The importance of dielectronic recombination in shock induced plasmas is demonstrated in Fig. 12, where coronal equilibria of

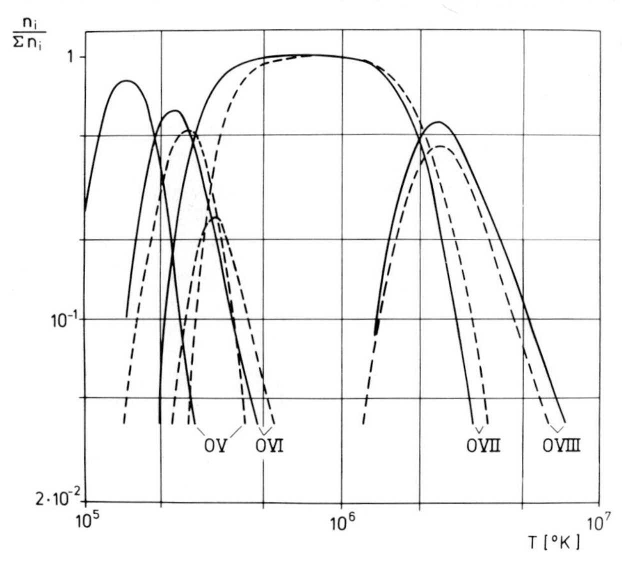

Fig. 12. Ionization equilibrium of oxygen in a low density plasma without (——) and with (---) dielectronic recombination (Pospieszczyk, 1975b).

oxygen are compared with and without consideration of dielectronic recombination.

An additional difficulty in the calculation of ion concentrations may be caused by the fast variation of the plasma parameters. This may require taking the ionization relaxation times into account. These times depend on the dominant ionization processes. At low densities, direct ionization from the ground state is the main process (Knorr, 1958). At high densities the ions are first excited to the $n = 2$ state and then ionized (Griem, 1964). The corresponding ionization times are called t_1 and t_2, respectively:

$$t_1 \approx 6 \cdot 10^5 \frac{\chi_z^{3/2}}{\zeta n_e} \exp\frac{\chi_z}{kT}, \tag{47}$$

$$t_2 \approx \frac{10^7}{f_{21} n_e} \frac{E_{21}(kT)^{1/2}}{\chi_H^{3/2}} \exp\left(\frac{E_{21}}{kT}\right) = \frac{2 \cdot 10^5}{f_{21} n_e} E_{21}(kT)^{1/2} \exp\left(\frac{E_{21}}{kT}\right), \tag{48}$$

where χ_z is the ionization energy in electron volts, E_{21} is the energy difference between levels 1 and 2, f_{21} is the oscillator strength of the resonance line, and ζ is the number of electrons in the outer shell.

For the calculation of line radiation intensities, excitation cross sections and/or transition probabilities [see the survey articles of Kunze (1972), Griem (1964), Wiese *et al.* (1966), McWhirter (1965)] must be known in addition to electron and ion densities and electron temperatures.

Since theta pinches can be operated in a large parameter range of T_e and n_e, they have often been applied to basic research on spectroscopy and on light scattering of high temperature plasmas. Soft x-ray spectra (lines as well as continua) were first observed in the Scylla device by Jahoda *et al.* (1960), Bearden *et al.* (1961), and Sawyer *et al.*(1963), using a Bragg crystal spectrometer. From the exponential decay of the free–free and free–bound continuum T_e has been deduced. Its intensity per frequency unit I_ν is given for hydrogenlike ions in the case of an optically thin layer of thickness l *by* (see e.g., Griem, 1964)

$$I_\nu = 1.36 \cdot 10^{-41} \left(\frac{\chi_H}{kT}\right)^{1/2} Z^2 n_{z+1} n_e l \cdot \left[g_{ff} + \sum_{\chi_{zn} < h\nu} \frac{2\chi_{zn}}{kT} \frac{g_{fn}}{n} \exp\left(\frac{\chi_{zn}}{kT}\right)\right] \exp\left(-\frac{h\nu}{kT}\right)\left(\frac{erg}{cm^2 sr}\right), \tag{49}$$

where χ_H is the ionization energy of hydrogen, $\chi_{zn} = Z^2 \chi_{Hn}$ is the ionization energy of the ion from the quantum level n, Z is the nuclear charge, $g_{ff}(\nu, T)$ the Gaunt factor for free–free transitions, and $g_{fn}(\nu, T)$is the Gaunt factor for free–bound transitions into shell n.

For details of the measurements, especially the calibration, see e.g., Stratton (1964), Schlüter (1966), and Bogen (1968).

Strong resonance lines up to SiXIII have been identified. A characteristic feature of these lines is that they are accompanied by a large number of weaker satellites. They can be regarded as x-ray lines (from inner shell transitions) of highly stripped ions which are excited either by collisional excitation or by dielectronic recombination. Most of these lines have been identified by comparison with theoretical results (Gabriel, 1972; Pospieszczyk, 1973, 1975a). In several cases, for which the plasma lifetime was longer than the ionization relaxation time, satellite intensities measured relative to the resonance line have been found to be in quantitative agreement with the dielectronic recombination theory.

The determination of collisional rate coefficients for ionization and excitation processes was made possible by measuring T_e and n_e by Thomson scattering of laser light. As summarized by Kunze (1972) several rate coefficients of He-, Li-, and Be-like ions could be determined and compared with theory. Although there are still discrepancies on the order of a factor two, the results are of great help for astrophysical and thermonuclear research. Whereas in the soft x-ray region the continuous spectrum is perturbed by numerous strong lines, the long wavelength region, especially the visible and infrared, shows a rather pure continuum under conditions where the particle density is high ($2 \cdot 10^{17}$ cm^{-3}) and the impurity concentration is low ($<0.2\%$). A measurement of the continuum between 0.27 and 22 μm (Bogen and Rusbüldt, 1968; Bogen *et al.*, 1972) showed that the wavelength dependence corresponds to pure bremsstrahlung as expected from Eq. (49) taking $Z = 1$ and $T_e = 100$ eV. At lower densities and higher impurity concentrations, strong deviations occur (Dippel, 1967; Engelhardt, 1973), presumably due to molecular hydrogen and impurity radiation. Therefore, care has to be taken in the interpretation of streak camera photographs.

The theta pinch has also been applied to research problems where dense plasmas of medium temperature are required. Line broadening measurements of HeII, CIII, and CIV spectra have been performed at densities up to about 10^{18} cm^{-3} and at temperatures of 5–20 eV (Eberhagen and Wunderlich, 1969; Bogen, 1970, 1972). Furthermore, several basic experiments on cooperative light scattering on plasma have utilized a theta pinch as a plasma source. For summarizing articles, see Kunze (1968), De Silva and Goldenbaum (1970).

V. PLASMA BEHAVIOR IN THE FRONT OF SHOCK WAVES

A. Structure of Collision Dominated Plasma Shocks

A summary of the shock wave structure in collision dominated plasma is given in the book by Zeldovich and Raizer (1967). Detailed calculations have been performed, e.g., by Shafranov (1957) and by Jaffrin and Probstein (1964), using hydrodynamic two fluid equations. Qualitatively the results of these calculations can be found in a heuristic manner as follows.

The kinetic energy $n(m_i + m_e)u_1^2/2$ of the plasma ahead of the shock consists predominantly of ion energy $n_1 m_i u_1^2/2$. This energy is converted into heat by the action of viscous forces on the ions. The thickness Δx_1 of the decelerating layer is determined by the condition that the energy flow $\frac{1}{2}n_1 m_i u_1^3$ into the layer is equal to the energy dissipation rate $u_1^2\mu/\Delta x_1$. With $\mu \approx n_2 k T_2 \tau_{i2}$ and $kT_2 \approx (1/f) m_i u_1^2$, one obtains

$$\Delta x_1 \approx u_1 \tau_{i2} \tag{50}$$

which corresponds to about one mean free path in the postshock plasma. The heating of the electrons by viscosity is normally negligible since the electrons have m_e/m_i times lower kinetic energies. The electrons are heated due to adiabatic compression ($T_{e2}/T_{e1} = [(\gamma + 1)/(\gamma - 1)]^{\gamma - 1} = 2.5$ for a strong shock) since their density is coupled to the ions through the condition of electric neutrality.

If the distance between the shock front and the piston is sufficiently long, thermal equilibrium will be established by electron–ion collisions. The characteristic length for this process is

$$\Delta x_2 = u_2 \tau_{eq2}. \tag{51}$$

Since $\tau_{eq} \approx (m_i/m_e)^{1/2}\tau_i$, Δx_2 is much longer than Δx_1, and therefore thermal equilibrium with $T_i = T_e$ will be achieved only if the separation between shock front and piston is large compared to Δx_2.

In case of $T_{e2} > T_{e1}$ a third process becomes effective, the electron thermal conductivity $\kappa_e \approx n_e k v_e^2 \tau_e$, which is larger than the ion thermal conductivity by a factor of about $(m_i/m_e)^{1/2}$. The thickness of the layer determined by κ_e can be estimated (Zeldowich and Raizer, 1967) by setting the electron heat flux $\kappa_e(dT_e/dx)$ equal to the energy flux $(3/2)kT_e(n_e u)$ due to convection:

$$\kappa_e(dT_e/dx) = (\tfrac{3}{2})kT_e(n_e u). \tag{52}$$

With $\kappa_e = n_e k(kT_e)\tau_e/m$, one obtains $\Delta x_3 \approx u_1 \cdot \tau_{eq2}$. A more exact integration, using $\kappa_e = aT^{5/2}$, gives

$$\Delta x_3 = \tfrac{4}{15}[\kappa_e(T_2) - \kappa_e(T_1)]/n_1 u_1 k. \tag{53}$$

The temperature gradient in the shock front not only produces a heat flux, but due to charge separation, also produces an electric field that reduces the heat flux (see, e.g., Spitzer, 1956). An effective thermal conductivity κ_{eff}

is therefore used that for $Z = 1$ is equal to

$$\kappa_{\text{eff}} = \frac{1.9 \cdot 10^{-5} T_e^{5/2}}{\ln \Lambda} \text{(erg/cm sec °K)}. \tag{54}$$

It is noteworthy that all three processes, viscosity, ion–electron energy transfer, and electron thermal conduction lead to characteristic lengths that are proportional to $u_1 \tau_2 \sim u_1 \cdot T_2^{3/2} \sim u_1{}^4$. The relative importance of them will therefore not be varied with increasing Mach number. However, equilibrium with $T_e = T_i$ will not be reached experimentally for high Mach numbers since then the separation between shock front and piston cannot be made sufficiently large.

Experiments on collision-dominated shocks in magnetic field free plasma suffer from the difficulty of producing a fully ionized, uniform initial plasma (see Section III.C). If the magnetic field in the preheating discharge has dropped to a value such that $\omega_{ce}\tau_e \ll 1$, the plasma has already started to recombine, and the neutral particle density is not negligible. Ionization times may become of the same order of magnitude as ion–electron equilibration times. A clear experimental demonstration of full ionization is a measured compression ratio $n_{e2}/n_{e1} = (\gamma + 1)/(\gamma - 1) = 4$. From the observed compression ratio and the jump in T the initial degree of ionization $\alpha = n_i/(n_i + n_0)$ can be derived. Using the conservation laws, one obtains

$$\alpha \approx [5(T_2 - T_1) + \chi_i - T_2]/[(n_{e2}/n_{e1})T_2 + \chi_i], \tag{55}$$

where χ_i is the ionization energy. Additional losses by radiation may be included in χ_i, but α is not very sensitive to the exact magnitude of χ_i. α approaches $4(n_{e1}/n_{e2})$ for large Mach numbers.

Figure 13 shows results on the structure of a shock with $M \approx 4$ in a plasma with $n_{e1} = 4 \cdot 10^{14}\ \text{cm}^{-3}$ and a degree of ionization of 50% (Siemsen, 1973). T_e and n_e have been measured by 90° Thomson scattering of laser

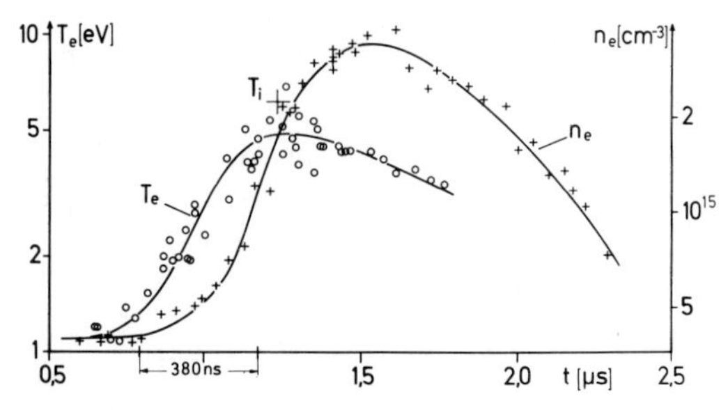

Fig. 13. $T_e(t)$ and $n_e(t)$ in a collision dominated shock front at $r = R_0/2$ (Bogen *et al.*, 1971, Siemsen, 1973).

light, T_i by 8.5° cooperative scattering. The viscous shock layer could not be resolved. A temperature increase ahead of the density front has been found at a distance expected from Eq. (53). The ratio $n_{e2}/n_{e1} = 8.5$ indicates the incomplete preionization, T_i and T_e have been found to be equal in the post shock layer. From a comparison of the scattered light amplitudes at 90° and 8.5°, respectively, it could be concluded that the fluctuation level corresponds to that of a thermal plasma, i.e., the plasma is collision dominated.

B. Resistive Dissipation in a Collisionless Shock

1. Laminar Shocks

For mhd shocks of small strength the magnetic field profile is given by (see Grad and Hu, 1967)

$$B - \frac{B_1 + B_2}{2} = \frac{B_2 - B_1}{2} \tanh\left(\frac{x}{L/\varepsilon}\right), \tag{56}$$

where L is a function of $\omega_{ce}\tau_{ei}$ and β. We restrict our attention to the parameter range $\omega_{ce}\tau_{ei} \gg 1$; $\beta \ll 1$ for which friction between electrons and ions is the main dissipation mechanism. The expression for L then reduces to

$$L \approx c^2/4\pi\sigma u_1 . \tag{57}$$

A rough estimate of the shock thickness gives a similar result. The penetration depth of the magnetic field of the post shock state into the upstream plasma is $\Delta = (c^2 t/4\pi\sigma)^{1/2}$. For the shock to be stationary t should be equal to the time that a fluid element needs to cross a shock front of width Δ, i.e., $t = \Delta/u_1$. Inserting this into the equation for the penetration depth leads to the above expression for L. Using $\sigma = ne^2/m_e \nu_{ei}$ one can also write

$$L \approx c^2 \nu_{ei}/\omega_{pe}^2 u_1 . \tag{58}$$

These results have been examined by measuring mhd-shock profiles at small ε (Fig. 14) and then plotting the measured Δ against $1/\varepsilon$. Figure 15 shows that there is a linear relationship between Δ and $1/\varepsilon$; the value of σ obtained from these measurements agrees within the accuracy of the measurements with the Spitzer conductivity (Bogen *et al.*, 1971).

With decreasing collision frequency $\nu_{ei} \sim n_e/T_e^{3/2}$ one will come across conditions where $\nu_{ei} < u_1/\Delta$ would occur; i.e., where electron–ion collisions are not possible in the shock front. The validity of Eq. (58) then breaks down. One can show that for shocks of medium strength this is the case if $\nu_{ei} < (\omega_{ce}\omega_{ci})^{1/2}$.

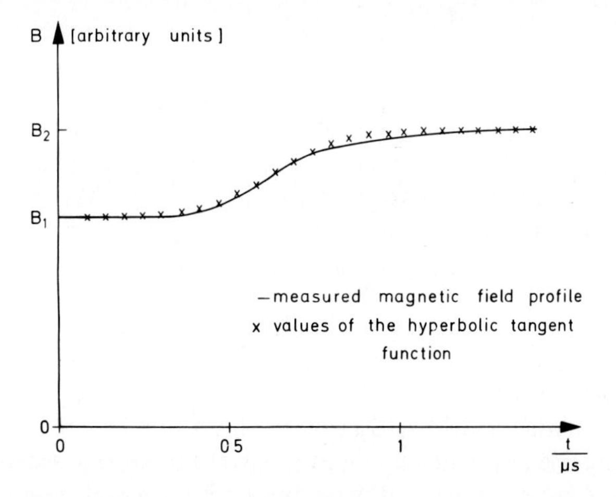

Fig. 14. Comparison between the observed magnetic field profile (x) and the expected tanh dependence for a weak shock (Bogen *et al.*, 1971).

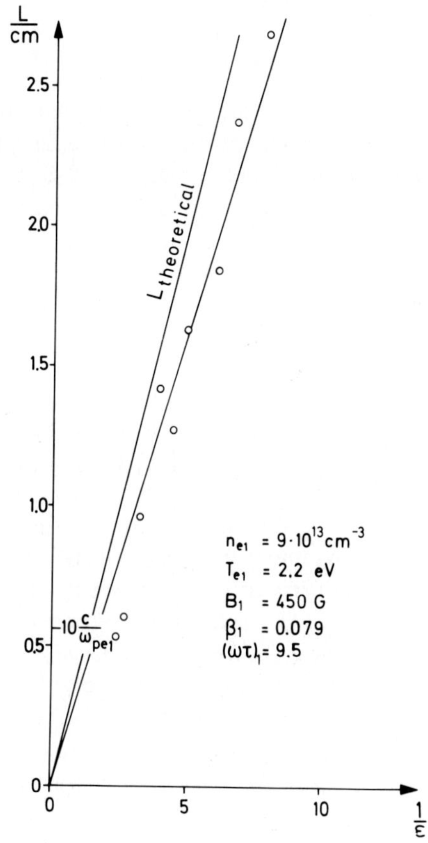

Fig. 15. Shock thickness Δ as function of the inverse shock strength $1/\varepsilon$ (Bogen *et al.*, 1971).

When electron collisions are negligible, electron inertia should become important in the formation of the shock front and should be taken into account in the generalized Ohm's law. If this is done, the phase velocity for small amplitude magnetoacoustic waves is no longer the Alfvén velocity V_A, instead one obtains

$$V_p^2 = V_A^2/[1 + (k^2c^2/\omega_{pe}^2)], \tag{59}$$

i.e., with decreasing wavelength the phase velocity drops. In a nonlinear wave this dispersion law can lead to a limitation of wave steepening at wavelengths of the order c/ω_{pe}.

Exact calculations yield an oscillatory shock profile with a characteristic length for the rise of B equal to about c/ω_{pe}; the nonlinear oscillation passes over into a damped sinusoidal wave train with a wavelength of the order $2\pi c/\omega_{pe}$ and a damping length u_1/ν_{ei}. With $\Delta \approx c/\omega_{pe}$ and $nev_d = (c/4\pi)\cdot(dB/dx)$ one can show that the electron drift velocity v_d necessary to support such a gradient is of order

$$v_d \approx (m_i/m_e)^{1/2}u_1, \tag{60}$$

i.e., $\frac{1}{2}m_e v_d^2 \approx \frac{1}{2}m_i u_1^2$, the ion flow energy is converted into drift energy of the electrons (see also the references at the end of Section II.A).

For the calculation of the shock profile it is further assumed that $\omega_{pe}^2/\omega_{ce}^2 \gg 1$. This is equivalent to $v_d \ll c$ and also implies that quasineutrality prevails in the shock front.

From a particle point of view one can describe the physical processes occurring in a resistive shock at $\nu_{ei} \ll (\omega_{ce}\omega_{ci})^{1/2}$ roughly as follows. With $\beta \ll 1$ the inequality $r_{ce} \ll c/\omega_{pe}$ holds. As an electron enters the shock front, it acquires a strong velocity component in the direction of current flow, i.e., perpendicular to B and to the plasma flow. The ion trajectories are hardly influenced by the jump in B. As a result, electrons and ions are separated and a space charge electric field E_r builds up which decelerates the ions: $ecE_r/\omega_{pe} = m_i u_1^2/2$. The drift velocity of the electrons should then be given approximately by $E_r c/B$. This turns out to be correct. A nonlinear oscillation is excited having a frequency of the order $(\omega_{ce}\omega_{ci})^{1/2}$ and being damped out because of friction between electrons and ions. For the case in which $(\omega_{ce}\omega_{ci})^{1/2} < \nu_{ei}$ the oscillation is aperiodically damped and the profile becomes monotonic again. Above a critical Alfvén Mach number $M_{A,c} \approx 3$ resistive dissipation is no longer sufficient to provide the dissipation for the shock to exist and viscous dissipation is required.

It should be a simple matter to generate oscillatory shock structures in an experiment where $\nu_{ei} \lesssim (\omega_{ce}\omega_{ci})^{1/2}$ or $\omega_{ce}\tau_{ei} > (m_i/m_e)^{1/2}$. Such profiles have indeed been observed as is shown in Fig. 16; however, their observation was erratic and was possible only in a narrow parameter range at small ε.

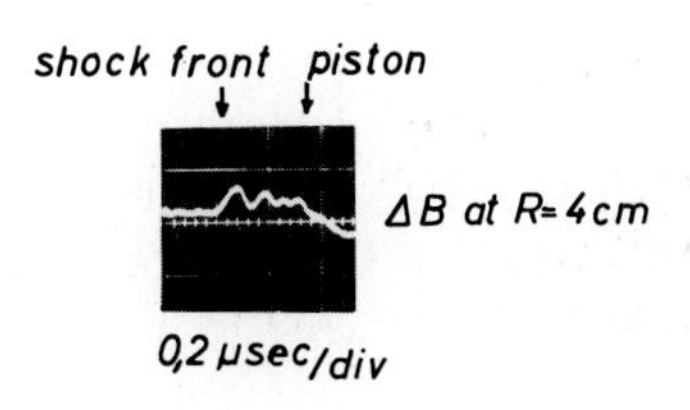

Fig. 16. Oscillatory structure for a shock at low density with $(B_1 - B_2)/B_1 \approx 0.2$ (Bogen *et al.*, 1971).

2. Turbulent Shocks

In case $(\omega_{ce}\tau_{ei})_1 < (m_i/m_e)^{1/2}$ one might expect that with ε increasing the plasma gets heated and with T_e rising in the front $\omega_{ce}\tau_{ei}$ increases, too. As a result, at sufficiently large ε, $(\omega_{ce}\tau_{ei})_2 > (m_i/m_e)^{1/2}$ will hold and oscillatory profiles should develop. Experimental observations, however, do not confirm this expectation. Measurements by a number of authors performed over a broad range of parameters have shown that in general shock profiles are monotonic even though $(\omega_{ce}\tau_{ei})_2$ is large. More important, however, the measured effective collision frequency ν_{eff} is large compared to that expected with binary collisions and is sufficient to provide aperiodic damping of the otherwise expected wavetrain (see, e.g., Alikhanov *et al.*, 1969; Chodura *et al.*, 1969; De Silva *et al.*, 1969; Hintz, 1969; Paul *et al.*, 1967; Babikhin and Smolkin, 1969).

Figure 17 shows density and magnetic field profiles as well as T_{e1} and T_{e2} of a low-β, mhd shock at $M_A = 2.4$ and $(\omega_{ce}\tau_{ei})_1 \approx 90$. Some of the physical properties of this shock are summarized in Table II. The measured values of n, B, and T_e behind the front are in good agreement with those derived from the shock relations if one assumes that all the energy dissipated in the shock front is transferred to the electrons and that $\gamma = 5/3$. With respect to the preceding discussion, the following observations are significant. At most two collisions can occur in the shock front. Nevertheless, the electrons are heated from 2.5 eV to 140 eV within $\tau_r = \Delta/u_1$, the rise time of the shock. Considering furthermore that $v_d/v_{e1} \approx 0.9$, it is obvious that collisions cannot even explain 10% of the observed temperature rise. The observation that $\sigma_{eff}/\sigma_{sp} \approx 10^{-3}$ indicates that binary collisions have been replaced by a very effective, but different mechanism.

One plausible explanation for the existence of collisionless shocks, such as those described above, assumes the existence of high level electric field

TABLE II

Typical Data of a Low-β, Turbulent MHD Shock

M_A	β_1	β_2	$(\omega_{ce}\tau_{ei})_1$	$(\omega_{ce}\tau_{ei})_2$	$\tau_r/\tau_{ei,1}$	σ_{eff}/σ_{sp}
2.4	$1.7\cdot10^{-2}$	$1.8\cdot10^{-1}$	90	$4.5\cdot10^{4}$	2	10^{-3}

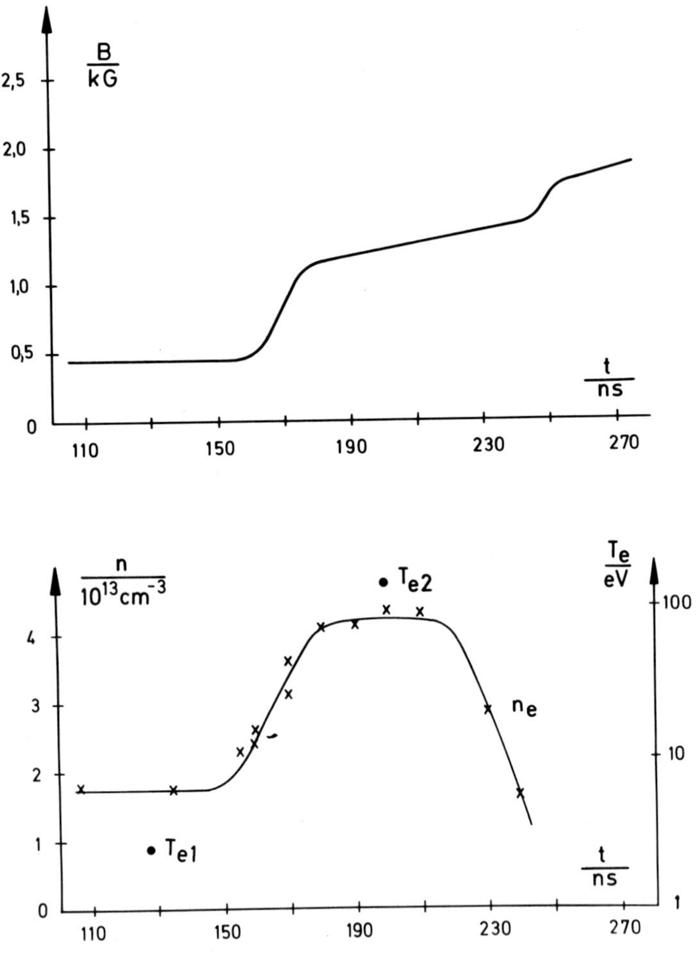

Fig. 17. Measured profiles of n_e and B in a resistive shock, T_e ahead of and behind the shock front (Bogen *et al.*, 1971).

fluctuations in the plasma, i.e., microturbulence. As a result of the interaction of electrons with these randomly phased fluctuations, they are scattered. In general, their straight trajectories suffer only small angle deflections by such scattering events. The time after which the root mean square deflection sums up to 90° is called the effective collision time $\tau_{\text{eff}} = 1/\nu_{\text{eff}}$. Long wavelength ($\lambda > \lambda_D$, where λ_D is the Debye length) coherent fluctuations of the electric charge or electric potential are always present in a plasma. However, in a thermal plasma their amplitude is low; the ratio of the total electric energy density of the fluctuations to the kinetic energy density is $W = E^2/8\pi nkT \approx 1/n\lambda_D{}^3$. With $n\lambda_D{}^3 \gg 1$, the contribution of collective fluctuations to particle scattering is small compared to that of binary collisions.

The reverse may be true if fluctuations are enhanced by some micro-instability. In plasmas with large deviations from thermal equilibrium, the conditions for the development of some instability are generally satisfied. The instability makes waves within certain wavelength ranges of the thermal spectrum grow exponentially. At large amplitudes when nonlinear effects become important, the growth stops and a quasi-stationary spectrum of enhanced fluctuations may develop. If the spectrum is known, the effective collision frequency ν_{eff} can be calculated. However, a complete theory of microturbulence does not exist for any of the instabilities. Although valuable results have been obtained by simulation experiments, further progress requires in particular comparison of existing theories with experimental results.

Which kind of instability can be responsible for the excitation of micro-turbulence? In the front of low-β mhd shocks, large magnetic field gradients do exist and high current densities are generated which provide large relative velocities between electrons and ions. Depending on the respective plasma parameters, this departure from local thermodynamic equilibrium can give rise to a number of instabilities belonging to the class of two stream instabilities. For example, with $v_d \approx (m_i/m_e)^{1/2}\beta^{-1/2}c_s \approx \beta^{-1/2}v_e$ [Eq. (60)] the instability condition for a Buneman type instability may easily be fulfilled. As a result of this instability electrons may heat up to a temperature at which the electron thermal velocity is larger than v_d and as a consequence, the instability is quenched. Although $T_{e1} = T_{i1}$ initially, T_e may now be sufficiently large compared to T_i for the ion sound instability to be excited. In general terms the conditions for the onset of this instability are $T_e/T_i \gg 1$, $v_d/c_s \gg 1$. More details about the onset and growth rates of the ion sound instability are given by Stringer (1964). The two types of instabilities mentioned shall only serve as examples. Various other instabilities have been proposed to explain the observed anomalous resistivities in perpendicular shocks. A survey on current driven instabilities and on the present status of the theory of turbulent resistivity is given by Sagdeev (1973).

Direct evidence for the existence of enhanced fluctuations in the front of resistive shocks has first been obtained in experiments at Culham (Paul *et al.*, 1969). There the level of density fluctuations at $k = 7.1 \cdot 10^5\ \text{m}^{-1} \approx \lambda_D^{-1}$ was determined by scattering the light of a ruby laser from these fluctuations and doing an absolute measurement of the scattered light intensity. Since spatial resolution was insufficient, only average values within the shock could be obtained. The measured level of fluctuation was about 230 times that of a thermal plasma. With some assumptions concerning the fluctuation spectrum, it could be shown that this amplitude of the fluctuations was consistent with the observed effective collision frequency. Figure 18 shows the frequency spectrum $S_K(\omega) \sim \langle \delta n_K{}^2(\omega) \rangle$ of the density fluctuations in case of the Culham experiment (Daughney *et al.*, 1970). The line center is shifted by about ω_{pi}. The direction of the shift shows that the scattering fluctuations propagate in the direction of the electron current. Further investigations did confirm that the frequency shift of the line is proportional to ω_{pi}. Apparently there are strong indications that the enhanced fluctuations are ion waves that are driven to instability by the electric current. For further results on the properties of plasma turbulence in resistive shocks, we refer to Perkin *et al.* (1974).

There have also been efforts to vary plasma and shock parameters over a broader range and to investigate thereby the dependence of ν_{eff} on n_e and on those parameters characterizing the deviation of the plasma state from equilibrium: v_d/v_e and T_e/T_i (Bogen *et al.*, 1971; Dippel *et al.*, 1972). Some results of these measurements are summarized in Table III.

To allow for a correct judgment of these data, we shall first give a brief description of the diagnostic methods employed. The magnetic field was measured with small magnetic induction probes. Their interaction with the surrounding plasma is only poorly understood and little is known about their actual temporal and spatial resolution. As a consequence, magnetic field gradients may be steeper than those measured and short wavelength fluctuations of the signal may have been lost. Electron temperatures and

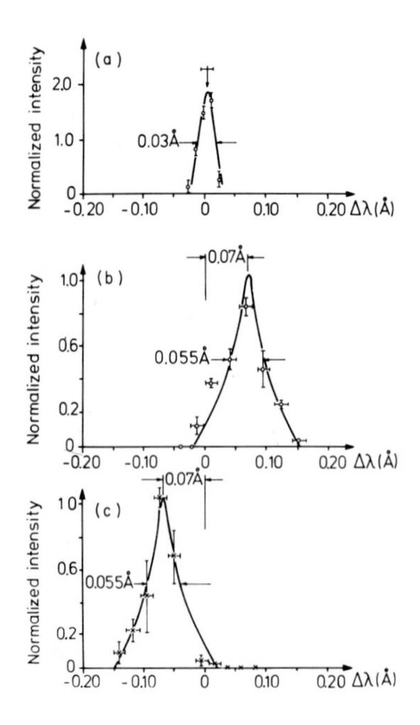

Fig. 18. Frequency spectrum $S_K(\omega)$: Spectral profiles of (a) incident laser light; (b) and (c) scattered light, i.e., $S_K(\omega)$ for opposite directions of B_1 and v_d as shown. Experimental points are mean of five measurements and error bars are standard deviation of the mean (Daughney *et al.*, 1970).

TABLE III

SUMMARY OF EXPERIMENTAL RESULTS ON LOW-β, TURBULENT MHD SHOCKS

	1 1	2 2	3 3	4 (Argon)
n_{e1}/cm^{-3}	$1.7 \cdot 10^{13}$	$4.4 \cdot 10^{13}$	9.10^{13}	5.10^{13}
β_1	$1.7 \cdot 10^{-2}$	$4.3 \cdot 10^{-2}$	$8.6 \cdot 10^{-2}$	6.10^{-2}
$(\omega_{ce}\tau_{ei})_1$	92	20	14	44
M_{A1}	2.4	3.2	1.7	
ν_{eff}/ν_{ei}	940	100	2	350
$\tau_r/(\tau_{ei})_1$	2.3	6.6	16	5.5
ν_{eff}	$8.1 \cdot 10^8$	$9.2 \cdot 10^8$	3.10^8	6.10^8
ω_{pe}	3.10^{11}	$4.5 \cdot 10^{11}$	$6.5 \cdot 10^{11}$	4.10^{11}
v_d/v_{e1}	0.9	0.72	0.36	0.8
v_d/v_e	0.17	0.18	0.21	0.19
T_e/T_i	22	11.4	2.3	17
ν_S/ν_{eff}	13	11	10	14
$(v_d/v_e) \cdot (m_i/m_e)^{1/4}$	1.3	1.4	1.6	3

electron densities were determined from the Thomson scattering of laser light. The electrical conductivity was derived by means of the energy balance equation for the electrons:

$$\frac{3}{2} n_1 u_1 \frac{k dT_e}{dx} = -nkT_e \frac{du}{dx} + \frac{j^2}{\sigma}. \tag{61}$$

By using the mean values of $dT_e/dx = (T_{e2} - T_{e1})/\Delta$, du/dx, T_e, n, and of the current density j, a mean value of the electrical conductivity in the shock front was calculated.

Within the set of measurements reported in Table III, the density is varied by a factor of 6, M_A by a factor of 2, the initial temperature is approximately constant: $T_{i1} \approx T_{e1} \approx 2$–3 eV. While T_e/T_i changes by a factor of 10, the ratio v_d/v_e is almost constant.

As pointed out before, most of the experimental evidence supports the assumption that ion sound turbulence is responsible for the observed turbulent resistivity. Let us examine the data in Table III from this point of view. For the ion sound instability to develop, in addition to $v_d/c_s \gg 1$, $T_e/T_i > 5$ is required. Since $T_{e1} \approx T_{i1}$, a mechanism is required that heats the electrons. This can be standard ohmic heating or, e.g., a Buneman instability. In case 1 of Table III, i.e., at low density, collisional heating can be excluded because $\tau_r \cdot (\nu_{ei})_1$ is close to one; with $v_d/v_{e1} = 0.9$ the instability condition $v_d \geq v_e$ is almost fulfilled. It is very likely that because of limited resolution of the magnetic probe the measured value of v_d is too low. At high density one would expect that ohmic heating could provide the required T_e. The fact that in case 3 the instability condition is not quite fulfilled ($T_e/T_i = 2.3$) may again be explained by the missing space resolution for the measurement of T_e.

The two lines at the end of the table serve for a comparison with theoretical results on turbulent resistivity. Here ν_s denotes a collision frequency calculated by means of a formula derived by Sagdeev (1965) for ion sound turbulence, assuming the growth of the waves to be limited by nonlinear ion Landau damping:

$$\nu_s = 10^{-2}(v_d/v_e)(T_e/T_i)\omega_{pe}. \quad (62)$$

The comparison with the experimental results is made by calculating the ratio ν_s/ν_{eff}. For the cases of Table III this ratio is nearly constant, indicating that the scaling proposed by Sagdeev is in agreement with experiments. There is, however, a discrepancy of about a factor of 10 with respect to absolute magnitude. This was also found by other authors (e.g., Paul *et al.*, 1967). Taking into consideration also that v_d/v_e is nearly constant, the comparison of theory with experiment is not quite conclusive.

Measurements as well as theory may need improvement. One might suspect the main experimental error to be connected with the use of mean values in Eq. (61) for the determination of ν_{eff}.

To obtain more meaningful measurements a spatial resolution of the shock front is required. Some progress in this direction was made by Zeyer (1975). In addition to the magnetic field profiles the complete n_e and T_e profiles were determined. Starting from these measurements the development of v_d/v_e, T_e/T_i, ν_{eff}/ν_{ei}, and ν_s/ν_{eff} in the shock front was calculated. These results are shown in Fig. 19. One important conclusion of these measurements is that v_{eff} calculated from mean values is about one order of magnitude smaller than the maximum value of ν_{eff} in the front. The results show further that ν_s/ν_{eff} is space dependent but approaches 1 at the end of the shock front.

It is still likely that the space resolution of the diagnostic methods employed is not sufficient to detect a possible fine structure of the shock. By using a special probe design, Eselevich *et al.* (1971) were able to show that such a fine structure does exist as illustrated in Fig. 20. The measured magnetic profile together with the mhd equations allowed the temporal development of v_d/v_e, ν_{eff}, and T_e/T_i in the shock front to be calculated. Detailed evidence about the development of microturbulence was obtained this way, in particular, the authors were able to show that for plasmas with $T_e = T_i$ ion sound instability can develop in a shock after T_e is first increased by ohmic heating or as the result of the Buneman instability.

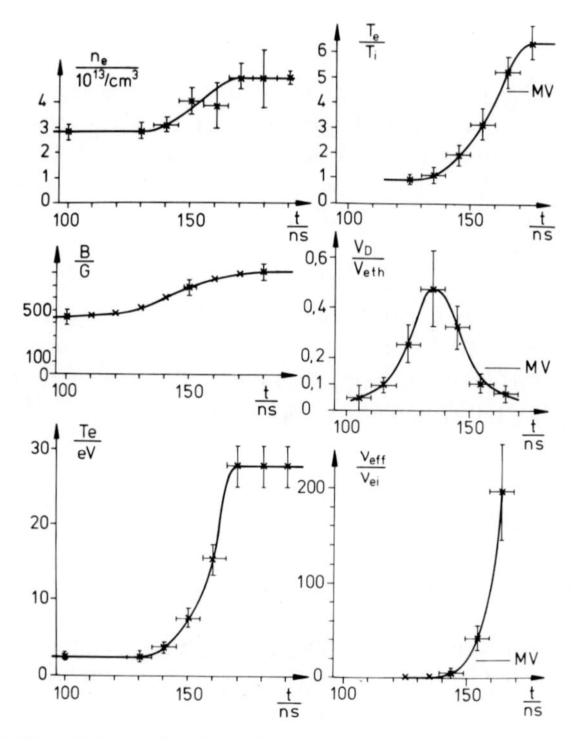

Fig. 19. Profiles of T_e, n_e, B, T_e/T_i, v_d/v_e, and ν_{eff}/ν_{ei} in a collision free shock wave with $M_A = 1.7$. The mean values are indicated by MV (Zeyer, 1975).

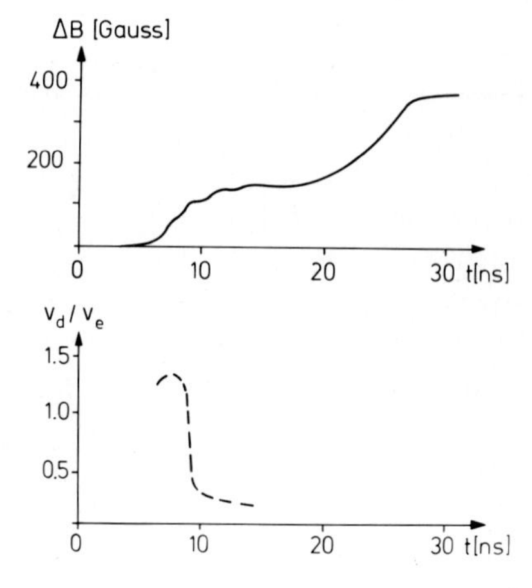

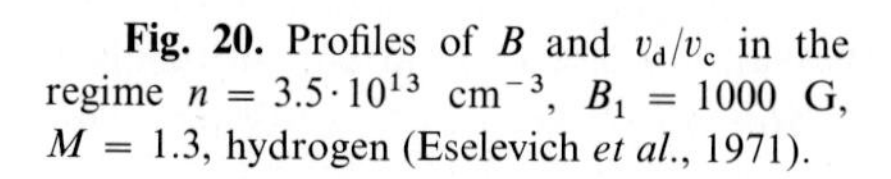

Fig. 20. Profiles of B and v_d/v_e in the regime $n = 3.5 \cdot 10^{13}$ cm^{-3}, $B_1 = 1000$ G, $M = 1.3$, hydrogen (Eselevich *et al.*, 1971).

According to the last line of Table III, the ratio v_d/v_e tends to a limiting value about equal to $(m_e/m_i)^{1/4}$. Only for the argon case a stronger deviation by a factor of 2 is observed. The observation $v_d/v_e \approx (m_e/m_i)^{1/4}$ is an important result; it agrees with the results of an improved theory on current driven ion sound turbulence by Vekshtein and Sagdeev (1970) and with the results of numerical computations by Dum *et al.* (1974). According to this theory an enhanced high energy tail of the ion velocity distribution is formed during the growth of the fluctuations, and the growth of the waves is limited due to linear Landau damping by the high energy ions. Strong evidence from various experiments (e.g., see Eselevich *et al.*, 1971; Höthker, 1975; Bogen *et al.*, 1975a) supports this theory. The discrepancy in the argon case disappears too, if one takes into consideration that the argon plasma at $T_e = 3.1$ eV most likely is doubly ionized.

VI. MISCELLANEOUS TOPICS

A. Theta Pinches with Strong Internal Fields

For shock heated plasmas with an internal magnetic field that is parallel to the external field, the following relation exists: $T_2 = T_1\eta\beta_2/\beta_1$. This formula makes use only of the fact that $\eta = n_2/n_1 = B_2/B_1$. High T_2 together with low β_2 is therefore difficult to realize, because $\eta < f + 1$. With the usual methods of plasma preparation, plasmas with low β, i.e., high internal magnetic field, show a diffusion profile for the density distribution. As a result the characteristic velocity for pressure perturbations may reach very high values in the vicinity of the walls. The short rise times of the magnetic field required for the generation of shock waves cannot be realized experimentally. It appears, therefore, that shock heating is not an appropriate method of producing hydrogen plasmas of high temperature and low β.

Pinches with trapped antiparallel fields show a behavior (cf. Fig. 21) strongly different from that with parallel fields. This begins at the start of the discharge. When voltage is applied to the coil, the initial plasma first moves outward since the magnetic pressure decreases until a sufficiently strong antiparallel field outside the plasma has been built up again. This behavior results in an enhanced wall contact that may lead to an increase of the impurity content of the plasma. Whereas the shock is quite similar in the parallel and antiparallel case, the boundary between plasma and piston is completely different. As a result of the strong magnetic field gradient high current densities occur and strong electron heating is observed. There is always a region where $B_i = 0$ and where for the pressure balance to be fulfilled $\tilde{\beta} = 1$ must hold in the plasma. Smear pictures mostly show a

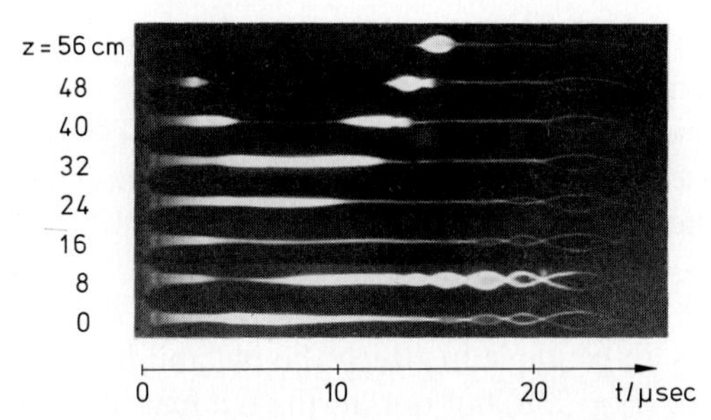

Fig. 21. Smear camera pictures of a theta pinch with trapped antiparallel bias field at different axial positions ($B_1 = -640$ G, maximum compression field 48 kG, quarter period 6.5 μs, coil length 128 cm, coil diameter 10 cm, filling pressure 0.05 Torr D_2). The tearing instability and rotation is clearly demonstrated (Kaleck *et al.*, 1969).

relatively narrow ring of high density, high temperature plasma and a "hole" in the center due to a density minimum which is normally also maintained in the adiabatic phase for some time. A great advantage of this configuration is that due to the closed field lines, thermal conduction losses and plasma losses at the ends of the coil are reduced. Furthermore, in addition to shock and adiabatic heating, joule heating is important because of the dissipation of the trapped antiparallel magnetic flux; this helps to achieve high temperatures (Griem *et al.*, 1962). However, strong axial contraction of the plasma, its decay into several plasmoids, and ejection of plasmoids out of the coil is observed, connected with rather unstable behavior of the plasma annulus (see Section VI.C and Fig. 21).

B. MHD Instabilities

Instabilities in high-β-plasmas have recently been discussed in a summarizing article by Bodin (1970). In the following only those phenomena will be described that are of special importance for the experiments. Since the equilibrium configuration of an infinitely long theta pinch is independent of ϑ and z, the perturbation of the equilibrium is usually taken in the form $\xi = \xi_0(r) \exp(\gamma t + im\vartheta + ikz)$.

Flutelike instabilities of the Rayleigh–Taylor type may occur if the plasma is accelerated toward the axis. They are similar to those observed in a gravitational field at the interface of a heavy and a light fluid. The growth rate for $2\pi r/m = \lambda_\theta \gg d$ is given by

$$\gamma = (gm/r)^{1/2}, \tag{63}$$

where g is the acceleration and d the sheath thickness $d = n/(dn/dr)$. For $\lambda_\theta \lesssim d$, γ can be approximated by $\gamma \approx (g/d)^{1/2}$.

In a theta pinch the acceleration at the beginning of the implosion and the inward acceleration during the radial oscillations (see Section IV) may excite these instabilities (Green and Niblett, 1960). They can be damped by the viscosity μ at high T_i. In this case maximum growth rate occurs at the wavelength

$$\lambda_\theta \approx 2\pi g^{-1/3}(\mu/\rho)^{2/3} \sim n^{-2/3}T_i^{5/3}. \tag{64}$$

Figure 22 shows an example in which the instability is observed during the oscillation phase (Fig. 22b) but not during the first implosion (Fig. 22a).

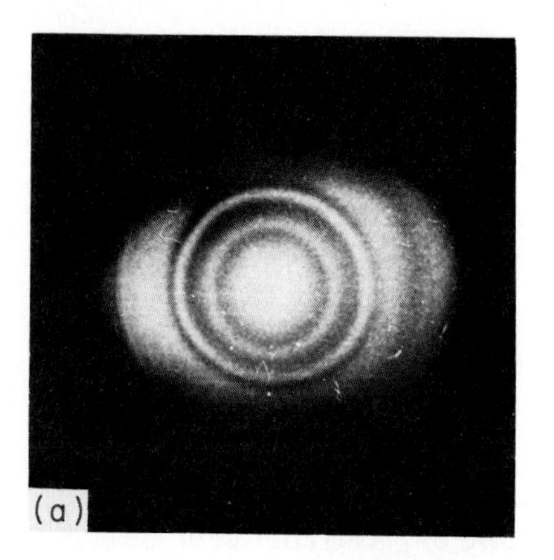

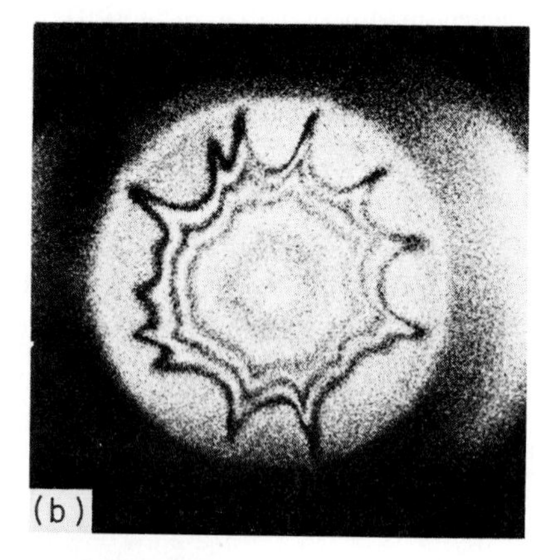

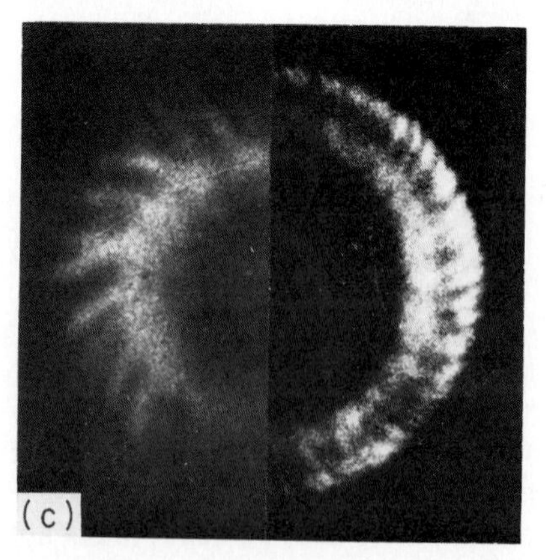

Fig. 22. Stability during the first compression (a) and Rayleigh–Taylor instability during inward acceleration of the oscillating plasma column (b) at 0.25 Torr H_2 filling pressure, coil diameter 4 cm; interferograms with zero phase shift (Dippel, 1967). Tilted striations at low initial pressure (0.01 Torr H_2) (c), coil diameter 20 cm. The two parts of the picture are successive photographs taken during the fast compression phase.

Under conditions for which the mean free path is long, instabilities in the form of tilted striations (Fig. 22c) are observed at the piston, even when it moves with a fairly constant velocity. These are interpreted (Shonk and Morse, 1968) as ion beam instabilities caused by counterstreaming ions in the piston region. Computer simulations gave azimuthal wavelengths of the order of c/ω_{pi}, in agreement with experiments.

In theta pinches with low internal magnetic fields or with strong antiparallel magnetic fields, plasma rotation, and rotational $m = 1$, and especially $m = 2$, instabilities have often been observed. The angular momentum is probably transferred by a contact with the wall during the preheating or during the main discharge. Several theories (see, e.g., Haines, 1965) try to explain how rotation could be induced by contact with the wall at the start of the magnetic compression pulse or at a later stage of the discharge. In the latter case a short circuiting of the radial electric field E_r at the coil ends is assumed which should cause torsional Alfvén waves to propagate along the axis toward the center plane of the coil, inducing a rotation of the plasma. Both mechanisms lead to a rotation in the direction of a diamagnetic ion current (direction of the gyrating ions).

Rotation has been measured by magnetic probes (Kolb *et al.*, 1965) by the Doppler shift of impurity lines (Keilhacker *et al.*, 1966); and by stereoscopic smear camera pictures (see, e.g., Rostocker and Kolb, 1961). Angular velocities of up to several times 10^6 revolutions per second have been observed, the rotation being in the direction of a diamagnetic ion current, as available theories predict. More recently, further investigations (Thomas, 1969; Kaufmann *et al.*, 1971b; Könen, 1973; Witulski, 1974) of the mechanisms leading to rotation have been carried out. So far, however, it seems impossible to give a unified description of the origin of rotation.

Stereoscopic smear camera pictures often show unstable rotation after more than an Alfvén wave transit time from the ends. Growth rates of these instabilities were calculated for $m \geq 2$ by Rostoker and Kolb (1961) and with an improved plasma model for $m \geq 1$ by Bowers and Haines (1971). This theory and experiments of Kaufmann *et al.* (1971b) and Thomas (1969) indicate that at high temperatures (large Larmor radius) the $m = 1$ rotational mode prevails (often called "wobble"), whereas at lower T the $m = 2$ mode is dominant. Higher modes are rarely observed. Haas and Wesson (1967) predicted $m \geq 1$ instabilities of the theta pinch without rotation. They found that the pinch is stable for $\tilde{\beta} = 1$ and $k > 0$, the mode $k = 0$ being marginally stable. For $\tilde{\beta} < 1$ a necessary condition for stability is $\tilde{\beta} > [1 + (r/R_w)^{2m}]^{-1}$, where r is the plasma radius, R_w the compression coil radius. These instabilities apparently have not been observed in straight theta pinches. However, an often observed drift of the plasma toward the slot in the compression coils shows that the theta pinch is sensitive to small

$m = 1$, $k \approx 0$ perturbations of the magnetic field. Field lines with a radius of curvature R give rise to a radial force $F = \tilde{\beta} B^2 r^2 / 4R$.

After fast compression, when the plasma is collision free, and a high degree of pressure anisotropy $p_\perp / p_\parallel$ exists, mirror instabilities ($m = 0$) have been observed (Kaufmann *et al.*, 1971a). They may occur if

$$\frac{p_\perp}{p_\parallel} > \frac{1}{\tilde{\beta}}\left(1 + \frac{1}{(R_w/r)^2 - 1}\right).$$

For an explanation consider a mirrorlike initial perturbation for a case in which $\tilde{\beta} \approx 1$: if $p_\perp / p_\parallel > 1$ particles tend to be reflected into the bulge. The particle density and pressure, therefore, increase, and the perturbation grows further, until nonlinear effects become effective and result in a decrease of the anisotropy.

C. Resistive Instabilities (Tearing Mode)

In short theta pinches with trapped antiparallel fields the plasma contracts axially, leading to a single plasmoid (see Fig. 21). This phenomenon can be understood by the fact that field lines close at the ends and tend to contract like a rubber band. A splitting into several plasmoids has been observed in pinches which are long compared to the radius. This is explained by the resistive tearing instability, caused by finite electrical conductivity which allows plasma and field lines to become detached from each other (Furth *et al.*, 1963).

For a cylindrical plasma the scaling of the growth rate γ and wavelength $\lambda = 2\pi/k$ of the tearing instability for the case $2\pi r/\lambda \ll 1$ is approximately

$$1/\gamma \sim r^{3/2} T_e^{1/2}, \qquad \lambda \sim r^{5/4} T_e^{1/2}. \tag{65}$$

For $r = 1, 1$ cm, and $T_e = 150$ eV, Kaleck *et al.* (1969) found by numerical methods $1/\gamma = 2.2$ μs and $\lambda = 19$ cm, whereas experiments gave $1/\gamma \approx 1$ μs and 12 cm $< \lambda <$ 25 cm, in rough agreement with the predictions.

REFERENCES

Adlam, J. H., and Allen, J. E. (1958). *Phil. Mag.* **3,** 448.

Alikhanov, S. G., *et al.* (1969). *In* "Plasma Physics and Controlled Nuclear Fusion Research," Vol. I, p. 47. IAEA, Vienna.

Allen, J. W., and Duprée, A. K. (1969). *Astrophys. J.* **155,** 27.

Andelfinger, C., Decker, G., Fünfer, E., Heiss, A., Keilhacker, M., and Sommer, J. (1966). *In* "Plasma Physics and Controlled Nuclear Fusion Research," Vol. 1, p. 249. IAEA, Vienna.

Andelfinger, C., *et al.*,(1967). *Proc. APS Topical Conf. Pulsed High Density Plasma, Los Alamos* Rep. LA-3770, G2.

Babykin, M. V., and Smolkin, G. E. (1969). *In* "Plasma Physics and Controlled Nuclear Fusion Research," Vol. I, p. 129. IAEA, Vienna.

Bearden, A. J., Ribe, F. L., Sawyer, G. A., and Stratton, T. F. (1961). *Phys. Rev. Lett.* **6,** 257.

Beerwald, H. (1965). Ber. Kernforschungsanlage Jülich, Jül-322 PP.

Beerwald, H. (1967). Ber. Kernforschungsanlage Jülich, Jül-480 PP.

Beerwald, H., and Hintz, E. (1960). *Proc. Int. Conf. Ionizat. Phenomena Gases, 4th, Uppsala* **2,** 468. North-Holland Publ., Amsterdam.

Beerwald, H., Bogen, P., El-Khalafawy, T., Fay, H., Hintz, E., and Kever, H. (1962). *Nucl. Fusion Suppl.* Part 2, 595.

Biskamp, D. (1973). *Nucl. Fusion* **13,** 719.

Bodin, H. A. B. (1970). *Methods Exp. Phys.* **9,** Part A, 395.

Bodin, H. A. B., Mc Cartan, J., Newton, A. A., and Wolf, G. H. (1969). *In* "Plasma Physics and Controlled Nuclear Fusion Research," Vol. II, p. 533. IAEA, Vienna.

Bogen, P. (1968). *In* "Plasma Diagnostics" (W. Lochte-Holtgreven, ed.). North-Holland Publ., Amsterdam.

Bogen, P. (1970). *Z. Naturforsch.* **25a,** 1151.

Bogen, P. (1972). *Z. Naturforsch.* **27a,** 210.

Bogen, P., and Hintz, E. (1961). *Proc. Conf. Ionizat. Phenomena Gases, 5th.* North-Holland Publ., Amsterdam.

Bogen, P., and Rusbüldt, D. (1966). *Phys. Fluids* **9,** 2296.

Bogen, P., and Rusbüldt, D. (1968). *Phys. Fluids* **11,** 2021.

Bogen, P., Lie, Y. T., Rusbüldt, D., and Schlüter, J. (1969). *In* "Plasma Physics and Controlled Nuclear Fusion Research," Vol. II, pp. 651–656. IAEA, Vienna.

Bogen, P., *et al.* (1971). *In* "Plasma Physics and Controlled Nuclear Fusion Research," Vol. III, p. 277. IAEA, Vienna.

Bogen, P., Lie, Y. T., and Rusbüldt, D. (1972). *Atomkernenergie* **19,** 217.

Bogen, P., Dietz, K. J., Hintz, E., Höthker, K., Lie, Y. T., and Pospieszczyk, A. (1975a). *In* "Plasma Physics and Controlled Nuclear Fusion Research," Vol. III, p. 349. IAEA, Vienna.

Bogen, P., Dietz, K. J., Hintz, E., Höthker, K., Lie, Y. T., and Pospieszczyk, A. (1975b). *In Proc. Eur. Conf. Controlled Fusion Plasma Phys., 7th, Lausanne* **I,** 53.

Bogen, P., Dietz, K. J., and Pospieszczyk, A. (1976). *In* "Pulsed High Beta Plasmas" (D. E. Evans, ed.), p. 197. Pergamon, Oxford.

Bowers, E., and Haines, M. G. (1971). *Phys. Fluids* **14,** 165.

Burgess, A. (1964). *Astrophys. J.* **139,** 776.

Chodura, R., Keilhacker, M., Kornherr, M., and Niedermayer, H. (1969). *In* "Plasma Physics and Controlled Nuclear Fusion Research," Vol. I, p. 81. IAEA, Vienna.

Chu, C. K., and Gross, R. A. (1969). *In Adv. Plasma Phys.* **2,** 139.

Cosler, A., Dietz, K. J., Dokopoulos, P., Schürer, M., and Wiegmann, W. (1974). *Proc. Symp. Fusion Technol., 8th,* p. 391.

Daughney, C. C., Holmes, L. S., and Paul, J. W. M. (1970). *Phys. Rev. Lett.* **25,** 497.

Davis, L., Lüst, R., and Schlüter, H. (1958). *Z. Naturforsch.* **13a,** 916.

De Silva, A. W., and Goldenbaum, G. C. (1970). *Methods Exp. Phys.* **9,** Part A.

De Silva, A. W., *et al.* (1969). *In* "Plasma Physics and Controlled Nuclear Fusion Research," Vol. I, p. 143. IAEA, Vienna.

Dietz, K. J. (1969). Private communication.

Dietz, K. J., and Hintz, E. (1970). *Proc. Eur. Conf. Contr. Fusion Plasma Phys., 5th, Rome,* p. 60.

Dippel, K. H. (1967). Ber. der KFA Jülich Nr. 493.

Dippel, K. H., Hintz, E., and Höthker, K. (1972). *Atomkernenergie* **19,** 225.

Dokopoulos, P. (1970). *Proc. Symp. Fusion Technol., 6th, Aachen,* p. 393. Commission of the European Communities, Centre for Information and Documentation, Luxembourg.

Dokopoulos, P., Göring, D., Kohlhaas, W., Liedtke, A., and Stickelmann, C. (1973). *Symp. Eng. Probl. Fusion Res., 5th,* IEEE Publ. No. 73, CH 0843-3-NPS, p. 537.

Dum, C. T., Chodura, R., and Biskamp, D. (1974). *Phys. Rev. Lett.* **32,** 1231.

Eberhagen, A., and Wunderlich, R. (1969). *Z. Physik* **232,** 1151.

Elwert, G. (1952). *Z. Naturforsch.* **7a,** 432.

Engelhardt, W. (1973). *Phys. Fluids* **15,** 2074.

Eselevich, V. G., Eskov, A. G., Kurtmullaev, R. Kh., and Malyutin, A. I. (1971). *Sov. Phys. JETP* **33,** 898.

Friedrich, F., and Hintz, E. (1966). *Proc. Symp. Eng. Probl. Controlled Thermonucl. Res.,* ORNL Rep. Conf. 66 1016, pp. 3–7.

Fünfer, E., Kaufmann, M., Lotz, W., Neuhauser, J., Schramm, G., and Seidel, V. (1975). *Nucl. Fusion* **15,** 133.

Furth, H. P., Killeen, J., and Rosenbluth, M. N. (1963). *Phys. Fluids* **6,** 459.

Gabriel, A. H. (1972). *Mon. Not. Roy. Astron. Soc.* **160,** 99.

Gardner, C. S., Goertzel, H., Grad, H., Morawetz, C. S., Rose, M. H., and Rubin, H. (1958). *Proc. U.N. Int. Conf. Peaceful Uses At. Energy, 2nd, Geneva,* **31,** 230.

Gouderley, G. (1942). *Luftfahrtforschung* **19,** 302.

Grad, H., and Hu, P. N. (1967). *Phys. Fluids* **10,** 2596.

Green, T. S. (1962). *Nucl. Fusion* **2,** 92.

Green, T. S., and Niblett, G. B. F. (1959). *Proc. Phys. Soc.* **74,** 737.

Green, T. S., and Niblett, G. B. F. (1960). *Nucl. Fusion* **1,** 42.

Green, T. S., Fisher, D. L., Gabriel, A. H., Morgan, F. J., and Newton, A. A. (1967). *Phys. Fluids* **10,** 1663.

Gribble, R. F., Little, E. M., Morse, R. L., and Quinn, W. E. (1968). *Phys. Fluids* **11,** 1221.

Griem, H. R. (1964). "Plasma Spectroscopy." McGraw-Hill, New York.

Griem, H. R., Kolb, A. C., Lupton, W. H., and Philipps, D. T. (1962). *Nucl. Fusion Suppl.* Part 2, 543.

Gross, R. A. (1971). *In* "Physics of High Energy Density" (*Proc. Int. School Phys., "Enrico Fermi"*), Course 48, p. 245. Academic Press, New York.

Gross, R. A. (1973). *In* "Recent Developments in Shock Tube Research" (D. Bershader and W. Griffith, eds.), p. 72. Stanford Univ. Press, Stanford, California.

Haas, F. A., and Wesson, J. A. (1967). *Phys. Fluids* **10,** 2245.

Hain, K., and Kolb, A. C. (1962). *Nucl. Fusion Suppl.* Part 2, 561.

Haines, M. G. (1965). *Adv. Phys.* **14,** 167.

Herppich, G. (1969). IPP-Rep. 4/68, Inst. für Plasmaphys. Garching, Munich.

Hintz, E. (1962). *Nucl. Fusion Suppl.* **2,** 601–605.

Hintz, E. (1969). *In* "Plasma Physics and Controlled Nuclear Fusion Research," Vol. I, p. 69. IAEA, Vienna.

Hintz, E. (1970). *Methods Exp. Phys.* **9,** 213.

Höthker, K. (1975). Dissertation Ruhr-Univ., Bochum.

Höthker, K. (1976). *Nucl. Fusion* **16,** 253.

Hundhausen, A. J. (1970). *In* "Intercorrelated Satellite Observations Related to Solar Events" (V. Manno and D. E. Perge, eds.). Reidel Publ., Dordrecht.

Jaffrin, M. Y., and Probstein, R. F. (1964). *Phys. Fluids* **7,** 1658.

Jahoda, F. C., Little, E. M., Quinn, W. E., Sawyer, G. A., Stratton, T. F. (1960). *Phys. Rev.* **119,** 843.

Kaleck, A., *et al.* (1969). *In* "Plasma Physics and Controlled Nuclear Fusion Research," Vol. II, p. 581. IAEA, Vienna.

Kantrowitz, A., and Petschek, H. E. (1966). *In* "Plasma Physics in Theory and Application" (W. B. Kunkel, ed.), p. 148. McGraw-Hill, New York.

Kaufmann, M., and Köppendörfer, W. (1974). *Nucl. Fusion Spec. Suppl.*, p. 285.
Kaufmann, M., Neuhauser, J., and Röhr, H. (1971a). *Z. Phys.* **244,** 99.
Kaufmann, M., Fünfer, E., Junker, J., Neuhauser, J., and Seidel, U. (1971b). IPP Rep. 1/123.
Keilhacker, M., and Herold, H. (1966). *In* "Plasma Physics and Controlled Nuclear Fusion Research," Vol. I, p. 315. IAEA, Vienna.
Kever, H. (1960). Berichte der KFA Jülich, Jül-2-PP.
Kever, H. (1962). *Nucl. Fusion Suppl.* Part 2, p. 613.
Kidder, R. E. (1971). *In* "Physics of High Energy Density" (*Proc. Int. School Phys.,* "Enrico Fermi") Course 48, p. 306. Academic Press, New York.
Knoepfel, K. H. (1970). "Pulsed High Magnetic Fields." North-Holland Publ., Amsterdam.
Knorr, G. (1958). *Z. Naturforsch.* **13a,** 441.
Kolb, A. C., and McWhirter, R. W. P. (1964). *Phys. Fluids* **7,** 519.
Kolb, A. C., Hintz, E., and Thonemann, P. C. (1965). *Phys. Fluids* **8,** 1005.
Könen, L. (1973). *Verhandl. DPG (VI)* **8,** 748.
Krall, N., *et al.* (1974). *Nucl. Fusion* **14,** 27.
Kunze, H. J. (1968). *In* "Plasma Diagnostics" (W. Lochte-Holtgreven, ed.). North-Holland Publ., Amsterdam.
Kunze, H. J. (1972). *Space Sci. Rev.* **13,** 565.
Lehner, G. (1967). *Z. Phys.* **206,** 284.
Lie, Y. T. (1973). *Appl. Phys.* **2,** 297–302.
Longmire, C. L. (1963). "Elementary Plasma Physics." Wiley (Interscience), New York.
Lupton, W. H., McLean, E. A., and Philips, O. T. (1961). *Proc. Int. Conf. Ionizat. Phenomena Gases, 5th.* North-Holland Publ., Amsterdam.
McKenna, F. C., Kristal, R. F., and Thomas, K. S. (1974). *Phys. Rev. Lett.* **32,** 409.
McLean, D. J. (1973). *In* "Coronal Disturbances" (G. Newkirk, Jr., ed.), p. 301. Reidel Publ., Dordrecht.
McWhirter, R. W. P. (1965). *In* "Plasma Diagnostic Techniques" (R. H. Huddlestone and S. L. Leonard, eds.). Academic Press, New York.
Meade, D. M. (1974). *Nucl. Fusion* **14,** 289.
Morse, R. L. (1972). Los Alamos Sci. Lab. Rep. LA-4930-MS.
Pai, S. (1959). *In* "Introduction to the Theory of Compressible Fluids." Van Nostrand–Reinhold, Princeton, New Jersey.
Patrick, R. M. (1959). *Phys. Fluids* **2,** 589.
Paul, J. W. M., Goldenbaum, G. C., Iiyoshi, A., Holmes, L. S., and Hardcastle, R. A. (1967). *Nature (London)* **216,** 363.
Paul, J. W. M., Daughney, C. C., and Hohnes, L. S. (1969). *Nature (London)* **223,** 822.
Perkin, R. M., Craig, A. D., Hohnes, L. S., and Paul, J. W. M. (1974). *Proc. Conf. Plasma Phys. Nucl. Fusion Res., 5th,* IAEA-CN-33-H6-2.
Pospieszczyk, A. (1973). Berichte der KFA Jülich, Jül-948 PP.
Pospieszczyk, A. (1975a). *Astron. Astrophys.* **39,** 357.
Pospieszczyk, A. (1975b). Private communication.
Ribe, F. L., Krakowski, R. A., Thomassen, K. I., and Coultas, T. A. (1974). *Nucl. Fusion Spec. Suppl.,* p. 99.
Rostoker, N., and Kolb, A. C. (1961). *Phys. Rev.* **124,** 965.
Sagdeev, R. Z. (1960). *Proc. Int. Conf. Ionizat. Phenomena Gases, 4th, Uppsala* **2,** 1081. North-Holland Publ., Amsterdam.
Sagdeev, R. Z. (1962). *Sov. Phys.–Tech. Phys.* **6,** 867.
Sagdeev, R. Z. (1965). *Proc. Symp. Appl. Math.* **18,** 281.
Sagdeev, R. Z. (1966). *Rev. Plasma Phys.* **4,** 23.
Sagdeev, R. Z. (1973). *Adv. Plasma Phys.* **5,** 153.

Savedoff, M. P., Hovenier, J. W., and van Leer, B. (1967). *Bull. Astron. Inst. Netherlands* **19,** 107.
Sawyer, G. A., Bearden, A. J., Henins, J., Jahoda, F. C., and Ribe, F. L. (1963). *Phys. Rev.* **131,** 1891.
Schlüter, J. (1966), Berichte der KFA Jülich, Jül-413 PP.
Shafranov, V. D. (1957). *Sov. Fiz. JETP* **5,** 1183.
Shonk, C. R., and Morse, R. L. (1968). *Proc. APS Topical Conf. Numerical Simulation Plasma* Paper C3.
Shore, B. W. (1969). *Astrophys. J.* **158,** 1205.
Siemsen, F. (1973). Berichte der KFA-Jülich, Jül-926 PP.
Spitzer, L. (1956). "Physics of Fully Ionized Gases." Wiley (Interscience), New York.
Stratton, T. F. (1965). *In* "Plasma Diagnostic Techniques" (R. H. Huddlestone and S. L. Leonard, eds.). Academic Press, New York.
Stringer, T. E. (1964). *Plasma Phys.* **6,** 267.
Taylor, J. B., and Wesson, J. A. (1965). *Nucl. Fusion* **5,** 159.
Thomas, K. S. (1969). *Phys. Rev. Lett.* **23,** 746.
Tidman, A., and Krall, N. A. (1971). "Shock Waves in Collisionless Plasmas." Wiley (Interscience), New York.
van Kessel, C. G. M., and Sigel, R. (1974). *Phys. Rev. Lett.* **33,** 1020.
Vekshtein, G. E., and Sagdeev, R. Z. (1970). *JETP Lett.* **11,** 194.
Vitkovitzky, J. M. (1966). *Proc. Symp. Fusion Technol., 4th, Frascati, Italy.*
Weniger, M. (1970). *Proc. Symp. Fusion Technol., 6th, Aachen,* p. 301. Commission Eur. Communities, Centre for Informat. and Documentation, Luxembourg.
Wiese, W. L., Smith, M. W., and Glennon, B. M. (1966). Atomic Transition Probabilities. Nat. Bur. Std., U.S., Vol. I (H through Ne), and Vol. II (Na through Ca).
Witham, G. B. (1958). *J. Fluid Mech.* **4,** 337.
Witulski, H. (1974). Dissertation Ruhr-Univ., Bochum.
Zeldovich, Ya. B., and Raizer, Yu. P. (1967). "Physics of Shock Waves and High Temperature Hydromagnetic Phenomena." Academic Press, New York.
Zeyer, G. (1975). Dissertation Ruhr-Univ., Bochum.

Index

A B C 8 D 9 E 0 F 1 G 2 H 3 I 4 J 5